AF479149

THE ERA OF MATERIALS

The Pennsylvania Academy of Science Book Publications

1. *Energy, Environment, and the Economy*, 1981. ISBN: 0-9606670-0-8.
2. *Pennsylvania Coal: Resources, Technology and Utilization*, 1983. ISBN: 0-9606670-1-6.
3. *Hazardous and Toxic Wastes: Technology, Management and Health Effects*, 1984. ISBN: 0-9606670-2-4.
4. *Solid and Liquid Wastes: Management, Methods and Socioeconomic Considerations*, 1984. ISBN: 0-9606670-3-2.
5. *Management of Radioactive Materials and Wastes: Issues and Progress*, 1985. ISBN: 0-9606670-4-0.
6. *Endangered and Threatened Species Programs in Pennsylvania and Other States: Causes, Issues and Management*, 1986. ISBN: 0-9606670-5-9.
7. *Environmental Consequences of Energy Production: Problems and Prospects*, 1987. ISBN: 0-9606670-6-7.
8. *Contaminant Problems and Management of Living Chesapeake Bay Resources*, 1987. ISBN: 0-9606670-7-5.
9. *Ecology and Restoration of The Delaware River Basin*, 1988. ISBN: 0-9606670-8-3.
10. *Management of Hazardous Materials and Wastes: Treatment, Minimization and Environmental Impacts*, 1989. ISBN: 0-9606670-9-1.
11. *Wetlands Ecology and Conservation: Emphasis in Pennsylvania*, 1989. ISBN: 0-945809-01-8.
12. *Water Resources in Pennsylvania: Availability, Quality and Management*, 1990. ISBN: 0-945809-02-6.
13. *Environmental Radon: Occurrence, Control and Health Hazards*, 1990. ISBN: 0-945809-03-4.
14. *Science Education in the United States: Issues, Crises, and Priorities*, 1991. ISBN: 0-945809-04-2.
15. *Air Pollution: Environmental Issues and Health Effects*, 1991. ISBN: 0-945809-05-0.
16. *Natural and Technological Disasters: Causes, Effects and Preventive Measures*, 1992. ISBN: 0-945809-06-9.
17. *Global Climate Change: Implications, Challenges and Mitigation Measures*, 1992. ISBN: 0-945809-07-7.
18. *Conservation and Resource Management*, 1993. ISBN: 0-945809-08-5.
19. *Biological Diversity: Problems and Challenges*, 1994. ISBN: 0-945809-09-3.
20. *The Oceans: Physical-Chemical Dynamics and Human Impact*, 1994. ISBN: 0-945809-10-7.
21. *Medicine and Health Care Into the 21st Century*, 1995. ISBN: 0-945809-11-5.
22. *Environmental Contaminants, Ecosystems and Human Health*, 1995. ISBN: 0-945809-12-3.
23. *Forests– A Global Perspective*, 1996. ISBN: 0-945809-13-1.
24. *The Era of Materials*, 1998. ISBN: 0-945809-15-8
25. *The Ecology of Wetlands and Associated Systems*, 1998. ISBN: 0-945809-14-X.

The Pennsylvania Academy of Science Publications • Books and Journal
Editor: Shyamal K. Majumdar • Professor and Head of Biology, Lafayette College
Easton, Pennsylvania 18042 • (610) 250-5464

THE ERA OF MATERIALS

EDITED BY

S.K. MAJUMDAR, Ph.D.
Marshall R. Metzgar Professor
and Head of Biology
Department of Biology
Lafayette College
Easton, PA 18042

RICHARD E. TRESSLER, Ph.D.
Professor and Head,
Department of Materials
Science and Engineering
Pennsylvania State University
University Park, PA 16802

E.W. MILLER, Ph.D.
Professor and Associate Dean (Emeritus)
Pennsylvania State University
University Park, PA 16802

Founded on April 18, 1924

**A Publication of
The Pennsylvania Academy of Science**

Library of Congress Cataloging in Publication Data

Bibliography
Index
Majumdar, Shyamal K. 1938-, ed.

Library of Congress Catalog Card No.: 97-76325

ISBN-0-945809-15-8
Copyright © 1998 By The Pennsylvania Academy of Science, Easton, PA 18042

A Publication of
The Pennsylvania Academy of Science, Easton, PA 18042

Printed in the United States of America by

Typehouse of Easton
Easton, PA 18042

PREFACE

The present economic prosperity and quality of life of industrial societies are linked to the dramatic development of advanced materials and processing technologies over the preceding several decades. Priorities for future investments to improve our quality of life and productivity – environmental quality, health-care, information and communication, infrastructure, transportation and efficient energy conversion – also depend strongly on improved materials and processes. Thus, we can expect robust research and development in all fields of materials and great demands for intellectual resources well into the 21st century.

Pennsylvania State University has played a major role in educating the leaders of the materials revolution and in pioneering an impressive list of research areas in its various laboratories. This book was conceived to provide a snapshot of the state-of-the-art at this time in the Era of Materials and to showcase the Penn State connections to the diverse segments of modern materials. The book has been written by the materials research faculty at Penn State, mostly from the Department of Materials Science and Engineering, the largest and most diverse department of its type in the U.S., and from the Intercollege Materials Research Laboratory, a pioneer in interdisciplinary materials research.

The book starts with an overview of the evolution of materials in our society with emphasis on the application and commercialization of new developments. The succeeding chapters are divided into two themes: the development of the fundamentals of materials science and the development of materials and devices for specific applications. The book is written for the technical layman who wants a broad perspective on the trends in materials research.

We greatly appreciate the cooperation and dedication of the contributors to this volume, a very impressive array of talent indeed. We are grateful to Lafayette College and Pennsylvania State University for providing facilities and resources to the editors.

Shyamal K.Majumdar
Richard E. Tressler
E. Willard Miller

Editors
February 1998

ACKNOWLEDGEMENT

The publication of this book was aided by contributions from the E. Willard and Ruby S. Miller book publication fund.

THE ERA OF MATERIALS

Table of Contents

Introduction

RECENT EVOLUTION OF MATERIALS IN OUR SOCIETY

R.E. TRESSLER

Department of Materials Science and Engineering
Steidle Building
Pennsylvania State University
University Park, PA 16802

INTRODUCTION

If Qin Shi Huang (the first emperor of China of Terra Cotta Warrior fame) lived today, with what highly prized, high technology material would he have chosen to be buried - a mechanically alloyed or directionally solidified nickel based turbine blade, a silicon carbide fiber reinforced silicon carbide ceramic aircraft engine exhaust flap, a graded index silica-based optical fiber cable, a yttrium-barium-copper oxide based superconducting magnet, a kevlar reinforced polymer body armor, a liquid crystal (polymer) display for a laptop computer, a 3-D silicon DRAM? What a bewildering array of "designer" materials and devices he would have to choose from today!

The historical trend of humans adapting or altering naturally occurring materials for use in structures and devices has been broken as our *scientific* understanding of processing-structure-property-performance relationships has developed. As a result, the evolution of materials has accelerated to meet the needs of design engineers for materials and devices with a broad range of properties. In parallel, the materials scientists have discovered new phenomena which have resulted in whole new classes of materials (such as the superconducting oxides with critical temperatures above the boiling point of liquid nitrogen) which enables new devices and machines. The acceleration of the evolution of materials technology is truly phenomenal, and some examples are cited in the following discussion.

Furthermore, in the U.S., stimulated by the publication in 1989 of a National Research Council (NRC) volume entitled "Materials Science and Engineering for the 1990s" (1) the agencies of the federal government teamed, in an unprecedented way, to develop the AMPP (Advanced Materials and Processing Program) Presidential initiative and focused the attention of the scientific community on "Materials." With increased federal spending for materials R&D, the field has become "sexy" with academic and national laboratory physicists and chemists turning their attention to basic scientific issues in materials. As a result, we can expect an acceleration of our fundamental knowledge of materials, particularly in the areas of synthesis and processing, which suggest exciting times ahead for the technologists who stay abreast of these developments.

At Penn State, our faculty, students and alums have played key roles in many of the most important developments. In this volume we've asked members of our faculty to review the developments in the fundamentals of materials science and in the applications of materials. Often this division is artificial so that scientific developments are discussed as they pertain to applications. What is striking to me, and also a great credit to our institution, is the extent to which Penn State has been a prolific source of ideas and key developments.

In this chapter, I attempt to highlight some of the key materials developments and their evolution into commercial successes which have had significant impact on our society. This is by no means intended as an all-inclusive menu of developments, but more as a "tickler" for the accounts in the following chapters. I have taken the approach of discussing materials developments in the context of their applications, because it is in the application of the scientific principles and new technology that society directly profits from their development.

MATERIALS IN COMMUNICATION AND INFORMATION SYSTEMS

In this century we've seen the development of the vacuum tube (which permitted electronic switching of electricity for the first time and amplification of electrical signals), the transistor (which was much faster than a vacuum tube and much more reliable), integrated circuits (transistors, resistors, capacitors and interconnects all fabricated together on a silicon chip), the laser (including solid state lasers which provide intense, coherent, monochromatic light) and low loss optical fibers (the optical signal carrying interconnect which has replaced the electrical signal carrying copper). These developments have had the most profound effects on our society by providing seemingly limitless computational power, fingertip access to enormous quantities of information, and instantaneous communication to any spot on earth. In all of these cases, fundamental scientific understanding permitted synthesis and fabrication of materials with precise properties, and inventions and technological breakthroughs led to cost-effective manufacture of these materials and devices.

For the vacuum tube refractory metals such as tungsten and thorium for the electron emitting cathodes had to be produced in reliable, cost effective forms. Glass envelope fabrication and glass-metal sealing were pushed to higher scientific and technological planes. With the invention of the bipolar transistor in 1948, intense efforts on growth of high purity and controlled purity single crystals of semiconductors such as silicon (and to a lesser extent, germanium) were launched. Recognizing the potential for miniaturization of the transistor and for assembly of many of them on one silicon wafer, Kelly and Hoerni and Noyce invented the integrated circuit in 1958 and 1959, respectively.

With a dizzying rate of design and materials invention and fabrication iterations, the level of IC sophistication has reached the point where a single silicon chip can now contain several million transistors using line widths (feature sizes of the circuit) of 0.35 μm (Figure 1). Thus, the electronic and computing aspect of the communications and information systems field has skyrocketed as an economic entity and is predicted to reach the one trillion dollar level in sales by the year 2000.

The photonic aspect, given birth by the invention of the laser in 1960, which produces coherent, monochromatic light, has enjoyed remarkable growth also. With the ability to rapidly modulate coherent monochromatic light to exceedingly rapidly transmit information, the critical need for a low loss waveguide was addressed with breakthrough inventions at Corning and AT&T Bell Labs in 1973 and 1974, respectively. The CVD deposition of the glass fiber optic performed by Dr. John MacChesney (Penn State Ph.D., 1959) has been called one of the ten most important inventions in AT&T history. The enormous improvement in transmittance of glass via this processing route is illustrated in Figure 2.

To convert electrical signals to light at the transmitter and from light to electrical signals at the receiver, a whole new set of semiconductor alloys were developed, namely the III-V semiconductors (referring to the columns in the periodic table of the elements). These materials are now fabricated into a variety of solid state lasers and light emitting diodes which generate the light signal, amplify it in repeaters periodically located along the length of the transmission cable, and convert it back to an electrical signal at the receiving end. Thus, photonics have become critical elements in our modern communication systems.

The perfection required of these electronic and photonic devices represent fundamental advances and breakthroughs which lead the overall field in materials science. The elegant methods and sophisticated instruments that have evolved to fabricate these devices represent an industry unto itself which will have (and is having) spin-offs into other materials technologies. An example is the use of photolithographic techniques to fabricate micromachines (ref. 2, Figure 3).

The interconnect technologies involved in an IC, in interconnecting IC's, in connecting the IC package to a printed wiring board, etc., are remarkably sophisticated and now present intrinsic limitations in the speed of communication and information systems. They also represent a significant part of the cost of modern electronic and photonic systems. Current research is concentrating on electronic

FIGURE 1. Intel Pentium™ Processor with 3.3 million transistors (Courtesy of Intel Corporation).

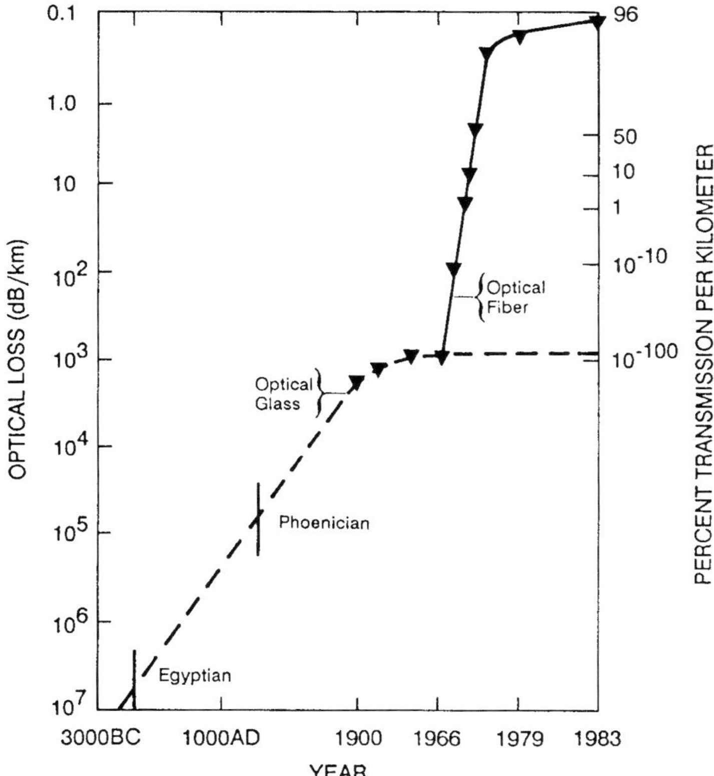

FIGURE 2. Historical improvement in glass transparency (Ref. 1).

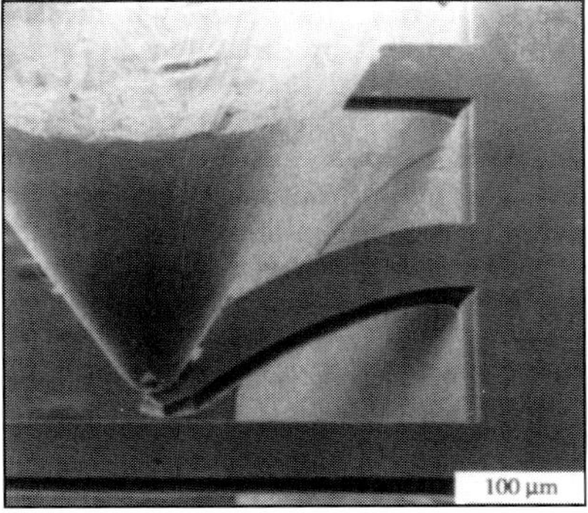

FIGURE 3. Cantilever beam bending experiment illustrating the remarkable elastic resilience of microsized silicon structures (Ref. 2).

packages that incorporate more of the circuit elements in the package close to the IC, on materials with low dielectric constant for higher speeds and lower losses at microwave frequencies, and at lower costs of fabrication (See Chapter 15 by Clive A. Randall and Joseph P. Dougherty).

The future in these fields is certain to be dynamic. Current research on silicon carbide (wide band gap) semiconductor devices is developing the technology for blue light lasers and light emitting diodes. Integrated circuits which can operate at 650°C have been fabricated in the research laboratory. In this field, the ability to manufacture high quality, large diameter ingots (and then wafers) is limiting progress. Plus, the materials science of defects and doping of a IV-IV compound semiconductor such as SiC is not well-developed. Similarly, GaN, which has a direct band gap, is being studied for LED and laser applications since conversion efficiencies to light should be 100X that of SiC which has an indirect band gap.

Diamond, with its outstanding carrier mobility, wide band gap, and extraordinary thermal conductivity is being researched for specialized semiconductor devices.

Photonics to date have been used primarily as the transmission media for information. Purely photonic devices that could assume some of the now electronic functions have not been fabricated in reliable, economic form. The ultimate goal is to have photonic transistors and logic gates which would lead to a photonic supercomputer which would be intrinsically faster and replace the electronic machines of today.

The idea of a superconducting supercomputer is still being pursued, particularly since the discovery of the oxide based superconductors with critical temperatures above the boiling point of liquid nitrogen.

MATERIALS FOR TRANSPORTATION

In this application area, the aerospace sector and land-based sector are discussed separately even though the driving force for improved materials and materials substitution is fundamentally the same-energy efficiency which translates to lighter weight in the vehicle and better efficiency in the propulsion system. In aerospace applications the life cycle cost benefit for weight savings is far greater than that for automotive, for example, so that much higher priced materials can be used in aerospace applications. Also, the structural performance and reliability requirements for aerospace applications are much greater than for land-based transportation, which again clearly draw distinctions between the two applications. However, there is considerable commonality in materials, particularly in the high performance automotive market, e.g., racing equipment. One good example is the use of carbon-carbon composite brakes on the latest commercial aircraft and on Indy cars.

Although there have been almost incredible development in avionics and a revolution in the use of electronics in land-based vehicles, they parallel the developments discussed in the previous section, and, therefore, are not discussed here.

Aerospace Applications

Advanced materials always appear first in military aircraft and then in helicopters and high performance business aircraft. the large commercial aircraft are more conventional in terms of materials used because of the large volumes of materials required and the extraordinary reliability requirements. Of course, space hardware, where the payoffs for weight savings are enormous, can be fabricated from very expensive materials, and thus, advanced materials, particularly composites, were first used there. The materials used in spacecraft in the final system cost about the same price as gold while those used in commercial aircraft are about the price of silver.

In discussing the developments in materials for aerospace, we must deal with the airframe separate from the engine since in the latter the advances are dedicated to increasing the operating temperature of the turbine, while in the airframe the objective has been to save weight while increasing structural performance (stiffness, strength, fatigue life).

Airframe

Aluminum alloys replaced wood and canvas as the primary airframe material some 75-80 years ago, and aluminum has been the dominant aircraft material since then. Clearly, advances in the microstructure via alloying and thermomechanical processing have led to alloys optimized for strength, for toughness, and for creep resistance (see Chapter 2 by Paul R. Howell). The most recent family of alloys is the Al-Li family which provides the lowest density alloys with improved stiffness.

Supersonic aircraft typified by the U.S. Air Force SR-71 or YF-12, introduced in the late 60's, created skin temperatures approaching 800°F so that titanium came to the fore as a material with higher temperature capability and greater structural efficiency for use in the airframe. At nearly the same time the Concorde was being built, and it was built primarily of aluminum alloys because of the high cost and low availability of titanium and the lack of performance data on titanium alloys. As a result, the Concorde was limited to Mach 2.0 because of the temperature limitations of aluminum alloys.

The major revolution in airframe design and construction came with the advent of high performance composites; first, graphite fiber and boron fiber reinforced polymers and later, metals (Al and Ti) reinforced with the same fibers. The most widely used composites in airframes are graphite reinforced epoxy. In Figure 4, a generalization of the use of materials in military and commercial airframes is depicted. On the order of 30% of the military airframe was composite in 1990, while the commercial airframe was much more conservative.

The planning for a High Speed Civil Transport (HSCT) which is a supersonic vehicle with a range of 5,000 miles is well underway with NASA being the lead agency. This aircraft will showcase the latest revolution in airframe design and materials utilization. The aircraft is projected to be a Mach 2.4 vehicle so that areas of the skin will be too hot for existing aluminum alloys or polymer matrix composites. Thus, titanium alloys will most likely be used selectively and in new hybrid laminates and sandwich structures which yield better specific properties (strength to weight and stiffness to weight ratios) and better impact, fatigue, and acoustic properties than monolithic materials. These hybrid laminate materials are already in use in secondary structures such as the cargo floor in the Boeing 777, etc. The specific laminate used for a specific application can be tailored for the conditions anticipated for the component. Automated fabrication of metal-polymer matrix composite laminates and sandwich structures of metal honeycomb and composite skin, for example, are being developed.

This HSCT aircraft is being designed for a 60,000 hr life under conditions for which we have very little durability data for the materials. A National Materials Advisory Board (NMAB) Committee has recently considered the designers' needs in this regard, and has recommended methods for the generation of the required data to assure safe use for this time period.

The remarkable gains in processing and fabrication of modern light metal alloys (superplastic forming, rapid solidification, etc.) which have been crucial developments are covered in Chapter 21 by Donald A. Koss).

Propulsion

The fact that the efficiency of a heat engine depends directly on the temperature at which the engine can operate has driven aircraft engine designers to remarkable

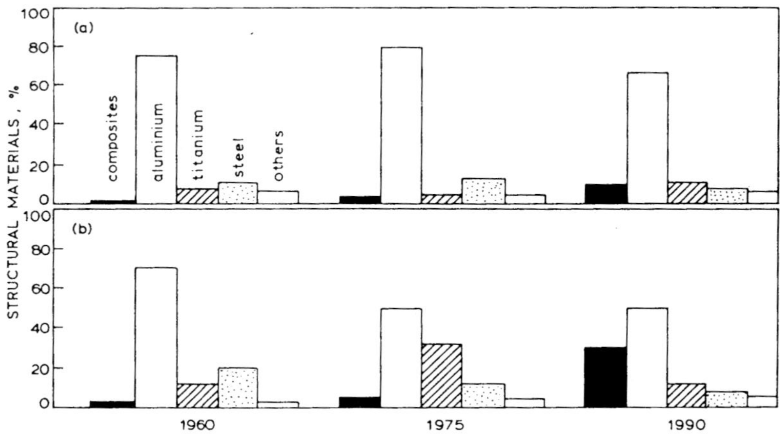

FIGURE 4. Materials used in (a) civilian, and (b) military aircraft, 1960-1990. (Courtesy of C. J. Peel, Materials Science and Technology, 1986, volume 2, pp. 1169-75).

ingenuity in the design of the engines and materials developers to alloy design at its highest level. The result has been remarkable gains in engine operating temperatures and thrust to weight ratios (Figure 5). Many of the gains in engine performance and efficiency have come from advanced design and advances in fluid mechanics. But the materials developments have been central in increasing the operating temperature of the structure. The developments in properties of superalloys (Ni and Co base alloys) have been spectacular. An example is the development of directionally solidified turbine blades (Figure 6) which are much more resistant to "creep," the gradual deformation of the blade under the stresses generated by the centrifugal force as the rotor spins at high speed. By directionally cooling the molten metal in the ceramic mold the grains grow elongated and aligned along the axis of the blade, thus causing orders of magnitude improvement in the creep resistance. This approach has been extended to the solidification of a blade which has no grain boundaries (single crystal) and is even stronger along the axis of the blade.

The use of zirconia-based oxide coatings with highly controlled microstructure (physical vapor deposited) on the surface of the turbine blades, so called "thermal barrier coatings," creates a significant temperature gradient from the hot gas to the metal surface, thus allowing the turbine to operate with an even higher

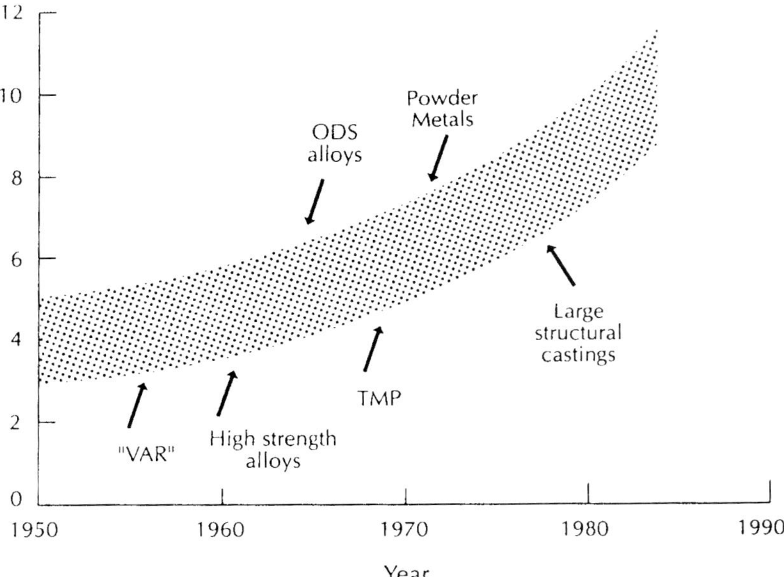

FIGURE 5. Engine thrust-to-weight ratio trends (Ref. 4).

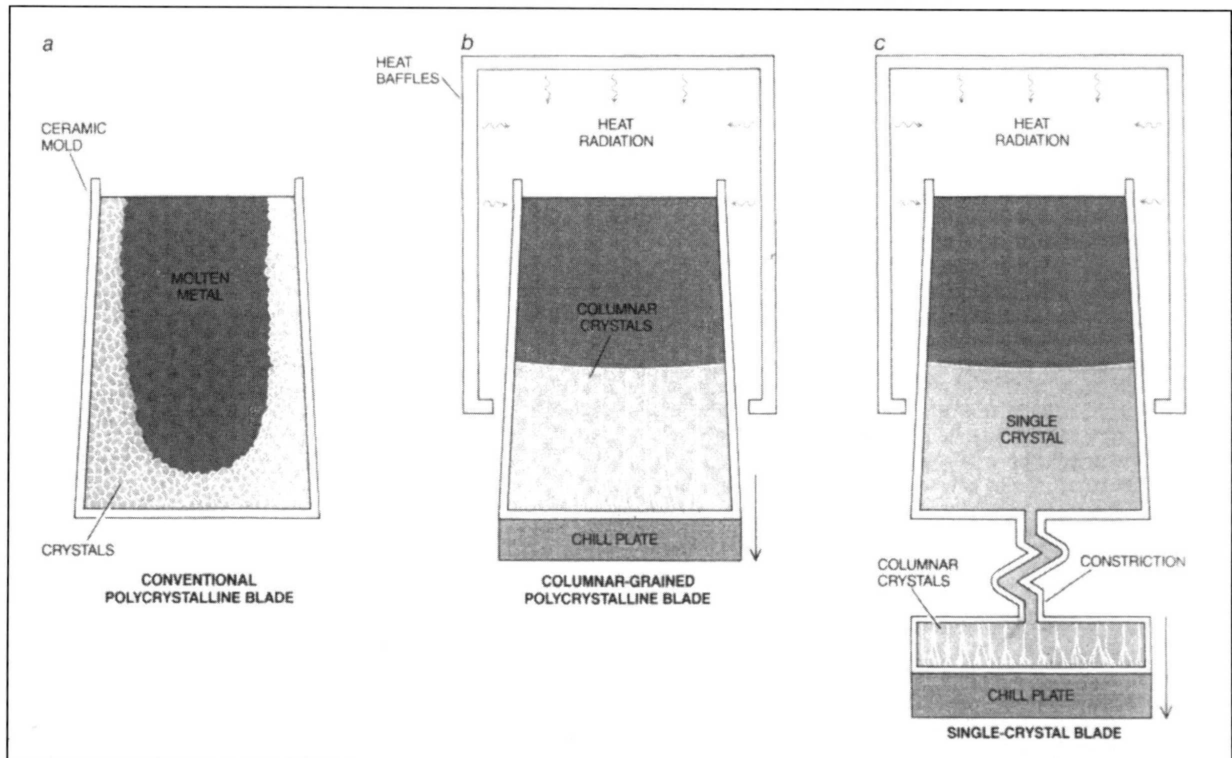

FIGURE 6. (a) A turbine blade fabricated by standard casting methods, (b) a directionally solidified turbine blade with elongated grains, and (c) a single crystal turbine blade produced by directional solidification (Ref. 5).

gas temperature. These coated blades have now experienced millions of hours of service and are being used in current production.

In the future, intermetallic materials such as titanium aluminides, iron and nickel aluminides are likely to be used in the hot end of the compression section and in the stator vanes because of their superior high temperature resistance. The engines for the HSCT will be massive structures which will require significant materials developments. For example, a ceramic or ceramic matrix composite liner for the combustor will allow the combustor to operate at 1540-1650°C versus approximately 1425°C which contributes to lower emissions and lower noise levels.

The use of ceramic matrix composites as panels on the exhaust nozzle of the engine will save 30% weight over the superalloys which would be required to withstand the temperatures. The French have already put C fiber reinforced SiC composites in the exhaust nozzle in their latest Mirage fighter, and these components were in the production model in 1996.

GROUND TRANSPORTATION

The production of land-based transportation vehicles is the largest component of the U.S. materials industry using ~60 million tons of metals, polymers, composites, glasses, and ceramics per year. Of course, the materials for infrastructure building and repair (concretes, asphalt, aggregate, etc.) dwarf these tonnages, but, of the advanced materials, cars, trucks, and trains use the largest total tonnage. The primary criteria for materials for automotive use has always been cheap, easy to process, and easy to secondarily form (weld, bend, etc.). The final system must be reliable and have long life. Since the 1970's, when our imported energy supply was in jeopardy, lighter weight has become a new criterion in the interest of fuel economy. As a result, significant changes in the mix of materials in the typical car have occurred. Also, new Federal regulation on increasing safety, reducing emissions and improving fuel economy along with foreign competition forced major changes in automobiles.

Focusing on the materials used, high strength steels that contain small concentrations of elements such as titanium and vanadium have replaced mild steels selectively. These micro-alloyed steels have strengths of 2-3 times the strength of conventional steels. Plastics, aluminum and glass fiber reinforced polymers have also made significant contributions to weight reduction. However, just comparing the densities and computing weight savings on an equal volume basis is not valid since the engineering properties must be compared to the design criteria for the component being replaced. For a stiffness critical application, for example, a much larger volume of aluminum must be used to replace cast iron because the elastic modulus of aluminum is less than one half that of cast iron. Thus, the savings in such a substitution is only 11%.

High strength steel in strength critical applications has a big advantage over mild steel, but in stiffness critical applications they are equal. Thus, the individual components must be examined in order to determine how much weight can actually be saved.

Light weight materials are generally more expensive than the materials that they replace, and thus, the saving of weight must more than compensate for the higher cost in order to be a cost effective substitution. In Figure 7, some substitutions are judged on this basis (in 1986, Ref. (6)), and, at that time, only high strength steel substituted for mild steel was truly cost effective on this basis. Yet light weight materials have been substituted in an industry where installed cost is the key parameter. The reason is that when *all* costs are included (fabrication cost,

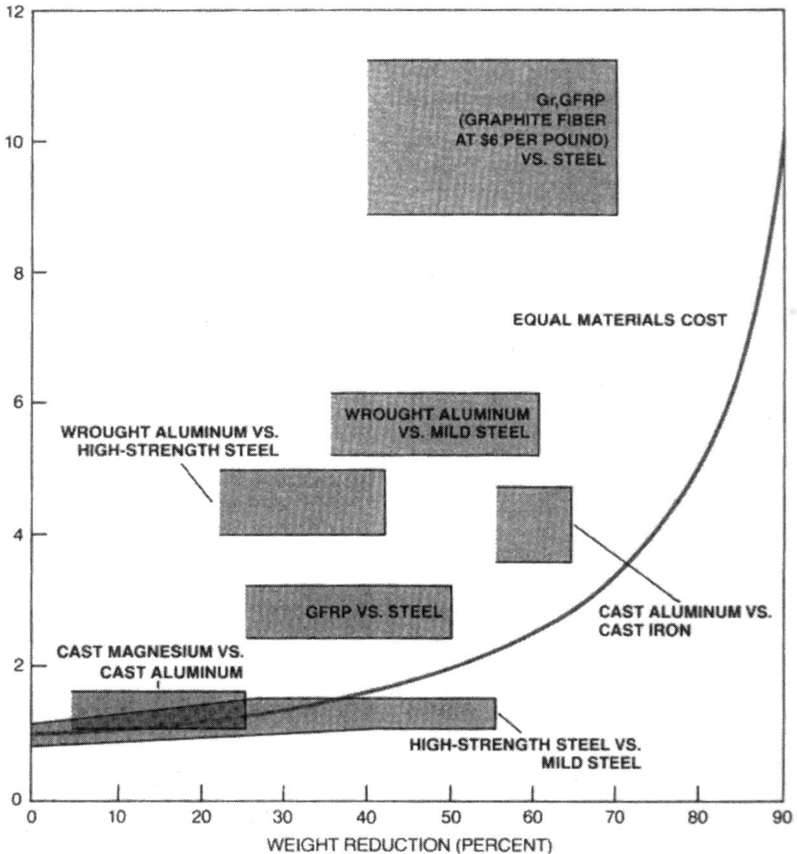

FIGURE 7. Cost of substituting a lightweight material in an automobile depends on the price of the new material relative to the conventional material and the weight savings (Ref. 6).

component integration advantages, tooling costs, facility costs, etc.) these materials are cost effective. For example, plastic parts can be molded in complex shapes which integrate into one part what would require several metal parts and an extra assembly process.

With the incorporation of various types of polymers, aluminum components, and fiberglass reinforced plastics, the recyclability of automobiles has actually decreased. The polymers essentially cannot be separated from each other and are not generally usable in the mixture found in automobiles, so they must be landfilled. Fiberglass cannot easily or economically be separated from the polymer matrix, thus limiting the extent to which the manufacturers are willing to use the material.

In Germany, automobiles must be 100% recyclable which puts serious constraints on the choice of materials. This type of regulation is expected to become global so that cost effective recycling processes must be developed.

Ceramics are being incorporated in key areas beyond the traditional sparkplug insulator and electrical and electronic devices. Turbocharger rotors are in mass production for automobiles in Japan, and are being introduced in diesel engines in the U.S. Critical wear resistant components in fuel injection systems are in production in silicon nitride ceramics. Ceramic valve train parts are being used in high performance engines and long life engines such as remotely sited diesel or natural gas engines for generators, pumps, etc.

The future will be one of advanced materials introductions in which the cost benefit ratio becomes favorable. Lightweight composite springs, crash worthy energy absorbing polymer blend body panels will become widely used as our fabrication experience and cost structure becomes favorable.

MATERIALS FOR ENERGY CONVERSION AND UTILIZATION

This application area is exceedingly diverse and includes nuclear energy conversion, fossil energy fueled power plants (coal, oil, natural gas), solar energy (large scale solar concentrator plants to solar cells), internal combustion engines, fuel cells, and batteries. The electric power generating industry is a conservative industry in terms of the introduction of new materials, primarily because of the enormous capital cost of new equipment, the long service lives required, and the requirement of reliable operation. The industry relies on conventional materials-steels, nickel-based alloys, oxide refractories, and concretes. Evolutionary improvements in materials which can be retrofit to existing plants is an emphasis. Maintenance, nondestructive testing and life prediction are all important aspects of assuring reliable operation.

However, new materials can have significant impact and are often borrowed from other sectors such as aerospace; for example, corrosion resistant coatings for turbine hot sections and creep resistant alloys such as single crystal turbine blades. New solar cell materials based on polycrystalline silicon have been developed so that efficiencies of 12% can be produced in mass and economically. New battery technologies which provide very high power densities are based on new materials such as lithium conducting polymers for the electrolyte in the lithium battery.

In the area of coal combustion, new ceramics and ceramic matrix composites are being developed for very high temperature heat exchangers and particulate filters for hot gas streams before the gas enters a turbine. The ceramics can withstand higher temperatures than metals and are more corrosion and erosion resistant.

The use of ceramics in land-based turbines as hot section parts is nearing reality with many successful demonstrations of small vehicular turbines. Current research is concentrating on larger engines which are used in the power industry for remote sites or for peak load generation.

In the area of fuel cells, the direct conversion of natural gas to electricity with efficiencies over 70% has been demonstrated using an oxide ion conducting zirconia-based electrolyte in a solid oxide fuel cell all of which operates at 1000°C and the exhaust powers a gas turbine. The Westinghouse Corporation has built two 27kw units which are now being rigorously tested for performance by Southern California Edison and a Japanese consortium. The target market for production machines is in the 3-25 mw range. The fabrication processes and reliability of these ceramic components must be improved to make this new technology fully competitive, economically, with advanced gas turbines.

MATERIALS FOR MEDICINE AND BIOMATERIALS

As in all of these general categories, this topic is extremely diverse. A National Research Council committee to survey the needs and opportunities in biomaterials segmented the field as follows (7):

- Synthetic biohybrid artificial organs
- Biosensors
- Biotechnology/growth factors
- Cardiovascular
- Commodity/disposals
- Drug delivery
- Genetic engineering (biomimetic materials, protein drugs, etc.)
- Maxillofacial, dental, EN&T, cranial
- Opthalmology
- Orthopedics
- Packaging
- Plastic and reconstructive surgery and
- Wound management

In all these areas materials are playing crucial roles and the active research is ongoing to develop new materials and to apply existing materials to the special needs of each segment.

In addition, the field of diagnostic instrumentation which is used external to the body has profited from materials research, and many opportunities exist for new and improved instruments. In the following paragraphs I've attempted to highlight some of the exciting new developments in several of these areas. An earlier summary covered many of the developments up to 1986 (Ref. 8).

One of the exciting developments in diagnostic materials was invented by one of our alums, Dr. Charles Greskovich of the G.E. Corporate R&D Center. Greskovich and his coworkers invented a transparent, polycrystalline x-ray scintillator material (a Nd-doped yttria material) which was much more efficient than its predecessor. This new x-ray detector is used in a large array in the latest CAT instrument, and it permits much more rapid imaging. This translates to much faster throughput and shorter residence time for the patient in this machine, thus reducing claustrophobic reactions. Also, the resolution and quality of images are much better with the new imaging system.

In the area of artificial organs, the popular press has kept us abreast of the artificial heart development and experimental use, but artificial kidneys and pancreas are also areas of active research and critical needs. In all of these cases, polymers play an important role in nondegradable formulations from sutures to vascular grafts to packaging and in bioabsorbable compositions from controlled release packages to sutures to absorbable implants. A major challenge continues to be fabrication of nonthrombogenic surfaces, surfaces which do not permit blood clots to form. The mimicking of the organ function, e.g., extracorporeal membrane oxygenator, is the goal of much polymer research. Membrane encapsulated living cells which are implanted into a diseased organ can secrete the desired product while filtering out antibodies which could attack the encapsulated cells. An artificial pancreas would contain insulin producing cells for treatment of diabetes. Polymers and sol-gel glasses are being researched for these applications.

In the drug delivery area Professor Langer's group at MIT has been a leading source of basic research and he has reviewed the various categories of controlled release processes (9). The systems either use pumps or polymers to deliver the drug. The pumps require surgery for implantation while the polymers generally do not. The polymers use one of the following mechanisms to release the drug: diffusion, solvent activation or chemical reaction. Some of these polymer carriers are implanted under the skin as in the Norplant[©] contraceptive which is a silicone rubber which in nondegradable and releases the steroid for over five years. Other drugs are delivered through the skin although the skin is a formidable barrier and drugs that are successfully delivered transdermally have a low molecular weight and appreciable solubility in water and oil. The most widely used example is the nitroglycerine patch for treatment of heart disease which was introduced in 1982. Current research is focused on using transdermal delivery more widely by using skin permeation enhancers, electric currents, ultrasonic excitation, and chemical modifications of the drug.

The Vietnam war, with its large number of young men with damaged skeletal structures, stimulated an activity in longer lived skeletal implants (orthopedic, maxillofacial, and dental). As a result, in the last three decades four basic methods for fixing load bearing implants to bone have been developed: mechanical fixation with screws and threaded devices, polymethylmethacrylate (PMMA) bone cement, porous coated implants for bone ingrowth and bioactive ceramic coatings or substrates such as metal alloys. All of these except PMMA cement have been used to fix dental implants.

In both the threaded implants and the porous-coated (or surface textured) implants, the fixation appears to involve bone formation within the threads and pores, respectfully, to form a mechanical interlocking. With the cemented implants, the cement appears to intrude into irregularities and pores on the inner surface of the bone, polymerize, thus creating a mechanical interlocking of cement and bone.

The "bioactive" ceramic coatings (for example Bioglass®) actually act as a substrate for growth of hydroxyapatite and collagen ("new bone"). The resulting bond is an intimate interdigitation of biological matrix and biomaterial which form a stronger bond than the substrate without the bioactive coating.

Recent work on hydroxyapatite as a dental and skeletal restorative has shown that the material can be installed as a slurry or plastic mass which sets rapidly at the body temperature. More details are presented in Chapter 18 by Paul W. Brown.

The progress in artificial skin development and artificial blood are very encouraging and the whole area of cultured organs and tissues is finding promising avenues for research. The interface between the biotechnology industry and the biomaterials industry will be a fertile one for developments which become rich markets.

SUMMARY

In this chapter I have tried to present a snapshot of materials developments in key application areas which provide the "pull" for new materials, new processes, and new assemblages. In the first set of chapters in the remainder of this volume, the scientific basis for new materials technology is reviewed across the broad spectrum of the materials research enterprise. In the second set of chapters the applications of the materials science is emphasized in the many niches of engineering materials technology. There are absences in our coverage since some key authors were not able to meet the publication deadline, but the breadth and richness of the field is certainly depicted with this selection of topics.

REFERENCES

1. NRC Committee on Materials Science and Engineering. 1989. *Materials Science and Engineering for the 1990's - Maintaining Competitiveness in the Age of Materials*. National Academy Press, Washington, D.C.

2. Schwertz, I.-A. 1992. Mechanical Characterization of Thin films by Micromechanical Techniques. *MRS Bull,* XVII(7):34-45.
3. NMAB Committee on Long Term Aging of Materials and Structures using Accelerated Test Methods. 1995. *Accelerated Aging of Materials Structures.* National Academy Press, Washington, D.C.
4. White, G. and D.W. Olson. 1990. *The New Materials Society: Challenges and Opportunities,* Vol. 2: New materials Science and Technology, pp. 5.1-5.77.
5. Kear, B.H. 1986. Advanced Metals. *Sci. Am.* 255(4):159-167.
6. Compton, W.D. and N.A. Cjostein. 1986. Materials for Ground Transportation, *Sci Am.* 255 (4):93-100.
7. Barenberg, S.A. 1991. Needs and Opportunities for the Biomedical Industry, *MRS Bull.,* XVI(9):26-38.
8. Fuller, R.A. and J.J. Rosen. 1986. Materials for Medicine, *Sci. Am.* 255: 119-125.
9. Langer, R. 1991. Drug Delivery Systems. *MRS Bull,* XVI(9):47-49.

Chapter One

THEORETICAL STUDIES OF PHASE TRANSFORMATIONS IN SOLIDS

L.Q. CHEN

Department of Materials Science and Engineering
Steidle Building
Pennsylvania State University
University Park, PA 16802

INTRODUCTION

Phase transformation is one of the most fundamental and important phenomena common to all kinds of materials including metals, ceramics, electronic and photonic materials, and polymers. Under a given thermodynamic condition: the temperature, pressure and overall composition, a material at thermodynamic equilibrium is made up of one or more phases with each having its unique crystal structure or composition. The physical properties of a material are determined by the number of phases and their volume fractions, their mutual spatial arrangement, the shapes, size and size distributions of phase particles, i.e., the *microstructure*. Changes in any of the thermodynamic parameters, temperature, pressure and composition, or application of external fields such as electrical, magnetic and stress fields, very often result in changes in the microstructure caused by phase transformations. Therefore, one of the efficient ways to obtain a desirable microstructure is by controlling phase transformations. As a matter of fact, phase transformations have been the underlying bases for the processing of many of the advanced materials (Please see Chapter 2).

A complete description of a phase transformation requires the understanding of two fundamental aspects: thermodynamics and kinetics. While the thermodynamics determines the possibility of a given phase transformation under a given thermodynamic constraint as well as the magnitude and the nature of the driving force, the kinetics addresses how, when, and in what sequence phase transformations will proceed. If we completely understood both aspects, then in principle, we should be able to predict the time-temperature dependencies under which a phase transformation occurs, and thus to predict the microstructure and properties of materials from fundamental scientific principles.

In 1963, a survey on phase transformations in solids was conducted by a group of distinguished materials scientists headed by Morris Cohen (1). It was summarized that "In light of this survey, it is evident that the science of phase transformations has a long way to go before it can be used to foretell the occurrence of solid-state reactions, or to predict their nature, kinetics and morphology". Indeed, we have come from a long way; for about thirty years since that survey, there have been many important advances in the theory of solid-solid phase transformations. Most of these new developments have been summarized in a number of excellent review articles and monographs. For example, various aspects of phase transformations are reviewed in the Materials Science and Technology Series: "Phase Transformations in Materials" (2); the static concentration wave theory for order-disorder transformations and the elasticity theory for coherent precipitates are extensively discussed in the monograph by Khachaturyan (3); and recent advances in the prediction of phase diagrams using electronic structure calculations in combination with statistical thermodynamics were emphasized in a recent review by de Fontaine (4) and in a monograph by Ducastelle (5). The proceedings from the solid-solid phase transformation conferences provide excellent references on new developments (6,7). The list is quite exhaustive and there exist many other excellent review articles and books emphasizing certain types of materials or certain kinds of phase transformations. In addition to the advances in the theory of phase transformations in solids, there has also been an explosion of research activities in the last decade using powerful computers to understand the science of phase transformations. These developments, coupled with increasingly sophisticated experimental techniques, have made materials science increasingly more quantitative and predictive. To a certain extent, now we are able to predict the possibility of phase transformations from first-principles, the rate of phase transformations and even the complex morphological evolution during various types of phase transformations in relatively simple systems. To simplify an analysis and reduce the computational time, still many approximations are made in both the theories and computational models, and we still have a long way to go in order to be able to predict from first-principles the phase transformations in complicated systems which are usually multicomponent and often have various structural defects such as vacancies or interstitials, dislocations, surfaces and grain boundaries. At this moment, it is fair to say that in the future the increasingly powerful computers will play a central role in realizing the dream of quantitatively predicting the structure and morphologies produced from phase transformations and the complexity of a system that can be dealt with will be directly correlated with the availability of computer resources.

This chapter is an attempt to highlight *some* of the major advances in the theory of phase transformations in crystalline solids during the last few decades. Because of the limit on the length of the manuscript, it is impossible to cover every aspect of and every new theory on such a broad subject, and only some aspects of solid-solid phase transformations will be briefly discussed, which reflect the author's interests and knowledge. Other important phase transformations such as liquid-solid transformations, glass transition and crystallization from the amorphous state, as well as ferromagnetic and ferroelectric transitions are not discussed.

THE CHEMICAL DRIVING FORCE FOR PHASE TRANSFORMATIONS

The first question to ask, with regard to a phase transformation, is whether or not a phase transformation is possible under a given thermodynamic condition. In principle, it can be simply answered by the classical Gibbs thermodynamics. The driving force for any phase transformation is the difference in the Gibbs free energies between the final and initial states; in the absence of any external fields, Gibbs free energy is the thermodynamic potential whose value has to be minimized at equilibrium for a given set of temperature, pressure and composition. As one of the parameters such as temperature, pressure, or composition changes, the originally equilibrium state will be brought out of equilibrium, i.e., to a nonequilibrium state. If sufficient time is allowed, a new equilibrium state will be established under the new specified thermodynamic condition. The question is then how to obtain the free energies of metastable and stable phases in a system.

Traditionally, the free energies of different phases as a function of temperature and composition are calculated by combining simple theoretical models such as the regular solution model with parameters obtained through experimental measurements such as calorimetry. There have been several computer softwares available for calculating phase diagrams using simple free energy models with computerized databases of thermodynamic properties of compounds and solutions (8). For essentially all systems of practical interest, this is still the most efficient way for obtaining the free energies of stable and metastable phases as a function of temperature and composition, thus the driving force for a given phase transformation. However, in the last decade, a rather powerful and desirable approach has emerged for calculating the free energies and thus the phase diagrams (4). It combines quantum mechanic electronic structure calculations with the statistical mechanics for solid solutions. Electronic calculations are performed to obtain the effective interaction energies among atomic species, which are then used as input to statistical mechanical calculations such as the Cluster Variation method (CVM) (9,10) to obtain the free energies. With the free energies of various phases as a function of composition and temperature, the transition temperatures among different phases can be determined.

In principle, such first-principle approaches can predict the free energies and the equilibrium phase diagrams for any system from the knowledge of only the atomic numbers of the species in the system. Currently, however, all these calculations are only possible for crystalline solids of single-component and binary alloy systems although some simple ternary alloys and even oxides have been studied (11, 12). The complexity increases as the number of component increases whereas alloys with commercial interest almost always contain many species. There are many approximations at each step of the calculations. For example, most of the calculations ignore the contribution from dynamic atom displacements, i.e., vibrational free energy contribution. It has difficulty incorporating long-range elastic interactions. The agreement between calculated and experimental phase diagrams are in most cases only qualitative; 50% difference between the experimentally measured transition temperatures and the calculated ones is not unusual. That is the reason why in all practical applications involving commercial systems, empirical methods using available thermodynamic data are being employed. Nevertheless, it is fair to say, the advances in the first-principle calculation of phase diagrams give us hope for the possibility of predicting the structure and properties *ab initio* rather than based on empirical methods. Improvement in electronic calculations and in the methods of obtaining the effective interaction energies must be achieved in order to have a real practical impact on materials design.

ELASTIC ENERGY CONTRIBUTION

The elastic strain energy contribution to solid-solid phase transformations is one of the areas which have received the most of attention in the last 10 to 15 years. Most solid-solid phase transformations involve a crystal lattice rearrangement which leads to a product phase whose lattice parameter is different from that of the parent phase (3). During the phase transformation and/or subsequent coarsening, the lattice mismatch between the new phase and parent phase must be accommodated to each other along the interphase boundary so as to minimize the total free energy of the muliphase mixture. An interphase boundary across which all lattice planes are continuos is called a coherent interface. For coherent interfaces, the lattice mismatch is accommodated entirely by elastic displacements of atoms from their regular lattice positions.

The study of elastic strain energy of inclusions was pioneered by Eshelby (13) who calculated the elastic strain energy of a coherent ellipsoidal inclusion in an elastically isotropic medium under the assumption that the inclusion and matrix have the same elastic moduli, i.e., homogeneous modulus assumption. It was advanced by Khachaturyan who proposed a general theory of strain energy in an arbitrary two-phase mixture with arbitrary crystal lattice rearrangement, arbitrary shape and mutual arrangement of inclusions in elastically anisotrophic media (14). However, in order to obtain analytical solutions, homogeneous modulus was assumed. Even with this assumption, Khachaturyan's theory has found tremendous success in analyzing precipitate morphology and habits in both alloys and ceramics, irrespective of the nature of phase transformations.

Our current understanding of the thermodynamics of coherent phase equilibria was primarily due to the work of Larche and Cahn (15), Williams (16), and Johnson and Voorhees (17), and was recently reviewed by Lache (18). An important and fundamental consequence of including the elastic energy contribution to the total free energy is the fact that the usual Gibbs thermodynamics is broken down as the free energy is no longer additive and the usual level rule is no longer valid. Moreover, the equilibrium compositions are functions of the overall composition. For example, if the elastic energy is sufficiently high, a two-phase mixture may become unstable within a certain, or in the extreme case, the entire composition range of the two-phase field of a phase diagram.

The morphological and microstructural evolution of a coherent mixture has been mainly investigated by numerical computer simulations due to the difficulty of obtaining any analytical solutions. The long-range elastic interactions have been shown to be responsible for a wide variety of rather unusual kinetic morphological phenomena such as strong directional alignments and spatial correlation of second-phase precipitates, particle shape transition and reverse coarsening, particle splitting and formation of tweed and polytwin structures (19-26). These phenomena cannot be understood in terms of the chemical driving force alone.

Most of the existing theoretical studies have made the assumption of homogeneous modulus. However, there are many examples in which the difference in elastic moduli is very important. For example, the effect of applied stress on the precipitate morphology with dilatational misfit can not be studied without including the elastic inhomogeneity. Theory of elastic strain energy in inhomogeneous solids have just started to emerge. For example, Khachaturyan and his coworkers have recently proposed a theory of elastic energy in elastically anisotropic and inhomogeneous systems employing a perturbation theory in combination with Green's functions (27). Computer simulations of the effect of elastic inhomogeneity on the multiphase morphology and evolution have been performed in model systems (19, 25). The possibility that coarsening kinetics of coherent precipitates may be totally frozen due to elastic inhomogeneity, suggested by Onuki (19), requires further investigation, since it may have important implications in alloy design.

Another important area is the effect of applied stress. Because of the difference in the elastic moduli, application of an external stress would affect crystal lattices of precipitates and matrix differently. As a result the applied stress will cause a crystal lattice mismatch in addition to the transformation-induced lattice mismatch. The total mismatch is accommodated by the elastic strain energy within the entire body of a multiphase alloy. Therefore, the microstructures of a coherent two-phase mixture can be modified by an applied stress. For systems containing several orientation variants of the precipitate, applied stress will result in certain variant(s) to drop out, and therefore modify the microstructure. This is an area which most likely will receive a lot of attention in the near future.

ORDER-DISORDER TRANSFORMATIONS

Although order-disorder transformations were primarily discussed in the context of alloys, they have been observed in ceramics and semiconductors as well. At sufficiently low temperatures, each species of atom in a given material occupies only a certain sublattice. A material in such a state is called completely ordered. As the temperature increases, some of the atoms will occupy wrong sublattice sties (sites which were occupied by other types of atoms in the completely ordered state) and the degree of order decreases; the corresponding material is called partially ordered. A disordered material is one in which the concentrations of atoms on different crystal lattice sites become equal at high temperatures but below its melting temperature. In this sense, we can view all compounds as ordered phases whether they are metals, ceramics or semiconductors. For most ceramic compounds, however, their order-disorder transition temperatures are above their melting temperatures, and hence one can not observe the order-disorder transformations. Assuming that we can obtain the fundamental interactions between atoms either from first principles, empirical potentials, or experimental measurements, there are two basic questions that need to be addressed with regard to order-disorder transformations: (1) what are the equilibrium order-disorder transition temperatures and the nature of these transitions? and (2) how fast does the ordering or disordering take place and how do the microstructures evolve during ordering and disordering?

The earliest theoretical model for order-disorder transformation is the Bragg-Williams model (28) considering only the nearest neighbor interactions and only the single-site point probability distribution for a given type of atom at a given site on a crystal lattice. In the last few decades, many different types of theoretical approaches have been proposed to study the thermodynamics of order-disorder transitions (5). The two which have been most often used are the static concentration wave method (3) and CVM (7,8). Of course, Monte-Carlo computer simulations (29) have also been employed to determine the order-disorder transformation temperatures. The main advantage of the static concentration wave method is the fact that it can take into account long-range interatomic interactions such as long-range electrostatic interactions and elastic interactions and it can predict the atomic structures of ordered phases, or the superlattices vectors, and the transition temperature without a priori assumptions. On the other hand, the static concentration wave method is a mean-field theory considering only the point probabilities although the accuracy of a point approximation increases with the range of interatomic interactions (30). CVM can take into account higher order atomic correlations such as pair, three-point, etc., correlations, and with the proper choice of the clusters, it can predict the order-disorder transition temperature with a very good accuracy as compared to those obtained from Monte-Carlo simulations. However, in order to predict the order-disorder transition temperature, CVM needs to know in advance what is the ordered structure. Furthermore, the complexity of carrying out a CVM calculation increases exponentially as the cluster size required or the range of interaction. Although Monte-Carlo can take into account all the correlation effects exactly, it is much more time consuming than CVM, and the computer time required increases rapidly with the range of interatomic interactions.

In summary, the concentration wave method is a very simple and elegant technique for predicting the ordered structures and their transition temperatures with known interatomic interactions. It should be employed in systems with very long-range interactions whereas CVM or Monte-Carlo computer simulations should be employed in systems with short-range interactions and when the accuracy of the ordering temperature is important.

An initially disordered phase annealed below its order-disorder transition temperature is expected to undergo ordering. For predicting the kinetics of ordering and the accompanying morphological evolution, various kinetic models have been developed, including microscopic diffusion equations (3), microscopic master equations (31-35), path probability methods (PPM) (36-37) and Monte-Carlo computer simulations (29). Microscopic diffusion equation describes the relaxation of the single-site occupation probability function using Onsager-type of diffusion equations. It has been applied not only to ordering kinetics but also to the microstructure evolution in much more complicated systems. It can easily incorporate long-range interactions such as long-range elastic interactions. Microscopic master equations characterize the ordering rates through the kinetic evolution equations for the cluster probability functions of different sizes. PPM is a kinetic version of the equilibrium CVM. Most existing applications of master equation methods and PPM are concerned with the ordering kinetics in homogeneous systems and the maximum cluster size which has been treated is a tetrahedron. Both of them become increasingly complicated as the cluster size increases. The master equation method has been applied to inhomogeneous systems for studying the morphological evolution during ordering and phase separation, but only in the point and pair approximation (35) and the inhomogeneous version of PPM is currently under development (38). In principle, Monte-Carlo may be applied to morphological evolution during ordering. However, there are a number of reasons that the kinetic equation approaches are, in many cases, preferred. First of all, the probability distribution functions generated by the kinetic equations are averages over a time-dependent nonequilibrium ensemble whereas in Monte-Carlo, a series of snapshots of instantaneous atomic configurations along the simulated Markov chain are produced. Therefore, a certain averaging procedure has to be designed in Monte-Carlo in order to obtain the information about local composition or local order. Secondly, while the time scale is clearly defined in kinetic equations, it is rather difficult to relate the Monte-Carlo time steps to real time. Finally, in most cases, simulations based on kinetic equations are computationally more efficient than Monte-Carlo. This is also the main reason that most of the equilibrium phase diagram calculations in alloys were performed using CVM instead of Monte-Carlo.

SPINODAL DECOMPOSITION

Decomposition of a homogeneous phase into a mixture of two phases with the same crystal structure but different compositions is called isostructural decomposition. As Gibbs realized more than 100 years ago that there exist two types of composition fluctuations that may eventually lead to decomposition: one is small in spatial extent but large in degree, which occurs when the initial homogeneous phase is metastable with respect to decomposition; and the other is large in spatial extent but small in degree, which takes place when the initially homogeneous phase is unstable with respect to decomposition. Decomposition through the first type of fluctuation is called nucleation-and-growth (discussed later) and that through the second type is referred to as spinodal decomposition.

The first kinetic theory of spinodal decomposition was developed by Cahn who analyzed the kinetics of spinodal decomposition by linearizing the nonlinear diffusion equation now known as the Cahn-Hilliard equation (39). It has been quite extensively applied to analyzing experimental data obtained from systems exhibiting spinodal decomposition (40). Since then, a non-linear theory has been put forward by Langer, Bar-On, and Miller (41), called the nonlinear spinodal theory of Langer-Baron-Miller. More recently, based on the nonlinear spinodal decomposition theory, Langer et al (42) have tried to develop a unified theory for nucleation and spinodal decomposition. On the other hand, Binder et al (43) developed a generalized nucleation theory using the cluster dynamics approach. Based on the cluster dynamics approach, Binder et al showed that there is no sharp change from the nucleation-and-growth regime to spinodal decomposition across the spinodal line in the binodal phase diagram, rather the change is gradual and smooth. However, except for theoretical analyses, these nonlinear theories as well as the so-called generalized theories have not been applied to interpreting experimental measurements.

There have been extensive studies of spinodal decomposition using computer simulations. Most of the computer simulation studies are based on either the Cahn-Hilliard equation or Monte-Carlo simulations, and focused on dynamic scaling of spinodally decomposed morphologies (44-46). Most of theories and computer simulation studies have been on binary systems, although the dynamics and morphological evolution in ternary model systems have been by Cahn-Hilliard equations and microscopic diffusion equations (47-50).

PRECIPITATION INVOLVING BOTH ORDERING AND DECOMPOSITION

Isostructural decomposition and ordering are two of the simplest diffusional phase transformations. Early theoretical works employing only the nearest neighbor interactions concluded that ordering and spinodal decomposition are two mutually exclusive processes. Only until recently it is realized that very often ordering and decomposition take place simultaneously or sequentially depending on a system. For example, it occurs during decomposition of a quenched homogeneous disordered phase into a two-phase mixture of ordered and disordered phases within an equilibrium two-phase field of a phase diagram. The dynamics is described by two order parameters; one is the composition characterizing the compositional difference between the ordered and the disordered phases and the other is the long-range order parameter which distinguishes the symmetry difference. Recent theoretical works based on thermodynamic stability analysis combining with some kinetic arguments (51-53) showed that when a homogenous disordered phase is quenched into a two-phase field below the ordering instability line, the first process that takes place during isothermal aging is a congruent ordering, resulting in appearance of a nonstoichiometric ordered single-phase. The congruently ordered single-phase has the same composition as the initial disordered phase. This corresponds to the much faster relaxation kinetics of the non-equilibrium state with respect to the long-range order than that with respect to the compositional clustering. The congruently ordered single-phase is, however, thermodynamically unstable with respect to phase separation. Therefore, it later decomposes through a spinodal mechanism into the equilibrium ordered and disordered phases. Of course, if the aging temperature is above the absolute ordering instability line but within the two-phase field, non-classical or classical nucleation will take place. These predictions are also confirmed by computer simulations of phase separation with ordering using TDGL continuum equations (54) and microscopic diffusion equations (55). Computer simulations showed that the subsequent phase separation of the congruently ordered single-phase occurs predominantly along the antiphase domain boundaries, i.e., it occurs heterogeneously.

The predictions made for the precipitation of an ordered intermetallic from a disordered matrix could be generalized for any systems whose kinetics are described by a conserved and a nonconserved order parameters. For example, if a cubic phase is quenched below its absolute instability line of transition into the tetragonal phase within a two-phase field of cubic and tetragonal phases, the first reaction should be the congruent transformation of the initial cubic phase into a single-phase tetragonal phase, followed by the decomposition of this tetragonal phase into the equilibrium two-phase mixture of cubic and tetragonal phases. However, nucleation and growth will occur if the initial quenched state is annealed above the absolute instability line but within the two-phase field. In the case of solidification of binary alloys, quenching a liquid below the instability line for crystallization, or called the to line, the first reaction will be the massive transformation of the whole liquid into a single-phase crystalline material. And the second stage of transformation involves the decomposition of the single-phase solid into a mixture of liquid and solid. It could be expected that liquid should reappear starting from grain boundaries.

NUCLEATION AND GROWTH

Nucleation and growth takes place when the initial uniform disordered phase is metastable, i.e., it is stable with respect to small fluctuations in the independent variables describing its free energy and unstable with respect to large fluctuations. It is the oldest problem, but seems to be the most difficult. In most cases, the classical nucleation theory is still used to estimate the nucleation rate based on available thermodynamic and kinetic data. In the classical nucleation theory, the nucleus is assumed to have the same structure and composition as the equilibrium precipitate phase and the interface between the nucleus and the matrix is sharp. The interfacial energy is assumed to be the same as a flat interface. The competition between the thermodynamic driving force and the creation of new surfaces results in the concept of critical nucleus size at which the total free energy increase (volume chemical energy + elastic strain energy + surface energy) is a maximum.

Based on their diffuse-interface theory Cahn and Hilliard (56) developed a continuum model taking into account the diffuse nature of interfaces, and studied the composition profiles of a critical nucleus, as a function of matrix compositions. They found that, for matrix compositions near the phase boundary, the composition within a critical nucleus was almost identical to that of the equilibrium precipitate phase. As the matrix composition increases, the profiles of a critical nucleus became increasingly diffuse, with the composition within the nucleus approaching that of the matrix. Based on this study, they concluded that classical nucleation, which required the nucleus composition to be fairly uniform and close to precipitate phase, was operative only when the matrix composition of the alloy was close to phase boundary. They also found that the radius of the critical nucleus diverged to infinity not only near the phase boundary, as predicted by the classical nucleation theory, but also near the spinodal line. More recently, LeGoues et al. (57) used a discrete lattice point model, to calculate the profiles of the occupation probabilities allowing interfacial energy anisotropy and seem to confirm most of the predictions by Cahn and Hilliard.

Recently, Poduri and Chen (58) extended the original non-classical nucleation theory of Cahn and Hilliard for isostructural decomposition, to the case in which the precipitate and matrix phases have different structures and compositions. They applied the non-classical theory to a particular example, precipitation of ordered particles from a disordered matrix in which the elastic energy contribution to nucleation is ignored. It was shown that it is only when the matrix composition is near the phase boundary of the disordered phase, that the composition and order parameter values inside the nucleus are close to those of the equilibrium ordered phase, and that the critical profiles become increasingly diffuse as the ordering instability line is approached. In contrast to the thermodynamic stability analysis, which predicted a region of congruent nucleation and growth, it is found that the critical nucleus consists of both composition and order parameter fluctuations through the entire composition range, from the disordered phase boundary to the ordering instability line.

In solids, nucleation of a new phase particle usually involves an elastic energy increase. In the classical nucleation theory, this elastic energy is simply added to the bulk driving force term so that the total driving force for nucleation is reduced. However, elastic energy is also dependent on the shape of a particle. Therefore, nucleation models which can predict not only the size but also the shape need to be developed. Furthermore, since elastic interactions are very long range, the nucleation at different locations is likely to be correlated. Such complicated cases have not been analyzed yet and may have to be solved using numerical simulations. Once the critical size and shape of a nucleus is determined, the rate of nucleation can be estimated (59).

DIFFUSIONLESS TRANSFORMATIONS

Both the order-disorder transformation and spinodal decomposition discussed above are diffusional transformations. Order-disorder transformations involve diffusion over distances comparable with interatomic separations whereas in spinodal decomposition the diffusion distance is on the order of the size of composition domains. Massive transformation, which is not discussed here, is a composition-invariant transformation involving diffusion only at the interfaces. Phase transformations which do not require diffusion at all are called displacive/diffusionless transformations. Cohen et al subdivided the displacive transformations into two main groups: shuffle transformations and lattice-distortive transformations (60). In shuffle transformations, the morphology is dominated by interfacial energy anisotropy caused by shuffle displacements which produce no net strain within the unit cell. On the other hand, lattice distortive transformations result in a homogeneous strain. The best known and most thoroughly studied diffusionless and lattice distortive transformation is the martensitic transformation which takes place by nucleation and growth and in which the morphology is dominated by shear strains (61). The phenomenological theory of martensitic transformation crystallography was developed more than 40 years ago (62) and has been applied to predicting the transformation morphology very successfully. The growth of a martensite is usually very fast with a speed on the order of sound velocity and hence the transformation kinetics is dominated by the nucleation process. The main theoretical question is centered on the issue how martensites are nucleated. Both classical and nonclassical nucleation theory have been applied to the nucleation of martensite (61). Since nucleation of a martensite involves lattice deformation which causes a large increase in elastic strain energy and it is concluded that homogeneous nucleation of a martensite is impossible because of the huge barrier for nucleation as compared to the thermal energy (61). As a result, it was proposed that defects must play a role in martensite nucleation. It is indeed shown that nucleation of martensite on defects, e.g., on dislocations reduces the barrier tremendously (63). Molecular dynamics computer simulations also demonstrate that defects can promote the nucleation of martensite (64). Very recently, Khachaturyan (65) proposed another mechanism for martensite nucleation. He showed that martensite can be homogeneously nucleated through an intermediate phase called the adaptive phase. An adaptive phase can be regarded as a microscopic limit of conventional accommodation twinning with twin thickness of one or several atomic layers. The main geometrical feature of the crystal lattice of the adaptive phase is that it is related to the parent phase lattice by an invariant plane crystal-lattice rearrangement, and the volume-dependent part of the elastic strain energy vanishes. The formation of adaptive phase as the first nucleation product seems to be confirmed by a recent computer simulation by Wang and Khacharturyan employing time-dependent Ginzburg-Landau equations (66).

In summary, direct homogeneous nucleation of martensite is impossible while heterogeneous nucleation on defects such as dislocations and indirect homogeneous nucleation of martensite through an intermediate adaptive phase are both possible. Which mechanism operates in a given system will depend on the system and experimental condition.

SUMMARY

In the last few decades, there have been many major advances in the theory of structural transformations in solids. The topics discussed in this chapter are just a few examples of those theoretical developments. There are others that the author is unaware of or does not understand well enough to discuss them. We still have a long way to go in order to be able to predict the thermodynamics and kinetics of phase transformations from first-principles in systems of practical interest. Some of the areas which are likely to receive a lot of attention in the near future include:

(1) Ab initio first-principle calculations of phase diagrams, and therefore the driving forces for phase transformations. Progresses are likely to be seen in the area of improving the accuracy of electronic calculations by removing some approximations and developments of techniques to incorporate the contributions of both static and dynamic displacements to the driving force.

(2) Most of the theories are proposed for phase transformations in bulk systems. However, the fundamental understanding of the thermodynamics and kinetics of phase transformations in finite systems are badly needed as modern devices become increasingly small. This should be true for both diffusional and diffusionless transformations. For example, can we apply the theories that we developed for bulk systems be applied to nanoscale bulk systems or thin films?

(3) Up to now, theoretical research in structural phase transformations relies on analytical theories and models which can be solved only for simple cases with many assumptions. Significant improvement on analytic theories will be difficult. Further major progress in the understanding of phase

transformations, in particular, the kinetics of phase transformations and the accompanying microstructural evolution, is likely to be achieved through extensive computer simulations based on physcially sound and realistic models.

REFERENCES

1. Cohen, M., Cahn, J.W., Christian, J.W., Flinn, P.A., Hillert, M., Kaufman, L. and T.A. Read. 1965. Perspectives in Materials Research, p. 309. Office of Naval Research, Washington.
2. Phase Transformations in Solids, Materials Science and Technology, Vol. 5, edited by R.W. Cahn, P. Hassen and E.J. Kramer, New York. 1991.
3. A.G. Khachaturyan. 1983. Theory of Structural Phase Transformations, Wiley & Sons, New York.
4. D. deFontaine. 1994. In "Solid State Physics", vol. 49, pp. 33-176, and references therein.
5. F. Ducastelle. 1991. Order and Phase Stability in Alloys, North-Holland, New York.
6. Proceedings of an International Conference on Solid-Solid Phase Transformations, edited by H.I. Aaronson, D.E. Laughlin, R.F. Sekerka and C.M. Wayman. 1982. Warrendale, PA: TMS.
7. Solid-Solid Phase Transformations in Inorganic Materials '94, edited by W.C. Johnson, J. M. Howe, D.E. Laughlin and W.A. Soffa. 1994. Warrendale, PA: TMS.
8. For a review of metallurgical thermodynamic databases, see C.W. Bale and N.G. Eriksson. 1990. Canad. Metall. Quart. *29*, 105-132.
9. R. Kikuchi, Phys. Rev. 1951. 81, 988.
10. Sanchez, J.M. and D. de Fontaine. 1978. Phys. Rev. *B 17*, 2926.
11. Burton, B.P. and R.E. Cohen. 1995. Phys. Rev. *B 52*, 792.
12. Tepesch, P.D., Garbulsky, G.D. and G. Ceder. 1994. Phys. Rev. Lett. *74*, 2272.
13. Eshelby, J.D. 1959. Proc. Roy. Soc. (A) *241*, 376; ibid 1959, *252*, 561.
14. Khachaturyan, A.G. 1967. Sov. Phys. Solid State *8*, 2163; A.G. Khachaturyan and G.A. Shatalov. 1969, ibid *11*, 118.
15. Lache, F. and J.W. Cahn. 1973. Acta metall. *21*, 1051; 1978, ibid *26*, 53; 1978, ibid *26*, 1579.
16. Williams, R.O. 198o. Metall. Trans. *11A*, 247; CALPHD 1978, 8,1.
17. Johnson, W.C. and P.W. Voorhees. 1987. Metal. Trans. *18A*, 1213.
18. Lache, F. 1990. Annu. Rev. Mater. Sci. *20*, 83.
19. Nishimori, H. and A. Onuki. 1990. Phys. Rev. *B 42;* A. Onuki and H. Nishimori. 1991, Phys. Rev. *B 43*, 13649; H. Nishimori and A. Onuki. 1991, J. Phs. Soc. Jpn. *60*, 1280.
20. Wang, Y., Chen, L.Q. and A.G. Khachaturyan. 1993. Acta metall. et mater. *41*, 279.

21. Johnson, W.C., Abinandanan, T.A. and P.W. Voorhees. 1990. Aca metall. mater. *38*, 1349.
22. Voorhees, P.W. and W.C. Johnson. 1988. Phys. Rev. Lett., *61*, 2225.
23. Lee, Jong, K. 1995. Scripta metall et mater. *32*, 559.
24. Chen, L.Q., Wang, Y.Z. and A.G. Khachaturyan. 1992. Phil. Mag. Lett. *65*, 15.
25. Semenovskaya, S. and A.G. Khachaturyan. 1991. Phys. Rev. Lett. *67*, 2223.
26. Fan, D.N. and L.Q. Chen. 1995. J. Am. Ceram. Soc *78*, 769.
27. Khachaturyan, A.G., Semenovskaya, S. and T. Tsakalakos. 1995. Phys. Rev. *B 52*, 15909.
28. Bragg, W.L. and E.J. Williams, Proc. Roy. Soc. 1934. *A145*, 699; 1935, *A152*, 231.
29. Binder, K. and D.W. Heermann,. 1988. Monte Carlo Simulation in Statistical Physics: An Introduction. Berlin: Springer.
30. Vaks, V.G., Larkin, A.E. and S.A. Pikin. 1968. Sov. Phys. Solid State *9*, 2249.
31. Van Baal, C.M. 1982. Physica *A111*, 591; 1982, *A113*, 117; 1993, *A196*, 116; 1985, *A129*, 601.
32. Fultz, B. 1989. Acta metall. *37*, 823; 1990, J. Mater. Res., *5*, 1419.
33. Martin, G. 1990. Phys. Rev. B *41*, 2279.
34. Penrose, O. 1991. J. Stat. Phys., *63*, 975.
35. Chen, L.Q. and J.A. Simmons. 1994. Acta metall et mater., *42*, 2943.
36. Kikuchi, R. 1960. Ann. Phys., Vol. 10, 127; 1969, Progr. Theor. Phys. Suppl., *35*, 1.
.37. Mohri, T. 1994. In *Solid-Solid Phase Transformations*, edited by W.C. Johnson, J.M. Howe, D.E. Laug Soffa, p. 52.
38. Kikuchi, R. 1995. Private communication.
39. Cahn, J.W. 1962. Acta metall *10*, 179; 1962, ibid *10*, 907.
40. Hilliard, J.W. 1970. In *Phase Transformations*, edited by H.I. Aaronson, Metals Par: Am. Soc. Metals, 219.
41. Langer, J.S., Baron, M. and H.D. Miller. 1975. Phys. Rev. *A11*, 1417.
42. Langer, J.S. 1975. In Fluctuations, Instabilities, and Phase Transisitons, edited by T. Riste, New York, Plenum Press, 19.
43. Binder, K. 1981. In Stochastic Nonlinear Systems in Physics, Chemistry and Biology: edited by L. Arnold and R. Lefever, Berlin:Springer, 62.
44. Gunton, J.D., San Miguel, M. and P.S. Sahni. 1983. In Phase Transitions and Critical Phenomena, Vol. 8 edited by C. Domb and J.L. Lebowitz, London: Academic Press, 267.
45. Kalos, M., Lebowitz, J.L., Penrose, O. and A. Sur. 1978. J. Stat. Phys. *18*, 39.
46. Bray, A.J. 1994. Adv. in Phys. *43*, 357.
47. Morral, J.E. and J.W. Cahn. 1971. Acta metall. *19*, 1037.
48. Hoyt, J.J. 1989. Acta metall. *37*, 2489.
49. Eyre, D.J. 1993. Systems of Cahn-Hilliard Equations, SIAM J. Appl Math. *53*, 1686.
50. Chen, L.Q. 1994. Acta metall. et mater. *42*, 3503.
51. Allen, S.M. and J.W. Cahn. 1976. Acta metall. *24*, 425.

52. Khachaturyan, A.G., Lindsey, T.F. and J.W. Morris, Jr. 1988. Met. Trans. *A19,* 249.
53. Soffa, W.A. and D.E. Laughlin. 1989. Acta metall. *37,* 3019.
54. Matsumura, S., Tanaka, Y., Tanako, K. and K. Oki. 1990. Proc. Int. Workshop on Computational Materials Science, Tsukuba, Japan, 103.
55. Chen, L.Q. and A.G. Khachaturyan. 1991. Acta metall. et mater. *39,* 2533.
56. Cahn, J.W. and J.E. Hilliard. 1959. J. Chem. Phys. *31,* 688.
57. LeGoues, F.K., Lee, Y.W. and H.I. Aaronson. 1984. Acta Met. *32,* 1837.
58. Poduri, R. and L.Q. Chen. 1996. Acta mater., 44,4253.
59. Russel, K.C. 1970. In *Phase Transformations,* edited by H.I. Aaronson, Metals Par: Am. Soc. Metals, 219.
60. Cohen, M., Olson, G.B. and P.C. Clapp. 1979. In Proc. ICOMAT-79, MIT, 1.
61. Olson, G.B. 1992. In martensite, edited by G.B. Olson and W.S. Owen, ASM Int., Metals Park, Ohio.
62. Wechsler, M.S., Lieberman, D.S. and T.A. Read. 1953. Trans. AIME *197,* 1053; J.S. Bowles and J.K. MacKenzie 1954. Acta metall. *2,* 129.
63. Olson, G.B. and M. Cohen. 1982. In Solid-State Phase Transformations, edited by H.I. Aaronson, D.E. Laughlin, D.F. Sekerka and C.M. Wayman, Warrendale, Met. Sco. AIME, 1145.
64. Clapp, P.C., Shao, Y. and J.A. Rifkin. 1993. MRS Symp. Proc. *246,* 1.
65. Khachaturyan, A.G., Shapiro, S.M. and S. Semenovskaya. 1991. Phys. Rev. *B 43,* 10832.
66. Wang, Y. and A.G. Khachaturyan. 1997. Acta mater., 45,759.

Chapter Two

EXPERIMENTAL STUDIES OF SOLID TO SOLID PHASE TRANSFORMATIONS IN METALS AND ALLOYS

PAUL R. HOWELL

Department of Materials Science and Engineering
Steidle Building
Pennsylvania State University
University Park, PA 16802

INTRODUCTION

The study of solid to solid phase transformations in metals and alloys is of vital importance in that it provides the ultimate link between microstructures and properties. Phase transformations have been studied for many decades and our understanding of the mechanisms of how one phase transforms into a mixture of one or more phases has increased as the sophistication of available experimental techniques has increased.

The development of the science of crystallography over the centuries has also been of vital importance to the study of phase transformations because in their lowest energy state, all solid metals are crystalline. One of the most fundamental laws of crystallography, the "Law of Constancy of Angle," had its genesis in the work of Nicolaus Steno[1]. He took numerous crystals of quartz, cut them perpendicular to what we now know is the triad axis (e.g., along ABCD in Figure 1), traced their outlines on paper and measured the angles between the various faces. No matter how irregular the crystal the angles were always 120°.

Unfortunately, metals rarely exhibit an external morphology which is remotely controlled by internal symmetry and crystal structure determination had to await the discovery of x-rays. Bragg's published the first crystal structure analysis (albeit of sodium chloride) in 1913 [2]. By the 1930's, the crystal structures of most metals and many alloys had been determined by x-ray techniques. In parallel with the x-ray work, some outstanding metallography, in the 1930's, 40s and 50s, using the light microscope, was beginning to elucidate the nature of various phase transformations, particularly in steels. The elegant work of Modin [3], using polarized light, even allowed the author to extract relative crystallographic information from the light microscope.

Perhaps the most important development in phase transformation studies occurred in the late 1950s and early 1960s, when the transmission electron microscope (TEM) was first employed to examine thin foils. For example, in 1962, Darken and Fisher [4] were able to show excellent images of pearlite (a mixture of two lamellar phases) in steels. At this point, it should be noted that TEM is not a relatively new technique; the first "electron diffraction camera" was built in the 1920s. This is perhaps what sets the TEM and its derivatives apart from other microstructural techniques: the TEM can, at the very least, give both crystallographic and structural information. Figure 2 is an electron diffraction pattern from an aluminum alloy which clearly shows that the "Law of Constancy of Angle," which was formulated from macroscopic crystal habits (and see Figure 1), is preserved on an internal/atomic scale. Figure 2 was recorded along a similar symmetry axis as Figure 1 and the regular hexagonal distribution of the diffracted discs is clearly visible.

In 1961, the first electron microprobe was produced, permitting the acquisition of chemical data from relatively small volumes. In the mid 1960s, the microprobe was modified to generate its cousin, the scanning electron microscope (SEM).

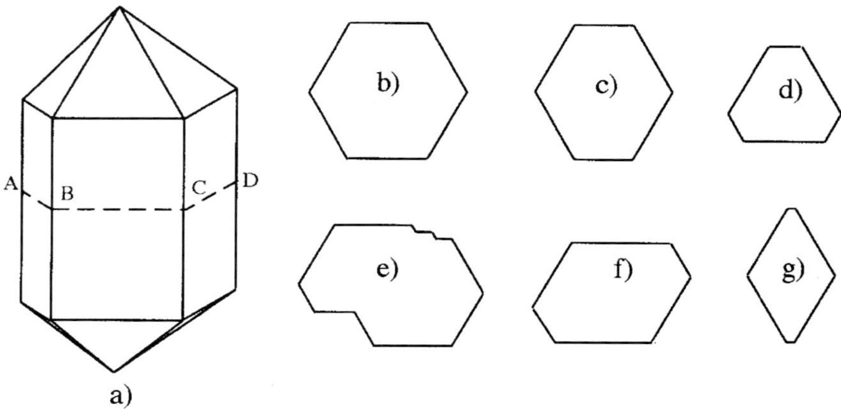

FIGURE 1. Schematic diagram of Steno's experiments on the interfacial angles in quartz (adapted from [1]). (a) A three dimensional drawing of a quartz crystal. Crystals to be analyzed were cut along plane ABCD which is perpendicular to the triad axis. (b) A two-dimensional section (along ABCD) of a regular quartz crystal. (c-g) Possible two-dimensional sections of irregular quartz crystals.

This "new" microscope was to revolutionize e.g., the study of fractography. Nowadays, the SEM is probably the most often used metallographic tool.

In the middle 1970s, the SEM and TEM were merged to form the scanning transmission electron microscope (STEM). This hybrid uses a scanning electron beam and a thin transmitting sample. Concurrently, the TEM was modified to form a second hybrid, the analytical electron microscope (AEM). This latter microscope can be used as a conventional TEM, a microprobe (with far superior resolution, see later) and an SEM.

In the 1980s and 1990s, the computer revolution radically changed all the techniques thus far described. For example, TEM images can now be stored on optical discs and the development of field emission electron guns has greatly broadened the compass of both SEMs and TEMs.

So much for a very brief history. In the following sections, examples of the application of the various microstructural techniques, to the study of phase transformations, will be presented. The examples are obviously, selective and predicated by my own research interests. Hence, steels and Al-Li based alloys will be rather prominent!

BANDED STEELS - ONE OF THE WORLDS' OLDEST, NATURALLY OCCURRING COMPOSITES

Most alloys are chemically inhomogeneous, i.e., a single phase region need not have a uniform composition. For example, when hypoeutectoid Fe-Mn-C steels solidify during processing, Mn is rejected into the melt, such that the last

FIGURE 2. An electron diffraction pattern of an Al-Li-Cu alloy. Note that the diffraction discs in Figure 2 create a geometric pattern which is very similar to the macroscopic morphology displayed by Figure 1b.

liquid to solidify is enriched in Mn. During hot rolling in the austenite (fcc) region, the Mn-rich pockets are "pancaked" yielding an almost sinusoidal variation in Mn concentration. When the austenite begins to transform to proeutectoid ferrite (the b.c.c. low temperature form of Fe), the reaction is initiated in the Mn-lean regions. Subsequent growth of the proeutectoid ferrite also occurs in the Mn-lean regions yielding bands of the product phase. Concomitantly, carbon is rejected into the Mn-rich regions which eventually transform to pearlite (the predominately lamellar mixture of ferrite plus cementite (Fe_3C).

Figure 3a is a light micrograph of the resulting microstructure. Alternate bands of proeutectoid ferrite (the light etching microconstituent) and pearlite (the dark etching microconstituent) are observed. Hence, we have "created" a lamellar composite consisting of a ductile "matrix" (ferrite) and a hard "reinforcement" (pearlite).

Two questions arise. (i) How do we know that it is Mn which is creating the microstructural banding? (ii) How do we know that the dark etching microconstituent in figure 3a is indeed a lamellar mixture of ferrite and pearlite?

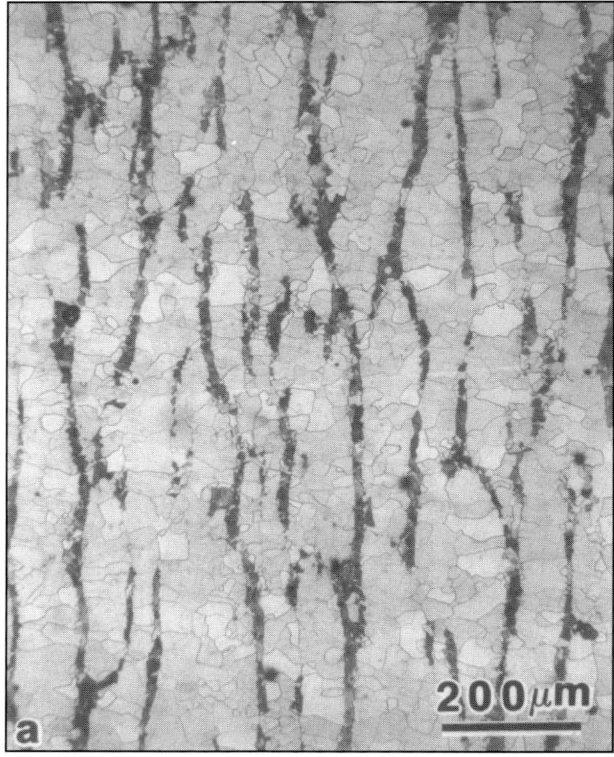

FIGURE 3. Ferrite/pearlite bands in an Fe-Mn-C alloy. (a) Light micrograph of alternate bands of ferrite (light) and pearlite (dark).

To answer these questions, we can make use of both the microprobe and the SEM. Figure 3b is a composition profile across several ferrite/pearlite bands. Low Mn levels correspond to ferrite (F) bands on Figure 3a and high Mn corresponds to bands of pearlite (P) on Figure 3a. Figure 3c is an SEM image of several pearlite colonies. The increased resolution of the SEM, as compared with the light microscope, shows that pearlite consists predominantly of alternate plates of ferrite (the dark etching phase) and cementite (the light etching phase).

Further information on banding in steels can be found in refs [5-8].

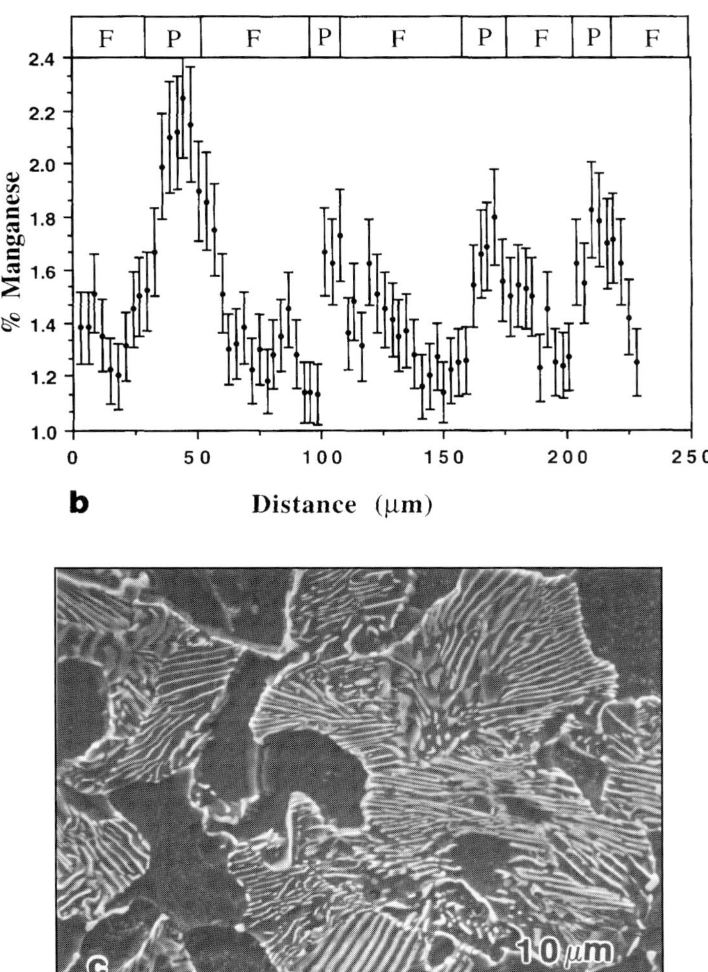

FIGURE 3. Ferrite/pearlite bands in an Fe-Mn-C alloy. (b) A Mn concentration profile taken perpendicular to the bands in Figure 3a. (c) Pearlite colonies within a pearlite band.

HIGH RESOLUTION MICROANALYSIS IN THE
AEM - DISCONTINUOUS PRECIPITATION

In certain precipitation reactions of the form

$$\alpha_{ss} \rightarrow \alpha + \beta$$

where α_{ss} is the supersaturated solid solution and β is the second phase, the β precipitates do not form uniformly throughout the matrix. Rather, the β forms as lamellae behind moving grain boundaries. (For a comprehensive review, see Butler and Williams [9]). Figure 4a is a TEM dark field image of VC lamellae in an Fe-Mn-V-C alloy. The discontinuous reaction front is arrowed A and the original position of the grain boundary is arrowed B. Figure 4b presents composition profiles, in the matrix phase, across a similar discontinuous colony. The data were obtained in the STEM mode of an AEM using energy dispersive spectometry (EDS) of x-rays. Again A represents the discontinuous front and B delineates the original position of the boundary. Note that there is a discontinuous change in composition at both A and B. There is also a residual supersaturation in the

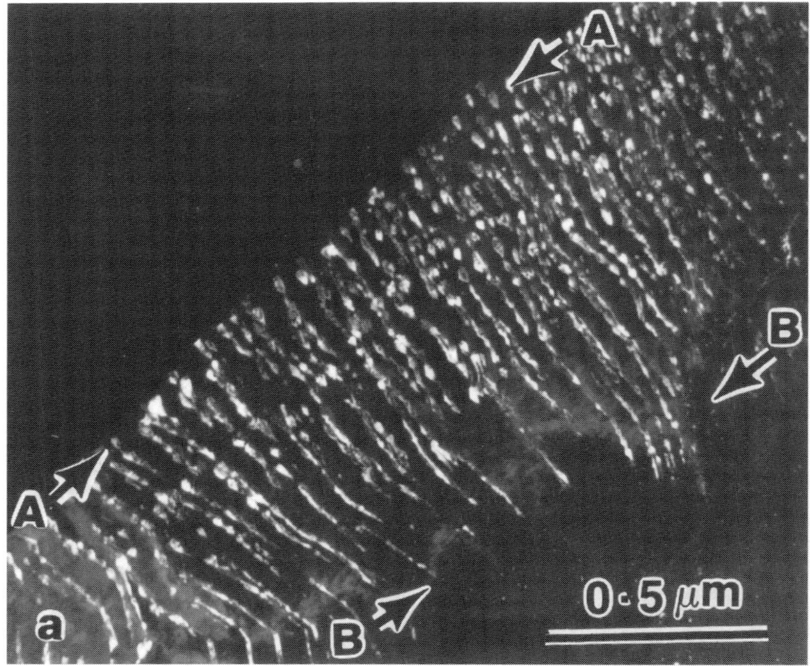

FIGURE 4. Discontinuous precipitation in an Fe-Mn-V-C alloy. (a) centered dark field (CDF) image of vandium carbide in a discontinuous colony. The position of the discontinuous front is arrowed A and the original position of the grain boundary is arrowed B.

discontinuous colony (the solubility limit for VC in the austenitic matrix is close to zero). This residual supersaturation was theoretically predicted by Cahn [10] who showed that a non-zero supersaturation, in the advancing discontinuous front, was needed for a finite growth rate.

Finally, a comparison between Figure 4b and Figure 3b shows the vastly improved spatial resolution of STEM-EDS when compared with the microprobe. This is a function of both the fine electron probes that can be generated in the AEM ($\cong$ 2nm diameter) and the thin nature of the sample ($\cong$ 100 - 2000 Å).

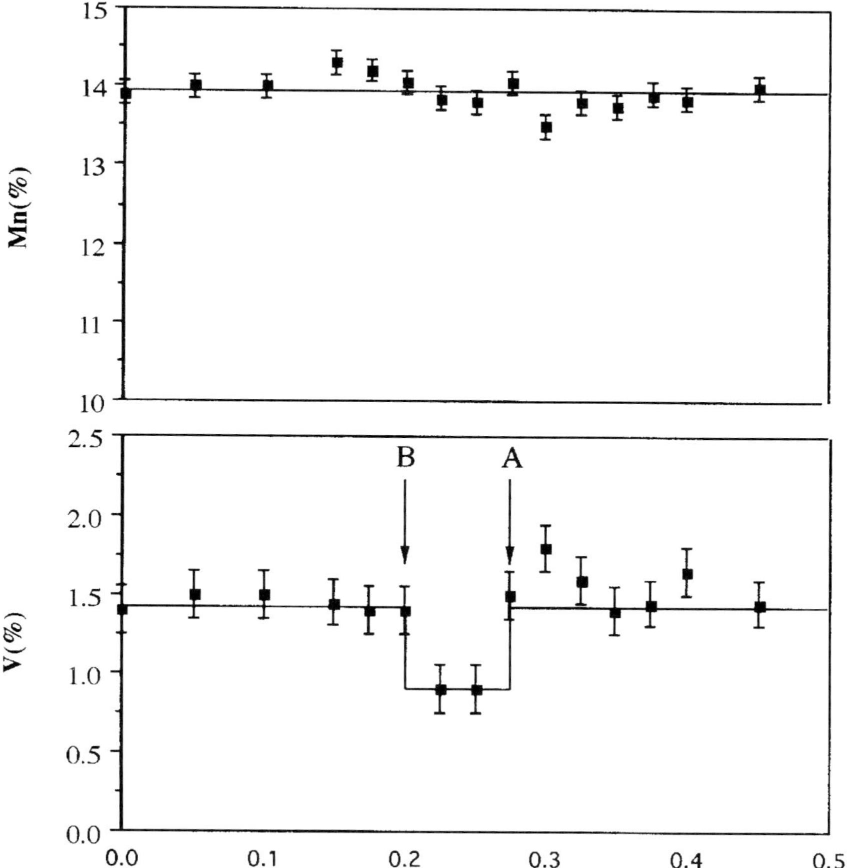

FIGURE 4. Discontinuous precipitation in an Fe-Mn-V-C alloy. (b) Concentration profiles across the discontinuous colony.

ALUMINUM - LITHIUM BASED ALLOYS FOR AEROSPACE

General

In the late 1970s, 1980s and continuing today, there was/is intense interest in developing Al-Li based alloys for aerospace applications. The driving force was to produce, low density, high stiffness and high strength materials which could compete with e.g., the lightweight carbon fibre composites. Table 1 lists some physical properties of binary Al-Li alloys. Note that as the weight % Li increases, the density decreases and the elastic modulus increases. A further benefit from alloying aluminum with lithium is the increase in yield strength, i.e., Al-Li alloys are age hardenable. The strengthening precipitate in binary Al-Li alloys is $\delta'(Al_3Li)$. δ' is ordered and is fully coherent with the α matrix. Figure 5 is a CDF image of the sperical δ' particles. Unfortunately, binary Al-Li alloys are prone to planar slip and they exhibit poor fracture toughness. The problem is exacerbated by the formation of brittle δ (Al Li) precipitates on the grain boundaries (Figure 6). This latter phase also leads to δ' precipitate free zones adjacent to the grain boundaries which can also lead to localized deformation.

In order to obviate the problems associated with planar slip, a number of approaches have been taken (for further details, see e.g., ref [11]. Most involve ternary or quaternary additions which promote the formation of other precipitate species, in addition to δ, or to replace the δ phase.

In an Al-3% Cu-2% Li alloy, the δ is retained and two additional strengthening precipitates, θ' (Al_2Cu) and T_1 (Al_2CuLi) form. Figure 7a is a CDF of the θ', which *appear* as needles. However, the θ' forms a plate on the $\{001\}_\alpha$ planes which is shown in the electron diffraction pattern of Figure 7b. The streaks in the <001> directions are indicative of thin precipitate plates on the $\{001\}$ matrix planes.

The T_1 precipitates form on the $\{111\}$ matrix planes and one T_1 variant is illuminated in the CDF image of Figure 8.

In Al-Li-Cu-Mg alloys, both θ' and T_1 can be replaced by the S' phase $(Al_2 Cu Mg)$. S' nucleates preferentially at dislocation loops and helices. In most aluminum alloys, dislocation loops form on the $\{110\}_\alpha$ planes. Figure 9 is a high magnification image of several S' precipitates on a dislocation loop. Lattice

TABLE 1.
Some Physical Properties of Al-Li Alloys.

%Li	Density (kg/m³)	% Lower Density*	Elastic Modulus (GPa)
1.1	2.71	3.6	76
2.1	2.61	7.1	79
2.9	2.51	10.7	83

* As compared with a "conventional" high strength aluminum alloy .

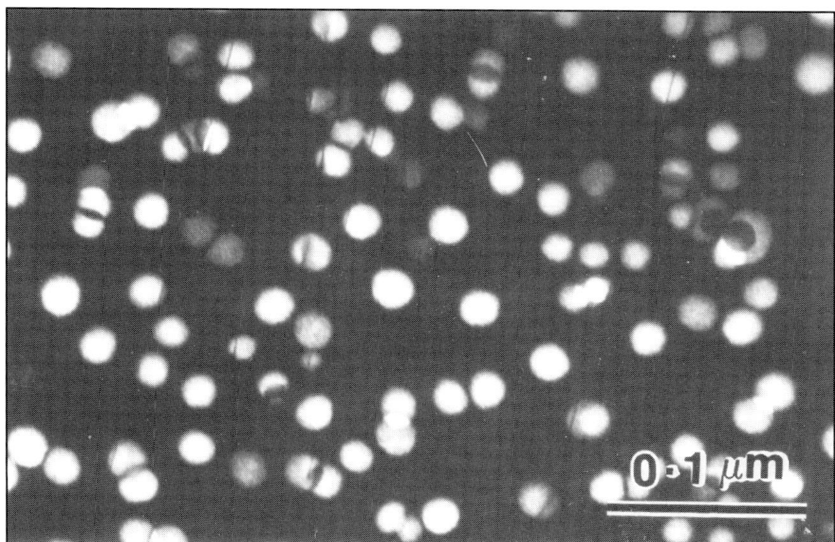

FIGURE 5. δ CDF image of an Al-Li alloy.

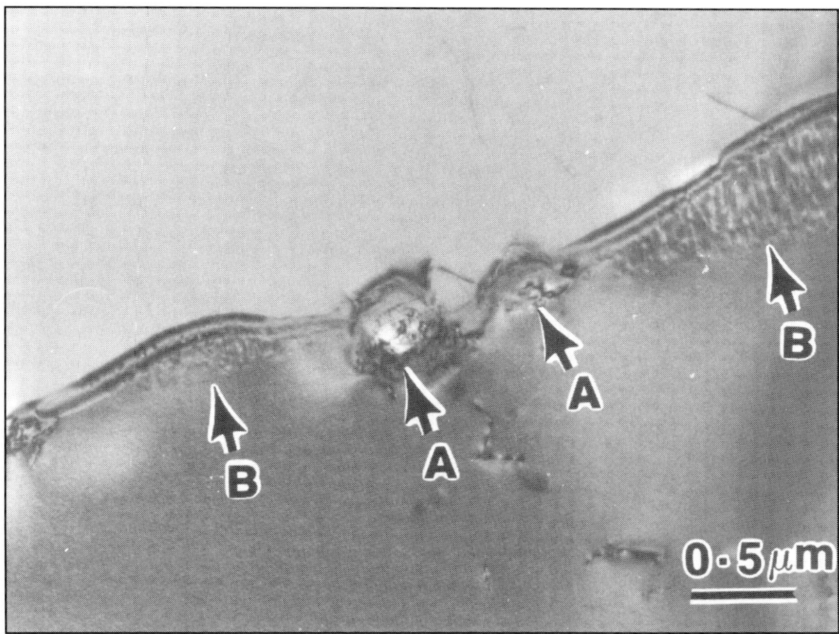

FIGURE 6. Bright field (BF) image of a grain boundary region in a binary Al-Li alloy. Two δ precipitates are arrowed A. Two colonies of discontinuous precipitation ($\alpha + \delta$) are arrowed B.

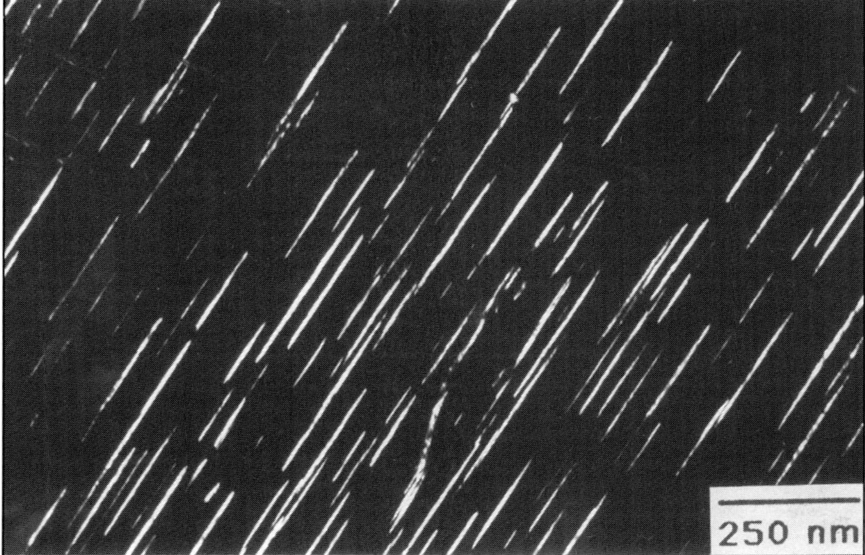

FIGURE 7. Precipitation of θ' in an Al-Li-Cu alloy. (a) X CDF of an "upright" variant.

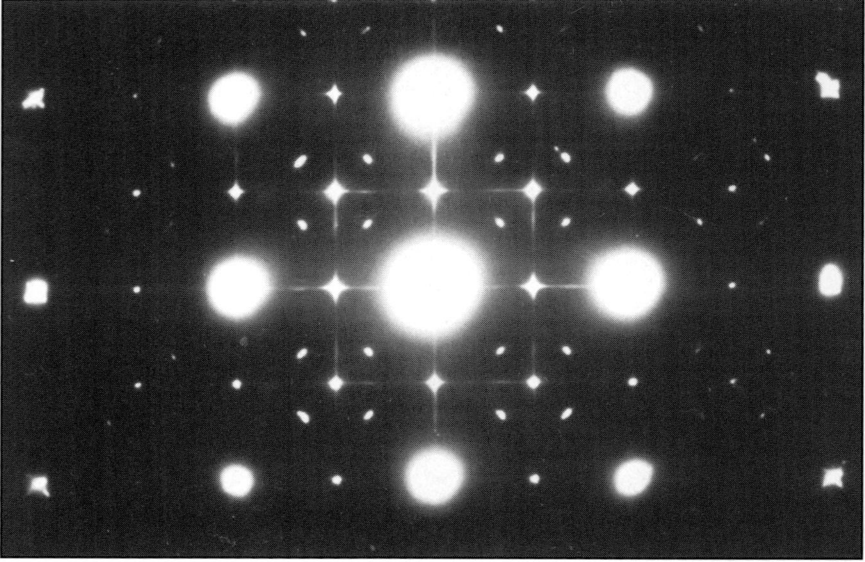

FIGURE 7. Precipitation of θ' in an Al-Li-Cu alloy. (b) Selected area diffraction pattern (SADP) of a region such as that shown in Figure 7a.

fringes can be seen in both the S precipitates and in the δ'. Using the lattice fringes in the δ' as a calibration, it is possible to perform a complete crystallographic characterization of both the precipitates and the loop. For example, the broad face of one of the variants was shown, unambigously to be $(001)_s \mid\mid (012)_\alpha$. In addition, the Burgers vector of the loop was found to be a/2[110] and the loop was in pure edge orientation.

Phase Identification

In modern AEMs, convergent beam electron diffraction (CBED) patterns and/or microdiffraction patterns can be routinely obtained with spot sizes of 2-3nm. Hence, it is possible to obtain single crystal diffraction patterns form

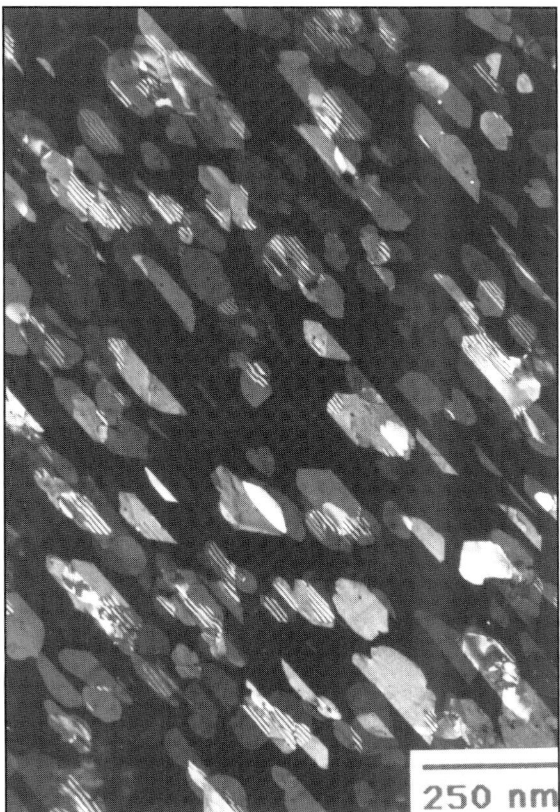

FIGURE 8. T_1 CDF. The T_1 plates form on the $\{111\}_\alpha$ phases. In this instance the T_1 variant is inclined to the electron beam.

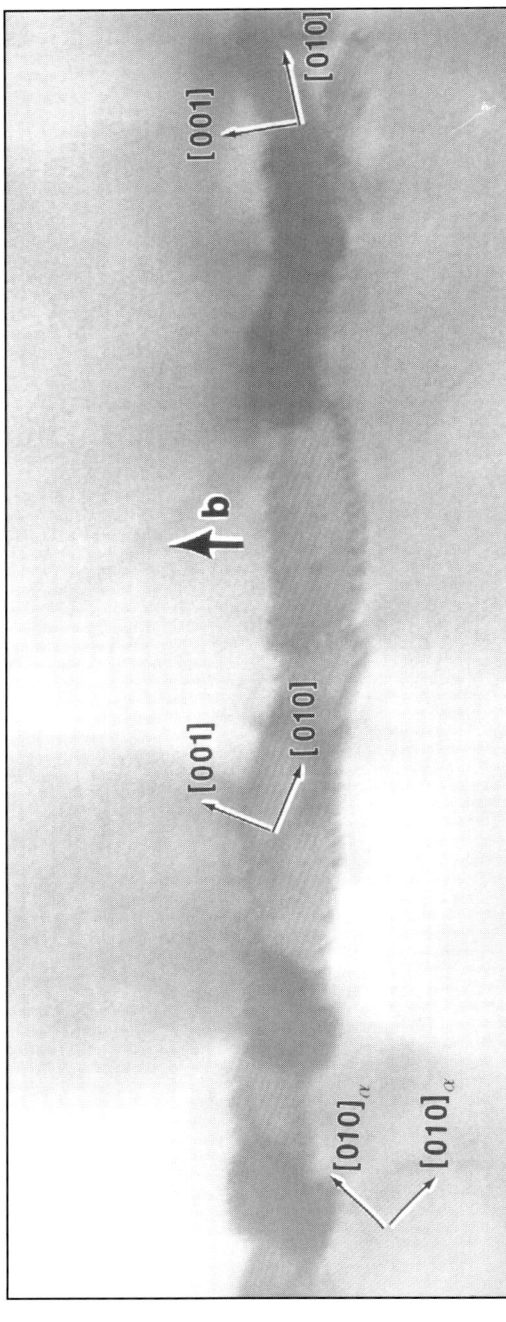

FIGURE 9. High magnification BF images of S' precipitates on a dislocation loop (see text for details).

individual precipitates. Figure 10a is a BF image of several T_2 precipitates on a low angle boundary (the arrowed precipitate is T_1). The T_2 phase was identified by obtaining CBED patterns of the type shown in Figure 10b. The T_2 precipitates exhibit the so-called "forbidden" icosahedral symmetry with a point group symmetry of $m\bar{3}5$. The fact that T_2 is icosahedral as opposed to decagonal (the pattern in figure 10b appears to show 10-fold symmetry) can be verified by looking along the same axis but at a low camera length (Figure 10c). In Figure 10c, reflections from the first order Laue Zone clearly shows 5-fold, and not 10-fold symmetry. Hence, by using whole-pattern symmetry, the AEM becomes a very powerful technique for crystal structure analysis.

When single crystal diffraction patterns can not be obtained, and a selected area diffraction pattern contains many reflections from multiple precipitate species and/or orientations, the patterns can be simulated. For example, Figure 11 is a computed diffraction pattern ([001] zone axis) and contains all the reflections that would arise from all twelve variants of the S' (Al$_2$ CuMg) phase.

The Disappearing T2 Phase

In 1988, Vecchio and Williams found that T_2 precipitates could display icosahedral symmetry when thick enough but appeared as a microcrystalline array when sufficiently thinned [12,13]. In the latter case, the selected area electron

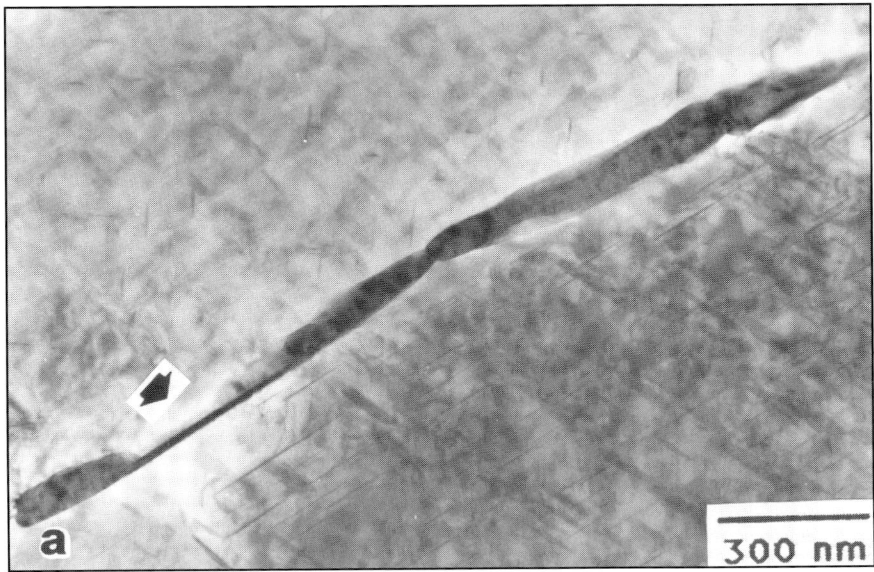

FIGURE 10. (a) BF image of precipitation on a low angle boundary in an Al-Li-Cu alloy. The arrowed precipitate is T_1; the virtually continuous grain boundary phase is T_2.

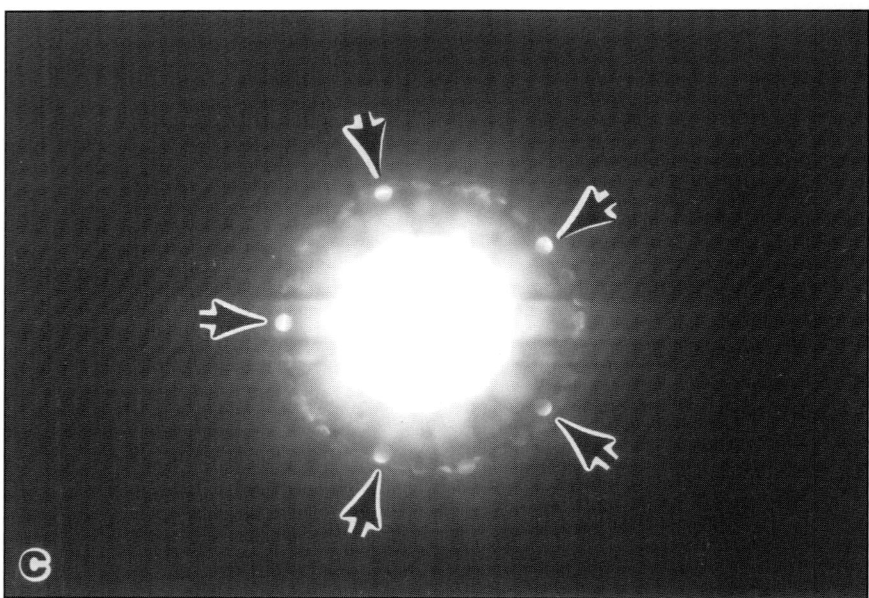

FIGURE 10. (b) Convergent beam electron diffraction (CBED) pattern from the T_2 phase. The pattern is recorded along a 5-fold axis. (c) Low camera length CBED pattern which shows that the symmetry is indeed 5-fold and not 10-fold.

diffraction patterns showed no vestige of the 5-fold symmetry. Hence, it was conjectured that T_2 was not icosahedral but was microcrystalline and only displayed apparent 5-fold symmetry when the particles were large enough. This above conjecture sparked a healthy debate between Penn State and Lehigh (and see refs [14,15] because we had already published a paper showing what we considered to be a transformation product of T_2 - a microcrystalline array [16]. A typical microcrystalline array is shown in Figure 12.

To solve the "debate", we employed both EDS in the AEM and CBED. A typical EDS spectrum from an untransformed T_2 particle is shown in Figure 13a. There is a significant CuK_α peak (T_2 has a stoichiometry of Al_6CuLi_3). However, when a EDS

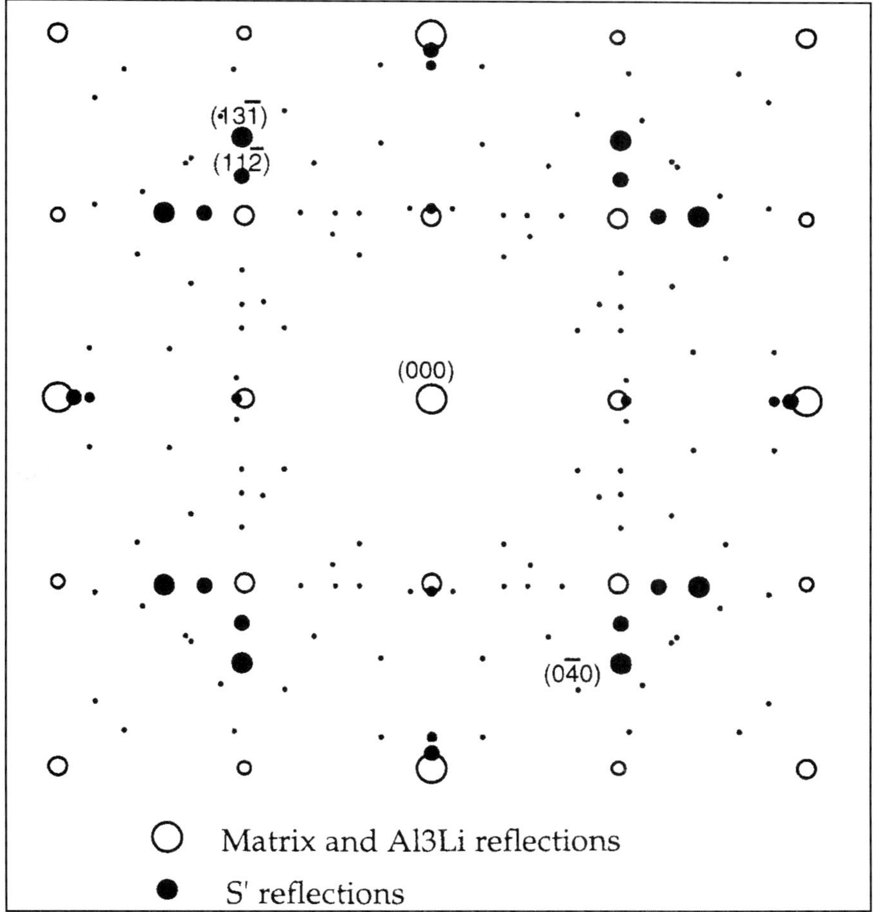

FIGURE 11. Computer generated diffraction pattern ([001] Zone axis) of the S^1 phase.

spectrum obtained from a microcrystal, virtually no "copper was found (Figure 13b). Hence, during the $T_2 \rightarrow$ microcrystalline array transformation, copper was being leached from the decomposing T_2. The above hypothesis was confirmed by obtaining single crystal diffraction patterns from many individual microcrystals and one such pattern is presented in Figure 14. This particular pattern can be indexed in terms of an fcc structure with a lattice parameter of $\cong 4.04$Å. Weak superlattice reflections can also be detected on Figure 14. Hence, it may be concluded that this particular microcrystal is an aluminum solid solution which contains δ.

A series of experiments [14,15,17] found that T_2 is unstable when in contact with a free surface and is:

(a) irradiated by the electron beam;
(b) irradiated by an ion beam;
(c) in-situ heated.

Other transformation products which were identified using CBED included T_1 and T_B.

Orientation Determination

Hardy and Silcock [18] first examined the T_B phase. T_B is cubic and has the CaF_2 structure with a = 0.583 nm. The orientation relationship quoted in [18] was

$$[100]_{T_B} \parallel [110]\alpha$$
$$[001]_{T_B} \parallel [001]\alpha$$

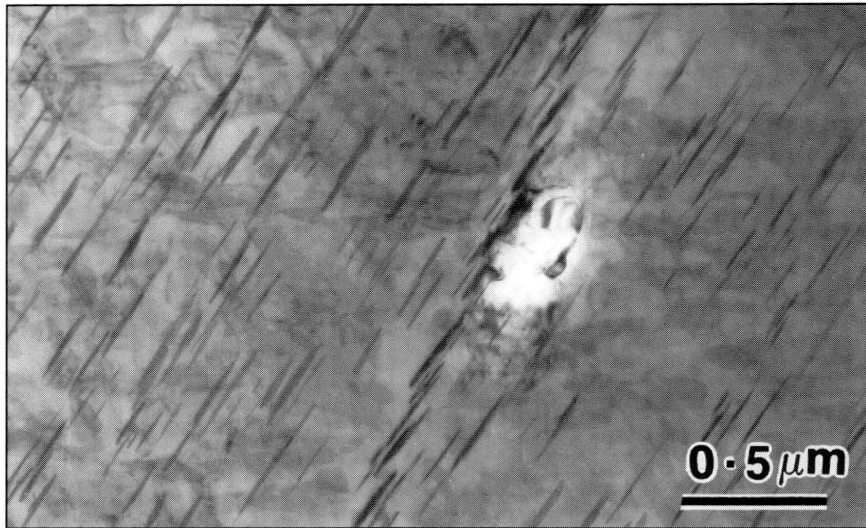

FIGURE 12. BF image of a microcrystalline array at a low angle boundary.

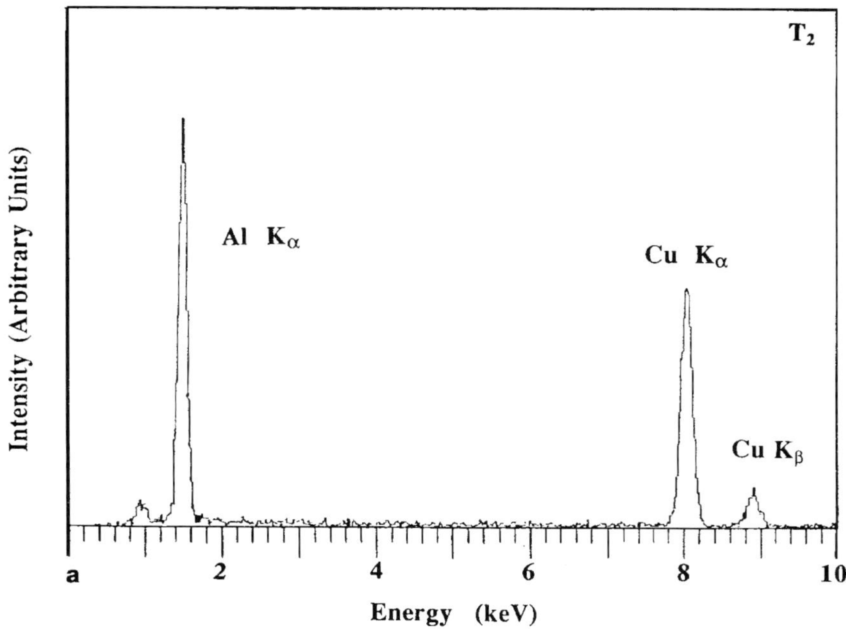

FIGURE 13. Energy dispersive spectra from T_2 and its transformation products. (a) Spectrum for a precipitate which displays 5-fold symmetry.

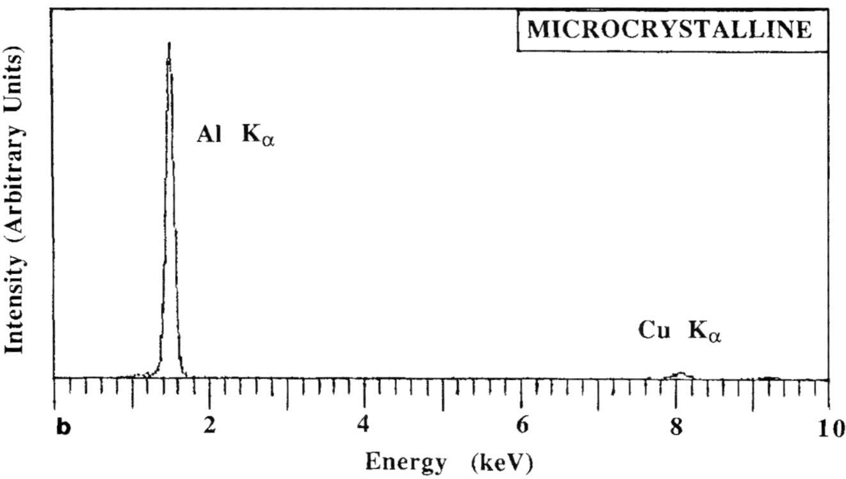

FIGURE 13. Energy dispersive spectra from T2 and its transformation products. (b) Spectrum from a microcrystalline region.

but with a rotation of 1 to 2° around the [100]α axes.

Figure 15a is a BF image of a 4% Cu - 1% Li alloy and shows a T_B precipitate on a high angle boundary. (The T_B was identified from CBED patterns). In this particular alloy, the strengthening phases in the matrix are T_1 and θ′ (no δ′ is present). Figure 15b is a CBED pattern of the T_B precipitate ([001] zone axis) whereas Figure 15c is a [110] pattern from the adjacent matrix. The [001] pattern is exactly on zone whereas the [110]α pattern is off-zone. Analyses of these, and other CBED patterns show that the [110]T_B axis is rotated by approximately 1° about [010]α which is fully consistent with the observations of Hardy and Silcock.

Stereology (Quantitative Microscopy)

In many studies, 'hard numbers" are required. For example, if yield strength is a function of precipitate spacing, then a method for obtaining these numbers as a function of e.g., aging time needs to be formulated.

The average precipitate spacing for plate shaped precipitate (h) is given by equation (1) where V_v is the volume fraction, $\bar{D}$ is the average precipitate diameter, $\bar{w}$ the average thickness and N_v is the number density.

The first problem is to measure the foil thickness. This can be achieved in a variety of ways:

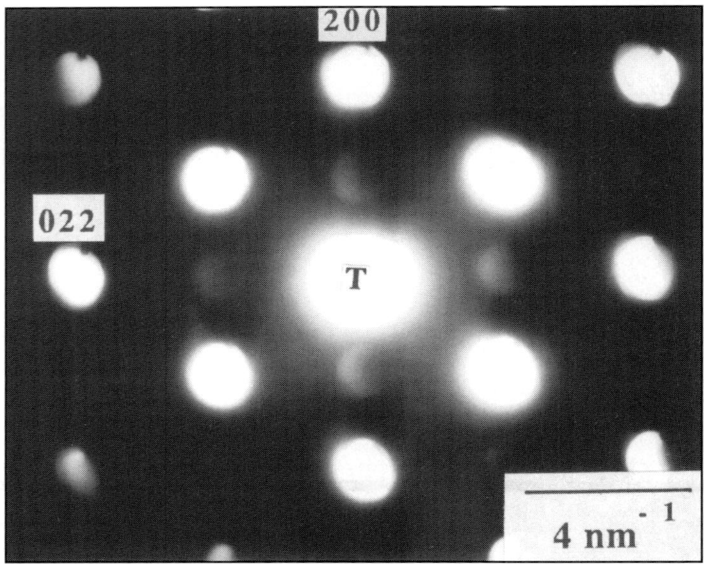

FIGURE 14. CBED pattern from a microcrystal which is indexed in terms of the aluminum solid solution.

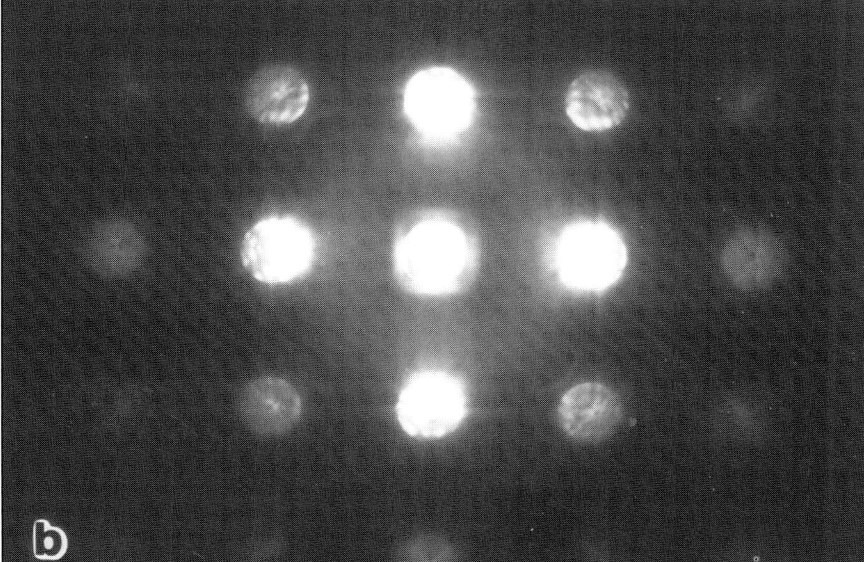

FIGURE 15a & 15b. The orientation relationship between T_B and the α matrix. (a) BF image of a T_2 precipitate on a high angle grain boundary. (b) CBED pattern from the T_B.

(i)   by using thickness fringes which are projected onto grain boundaries (e.g., see Figure 6);
(ii) by using the projected widths of plate-shaped precipitates (e.g., see Figure 8)
(iii) by using contamination spots;
(iv) by using Kossel Mollenstedt fringes.

A review of the ways of obtaining foil thicknesses is contained in [19].

$$h = \frac{8(1 - V_v)}{\pi \overline{D} N_v (\overline{D} = 2\overline{w})} \tag{1}$$

The apparent diameter distribution can be obtained from images such as that shown in Figure 7a by simple measurement. However, the data need to be corrected for the fact that many of the precipitates which are imaged, have their centers located outside the confines of the foil. This can lead to serious errors in e.g., number density. For example, Figures 16a and b compare the T_1 size distributions both before and after the correction procedure has been applied. Note that the number density has been reduced by a factor of approximately 4 after correction. Figures 17 a-e show the diameter distribution as a function of aging time at 190° C, the width distribution, the number density, the volume fraction and finally the precipitate spacing. As may be seen, the precipitate spacing is a minimum after 18 hour aging which corresponds to the peak aging time for this temperature.

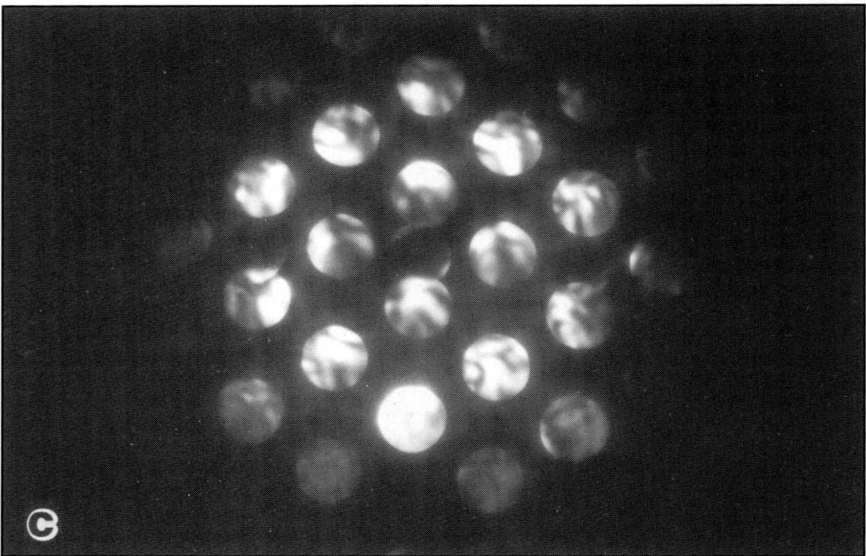

FIGURE 15. The orientation relationship between T_B and the α matrix. (c) CBED pattern from the lower aluminum grain.

Welding of an Al-Li based Alloy

Welding of aluminum alloys is becoming increasingly important. In particular, it is necessary to know what is happening in the heat affected zone (HAZ) as a function of position and peak temperature.

Figure 18 is a schematic diagram of an actual weld together with the hardness

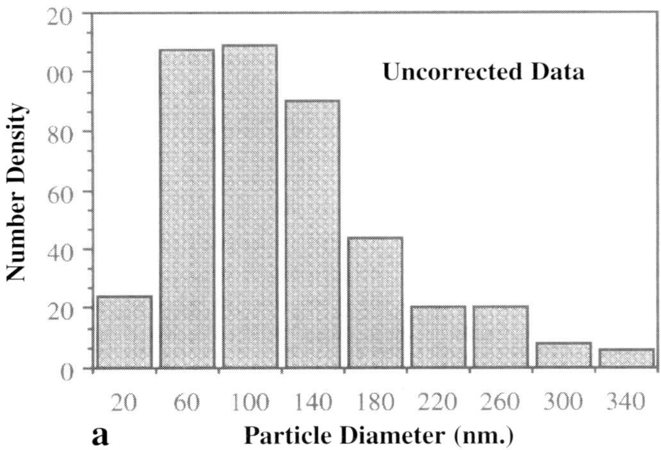

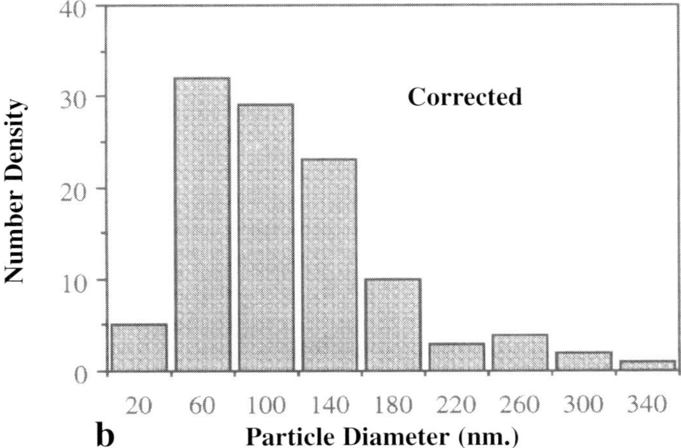

FIGURE 16. Histograms of particle diameters for the T_1 precipitate in an Al-Li-Cu alloy. Note that the number densities should be multiplied by 10^{19}. (a) Uncorrected data. (b) Corrected data (see text for details).

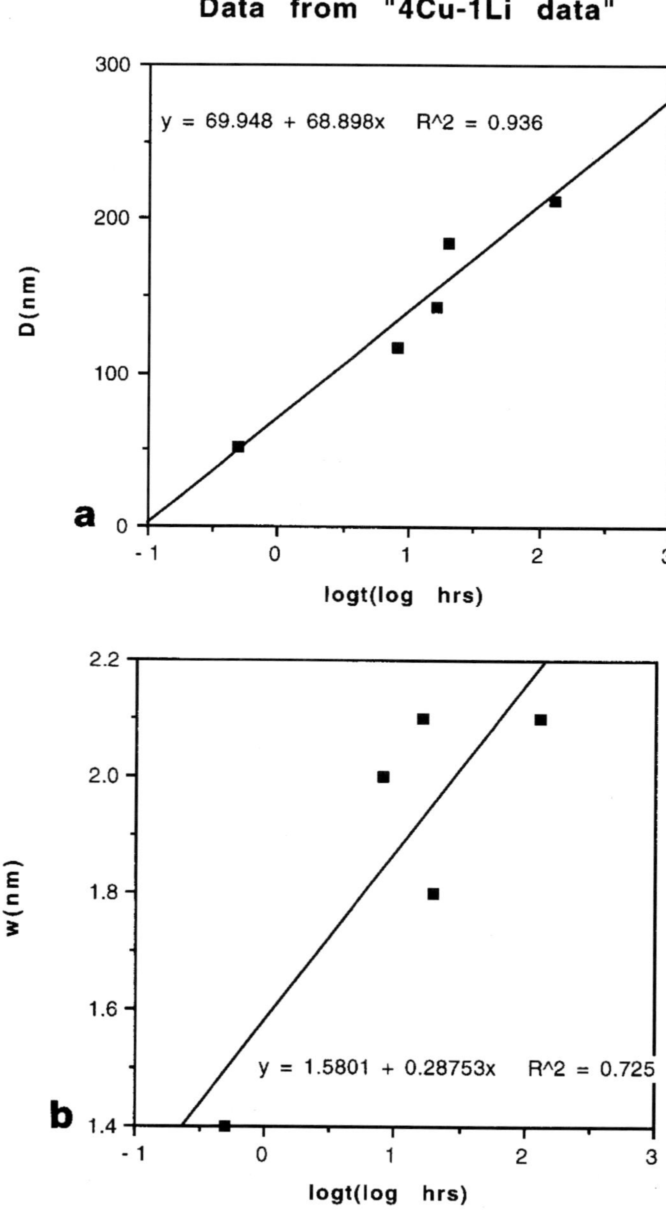

FIGURE 17. Stereological data for the alloy shown in Figure 16 (aging temperature is 190°C). (a) Variation of average diameter of T_1 with time (note the log scale). (b) Variation of average precipitate width with time. Note the very large scatter in the data.

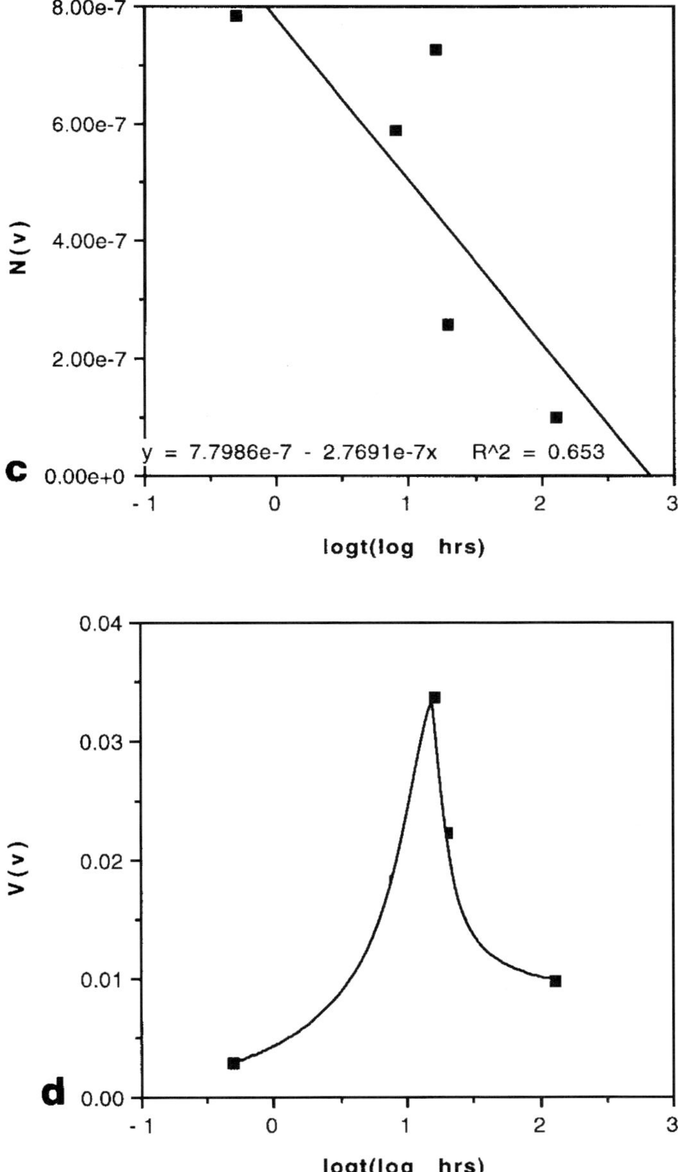

FIGURE 17. Stereological data for the alloy shown in Figure 16 (aging temperature is 190 °C). (c) Variation in number density with time. (d) Variation in volume fraction with time.

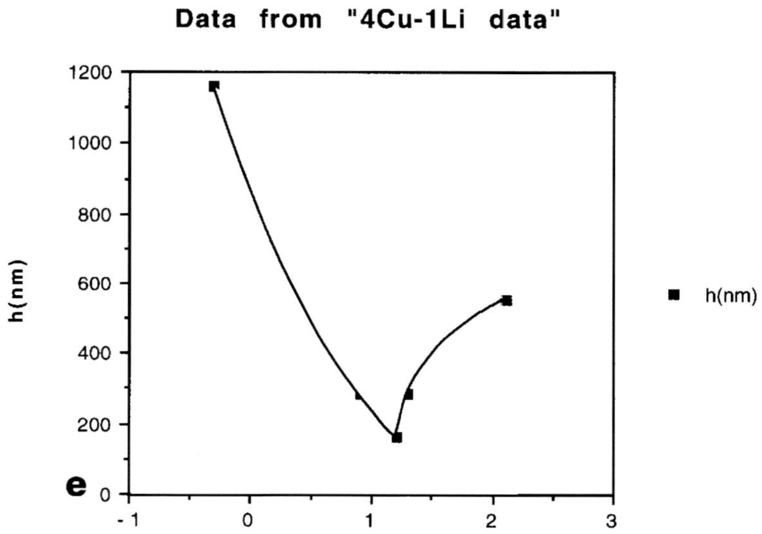

FIGURE 17e. Stereological data for the alloy shown in Figure 16 (aging temperature is 190 C). (e) Variation of precipitate spacing with time.

in the HAZ . The numbers on the schematic diagram represents positions that have currently been evaluated[20]. Figures 19a, b are θ' are T_1 CDFs respectively from the base metal (BM). This will provide the quantitative data against which all other data can be compared. Progressing into the HAZ, the θ' is dissolved at position 7 and there is a marked decrease in the T_1 number density. At position 1, the T_1 has virtually disappeared as can be seen in Figure 20a. The only remaining phase is the β (Al_3Zr) dispersoid (the spherical particles). However, close examination of the background reveals mottling which could be indicative of the presence of GP zones. This was confirmed by recording the selected area diffraction pattern and noting the diffuse streaks in the <001>α directions (Figure 20b).

SUMMARY

This chapter has demonstrated the wealth of information that can be obtained through the use of various microstructural techniques. Again, I would like to emphasize that the topics chosen have been very selective. In addition, much of the focus has been on TEM investigations. Hence, a word of caution. Figure 3 is a light micrograph of a banded microstructure with a banding wavelength of 50 μm. The TEM would be of little value in a study of this phenomenom because of the limited volume that can be analyzed. Indeed, bypassing the light microscope and going directly to the TEM could give erroneous results. Depending on the location of the perforation in the thin foil, one might conclude that the sample was either 100% ferrite or 100% pearlite!

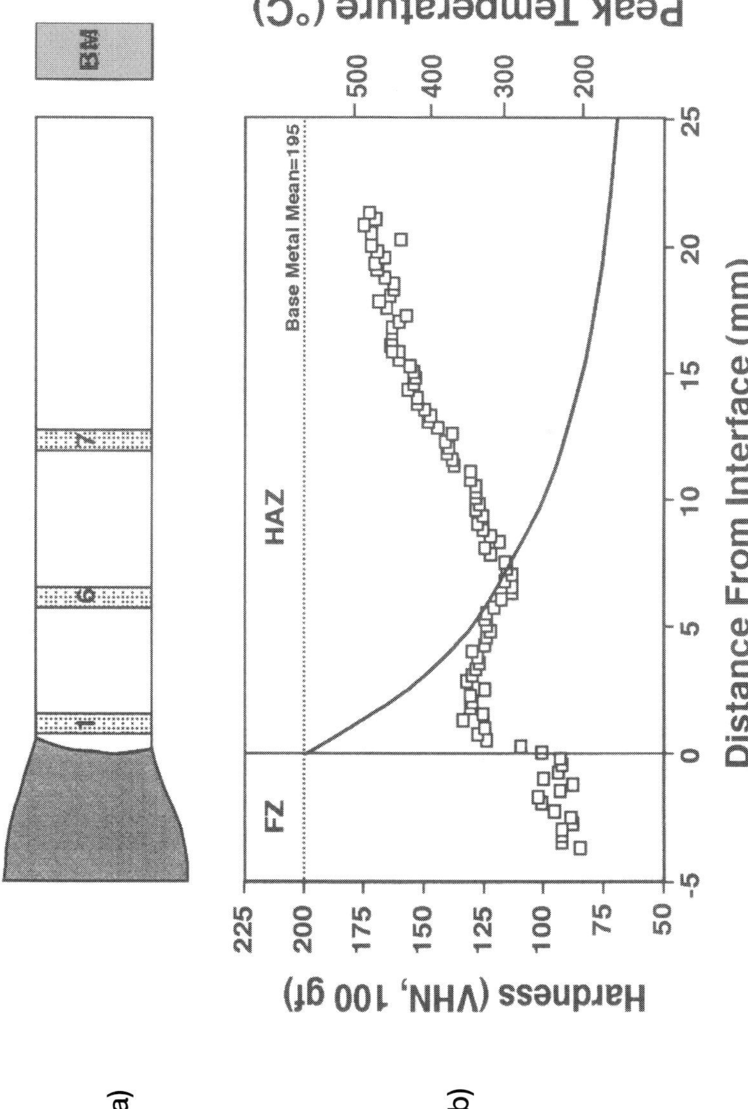

FIGURE 18. VPPA welding of an Al-Li-Cu-Mg-Ag alloy. (a) Schematic diagrams of the weld and positions that have been examined in the TEM. (b) Hardness profile in the weld and heat affected zone.

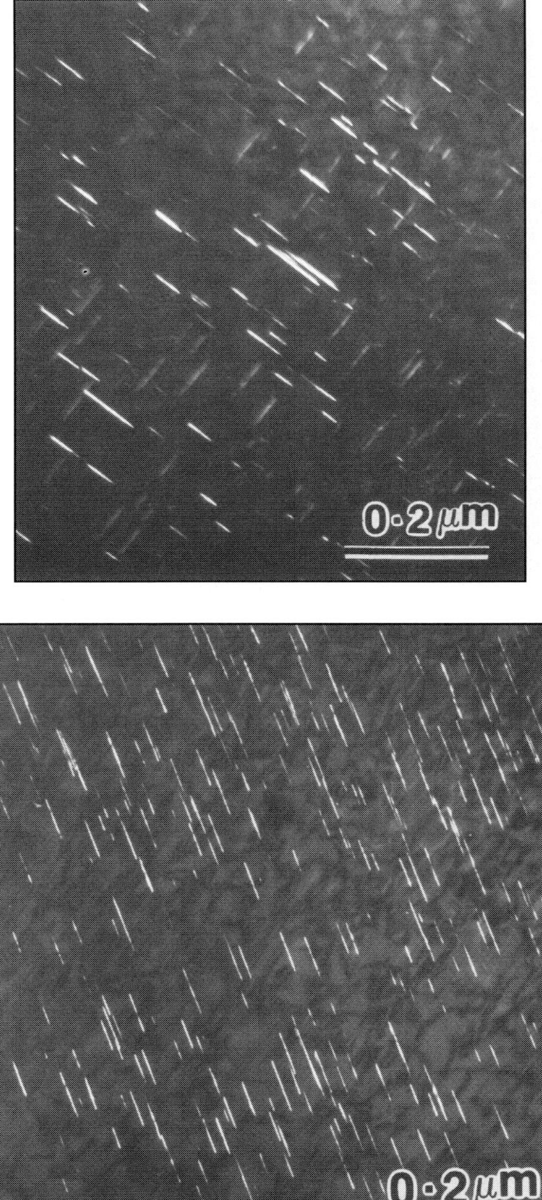

FIGURE 19. CDF images of the base metal. (a) θ' CDF. Two variants are illuminated. (b) T_1 CDF.

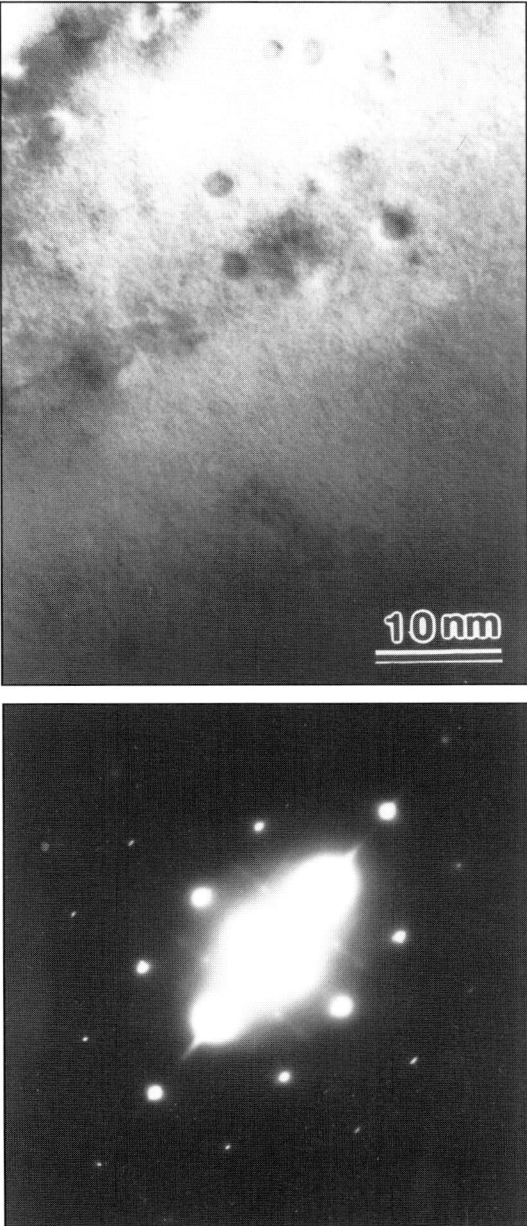

FIGURE 20. Microstructure at region 1 (Figure 18). (a) BF image. Note the mottling in the background. (b) SADP showing streaking in the <001> directions.

REFERENCES

1. Steno, N. De Solido intra Solidum Naturaliter Contento Dissertationis Prodromus. As quoted in F.C. Phillips, 1970 *An Introduction to Crystallography,* 4th Ed. Longman, p. 13.
2. Bragg, W.H. and W.L. Bragg. 1913. Proc. Royal Soc., A88:428 and 89:246.
3. Modin, S. 1951. *Jernkontorek Ann.,* 135:169.
4. Darken, L.S. and R.H. Fisher. 1962. In: V.F. Zackay and H.I. Aaronson (Eds.) *Decomposition of Austenite by Diffusional Processes.* Interscience Publishers, p. 249.
5. Kirkaldy, J.S., Destinon-Forstman, J. and R.J. Brigham. 1962. *Can. Metall.* Q. 1:59.
6. Grange, R.A. 1971. *Met. Trans.* 2:417.
7. Thompson, S.W. and P.R. Howell. 1992. *Mat. Sci. Tech.* 8:777.
8. Sychterz, J.B., Marino, M.P. and P.R. Howell. 1995. In: E.B. Halbolt and S.Yue (Eds.). *Phase Transformations during the Thermal/Mechanical Processing of Steel.* The Metallurgical Society of CIM, pp. 135.
9. Williams, D.B. and E.P. Butler. 1981. *Int. Met. Rev.* 3:153.
10. Cahn, J.W. 1959. *Acta Met.* 7:18.
11. Rioja, R.J., Bretz, P.E., Sowtell, R.R, Hunt, W.H. and E.A. Ludwiczak. 1986. In: E.A. Starke Jr. and T.H. Sanders Jr. (Eds.) *Aluminum Alloys, Their Physical and Mechanical Properties,* Vol. 3, EMAS, Worley, U.K., p. 1781.
12. Vecchio, K.S. and D.B. Williams. 1988. *Phil. Mag. B.* 57:535.
13. Vecchio, K.S. and D.B. Williams. 1988. *Met. Trans. A.* 19A:2875.
14. Tosten, M.H., Ramani, A., Bartges, C.W., Michel, D.J., Ryba, E.R. and P.R. Howell. 1989. *Scripta Met.* 23:829.
15. Howell, P.R., Michel, D.J., and E.R. Ryba. 1989. *Scripta Met.* 23:825.
16. Bartges. C., Tosten, M.H., Howell, P.R., and E.R. Ryba. *J. Mat. Sci* 22:1663.
17. Howell, P.R., Michel, D.J., and J.R. Reed. 1990. *Scripta Met.* 24:1033.
18. Hardy, H.K. and J.H. Silcock. 1955-56. *J. Inst. Met.* 84:423.
19. Williams, D.B. 1984. *An Introduction to Analytical Electron Microscopy.* Philips Electronic Instruments Inc., Electron Optics Publishing Group, Mahwah, NJ.
20. Martukanitz, R.P. and P.R. Howell. 1996. *4th Intl. Conf. on Trends in Welding Research.* In press.

The Era of Materials. Edited by S.K. Majumdar, R.E. Tressler, and E.W. Miller © 1998, The Pennsylvania Academy of Science.

Chapter Three

CRYSTALLINE VS. AMORPHOUS STATE: METALS

EARLE RYBA

Department of Materials Science and Engineering
Steidle Building
Pennsylvania State University
University Park, PA 16802

INTRODUCTION

The arrangement of atoms in space is the basis for understanding the nature and behavior of solids. As in other classes of materials, there is a rich variety of these structural arrangements in metals and alloys. Perhaps unique to metallic materials, however, is the extensive occurrence of isostructural atomic arrangements in both pure metals and alloys. Thus, most metallic materials can be assigned to structural classes. Also unique to metallic materials is the ability to produce amorphous and near-amorphous structural arrangements in a wide variety of alloys.

Metallic materials can be either pure metals or mixtures (alloys) of two or more metals. Some alloys contain non-metals such as boron, carbon, or silicon as alloying elements. The phases present in alloys are of three types: (a) the pure metal, (b) solid solutions of other elements in the pure metal, and (c) intermediate phases, or intermetallic compounds. We begin by discussing the crystal structures of materials in each of these categories.

PURE METALS

Generally, the atomic arrangements in these types of phases are very regular and periodic in nature (crystalline). However, while there are approximately 80 pure metals in the periodic table, there are not 80 different atomic arrangements to be considered. Most of the pure metals crystallize in one of three simple crystal structure types: cubic closest packing, hexagonal closest packing, and "body centered cubic". 45 metals exhibit closest packed structures at room temperature. These structures can be easily visualized as the packing of equal-sized hard sphere atoms, in which the atoms are placed as close together as possible. As shown in Figure 1, these closest packed structures can be formed by stacking together close packed layers of hard sphere atoms. There are only three possible positions (A, B, and C) for these layers.

16 of the metals with closest packed structures have their closed packed layers stacked in the sequence ABCABC.....(cubic close packed) and 28 have the stacking sequence ABAB.....or ABACABAC..... (hexagonal close packed). 15 more exhibit a non-closest packed structural arrangement termed "body centered cubic" at room temperature, in which the metal atoms are located at the corners and centers of a cubic structural building block or unit cell. The rest of the pure metals exhibit a variety of crystal structures, most of which are rather different from the three common crystal structure types just described. Note that there are no molecules formed in any of these structures.

SOLID SOLUTIONS OF THE PURE METALS

A solid solution of a pure metal exhibits the same crystal structure as the pure metal, except that the periodicity is slightly interrupted and some minor displacements in the positions of the atoms are introduced by the substitution of alloy elements (solute) atoms into the sites otherwise occupied by the solvent

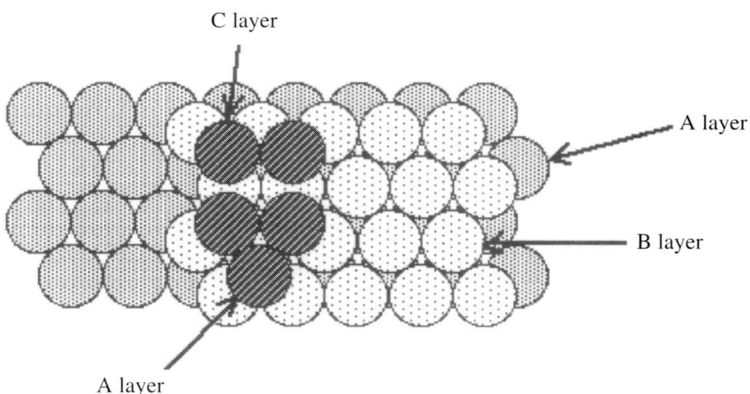

FIGURE 1. The stacking of close packed layers of hard sphere atoms to form the various types of closest packed crystal structures for pure metals

atoms, or by the insertion of much smaller alloy element atoms into the interstices of the structures. Thus, the formation of solid solutions does not really result in any new crystal structure types.

INTERMETALLIC COMPOUNDS

When two or more metals are combined to form alloys, intermediate phases or intermetallic compounds frequently form. Some intermetallic compounds form only at a specific compositions (line compounds) (see Figure 2), whereas others form over a composition range due to solid solution.

For either type, the common chemical valence rules of combination are not followed, and the stoichiometries of intermetallic compounds take on a wide variety of values, sometimes non-integer values. The occurrence of intermetallic compounds is widespread; for instance, approximately 3/4 of the binary alloy systems exhibit at least one intermetallic compound, and a few have as many as 11-13 compounds. True ternary and higher compounds, those with crystal structures different from the binary compounds in the alloy system, are

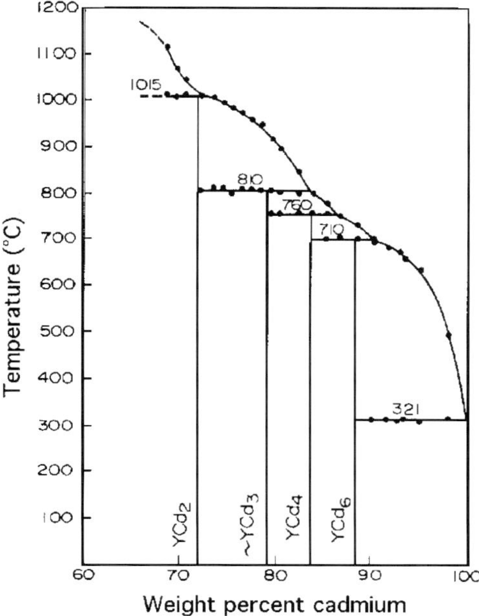

Figure 2. The occurrence of line compounds in the yttrium-cadmium alloy system. From Ryba, Kejriwal & Elmendorf [1]

relatively uncommon. Additional elements are usually accommodated in binary intermetallic compounds by solid solution.

STRUCTURAL CLASSES IN INTERMETALLIC COMPOUNDS

Because there are so many intermetallic compounds with different component atoms and a wide variety of stoichiometries, the description of their crystal structures is complicated. Again, the hard sphere atom model is used to visualize the structural arrangements, but sphere size varies from one element to the next. Structural classifications are possible. An example of a classification scheme for a few groups of intermetallic compounds based upon composition/stoichiometry is shown in Figure 3.

The intermetallic compounds denoted tetrahedrally close packed are unique in that, while they are closely structurally related, they occur at different compositions in different alloy systems. These compounds exhibit variations of a structural formation in which tetrahedra with atoms at each vertex are packed together to fill the crystal space. The tetrahedrally close packed phases are formed by combinations of transition metals, do not have fixed stoichiometries from one alloy system to another, and usually exist over a composition range. A primary example is the σ phase; the structure of the well-known σ (Fe,Cr) phase is shown in Figure 4.

Another class of structures occurs for combinations of metals whose atom sizes and electronegativities are similar, wherein the resulting structures of the intermediate phases are essentially ordered versions of the closest packed structures of the pure

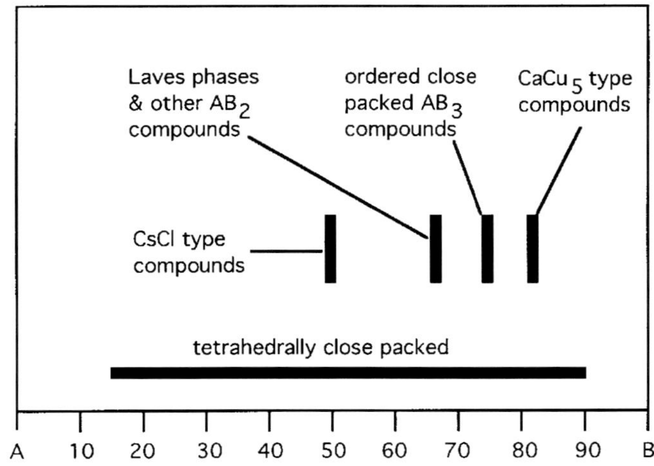

FIGURE 3. A structural classification of a few of the more prominent groups of intermetallic compounds according to composition/stoichiometry.

metals. Examples are shown in Figure 5; six common crystal structure types are observed to form from these two basic layers through different stacking sequences.

A class of structures based on the pure metal "bcc" structure also exists, in which various ordered arrays of atoms on the structure sites, such as that shown in Figure 6, can be formed.

A simple form of this type of ordered structure is one in which one metal atom sits on the cubic unit cell corner and a second at the cell center. A large number of intermetallic compounds exhibit this CsCl type structure.

On the other hand, when the component atoms have rather different sizes and/or when their electronegativities are quite different, intermetallic compounds have crystal structures which bear little resemblance to the pure metal structures. Compounds with the stoichiometry AB_2 are very prominent in alloy systems. Hundreds of these compounds exhibit what are known as the Laves phase structures,

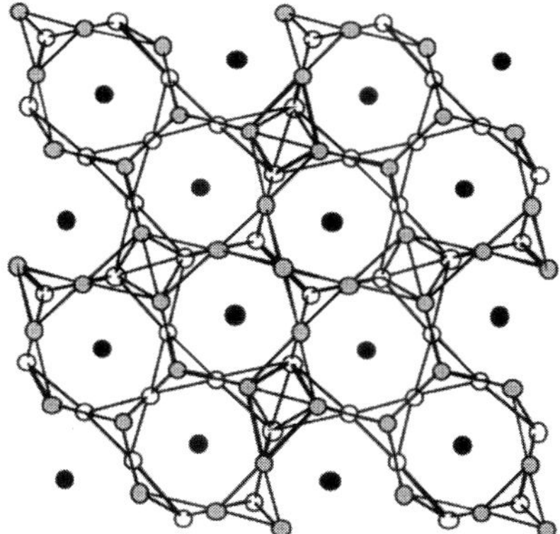

FIGURE 4. The crystal structure of the σ (Fe,Cr) phase shown as a projection of three planar layers (indicated by the three different shadings) perpendicular to those layers. There is a slight tendency toward preferred occupancy of the different atom sites by the Fe and Cr atoms. From Bergman & Shoemaker [2].

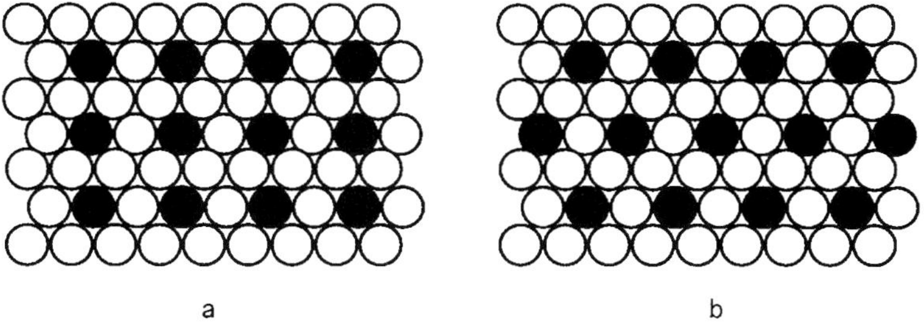

a   b

FIGURE 5. The two basic ordering schemes in layers of the close packed AB_3 intermetallic compounds. a) in-layer ordering for $TiCu_3$ and $TiAl_3$ type structures; b) in-layer ordering for the $AuCu_3$, $TiNi_3$, $MgCd_3$, and $PuNi_3$ type structures.

one of three very closely related crystal structures made up of the packing of truncated tetrahedra with the smaller B atom at each vertex of the polyhedra and a larger A atom located at their centers (see Figure 7). The structural building block is common in other structures also. An example is shown in Figure 8 for the rather complex crystal structure of Cu_4Cd_3.

There are a number of other types of crystal structures for AB_2 compounds. For example, the KHg_2 type structure (see Figure 9) occurs in several dozen binary alloy systems. This structure type is useful for the illustration of the importance of the role atomic size plays in the positions of the atoms in the structure. It is possible to predict the positions of the atoms, based upon packing of hard sphere atoms. For example, the angle of the zig-zag A atom chain (see Figure 9) depends upon the relative sizes of the A and B atoms. This is illustrated in Figure 10 for a series

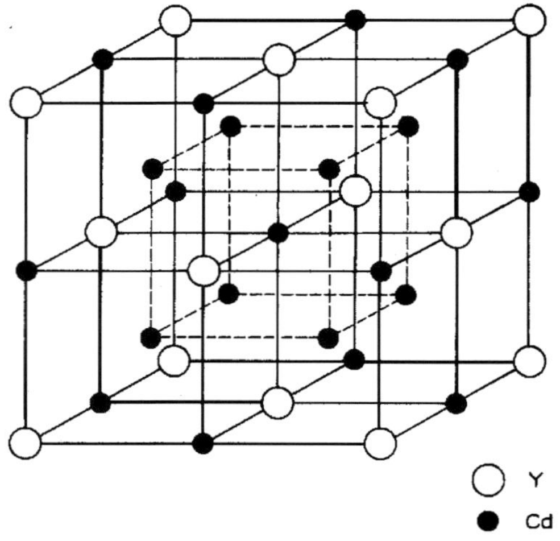

FIGURE 6. The crystal structure of YCd_3. The atom arrangement is an ordered form of the "bcc" pure metal structure. From Elmendorf [3].

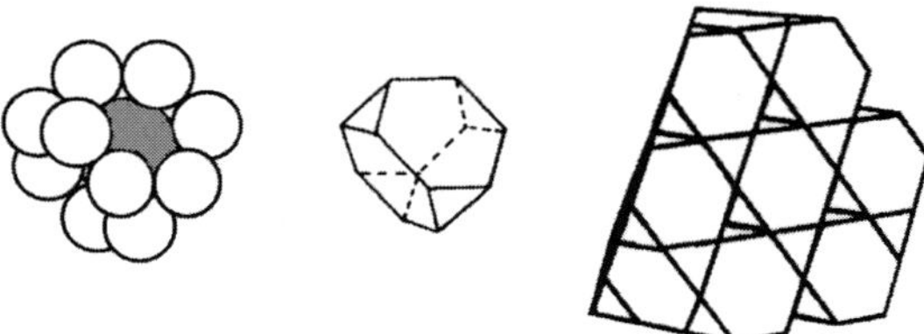

FIGURE 7. The crystal structure of the cubic form of the AB_2 Laves phases as represented by the packing of truncated tetrahedra. Each tetrahedron has an A atom at its center an B atoms at its vertices, as shown on the left.

of KHg$_2$ type compounds for which the crystal structures have been experimentally determined. Most of the chain angle values lie within about 2° of the angles predicted from packing of the two sizes of hard spheres in this type of arrangement.

In the Laves phases, on the other hand, the positions of the atoms are not variable. Instead, the sizes of the atoms change when incorporated into the Laves phase structures.

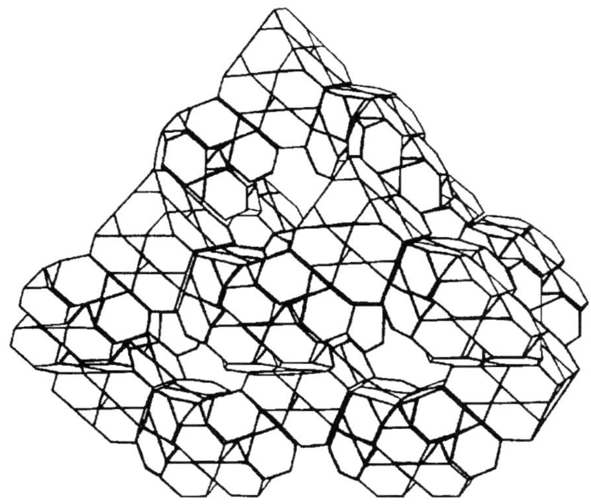

FIGURE 8. The complex structure of Cu$_4$Cd$_3$ represented as an assembly of truncated tetrahedra. The truncated tetrahedra have Cu atoms at their vertices and Cd atoms at their centers. From Samson [4].

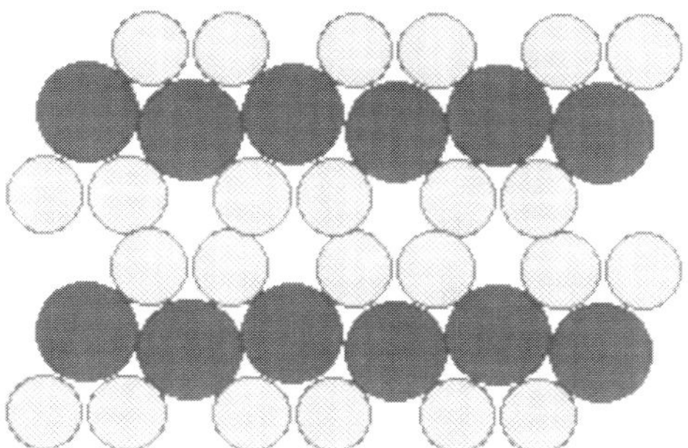

FIGURE 9. Planar basic layer of the crystal structure of YZn$_2$. Note the zig-zag chain of Y atoms. The layers are stacked so that the Y atoms (larger) fit over the open spaces in the adjacent layers. From Sree Harsha [5].

A large group of intermetallic compounds of various stoichiometries exhibit structures which are based upon the $CaCu_5$ type structure. A number of compounds of this type which contain certain rare earth and transition metals are important permanent magnet and hydrogen storage materials. The basic layers of atoms in the AB_5 structure, of which these structures are built, are shown in Figure 11. For the structures of the related compounds with A_2B_{17}, AB_{12}, and other stoichiometries, a pair of smaller B atoms is systematically substituted for selected A atoms.

An example of a true ternary intermetallic compound with a complex structure is Al_5CuLi_3 (see Figure 12). There are about a dozen representatives of this structure type. The structure is described as an 80 atom array packed together in a body centered cubic arrangement. The 80 atom complex is built up of the truncated tetrahedra described above, with the larger lithium atoms usually at their centers, and the vertices are occupied by both aluminum and copper atoms in a disordered fashion.

Thus, it can be seen that there are a large number of widely different atom arrangements for metals and intermetallic compounds. While the crystal structures of most of the pure metals and a number of intermetallic compounds are relatively simple in nature, most intermetallic compounds exhibit crystal structures of considerable complexity. However, there is as yet little real understanding of the relationship of chemical bonding to structural arrangement in either the pure metals or intermetallic compounds. Sphere packing cannot predict these structural arrangements, stacking sequences, or ordering *ab initio*. It seems apparent that there must be at least some influence of covalent bonding and second NN interactions. But no theoretical treatments have been advanced which successfully predict the simple crystal structures of the pure metals. It would seem that considerations of directional bonding and

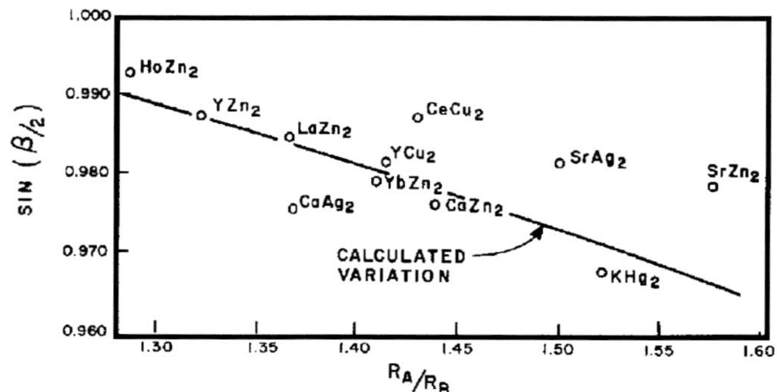

FIGURE 10. Variation of the A atom chain angle β with relative atom size for the KHg_2 type crystal structure. Goldschmidt CN12 values are used for the metallic radii of the A and B atoms. From Ryba & DebRay [6].

orbital overlap are necessary. The tetrahedral character of the connections between metal atoms in pure metals and tetrahedrally close packed inter-metallic compounds suggests a strong influence of directional bonding. The atom coordinations in the Laves phases are reminiscent of hybrid sp^3 orbitals. Trost [8] proposed a scheme involving orbital overlap and hybrid orbitals as an explanation of the structures of the transition metals. Others [e.g., 9-11] have proposed similar bonding schemes for several intermetallic compound structure types. However, these approaches have not been popularly accepted, nor are they well-developed.

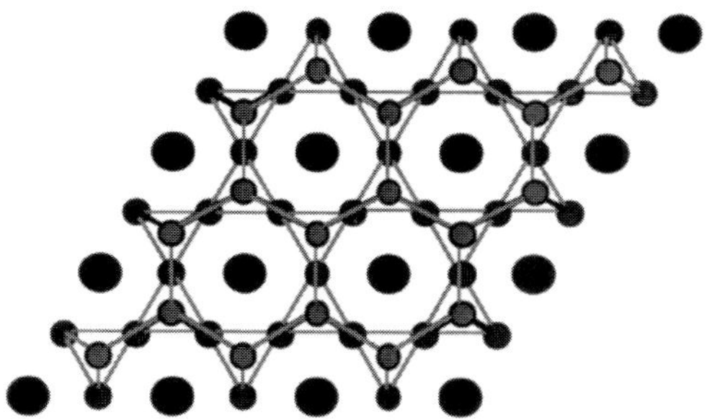

FIGURE 11. The CaCu$_5$ crystal structure type. Two planar layers are shown in projection; the larger A atoms center the larger hexagons of B atoms in the main layer of the structure, while smaller B atom hexagon layers appear above and below the main layer.

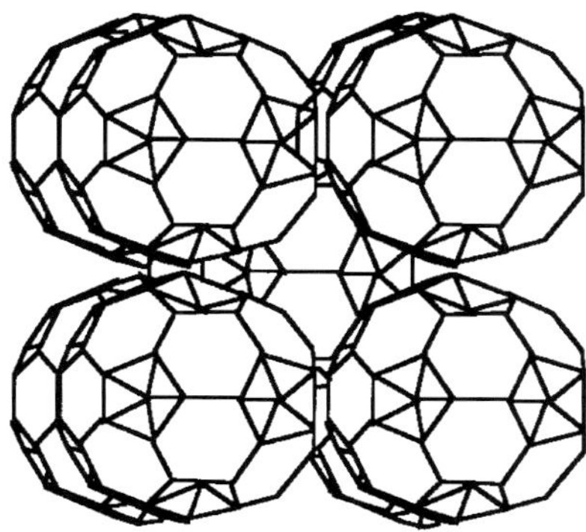

FIGURE 12. The crystal structure of Al$_5$CuLi$_3$ shown as a body centered cubic packing of super complexes of 80 atoms arranged in truncated tetrahedra. From Bartges & Ryba [7].

AMORPHOUS METALS

While metals and alloys usually exhibit crystalline structures, it is possible to prepare a large number of metallic alloys in the non-crystalline or amorphous state, usually through the use of a technique involving rapid solidification, (cooling rates of about 10^6 K/sec.) from the melt. In 1959, Duwez and his colleagues [12] began rapid solidification studies with the aim of extending solid solubility beyond the equilibrium limits, and found that amorphous structures resulted for a large group of alloys. These materials are called metallic glasses. Since that time, a wide variety of techniques have been shown to produce metallic glasses and similar amorphous metallic materials. Earlier techniques included vapor deposition, sputtering, electrodeposition, gas atomization, splat cooling, and melt spinning. More recently, amorphous metallic alloys have been formed by mechanical alloying, hydrogen absorption, irradiation, and interdiffusion between metallic layers. The resulting amorphous structures are very unstable and transform to crystalline materials when heated above a critical temperature.

Early metallic glasses contained "glassifiers" which increased the glass forming tendency of a metal or alloy. These glassifiers are metalloids such as boron, silicon, and phosphorus. In some cases, the glassifier has a very potent effect. For example, the alloy $Pd_{40}Ni_{40}P_{20}$ can be made glassy in 1 cm sections at cooling rates as low as 1 K/sec. Subsequently, it was found possible to form metallic glasses without glassifiers in many alloy systems. As a result, there are considered to be two general classes of amorphous alloys: those consisting of metals and metalloids, and those consisting of two or more metals. For the latter, there are subclasses of materials depending upon the types of metals (an early transition metal with a late transition metal, aluminum with a rare earth metal and another transition metal, etc.) involved. Metallic glasses form only in cases where the sizes of the constituent atoms differ by at least 10%.

X-ray, neutron, and electron diffraction measurements provide the major evidence for the amorphous character of metallic glasses. While crystalline materials exhibit many sharp diffraction maxima, diffraction patterns for amorphous alloys consist of only a few very broad peaks. Information concerning their disordered structures can be extracted from the diffraction patterns with difficulty. Pair distribution functions are extracted from the diffraction data and compared with those calculated for a model structure.

Some of the models which have been used are:

   a. dense random packing of hard or soft spheres
   b. close packing of irregular tetrahedra
   c. assemblies of trigonal prisms (see Figure 13)
   d. polyclusters - nano-sized clusters of locally ordered atoms

If strong short range chemical interactions are not present, then the dense random packing of spheres or packing of regular tetrahedra are appropriate models. If there is local ordering, then the trigonal prism or the polycluster models may be used. There is experimental evidence that chemical and other types of local ordering exist in many metallic glasses.

A recent study by Petkov and Apostolov [14] on the structures of $TbCu_2$ and $DyCu_2$ metallic glasses illustrates the importance of local ordering. Models for the glass structure were based upon local atomic arrangements in selected crystal structures for AB_2 compounds. The Laves phase, KHg_2, and $A1B_2$ type crystal structures were considered. Local order in the structure of the metallic glasses was found to correspond closely to that in crystalline $A1B_2$ (see Figure 14). The trigonal prisms of the $A1B_2$ structure are disordered by vertex and edge sharing in the glassy state. Complete face sharing would give a crystalline arrangement.

The $A1B_2$ crystal structure is closely related to the KHg_2 structure , exhibited by both $TbCu_2$ and $DyCu_2$ in the crystalline state. A high temperature martensitic phase transformation has been detected in some crystalline $REZn_2$ compounds, which also have the KHg_2 crystal structure. The structure of the high temperature phase has not yet been determined. The transformation probably exists for the $RECu_2$ compounds as well, based upon the extensive twinning observed in both $REZn_2$ and $RECu_2$ compounds. This twinning is well documented and has been quantified for $HoCu_2$ by Seidensticker & Ryba [15]. The supposition that the local order in the glassy state and the liquid are very similar strongly suggests that the crystal structure of the high temperature phase may well be the $A1B_2$ type.

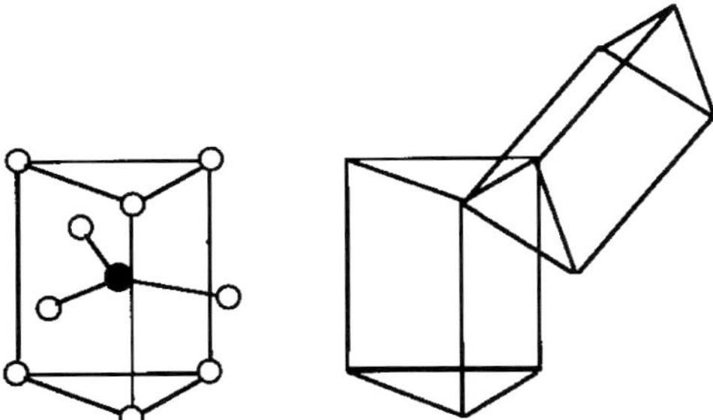

FIGURE 13. Trigonal prism atom arrangement (left) and prism packing by edge sharing as a model for the structure of metallic glasses. From Gaskell [13]

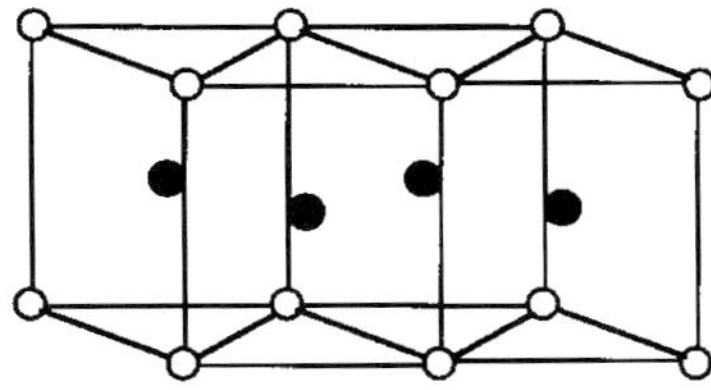

FIGURE 14. A fragment of the $A1B_2$ crystal structure.

QUASICRYSTALS

Rapid solidification of many liquid alloys can produce another type of structural ordering which is neither crystalline nor amorphous. The structure has no periodic character, in the usual sense, in at least one direction, but yet its atom connections are not random, as in amorphous materials. These recently discovered materials, dubbed quasicrystals, exhibit bond orientation order. This type of order is unique to metallic materials. As a result, the diffraction patterns of quasicrystals exhibit sharp maxima. However, the symmetries observed in these diffraction patterns (5-fold, 8-fold, 10-fold, and 12-fold rotation axes) belie atom arrangements which are not crystalline (i.e., periodic) in nature. For example, for the icosahedral phases, which exhibit 5-fold symmetry, the interatomic bonds all lie essentially along directions parallel to the five axes defining an icosahedron. For the decagonal phases, which exhibit 10-fold symmetry, the bonds in any plane perpendicular to a particular direction all lie essentially along the five axes defining a pentagon; the structure is periodic in the direction which is perpendicular to these planes.

Quasicrystals are produced in many alloys by melt spinning and other rapid solidification techniques. However, just as in the production of metallic glasses, there is a considerable number of quasicrystalline phases which form under rather slow cooling conditions. Some are thought to be equilibrium phases, and others transform to crystalline phases upon cooling slowly. On the other hand, quasicrystalline Al_6CuLi_3 has been found to transform to an amorphous material in an electron beam [16].

Because of the lack of periodicity, it has been difficult to extract structural information from the diffraction patterns of quasicrystals. However, through the gradual development of specialized structure determination techniques which make use of the property for quasicrystals of periodicity in spaces higher than three dimensions, the nature of the structures of a number of these materials is being revealed. For example, the decagonal phase $Al_{65}Co_{15}Cu_{20}$ has been studied extensively by x-ray and electron diffraction and other techniques. Steurer [17] showed that the atom arrangement for this quasicrystalline material is made up of a periodic stacking of layers of polyhedra with 5- and 10-sided faces quasiperiodically packed together, as shown in Figure 15. Clusters of columns of these pentagonal polyhedra are a major characteristic of the structure.

Another approach was taken by Song & Ryba [18], who constructed a structure model from an extensive examination of high resolution transmission electron microscope and scanning tunneling microscope images. They show that the structure of the quasicrystalline phase and its closely related large unit cell "crystalline approximants" can be represented by the packing of two basic pentagonal polyhedra (see Figure 16). The two structure models have many similar characteristics; this model also results in clusters of columns of pentagonal polyhedra.

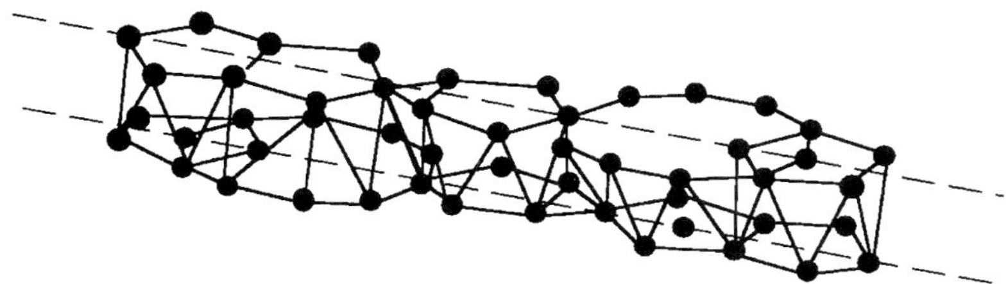

FIGURE 15. A portion of one layer of the structure of quasicrystalline $Al_{65}Co_{15}Cu_{20}$. Layers of these quasiperiodically packed polyhedra are stacked periodically in the vertical direction. From Steurer [17].

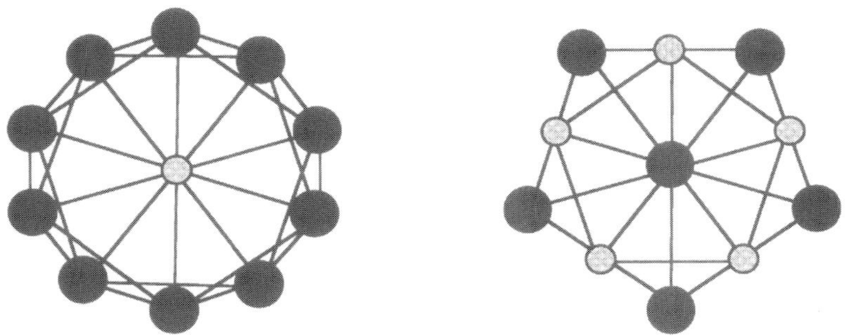

FIGURE 16. The two basic polyhedra for the structures of quasicrystalline $Al_{65}Co_{15}Cu_{20}$ and its crystalline approximants derived by Song & Ryba [18] from high resolution transmission electron microscope and scanning tunneling microscope images.

REFERENCES

1. Ryba, E., Kejriwal, P.K. and Elmendorf, R. 1969. J. Less-Common Metals *18*, 419.
2. Bergman, B.G. and Shoemaker, D.P. 1954. Acta Cryst. *7*, 857.
3. Elmendorf, R. 1967. M.S. Thesis, The Pennsylvania State University.
4. Samson, S. 1967. Acta Cryst. *23*, 586.
5. Sree Harsha, K.S. 1964. Ph.D. Thesis, The Pennsylvania State University.
6. Ryba, E. and DebRay, D.K. 1966. Winter Meeting of the American Crystallographic Association, Austin, TX.
7. Bartges, C.W. and Ryba, E. 1989. *Aluminum-Lithium Alloys.* Vol. II. T.H. Sanders and E.A. Starke, Jr., Eds., Materials and Component Engineering Publications Ltd., Birmingham, England, p. 711.
8. Trost, W.R. 1959. Can. J. of Chem. *37*, 460.
9. Carter, F.L. 1964. Westinghouse Research Laboratories Paper 64-929-442-P3.

10. Hansen, D.A. and Smith. J.F. 1967. Acta Cryst. *22*, 836.
11. Mulokozi, A.M. 1977. J. Less-Common Metals *53*, 205.
12. Klement, K., Willens, R.H., & Duwez, P. 1960. Nature *187*, 869,
13. Gaskell, P.H. 1979. J. Non-Cryst. Solids *32*, 207.
14. Petkov, V. and Apostolov, A. 1994. J. Non-Cryst. Solids *167*, 255.
15. Seidensticker, J.R. and Ryba, E. 1993. Acta Cryst. *B49*, 213.
16. Michel, D.J., Reed, J.R., Tosten, M.H., Bartges, C.W., Ramani, A., Ryba, E., and Howell, P.R. 1990. Proceedings of the International Conference on Advanced Aluminum and Magnesium Alloys, Amsterdam.
17. Steurer, W. 1994. *Quasicrystals and Imperfectly Ordered Crystals*, K.H. Kuo and S. Takeuchi, Eds., Materials Science Forum 150-151, Trans Tech Publications, Switzerland, p. 15.
18. Song. S. & Ryba, E. 1994. Phil. Mag. 69, 707.

Chapter Four

CRYSTALLINE AND AMORPHOUS STATES OF POLYMERS AND POLYMER BLENDS

JAMES P. RUNT and SAPNA TALIBUDDIN

Department of Materials Science and Engineering
Steidle Building
Pennsylvania State University
University Park, PA 16802

INTRODUCTION

The solid-state microstructure of semi-crystalline polymers has been studied for many years and a relatively clear picture of the details has developed [e.g. 1-3]. Very briefly, crystalline polymers generally consist of chain-folded lamellar crystals which are separated by regions of disorder. These crystals are ca. 5-20 nm thick and the polymer chain axis is more or less normal to the lamellar faces. Lamellae are organized into fibrils which form a component part of a spherulitic (or more complex) superstructure. Generally speaking, chain segments residing in amorphous (disordered) regions have restricted mobility as a consequence of connections to the crystallites. In addition, the amorphous regions are not expected to be uniform. It has been argued that the transition from the highly ordered crystalline state to the isotropic amorphous state cannot occur abruptly and, hence, the existence of crystal-amorphous interphases has been predicted [e.g. 4]. The interphase is envisioned as a region in which the chain segments retain some traits of crystalline order while lacking rigid lattice constraints. A variety of experimental studies on neat semi-crystalline polymers have confirmed the existence and generality of crystal-amorphous interphases and have estimated their size to be on the order of 0.5-3.5 nm, depending on crystallization conditions, molecular weight, etc. [e.g. 5, 6].

Polymer blends — physical mixtures of two or more polymers — have been under intense scrutiny in both industry and academia during the past two decades. This activity has resulted in advances in our fundamental understanding of this group of materials as well as a large number of important industrial products. In fact, as of 1992, it has been estimated that 30% of all polymers are sold in mixed form [7]. Interest has burgeoned in polymer blends as a result of one's ability to readily tailor properties to specific applications, their great versatility and the promise of enhanced performance. A significant portion of this activity has been focused on enhancement of mechanical properties, particularly toughness. The vast majority of polymer blends (~90%) have been found to be multiphase — either completely immiscible or multiphase with some limited mixing between the component polymers. This behavior is primarily the result of the low combinatorial entropy of mixing of the dissimilar polymer chains. The remaining fraction are single phase, melt-miscible mixtures.

Since roughly 1/2 to 2/3 of all useful polymers are crystalline or crystallizable, blends containing crystalline polymers are of particular importance. However, the microstructure and properties of such mixtures are less well understood at the present time than their neat semi-crystalline counterparts. In the next sections we present a brief overview of some of our recent research on binary mixtures containing crystalline polymers. The first portion focuses on homopolymer - copolymer blends, where both components are semi-crystalline. Of particular interest is the important role that the homopolymer plays in crystallization of the copolymer (i.e. nucleation). In addition, the possibility of co-crystallization in these mixtures (i.e. crystallization of homopolymer and copolymer chains in close proximity in the same lamellar crystal) is addressed. The second portion of this article summarizes ongoing research on the fundamental factors that influence the microstructure in melt-miscible, crystalline blends.

SEMI-CRYSTALLINE POLYMER BLENDS

Crystalline Homopolymer - Copolymer Blends

Relatively strong intermolecular interactions are often required to attain melt miscibility of polymer mixtures, although there are a number of cases where the components' chemical structures are similar enough so that, although small, the combinatorial entropy is apparently sufficient to overcome unfavorable dispersive forces. Some examples of this situation are blends of polyethylenes [8], poly(vinyl fluoride-trifluoroethylene) copolymers [9] and the homopolyester - segmented block copoly(ester-ether) blends [10, 11]. In addition, in each of these cases the component polymers are both crystallizable and the unit cell structures and crystal compositions are very similar or identical, providing the opportunity for co-crystallization.

In an extension of our earlier work on the homopolyester - segmented block copoly(ester-ether) blends [10,11], we have explored the miscibility and crystallization behavior of mixtures of two crystalline fluoropolymers that also have similar chemical structures: poly(tetrafluoroethylene) (PTFE) and a random copolymer of tetrafluoroethylene (TFE) and a small amount (~1-2 mole %) of perfluoropropylvinyl ether, PFA [12]. The addition of the perfluoroalkoxy, even at such low concentration, renders the copolymer melt-processible, lowers the melting point by some 20°C, but only slightly reduces the material's thermal stability.

It is well established that the mechanical β-process (observed at low temperatures) of PTFE and PFA involves oscillations of chain segments within the crystalline phase and is associated with the first-order crystal disordering transitions seen near room temperature in differential scanning calorimetry (DSC) experiments [13]. We observe two β-processes for as-prepared PTFE/PFA blends and the character of the parent polymers is largely retained (modified by the blend composition). This behavior suggests separate PTFE and PFA crystalline regions. We observe a single α-transition for the blends, which is close to that of PFA or intermediate to that of the neat polymer transitions. The α-relaxation of PTFE and copolymers has been associated with cooperative, segmental (i.e. Tg-like) motion but this assignment remains controversial [14]. Thus, the question of amorphous miscibility of these blends is not readily addressed by analysis of the number and location of relaxations associated with segmental motion, due to the uncertain origin of the α-transition. However, although largely circumstantial, results of DSC experiments, including those conducted on physically mixed powders directly in the DSC, point to a significant degree of mixing of PTFE and PFA in the temperature range of interest.

Since only dispersive forces are expected between PTFE and PFA, their miscibility was modeled using a Flory-Huggins formalism and estimating the interaction parameter from the component solubility parameters [12]. This treatment predicts that relatively high molecular weight PTFE and PFA are expected to be miscible when PFA contains ≤~4-5 wt % (1-2 mole %) FPVE comonomer.

Degrees of crystallinity of the as-prepared neat polymers and blends range from ~30-50%. The dominant melting endotherm for the blends is seen at an intermediate temperature compared to that of the neat polymers although there is evidence of a second small endotherm located at a temperature below that of the PFA melting peak (but at a temperature that corresponds to the lower temperature portion of the PFA endotherm). Likewise, there is one dominant crystallization exotherm and also indication of a lower temperature crystallization process in the blends.

The significant influence of PTFE on the crystallization of PFA is seen particularly well in Figure 1 when the blends and neat polymers are crystallized at a relatively low cooling rate. Heats of crystallization of all blends fall between those of the neat polymers, with the overall heat of crystallization decreasing with increasing PFA content in the blend. Taking as an example the 50/50 blend, it is seen that much (but not all) of the PFA in the mixture has concluded crystallization at a temperature well above the onset of crystallization of PFA in the unblended state. Although the dominant crystallization process in the blends occurs at a lower temperature than neat PTFE (i.e. PFA inhibits the crystallization of the PTFE chains), the role of PTFE in nucleating the crystallization of at least a portion of the PFA is clear. The fact that we observe one melting point (T_m) from the primary crystallization process is consistent with the idea of co-crystallization of PTFE with a portion of the PFA (we would anticipate two significant melting endotherms if PTFE simply nucleated PFA crystals), although the behavior of the crystal-crystal transitions strongly supports the existence of separate crystal populations.

In DSC experiments, PTFE was found to exhibit two transitions near room temperature (one at ~20°C and another at ~32°C) and these have been assigned to solid-solid phase transformations and result in an increase of crystalline disorder [e.g. 15]. PFA exhibits one crystalline transition in this temperature range, consistent with previous observations. The blends, either on heating or cooling, all exhibit multiple transitions— the behavior appears to be a weighted average of the thermograms of the components. This is in agreement with the mechanical relaxation behavior presented earlier, although the additive-type

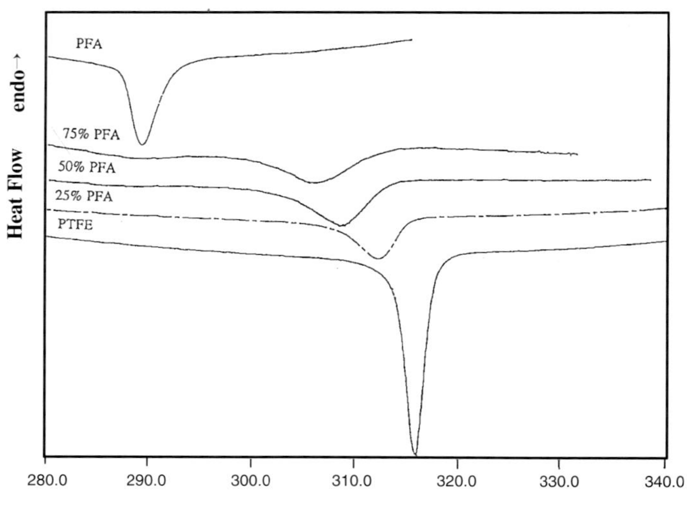

Temperature (°C)

FIGURE 1. Crystallization behavior (cooling at 2°C/min) after 1 min. in the melt at 400°C of PTFE, PFA and 3 blends [from 12].

behavior is in distinct contrast to the melting and crystallization behavior. The character of the thermograms of the as-prepared blends in this temperature interval is consistent with the idea that there are sufficiently large spatial regions of crystalline PFA and crystalline PTFE so that separate crystalline transitions are observed. Yet these regions must melt at similar temperatures as only one dominant T_m is observed.

There are crystallization conditions under which a single, intermediate-temperature transition is observed. DSC thermograms of the blends in the low temperature region after rapid cooling from the melt (from 380°C) into liquid nitrogen exhibit one (albeit relatively broad) intermediate transition, which moves to higher temperatures with increasing PTFE content. In addition, a single melting endotherm is observed for these rapidly cooled samples. Co-crystallization is well known to be favored under rapid crystallization conditions and we speculate that this is the origin of the behavior observed here – that is, PFA and PTFE chains are well mixed on the local level in the crystals.

Microstructure of Crystalline Polymer Blends:
Model Poly(ethylene oxide) Blends

An intriguing subset of melt-miscible polymer blends are those that contain one or more polymers that are capable of crystallizing. The influence of the second polymer on crystallization and crystalline microstructure is expected to be quite significant. In general, the T_g of the mixture can be raised or lowered, thereby contracting or expanding the temperature window over which the mixture may crystallize. A reduction in the equilibrium melting point would also be expected, the magnitude of which will depend, among other factors, upon the polymer-polymer interaction parameter (χ) and blend composition [e.g. 16]. In addition, one generally expects a reduction in crystallization rate with increasing diluent concentration (i.e. proportional to the volume fraction of the crystalliz-able polymer in the mixture).

One of the important considerations in the microstructure of these crystalline, melt-miscible polymer blends is the location of the amorphous polymeric diluent in the microstructure. As illustrated in Figure 2, the diluent molecules can be located in interspherulitic regions, interfibrillar regions (i.e. between the fibrils or lamellar stacks), interlamellar regions (between lamellar crystals) or some com-bination of these, yielding different microstructures, which in turn give rise to different materials properties. It is important, therefore, to determine the factors that influence the extent of diluent segregation. It is also interesting to note that multiple locations for the amorphous diluent, depending on the dispo-sition of the amorphous polymer in the final microstructure, lead naturally to multiple T_g-like transitions although the polymers themselves are miscible at this temperature [18].

What controls the placement of the amorphous polymer in the microstructure of such melt-miscible systems? From the point of view of interlamellar incorporation, the diameter of gyration of the diluent polymer is often comparable to or greater than the separation between crystalline lamellae (particularly at high crystalline polymer concentrations) [19]. Confinement of the diluent would therefore lead to an entropic driving force for escape from the interlamellar zones. In competition with this, it has been proposed that the favorable intermolecular interactions gained by mixing the diluent with the amorphous portion of the crystalline polymer will promote interlamellar incorporation.

In the literature (where most citations are for relatively weakly-interacting mixtures), one generally finds that the amorphous polymer resides in interlamellar regions when its glass transition temperature (T_g) is high compared to the crystallization temperature (T_c), whereas when its T_g is low relative to T_c at least some of the amorphous polymer is located outside of the interlamellar regions. A case in point is the poly (vinylidene fluoride) and poly (methyl methacrylate) (PMMA) blend system for which completely interlamellar [20] as well as extralamellar [21] incorporation of PMMA are obtained at different T_cs. Thus, mobility of the amorphous polymer, as determined by its T_g and by the mixed phase environment at and away from the crystal growth front, appears to be one factor that plays an important role in diluent segregation. The role of the crystal growth rate is also expected to be important [22].

We have recently designed a series of 'model' melt-miscible blends and examined their microstructure using small-angle x-ray scattering [SAXS] (along with supplementary studies of the supermolecular structure and crystallization kinetics) with a view to targeting the effects of glass transition temperature and intermolecular interactions on phase segregation in crystalline blends [23]. Poly (ethylene oxide) was used as the crystalline component in each model blend system while the melt-miscible amorphous polymers were selected with regard

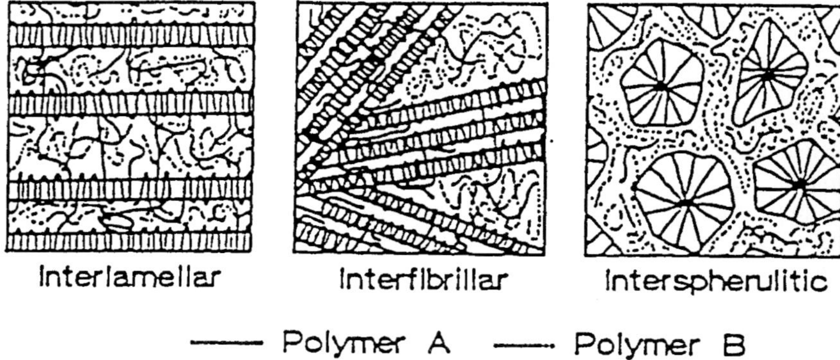

FIGURE 2. Possible solid-state microstructures of a melt-miscible binary polymer blend containing a crystalline component (from [17]). Polymers A and B represent the crystallizable and amorphous polymers, respectively.

to their T_gs relative to the crystallization temperature and the strength of their interactions with PEO. Poly(methyl methacrylate) [PMMA] and poly(vinyl acetate) [PVAc] represented weakly-interacting high and low T_g diluents (T_g = 113°C and 31°C, respectively). Two random copolymers, one composed of ethylene and methacrylic acid containing 55% by weight acid units [EMAA55], and another of 50 weight percent styrene and hydroxystyrene [SHS50], comprised corresponding strongly-interacting low and high Tg diluents (Tg = 36°C and 150°C, respectively). A summary of the results of our experiments on blends crystallized isothermally at 45°C are provided below.

Optical microscopy on the weakly-interacting systems reveals linear spherulitic growth over the entire composition range studied. The lower T_g mixtures, PEO/PVAc, crystallize rapidly (i.e. in a few minutes) while corresponding growth rates for PEO/PMMA are somewhat lower, as would be expected for a higher T_g diluent. In both cases, volume-filling spherulites are observed at all compositions, indicating that the diluents reside completely within the spherulites. On the other hand, the two strongly-interacting systems, PEO/EMAA55 and PEO/SHS50, exhibit volume-filling spherulites up to a diluent concentration of about 20 weight percent. Furthermore, crystallization rates for these systems are much lower than those for their weakly-interacting counterparts. For example, since PVAc and EMAA55 have similar glass transition temperatures, their blends with PEO might initially be expected to crystallize at comparable rates. However, PEO/EMAA55 crystallizes much more slowly than PEO/PVAc: at 20% diluent concentration, the growth rate of PEO/EMAA55 is an order of magnitude lower than that of 80/20 PEO/PVAc. At 30% EMAA55 concentration, the growth rate not only diminishes further but growth ceases before the spherulites impinge. On cooling to room temperature, the spherulites continue to grow, eventually becoming volume-filling. Such behavior is characteristic of a decrease in the degree of supercooling brought about by the addition of a strongly- interacting diluent. PEO/SHS50 blends are found to crystallize at even slower rates than the corresponding PEO/EMAA55 blends. Given the much higher T_g of SHS50, this is understandable. 70/30 PEO/SHS50 behaves much like 70/30 PEO/EMAA55. However, growth does not resume on cooling to room temperature, implying sufficient build-up of the high-T_g SHS50 at the spherulitic boundaries to immobilize the crystallizable PEO.

The two weakly-interacting blends, PEO/PMMA and PEO/PVAc, display strikingly different behavior SAXS behavior when crystallized at 45°C. For PEO/PMMA, Lorentz-corrected scattering maxima move progressively to lower values of the scattering vector ($q = (4\pi/\lambda)\sin(\theta/2)$, where θ is the scattering angle and λ the x-ray wavelength) or to larger long periods, with increasing PMMA content. For a weakly interacting mixture, it is reasonable to expect the crystal thickness to be independent of blend composition. The average thickness of the PEO lamellae in the PEO/PMMA blends may then be assumed to equal that of neat PEO lamellae crystallized at the same temperature and this value can be used to compute the average long period of the all-interlamellar case. The long periods thus obtained were found to agree with those derived experimentally, suggesting complete interlamellar incorporation of PMMA at all compositions studied. The excellent agreement between the experimental and calculated invariants (i.e. the total scattered intensity — as derived from the correlation function) lends further credence to the conclusion that PMMA resides completely within interlamellar regions. In the case of the PEO/PVAc blends, however, although the scattering peaks broaden somewhat with increasing PVAc content, the maxima shift only slightly to lower q values. In addition, the experimental invariants do not correspond to all-interlamellar model values. These findings, in conjunction with optical microscopy results, suggest at least partial exclusion of PVAc from interlamellar to the interfibrillar regions.

On the other hand, the results of SAXS and optical microscopy experiments showed both strongly-interacting diluents to be largely extralamellar: interfibrillar up to a concentration of *ca.* 20% and at least partially interspherulitic at 30% diluent concentration. As an example, Figure 3a presents the Lorentz-corrected SAXS intensity as a function of scattering vector for PEO and several PEO/EMA55 blends. A dramatic increase in long period is observed as the EMA55 content is increased up to 20%. One may at first take this as evidence that the EMAA is located in interlamellar regions but comparison of the expected SAXS invariant for the case of interlamellar incorporation vs. that measured experimentally leads to a quite different conclusion (Figure 3b). EMAA55 has an electron density that is substantially smaller than those of neat amorphous and crtystalline PEO and even modest interlamellar inclusion of the amorphous polymer would be expected to lead to a significant change in the invariant for PEO/EMAA55 blends. Thus this data strongly suggests displacement of EMAA55 to extralamellar regions. Consequently, the origin of the increase in long period with increasing EMAA55 content is an increase in lamellar thickness, which is consistent with a lower degree of supercooling (i.e. a significant equilibrium melting point depression), as expected for a strongly-interacting blend. In fact, to a first approximation, we can estimate the degree of supercooling at which the EMAA blends were crystallized by comparing the experimental long periods with those determined by Arlie, et. al. [24] for neat PEO. It appears that the addition of EMAA55 to PEO can effectively depress the supercooling by as much as 9 - 14°C in the range of 10 - 20% EMAA55. This estimate seems plausible in light of the drastic reduction observed in crystallization rates of these blends with increasing EMAA55 content. The behavior of the SHS50 blends is similar although the changes in crystal thickness with SHS50 composition are not as large as for PEO/EMAA55, suggesting that the melting point depression is not as significant for this blend.

A number of inferences can be drawn from the above analysis. The contrasting behavior of the two weakly-interacting blends seems to suggest that the mobility of the amorphous component at T_c determines diluent location: relatively high T_g PMMA is trapped between crystal lamellae whereas low T_g PVAc can diffuse away from the crystal growth front and reside partially in interfibrillar regions.

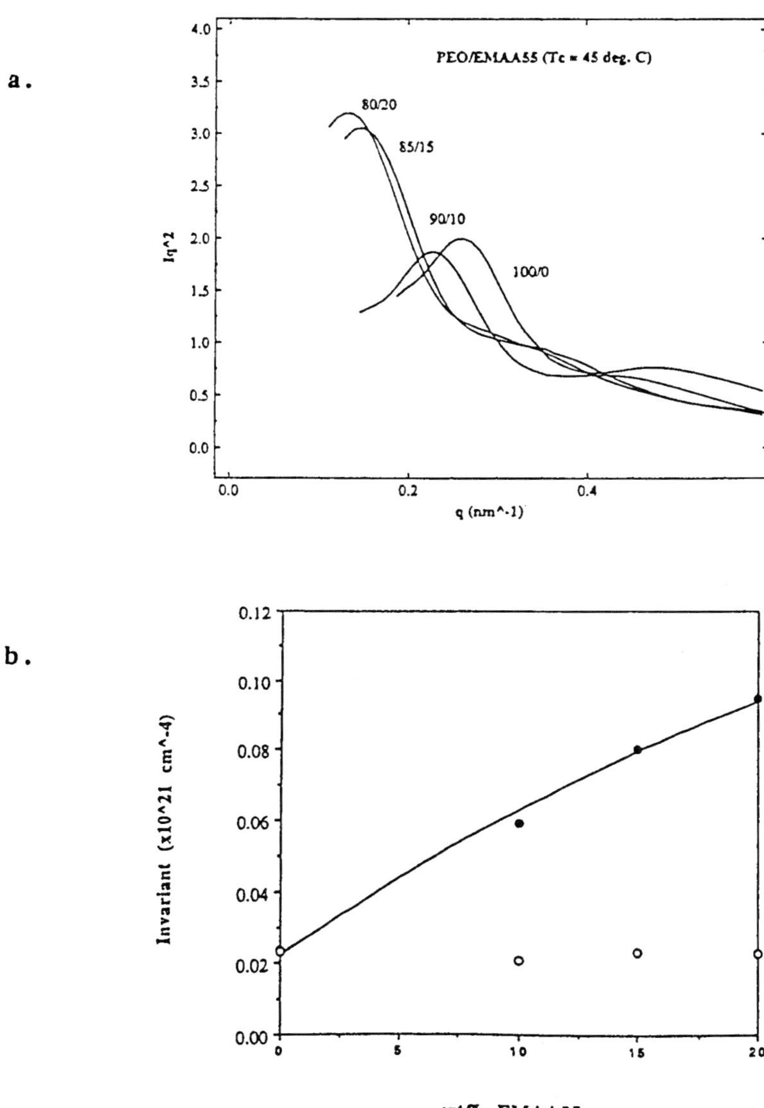

FIGURE 3a: Lorentz-corrected SAXS intensity as a function of scattering vector for PEO and three PEO/EMAA55 blends crystallized at 45°C.

FIGURE 3b. SAXS invariants for the same materials. The filled circles (and solid line) represent the predicted invariants for the case of complete EMAA55 inclusion in interlamellar regions. The open circles are the experimental invariants (from [23b]).

However, an examination of the microstructures of the two high T_g systems, PEO/PMMA and PEO/SHS50, clearly demonstrates that diluent mobility is not the sole criterion for amorphous phase segregation. Although SHSS50 has a significantly higher T_g (and molecular weight) and therefore reduced mobility in the mixed environment compared to PMMA, SHS50 is found to reside in interfibrillar regions at blend compositions up to 20%, becoming at least partially interspherulitic at higher concentrations, whereas PMMA is completely interlamellar at all blend compositions studied. The appreciably lower crystal growth rates for the PEO/SHS50 blends allow more time for SHS50 diffusion over larger distances from the growth front. Such a relationship between the scale of diluent segregation and relative diffusion and growth rates has been proposed previously by Keith and Padden [22].

A comparison of the two low T_g blends, PEO/PVAc and PEO/EMAA55, further illustrates the influence of crystallization rate on diluent migration. The diluent is extralamellar in both cases, but diffuses over larger distances for the more slowly crystallizing PEO/EMAA55 blends, becoming partly interspherulitic at ~30% EMAA55 content. Despite the lower molecular weight of EMAA55 compared to PVAc, the fact that the amorphous polymers in both slow-crystallizing systems (PEO/EMAA55 and PEO/SHS50) exhibit the same degree of segregation with increasing diluent content, irrespective of relative diluent T_gs and molecular weights, strongly favors crystal growth rate as a crucial factor in the extent of migration of the amorphous component.

For these systems it follows therefore that while the effect of diluent mobility cannot be ignored, it is the growth rate of the PEO crystals, and hence the factors that influence the growth rate, which dominate the length scale of diluent segregation at the crystallization temperatures explored in our studies.

Quenched Semicrystalline/Amorphous Blends

While diluent segregation in melt-miscible crystalline blends may give rise to regions of varying concentration (and consequently, multiple glass transitions) as noted earlier, single composition-dependent T_gs would no doubt be expected in the quenched, amorphous state. Indeed this has been found to be the case for several miscible semicrystalline/amorphous blends such as poly) (ε-caprolactone)/ poly(vinyl chloride) [25] and isotactic poly(styrene)/poly(phenylene oxide) [26]. However, recent DSC experiments on our weakly-interacting PEO blends after quenching from the melt [27] have shown a T_g close to that of pure PEO for all compositions studied. A mixed-phase T_g was additionally observed for the PEO/PVAc blends but not in the case of the higher T_g PEO/PMMA blends, as it was obscured by the melting of the PEO crystals formed on quenching. Both strongly-ineracting blends only exhibited single, composition-dependent T_gs.

The above behavior may be explained by local concentration fluctuations that are known to occur around the mean blend composition in nominally miscible ($\chi = 0$) polymer blends [28]. For strongly interacting systems, these fluctuations would understandably be more damped. Recently, Kumar et al. [29] proposed a concentration correlation model for binary homogeneous polymer systems with components of widely differing T_gs. It predicts two distinct "dynamic microenvironments" for weakly-interacting systems, one associated with the low T_g component and the other with the mixed phase, which is consistent with our findings for PEO/PVAc and PEO/PMMA blends. For the case of strong intermolecular interactions, the model predicts a dynamic response corresponding to the average blend composition, which once again is in accordance with our results for the strongly-interacting PEO/EMAA55 and PEO/SHS50 blends.

REFERENCES

1. Wunderlich, B. 1973 and 1976. *Macromolecular Physics* (Vols 1 & 2) Academic Press, New York.
2. Bassett, D.C. 1981. *Principles of Polymer Morphology.* Cambridge University Press, Cambridge, UK.
3. Phillips, P.J. 1990. Polymer Crystals. *Rep. Prog. Phys.* 53: 549-604.
4. Flory, P.J., Yoon, D.Y. and D.A. Dill. 1984. The Interphase in Lamellar Semicrystalline Polymers. *Macromolecules* 17:862-868.
5. Muira, H., Hirschenger, J. and A.D. English. 1990. Segmental Dynamics in the Amorphous Phase of Nylon 66: Solid-State ^{2}H NMR. *Macromolecules* 23:2169-2182.
6. Mandelkern, L., Alamo, R.G. and M.A. Kennedy. 1990. Interphase Thickness in Linear Polyethylene. *Macromolecules* 23: 4721-4723.
7. Utracki, L.A. 1990. *Polymer Alloys and Blends. Thermodynamics and Rheology,* Oxford University Press, Hanser, New York.
8. Alamo, R.G., Glaser, R.H. and L. Mandelkern. 1988. The Cocrystallization of Polymers: Polyethylene and Its Copolymers. *J. Polym. Sci. Polym. Phys. Ed.* 26: 2169-2195.
9. Tanaka, H., Lovinger, A. and D.D. Davis. 1990. Miscibility and Isomorphic Cocrystallization in Blends of Ferroelectric Copolymers of Vinylidene Fluoride and Trifluoroethylene. *J. Polym. Sci. Polym. Phys. Ed.* 28: 2183-2198.
10. Runt, J., Du, L., Martynowicz, L.M., Brezny, D.M., Mayo, M. and M.E. H a P ncock, 1989. Dielectric Properties and Cocrystallization of Mixtures of Poly(butylene terephthalate) and Poly(ester-ether) Segmented Block Copolymers. *Macromolecules* 22:3908-3913.
11. Gallagher, K.P., Zhang, X., Runt, J., Huynh-ba, G. and J.S. Lin. 1993. Miscibility and Cocrystallization in Homopolymer - Segmented Block Copolymer Blends. *Macromolecules* 26: 588-596.
12. Runt, J., Jin, L., Talibuddin, S. and C.R. Davis. 1995. Crystalline Homopolymer-Copolymer Blends: Poly(tetrafluroethylene) - Poly(tetrafluroethylene-co-perfluoroalkylvinyl ether). *Macromolecules* 28: 2781-2786.
13. Starkweather, H.W. 1986. Dynamic Mechanical Study of the β-Relaxation in Poly(tetrafluoroethylene). *Macromolecules* 19:2541-2544.

14. Lau, S.F., Wesson, J.P. and B. Wunderlich. 1984. Glass Transition of Poly(tetrafluoroethylent) *Macromolecules* 17: 1102-1104.

15. Kimmig, M., Strobl, G. and B. Stuhn. 1994. Chain Reorientation in Poly(tetrafluoroethylene) by Mobile Twin-Helix Reversal Effects. *Macromolecules* 27: 2481-2495.

16. Runt, J. and K.P. Gallagher. 1991. Polymer-Polymer on Parameters via Melting Point Depression - A Critical Analysis. *Polym. Comm.* 32: 180-182.

17. Crevecoeur, G. and G. Groeninckx. 1991. Binary Blends of Poly(ether ether ketone) and Poly(ether imide). Miscibility, Crystallization Behavior, and Semicrystalline Morphology. *Macromolecules* 24:1190-1195.

18. For a discussion of the relaxation behavior of crystalline polymer blends see: Runt, J. Dielectric Studies of Polymer Blends. In J. Runt and J.J. Fitzgerald (Ed.) 1997. *Dielectric Spectroscopy of Polymeric Materials.* American Chemical Society. Washington.

19. Russell, T.P., Ito, H. and G.D. Wignall. 1988. Neutron and X-ray Scattering Studies on Semicrystalline Polymer Blends. *Macromolecules* 21: 1703-1709.

20. Hahn, B., Hermann-Schonherr, O. and J. Wendorff. 1987. Evidence for a Crystal-Amorphous Interphase in PVDF and PVDF/PMMA Blends. *Polymer* 28: 201-208.

21. Saito, H. and B. Stuhn. 1994. Exclusion of Noncrystalline Polymer from the Interlamellar Region in Poly(vinylidene fluoride)/Poly(methyl methacrylate) Blends. *Macromolecules* 27: 216-218.

22. Keith, H.D. and F.J. Padden. 1987. Influence of Reptation on Localized Diffusion in Crystallizing Polymers. *J. Polymer Sci., Polym. Phys. Ed.* 25: 229-242.

23. a. Barron, C.A. 1994. Ph.D. Thesis, The Pennsylvania State University.
b. Talibuddin, S., Wu, L., Runt, J. and J.S. Lin. 1996. Microstructure of Melt-Miscible, Semi-Crystalline Polymer Blends. *Macromolecules* 29:7527-7535.

24. Arlie, J.P., Sept, P. and A. Skoulios. 1967. Etude de la cristallisation des polmeres II. Structure lamellaire et repliement des chaines du polyoxyethylene. *Makromol. Chem.* 104: 212-229.

25. Ong., C.J. and F.P. Price. 1978. Blends of Poly(ε-caprolactone) with Poly(vinyl chloride) I. Morphology. *J. Polym. Sci.:Polym. Symp.* 63; 45-58.

26. Plans, J., MacKnight, W.J. and F.E. Karasz. 1984. Equilibrium Melting Point Depression for Blends of Isotactic Polystyrene with Poly(2,6-dimethyl phenylene oxide). *Macromolecules* 17: 810-814.

27. Talibuddin, S. and J. Runt. Phase Behavior in Quenched Semicrystalline/ Amorphous Polymer Blends. *Polymer* Submitted for publication.

28. de Gennes, P.G. 1980. Dynamics of Fluctuations and Spinodal Decomposition in Polymer Blends. *J. Chem. Phys.* 72: 4756-4763.

29. Kumar, S.K., Colby, R.H., Anastasiadis, S.H. and G. Fytas. 1996. Concentration Fluctuation Induced Dynamic Heterogeneities in Polymer Blends. *J. Chem. Phys.* 105:3777-3788.

Chapter Five

THE ROLE OF POLYMER SURFACES IN THIN FILM DISPLAY TECHNOLOGY

SANAT K. KUMAR[1], DAVID L. ALLARA[1] and THOMAS P. RUSSELL[2]

[1]Department of Materials Science and Engineering
Steidle Building
Pennsylvania State University
University Park, PA 16802
and
[2]Department of Polymer Science
University of Massachusetts
Amherst, MA 01003

ABSTRACT

A technologically important application of polymeric surfaces is in the context of twisted nematic flat panel displays where thin [≈0.2 μm] uniaxially rubbed polymer films are utilized as template layers for controllably orienting the optically active liquid crystal molecules. While the molecular origins of this transfer of orientation from a rubbed polymer layer to an over-layer of liquid crystal molecules are not understood at this time, it must be emphasized that this forms the basis of a $8 billion industry. To establish the scientific underpinnings of this technological application we have begun a detailed study, using a variety of surface sensitive tools, to characterize the structure of thin polymer films, especially the role of the commercially employed uniaxial buffing. We find that the buffing process affects the near surface [i.e., the top 2-5 nm] polymer molecules and serves to increase the effective orientation of polymer molecules in the direction of buffing. The chain stems oriented along the buffing direction then serve as a molecular template for the LC overlayer molecules, which then align with their long axes parallel to the buffing direction.

TECHNOLOGY ISSUES

Thin film displays have become ubiquitous in a variety of electronic devices, such as in quartz watches and in portable computers, and it has been estimated that the flat panel display market in the US in 1995 alone was $8 billion. The basic structure of a thin film display, which is sketched schematically in Figure 1 in the case of a quartz watch application [1]-[3], consists of an organized thin film assembly of liquid crystalline [LC] molecules, which is the active electro-optic component, sandwiched between glass sheets coated with conductive indium-tin oxide layers. The whole setup is then placed between cross polarizers. In the "off" state [left cartoon in Figure 1] the LC molecules are in an orientation where the polarization of a light beam is rotated by 90° when it passes through the display. Consequently, the light beam is able to pass through both polarizers and the device appears illuminated. The device is then switched to its "on" state by the passage of an electric current. The LC molecules orient in a direction parallel to the applied field, i.e., perpendicular to the glass sheets in this

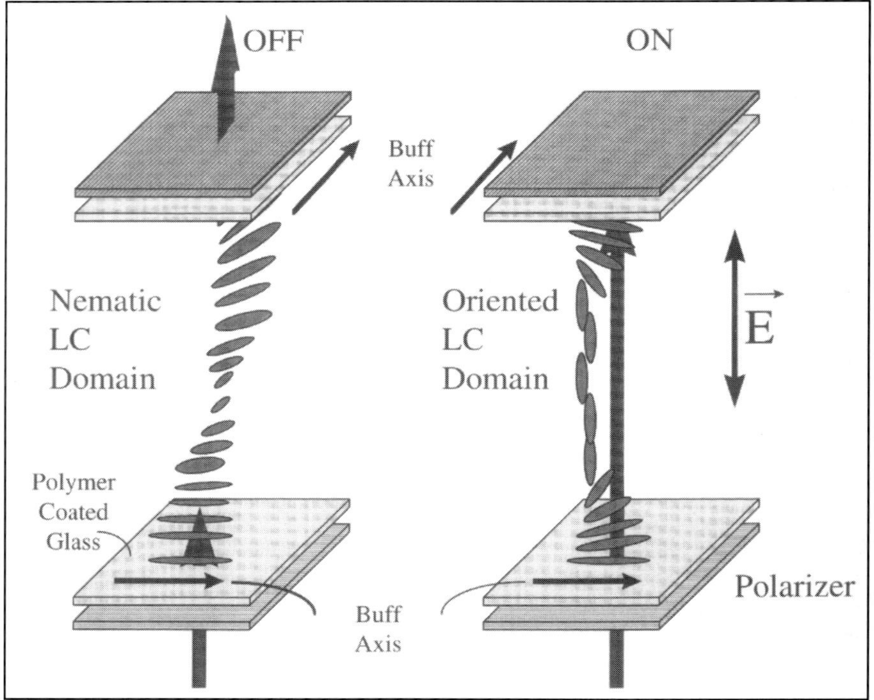

FIGURE 1. Cartoon representation of a thin film display as utilized in a quartz watch. Left cartoon is the "off" state, while the right state is "on". Adapted from reference [3].

case [right cartoon in Figure 1]. Due to this orientation of the LC molecules, the polarization of a light beam is no longer rotated when it traverses the device, which consequently appears dark. The most important issue in these devices is the ability to control the initial orientation of LC molecules, as well as to ensure the stability of the assembly so switching rapidly between the off and on states occurs in a reproducible manner.

The industrially utilized method to ensure that the LC molecules will return reproducibly to the "off" state is to deposit a thin film of a polyimide layer on the electrode and then uniaxially rubbing the polymer surface under ambient conditions [1]-[3]. The long axis of the LC molecules are then found to orient along the buffing direction. In this manner the uniaxially buffed polymer surface acts a molecular template for LC alignment. While the capability of rubbed surfaces to orient LC molecules in their vicinity has been empirically known from 1911 following the pioneering work of Mauguin [4], the underlying role played by the polymer layer, especially the buffing process, is not clearly understood at this time.

We assert that improvements in the manufacture of thin film displays, which have a 80-90% rate of success based on current buffing techniques used industrially, can only be achieved if the scientific underpinnings of this technology are established in an unequivocal fashion. This paper addresses recent findings in this area which address the molecular basis for this important technological application. In addition, these results provide new insights into the mechanical behavior of thin films, which are shown to be very different from their bulk counterparts. In this context it must also be emphasized that textured polymer surfaces are relevant in a variety of other applications such as in paints and floor finishings where the interest is not in these effects, but rather in minimizing them, so as to maximize, for example, the optical "finish" of a floor coating.

The rest of this paper is organized as follows. The next section reviews past work and establishes scientific issues that need to be resolved in this area. Section 3 then discusses experimental results which have emerged from our group in the last 2 years, while Section 4 presents a brief summary of our findings as well as directions for future research.

PAST WORK AND UNRESOLVED SCIENTIFIC ISSUES

Mauguin [4] showed in 1911 that buffed surfaces could be utilized to orient LC molecules. Following this pioneering work the study of the effects of buffing on thin polymer layers has been the focus of several studies, e.g., references [5]-[11] Berreman [5,6] suggested that the primary role of buffing was in the creation of grooves on the surface of the polymer layer. A LC layer placed on this surface will align the long axis of the molecules parallel to the grooves so as to minimize the elastic deformation free energy. While the grooving of polymer surfaces under the action of light buffing has been verified by the use of atomic force microscopy [10], however, we and others [11] have found that, while the grooving cannot be ruled out as a mechanism for the orientation of LC molecules, the extent of orientation is inversely proportional to this quantity. Consequently, a correct molecular mechanism must contain other fundamental factors.

Patel [7] and van Aerle [8,9] have suggested, based on measurements of sample birefringence, that the polyimide molecules in the near surface region of the template layer are oriented in the direction of the buffing. These workers then propose that the oriented polymer stems "epitaxially" align the molecules in the overlayer LC layer. This conclusion appears consistent with findings of Smith and coworkers [12] on the orientation of LC molecules on polymer layers deposited through frictional transfer during rubbing of a polymer on an abrasive surface. While this surface orientation hypothesis appears reasonable, there are two major problems. First, the probes utilized in these experiments are not surface sensitive. Hence, they cannot unequivocally determine that the orientation of the polymer molecules under the action of rubbing is at the surface of the film. Probably more importantly, the typical polyimide molecules used in these experiments [2,11] have a bulk glass transition temperature, T_g, which is in excess of 400°C. Since the buffing is conducted at room temperature under relatively mild conditions [typically the buffing load is 2 gm/cm^2] it is hard to explain why the molecules in the surface layer are reoriented. While several possible scenarios, such as local frictional heating due to the rubbing process [13], the depression of the glass transition temperature, T_g, of the surface layer [14]-[16] or the reduced yield stress for a material near the surface [11,17] can be utilized to rationalize these findings, it still remains to be established conclusively that it is the surface of the films that are reoriented under the action of buffing.

It is clear from the above discussion that, in spite of previous efforts, important scientific issues critical to the improvement of thin film display technology remain unresolved. Below, we summarize some of the experimental work conducted in our group that sheds light on these issues, and helps to establish the scientific underpinnings of this important technological application.

EXPERIMENTAL RESULTS

To clarify the unresolved issues discussed above we have recently begun a multi-technique, surface sensitive study of buffed polyimide layers. The primary tool we have employed in these studies is Grazing Incidence X-ray Scattering [GIXS] [2], a variation of the standard x-ray scattering technique which can be tuned to be highly depth sensitive. While more detailed descriptions of this method are reported elsewhere [18] we present a brief summary of the technique here. The GIXS geometry is presented in Figure 2. We consider a thin polyimide film on a flat Si[100] wafer. Since the refractive index for this material is smaller than for air it possesses a critical angle, denoted by α_c. Consequently, the depth sensitivity of the GIXS probe can be altered by varying the angle of incidence of the x-rays from below the critical angle [$\alpha < \alpha_c$], where it probes the top 5 nm through an evanescent wave, to bulk penetration when the incidence angle is larger than the critical angle of the medium [$\alpha > \alpha_c$]. Further, since we employ a four circle goniometer at NSLS we can vary the scattering wavevector, q [$\equiv (4\pi/\lambda)\, sin\theta$, where λ is the x-ray wavelength, 1.703Å in this case] by varying θ but can keep the scattering vector in a fixed direction [e.g., parallel or perpendicular to the buffing direction].

In addition to GIXS, we have also employed Near edge x-ray absorption fine structure [NEX-AFS] and high signal/noise transmission infrared spectroscopy [IRS] to obtain chemical information on the systems of interest. The NEXAFS spectroscopy is based on the principle that near the K-shell absorption threshold these spectra are dominated by transitions arising from the 1s core level to unfilled molecular orbitals of π^* and σ^* symmetry [19]. Since these states are specific to the bonding within different functional groups, NEXAFS can be utilized to probe, for example, compositional heterogeneities in thin polymer films with high spatial resolution. By using electron yield detection we can ensure that the NEXAFS signal originates close to the surface. Two commonly encountered methods are Auger electron [AEY] or total electron yield [TEY], which provide sampling of the top 1 and 10 nm of the film, respectively [19]. Further, by utilizing polarized x-rays in the incident beam one may obtain information on near surface orientation of polymer moieties in a manner that complements the GIXS measurements, i.e., parallel or perpendicular to the buffing direction.

We have also conducted high signal/noise transmission mode infrared spectroscopy [IRS] [21]. The reason for selecting IRS is that, similar to NEXAFS, it is a molecular spectroscopy and can provide important complementary information to the GIXS studies. While GIXS can yield depth sensitive information, it must be stressed that the transmission IR data are not capable of providing this information. However, IRS is very sensitive to chemical environments and can,

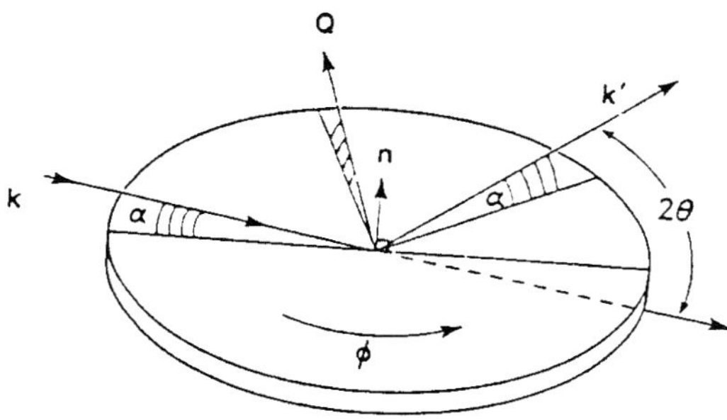

FIGURE 2. Schematic representation of the GIXS setup. α represents the incidence angle, as well as the angle at which the scattering is monitored. The wavevector is changed by altering θ using a four circle goniometer at the National Synchroton Light Source [NSLS] at the Brookhaven National Laboratories [2].

therefore, distinguish between regions which are ordered vs. disordered. Consequently, a combination of GIXS, NEXAFS and IRS provide complementary information on the order and the preferential orientation of chemical groups in the films. We conduct IRS measurements with the electric field in the incident IR beam polarized either parallel or perpendicular to the buffing direction. Polarization dependent [PD] spectra as well as the IR dichroism of the buffed samples (i.e., the difference of the IR absorption in the direction parallel to and perpendicular to the buffing) can therefore be obtained. While this work is similar in spirit to that of van Aerle [8,9] and Patel [7], the advantage of our procedure, however, is that these samples have also been characterized by GIXS. Further, the IR techniques available to us in our laboratory are of extremely high resolution and permit us to resolve structures on the scale of a single monolayer.

The polyimide utilized in this work was obtained by spin coating the polyamic acid formed from bisphenylenedianhydride and paraphenylenediamine [BPDA/PDA] on a Si[100] surface. The resulting layer was then "cycloimidized" at elevated temperatures [200-400°C] where the acid precursor is converted to the more rigid polyimide film, which has been deduced as being liquid crystalline in nature and has a glass transition in excess of 400°C, is then uniaxially buffed by pulling a velor cloth under fixed load over a fixed length at a fixed rate [typically 1 cm/s] [2]. The imidization temperature, the buffing load and the buffing distance were then varied systematically in a series of experiments.

EFFECTS OF BUFFING

The samples considered in these series of experiments were imidized at 300°C, and then buffed by pressing the sample against a velor cloth under a load of 200 Pa (2 gm/cm^2) and pulling the cloth unidirectionally over 300 cm at a rate of 1 cm/s [2]. Figure 3 shows the GIXS obtained from a 6 nm thick sample with $\alpha < \alpha_c$ to minimize any scattering from the Si substrate. The only peaks present in the buffing direction are sharp Bragg peaks, which can be attributed to intramolecular order in these stiff polyimide chains [11]. These sharp peaks are completely absent in the perpendicular direction, and the only two observable diffuse reflections at $q = 13$ nm^{-1} and 30 nm^{-1}, arise from intermolecular disorder. It is important to note that the sample was isotropic before buffing, and showed the evidence of Bragg peaks and diffuse reflections in all directions. As a consequence of these findings we assert that chain stems in this film are aligned almost completely in the direction of the buffing. Similar studies have also been conducted on samples with a range of thicknesses up to 300 nm. These studies show that although most of the molecules in the top 5 nm of this thick polyimide layer are also oriented in the buffing direction, the bulk of the film is isotropic, and unaffected by the rubbing. Consequently, we stress that the buffing operation serves to only orient the chain stems in ear surface region of the film, and that the alignment of the LC molecules in the thin film display are dictated by this extremely thin layer of oriented molecules. We have recently verified these results through the use of NEXAFS spectroscopy [20] which also suggest that the orientation created by the buffing process is isolated to less than the top 10 nm of the film.

BUFFING LOAD AND DISTANCE DEPENDENCE

To understand the role playing by the buffing load and distance we have conducted experiments utilizing GIXS, NEXAFS and IRS. Here we focus on results obtained by NEXAFS which is also a highly surface sensitive probe, and in addition yields chemical information. By using the Auger electron yield [AEY] from a buffed polymer sample, and by carefully controlling the polarization of the incoming x-ray beam we can obtain orientation dependent information from the top 1 nm of the polyimide film. Figure 4 reports results obtained from NEXAFS on 100 nm thick films imidized at 300°C and buffed under a load of 2 *gm/cm²* at 1 cm/s as a function of buffing distance. The results are shown for all of the different π^* reso-

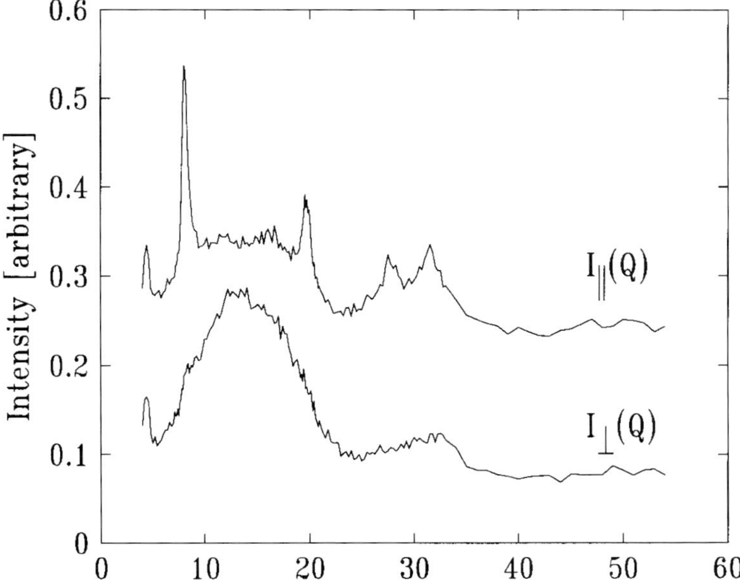

FIGURE. 3. GIXS data obtained from a 6 nm thick BPDA/PDA polyimide imidized at 300°C. The sample was buffed under a load of 2 gm/cm² for a length of 300 cm. The GIXS was conducted below the critical angle, and the two curves correspond to scattering vector parallel and perpendicular to the buffing direction, respectively .

nances in the BPDA/PDA molecule. Clearly, all of the spectra show the same behavior and demonstrate that the extent of surface orientation, as characterized by the second Legendre polynomial, increases with increasing buffing length. The line shown in this figure is an exponential fit to the data, which suggests that the (1/e) buffing distance for this data is 67 cm.

In parallel, we have considered 150 nm thick films which were buffed under varying loads for 50 cm at 1 cm/s. Figure 4b suggests that the extent of orientation increases with increasing load, with a (1/e) load of 1.2 *gm/cm²*. These results are qualitatively consistent with GIXS data and illustrate that the difference in order between the direction parallel to and perpendicular to the buffing direction increases with buffing load and distance.

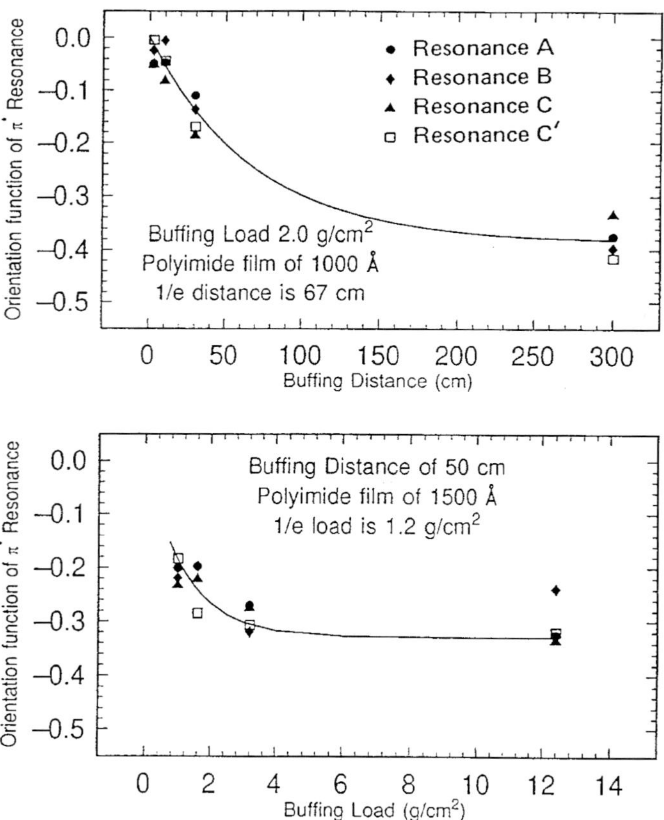

FIGURE 4. Plot of the orientation function obtained from AEY results as a function of buffing distance and buffing load. The data points include all of the π^* resonances. The solid curve is a fit to the data using an exponential decay curve.

FILM THICKNESS EFFECTS

In Figure 5 we plot IR dichroism as a function of wavenumber for BPDA/PDA samples of five different thicknesses imidized at 400°C and then buffed with a load of 2 gm/cm^2 over a length of 50 cm. In agreement with the GIXS we find that buffing creates orientation in the sample. However, the orientation, as measured by IR, decreases with increasing film thickness. The underlying reason for this change with film thickness is not understood at this time and remains to be studied further. In addition, the intensity of some of the dichroism peaks are positive while some others are negative thus helping to assign the orientation of different moieties in the surface layer of the film. While we shall not pursue this issue in more detail here, we conclude by considering the effective thickness of the layer that is oriented by the buffing process. We create a plot of the IR absorption for a given peak position as a function of film thickness for several unbuffed samples. The resulting plot is an exact straight line for all thicknesses considered ($< 300nm$) with a zero intercept, verifying the applicability of Beer's law in this context. When the measured dichroism for the thin films at the same peak position is mapped on this plot it suggests that the net orientation achieved by the buffing process is equivalent to a bulk sample with a thickness

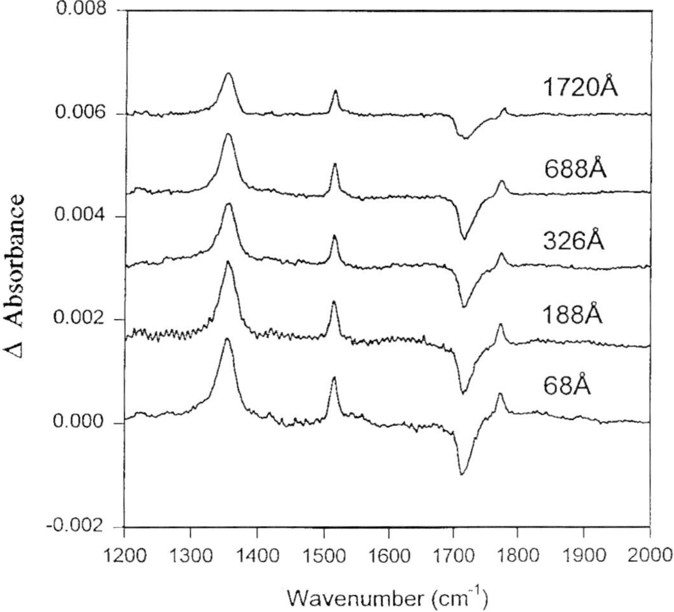

FIGURE 5. IR dichroism plot as a function of wavenumber for several BPDA/PDA polyimide films with thicknesses shown in the figure. The different curves have been offset vertically for clarity.

of $\approx$ 2 nm. This result is consistent with the GIXS and NEXAFS data, which suggest that only the top 5 nm of the film are oriented by the buffing, and verify that the IR, NEXAFS and GIXS are seeing one and the same thing.

DISCUSSION AND SUMMARY OF RESULTS

To summarize, we have shown that polyimide chains near the air surface are reoriented to lie along the buffing direction, although these systems are several hundred degrees below their nominal bulk glass transition temperature, and the buffing conditions are relatively mild. This result establishes that the orientation of a LC overlayer in a thin film display is driven by the underlying alignment of polyimide molecules.

Several issues remain unresolved at this time. The most important result which is unexplained is the realignment of polymer chains by buffing at temperatures several hundred degrees below their glass transition temperatures. It has been proposed [11] that this reorientation can be rationalized by postulating that the material yields under the applied load, and locally reorients in the buffing direction. It must be noted, in this context, that while the yield stress of the BPDA/PDA polyimide is 280 MPa [11] the applied load is only 200 Pa. We stress, however, that the contact between the buffing cloth and the polyimide layer can only be achieved through the fibers in the cloth. This will, effectively, reduce the area of contact quite dramatically. We also note that, under these stresses, the sample has to be rubbed over distances of $\approx$ 2 *cm* before any noticeable reorientation of the molecules result. If we assume that the reorientation, at the molecular scale, occurs over a distance comparable to the c-axis dimension of the unit cell of the BPDA/PDA polyimide [120Å, then, we can argue that the improper contact between the cloth and the polyimide leads to a local stress concentration, which to a zeroth order is:

$$\frac{2}{120 \text{ x } 10^{-8}} \approx 1 \text{ x } 10^{6} \tag{1}$$

From a molecular point of view, therefore, an applied strss of 200 Pa, can therefore be effectively a 200 MPa due to this amplification mechanism. Since this number is very close to the yield stress of the BPDA/PDA polyimide we conjecture, in agreement with past speculations, that the orientation of polymer molecules under the action of buffing occurs by a yielding mechanism.

Another aspect that remains unexplained at this time is the unexpected thickness dependence of the orientation created by buffing. In Figure 5, for example, we had found that the extent of orientation induced by buffing decreased with increasing film thickness. While we do not have conclusive proof, we speculate that this result bears significant impact on the mechanical properties of thin polymer films. Following the proposal of Brown and Russell [17] we conjecture that, in the case of very thin films, the polymer chains are effectively two dimensional and hence not topologically constrained by chains in layers which are some distance away from the surface. These systems should reorient with ease under the action of buffing. As one increases the film thickness, however, these topological constraints should increase thus leading to a decrease in the extent of orientation created at the surface by the buffing process. These results are in qualitative accord with experimental findings, and represent, to our knowledge, the first concrete evidence of the role of topological constraints in the dynamics of thin polymer films. It must be stressed that these ideas are very preliminary and need to be verified by additional experiments on the same polymer, as well as different polymers so as to establish the generality of the phenomenon discussed. Nevertheless, we stress that the experiments we have conducted appear to bear real promise in resolving an extremely important fundamental problem, which could have ramifications, not only in thin film display technology, but in other areas such as in lubrication, adhesion etc.

Finally, we note that none of our work has focused on understanding the transfer of orientation from this rubbed polyimide layer to an overlayer of LC molecules. This issue which is critical to the proper operation of a flat panel display will focussed on in future work.

REFERENCES

1. Depp, S.W. and W.E. Howard. 1993. *Sci. Am., 268,* 90.
2. Toney, M.F., Russell, T.P., Kikuchi, H., Logan, A., Sands, J. and S.K. Kumar. 1995. *Nature, 374,* 709.
3. Patel, J.S. 1993. *Ann. Rev. Mat. Sci., 23,* 269.
4. Mauguin, C. 1911. *Bull. Soc. fr. Min., 34,* 71.
5. Berreman, D.W. 1973. *Mol. Cryst. Liq. Cryst., 23,* 215.
6. Berreman, D.W. 1972. *Phys. Rev. Lett., 28,* 1683.
7. Geary, J.M. Goodby, J.W., Kmetz, A.R., and J.S. Patel. 1987. *J. Appl. Phys., 62,* 4100.
8. van Aerle, N.J.A.M., Barmentlo, M. and R.W.J. Hollering. 1993. *J. Appl. Phys., 74,* 3111.
9. van Aerle, N.A.J.M. and A.J.W. Tol. 1994. *Macromolecules, 27,* 6520.
10. Y.-M. Zhu, Wang, L., Lu, Z.-H., Wei, Y. Chen, X.X. and J.H. Tang. 1994. *Appl. Phys. Lett., 65,* 49.
11. H. Kikuchi, Logan, J.A. and D.Y. Yoon. 1994. *Mat. Res. Soc. Symp. Proc., 345,* 247.
12. Wittman, J.C. and P. Smith. 1991. *Nature, 352,* 414.
13. F.P. Bowden and D. Tabor. 1950. *The Friction and Lubrication of Solids,* Oxford University Press, London.
14. W.J. Orts, van Zanten, J.H., Wu, W.L. and S.K. Satija. 1993. *Phys. Rev. Lett., 71,* 87.

15. Keddie, J.L., Jones, R.A.L., and R.A. Corey. 1994. *Europhys. Lett., 27,* 59.

16. Ray, P. and K. Binder. 1994. *Europhys. Lett., 27,* 53.

17. T.P. Russell and H.R. Brown *Macromolecules,* in press.

18. Factor, B. 1992. *Ph. D Thesis,* Stanford University.

19. Stohr, J. 1992. NEXAFS *Spectroscopy* , Spriner-Verlag, Berlin.

20. Samanth, M., Stohr, J., Brown, H.R., Russell, T.P., Sands, J.M. and S.K. Kumar. 1995. Submitted Macromolecules.

21. G. Heitpas, Sands, J., Kumar, S.K. and D.L. Allara. 1996. Mauscript in preparation.

The Era of Materials. Edited by S.K. Majumdar, R.E. Tressler, and E.W. Miller © 1998, The Pennsylvania Academy of Science.

Chapter Six

COLLOIDS AND SURFACES IN AQUEOUS SYSTEMS

K. OSSEO-ASARE

Department of Materials Science and Engineering
Steidle Building
Pennsylvania State University
University Park, PA 16802

INTRODUCTION

Physicochemical processes at interfaces (1-4) underlie many of the techniques exploited in materials processing in aqueous systems, e.g., leaching and precipitation in hydrometallurgy and materials synthesis; etching in semiconductor device microfabrication; dispersion, coagulation, and flocculation in ceramic powder processing and hydrometallurgy; electroplating in surface finishing. Also, under service conditions, materials frequently encounter aqueous environments where interfacial physicochemical processes can lead to undesirable effects, such as corrosion and failure of components, devices, equipment, and structures. Figures 1 and 2 illustrate some of the basic interfacial processes of relevance in materials/aqueous systems. Ions and molecules at interfaces and in solution can participate in dissolution, precipitation, adsorption, membrane transport, and partitioning processes, as shown in Figure 1. As a result of interfacial interactions, particulate materials in aqueous solutions can be maintained as stable dispersions (Figure 2, A and C) or they can be destabilized to form aggregates (Figure 2, B).

Depending on the circumstances, the interface of interest may be of the type solid/aqueous, oil/aqueous, or gas/aqueous. In the discussion that follows, the emphasis will be on solid/aqueous systems. Aspects of oil/aqueous interfacial effects in materials synthesis and in metal separations are discussed elsewhere

(5,6). A major concern in this chapter is the quantitative description of the electrified interface and the associated adsorption processes. The roles of interfacial reactions in materials/aqueous systems are illustrated with the aid of examples based on dissolution processes.

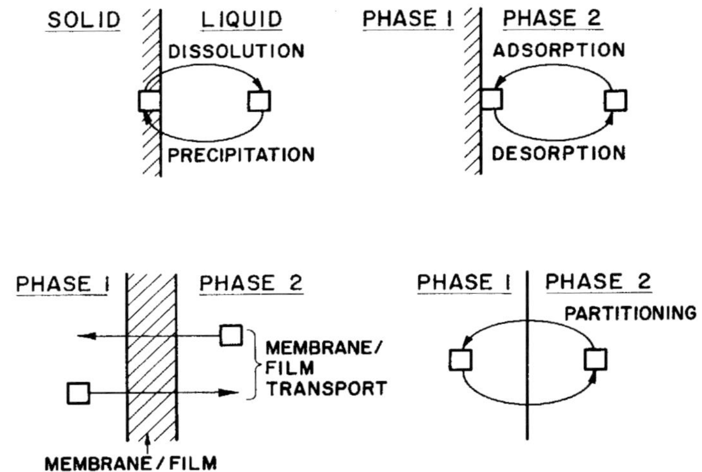

FIGURE 1. Basic interfacial processes involving ions and molecules.

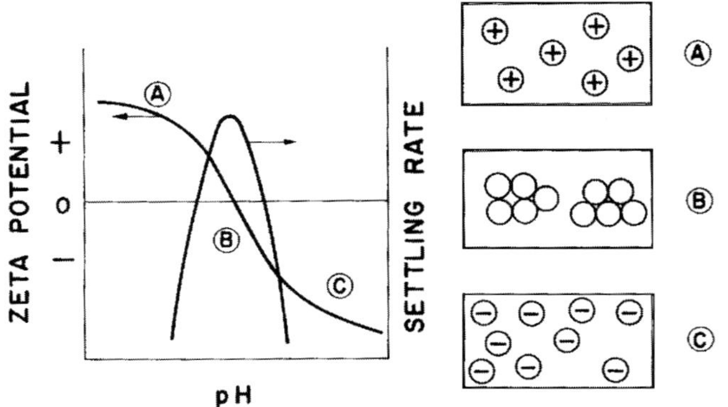

FIGURE 2. A schematic illustration of the effects of zeta potential on particle stabilization (A, C) and destabilization (B). The zeta potential is the electrical potential at the boundary between a surface and the surrounding bulk liquid when the two phases are in relative motion. A high (low) surface charge gives a high (low) zeta potential. Repulsive electrostatic interactions keep highly charged particles dispersed, whereas particles with relatively low surface charges tend to coagulate.

THE ELECTRIFIED INTERFACE ON COLLOIDAL SURFACES

Origins of Surface Charge

When materials such as oxides, sulfides, insoluble salts, and activated carbons are immersed in aqueous solution, they acquire surface charges, the presence of which controls much of their interfacial behavior (1-4). These charges arise from the dissociation of surface functional groups (7-10). In the case of metal oxides, hydroxyls constitute the surface groups and, as illustrated in Figure 3, their formation involves hydration followed by O-H bond rupture.

Subsequent protonation and dissociation (i.e., deprotonation) of the surface hydroxyls produces a charged oxide surface:

$$-MOH \text{ (surf)} + H^+ = -MOH_2^+ \text{ (surf)} \tag{1}$$
$$-MOH \text{ (surf)} = -MO^- \text{ (surf)} + H^+ \tag{2}$$

It can be seen that the surface hydroxyl is amphoteric, i.e., it can act both as a basic site (Equation 1) and an acidic site (Equation 2). When the hydroxyl group acts as a base, adsorption of H^+ yields a positively charged site. On the other hand, when the hydroxyl group behaves as an acid, it can deprotonate to give a negatively charged site. A similar process is responsible for the ionization of carbon surfaces. Here, the surface charge arises from the acid-base reactions of hydroxyl groups present on the original material or produced by interaction of oxygen-containing surface functional groups with water. The situation may be represented symbolically as:

$$-CO \text{ (surf)} + H_2O = -COH \cdot OH \text{ (surf)} \tag{3}$$

The charge on a solid surface may also result from the selective desorption (or dissolution) of the constituent surface atoms:

$$-MA \text{ (surf)} + -M^+ \text{ (surf)} + A^- \tag{4}$$
$$-MA \text{ (surf)} = -A^- \text{ (surf)} + M^+ \tag{5}$$

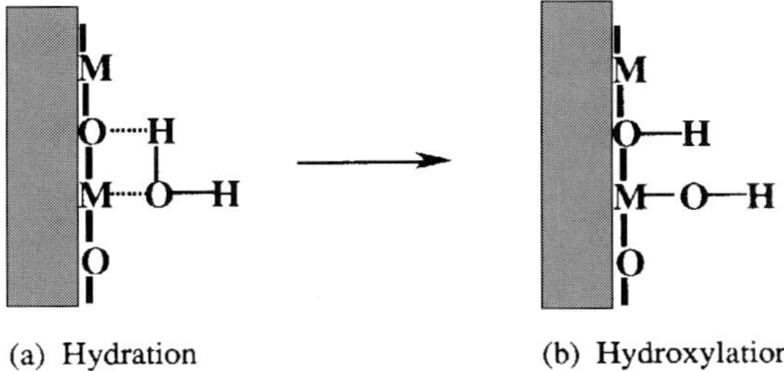

FIGURE 3. Formation of a hydroxylated oxide surface.

Alternatively the surface charging of MA may result from the preferential adsorption of the constituent ions from the bulk aqueous phase:

$$-MA \ (surf) + M^+ = -M_2A^+ \ (surf) \tag{6}$$
$$-MA \ (surf) + A^- = -MA_2^- \ (surf) \tag{7}$$

The surface ionization processes described by Equations 4 - 7 are characteristic of the interfacial behavior of salt-type materials, e.g., $CaCO_3$, $CaWO_4$, and AgI.

In the case of metal sulfides, the surface ionization is attributable to a combination of oxide-type and salt-type behavior. The interaction of a sulfide, MS, with water may be depicted as:

$$-MS \ (surf) + H_2O = MS \cdot H_2O \ (surf) = MOH \cdot SH \ (surf) \tag{8}$$

That is, a surface metal atom binds a hydroxyl group provided by the adsorbed water molecule, while the sulfur atom binds the corresponding proton to give a thiol group. The resulting MOH group can undergo acid-base ionization reactions as described by Equations 1 and 2. The thiol group can also participate in protonation/deprotonation reactions:

$$-SH \ (surf) + H^+ = -SH_2^+ \ (surf) \tag{9}$$
$$-SH \ (surf) = -S^- \ (surf) + H^+ \tag{10}$$

Crystal lattice defects provide another mechanism for charge generation at the solid/aqueous interface. This case is especially important for clay minerals. Here, substitution of Al^{3+} for Si^{4+} results in the generation of negative charge in the basal planes.

The Colloid-Ion Analogy

The charge on a surface will attract oppositely charged ions from the aqueous phase to the surface. The resulting system of a surface charge and a corresponding region of counter-charge on the aqueous side of the interface is called the *electric double layer* (1-4). A useful analogy can be drawn between the double layer associated with a colloidal particle and the system of a central ion and its ionic atmosphere.

An approximate expression for the potential distribution around a central j-ion carrying a charge z_j, as derived from Debye-Huckel theory, is (11):

$$\psi_j \ (r) = \frac{z_j e}{\varepsilon} \cdot \frac{e^{\kappa a}}{1+\kappa a} \cdot \frac{e^{-\kappa r}}{r} \tag{11}$$

where ε is the dielectric constant of water, r is the distance from the central j-ion, e is the electronic charge, and κ has dimensions of reciprocal length. The quantity $1/\kappa$ is called the *thickness of the ionic atmosphere*. At room temperature (25°C), the thickness of the ionic atmosphere is given by

$$1/\kappa = 3.04 \ (I)^{-1/2} \ \text{Å} \tag{12}$$

where, for an electrolyte solution containing different ionic species identified by the subscripts 1, 2, ...s, with molar concentrations C_i , I is the ionic strength, defined as:

$$I = (1/2) \sum_i C_i z_i^2 \tag{13}$$

Thus, for a 1:1 electrolyte, the thickness of the ionic atmosphere is 9.6Å and 304Å respectively for ionic strengths of 10^{-1} and 10^{-4} molar.

Referring to Equation 11, the potential on the particle surface (r = a, where a is the particle radius) can be expressed as:

$$\Psi_a = \frac{ze}{ea} \cdot \frac{1}{1 + \kappa a} \tag{14}$$

Substitution of Equation 14 into Equation 11 gives the potential around the particle as:

$$\Psi(r) = \Psi_a \, (a/r) \, \exp[-\kappa(r-a)] \tag{15}$$

Based on the colloid-ion analogy, the charge of the central ion, ze, corresponds to the charge (σ) of the colloidal particle. Thus, Equation 14 can be rewritten for the colloidal particle as:

$$\sigma = a\varepsilon \, (1 + \kappa a) \, \Psi_a \tag{16}$$

The parameter $1/\kappa$, previously identified as the thickness of the ionic atmosphere, now represents the *thickness of the electric double layer.*

The Gouy-Chapman-Stern-Graham (GCSG) Model

Various models of the electric double layer have been presented, the main differences in the models being related to the manner in which the spatial distribution of the counter-charge is visualized (1-4, 7-10,12,13). The simplest model, The *Gouy-Chapman model*, considers that the electric double layer consists of a surface layer of charge on the solid surface (σ_o) balanced by a *diffuse layer* of counter-ion charge (σ_d) located at the aqueous side of the interface. The surface charge (σ_o) results in a surface potential (ψ_o) and a potential distribution ($\psi(x)$) in the adjacent aqueous phase. This model of the electric double layer corresponds directly to the view of the colloidal particle as a large ion.

The Gouy-Chapman model, however, breaks down at small values of κx and high ψ_o. This difficulty arises because the model assumes that the ions in the aqueous side of the interface are point charges. A more realistic model is the Gouy-Chapman-Stern-Graham model, illustrated in Figure 4a. Here, the ions closest to the solid surface are considered to have lost their waters of hydration. This layer of dehydrated ions is called the *Inner Helmholtz Plane (IHP)* and the dehydrated ions are said to be specifically adsorbed. The potential and charge of the IHP are represented by ψ_β and σ_β respectively. The first layer of hydrated ions is termed the *Outer Helmholtz Plane (OHP)*. The OHP marks the beginning of the diffuse layer; the ions in this layer retain their hydration shells.

The potential distribution associated with the GCSG model is illustrated in Figure 4b. The electroneutrality condition is:

$$\sigma_o + \sigma_\beta + \sigma_d = 0 \tag{17}$$

The region between the surface plane ($x = 0$) and the IHP may be treated as a condenser with an integral capacitance of C_1. Similarly, the region between the IHP and the OHP may be treated as a condenser with an integral capacitance of C_2. Thus, the following potential-charge relationships are obtained:

$$\psi_o - \psi_\beta = \sigma_o/C_1 \tag{18}$$

$$\psi_\beta - \psi_d = (\sigma_o + \sigma_\beta)/C_2 = -\sigma_d/C_2 \tag{19}$$

If ψ is in volts and the ionic strength (I) is in mol dm^{-3}, the diffuse layer charge at 298 K is given by:

$$\sigma_d \ (\mu C \ cm^{-2}) = -11.74I^{1/2} \sinh(19.46z\psi_d) \tag{20}$$

SURFACE IONIZATION EQUILIBRIA

One-site Amphoteric Surfaces

As already noted above, an amphoteric site is bifunctional, i.e., it can serve both as an acidic and as a basic functional group. Consider a surface occupied by XH groups which can undergo protonation and deprotonation (7-10,12,13):

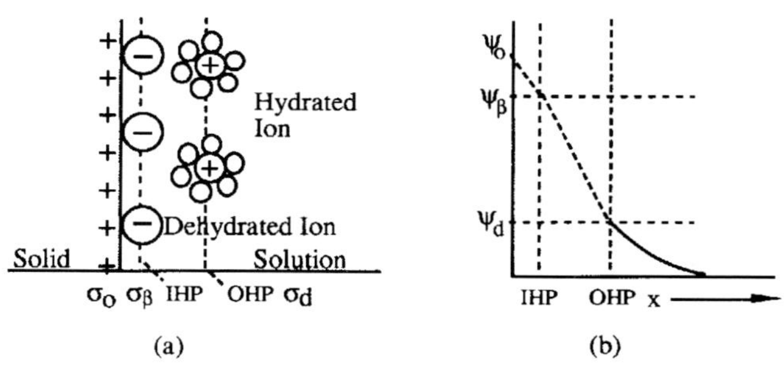

FIGURE 4. The Gouy-Chapman-Stern-Graham (GCSG) electric double layer: (a) Charge distribution, (b) Potential distribution.

$$XH_2^+ = XH + H^+ \tag{21}$$

$$XH = X^- + H^+ \tag{22}$$

The corresponding surface equilibrium constants are:

$$K_1 = [XH] \, \{H^+\}_s/[XH] \tag{23}$$

$$K_2 = [X^-] \, \{H^+\}_s/[XH] \tag{24}$$

The terms [XH] etc., represent the concentrations of the surface groups (e.g., moles per unit volume of solution); concentrations have been used in the equilibrium expressions, with the understanding that the corresponding surface activity coefficients are incorporated into the equilibrium constants. The surface proton activity $\{H^+\}_s$ is related to the bulk aqueous phase activity $\{H^+\}_a$ via a Boltzmann distribution function:

$$\{H^+\}_s = \{H^+\}_a \exp \, (-e\psi_o/kT) \tag{25}$$

A mass balance on the surface sites can be written as:

$$[X]_T = [XH] + [XH_2^+] + [X^-] \tag{26}$$

where $[X]_T$ is the total concentration of the surface groups. If N_s (number of sites per unit area) is the total density of the surface groups, then it follows that

$$[X]_T = (S/N_{Avog})N_s \tag{27}$$

where S is the total surface area of the solid per unit volume of solution and N_{Avog} is the Avogadro number. The surface charge corresponding to a particular pH is given by

$$\sigma_o = (eN_{Avog}/S) \, ([XH_2^+] - [X^-]) \tag{28}$$

The surface concentrations as well as the surface potential and surface charge can be computed as a function of pH, provided the interfacial parameters S, N_s, K_1, and K_2 are known. There are then five unknowns, i.e., $[XH_2^+]$ $[XH]$, $[X^-]$, σ_o, and ψ_o. Four of the necessary five independent equations are provided by Equations 23, 24, 26, and 28. A relationship between σ_o and pH gives the fifth equation. Depending upon the experimental conditions, a choice can be made between the various available double layer models (7-10,12,13). Also, these same equations can be used to determine the surface protonation constants (K_1 and K_2) from experimental σ_o vs pH data. The surface charge can be determined experimentally by the potentiometric titration of colloidal dispersions with the appropriate potential-determining ions (7-10,12,13). Tabulations of surface protonation constants for selected metal oxides, based on various models of the electric double layer can be found in the literature (7-10,12,13).

The Point of Zero Charge and the Isoelectric Point

The *point of zero charge (pzc)* is an important interfacial parameter which is used extensively in characterizing the ionization behavior of a surface. The pzc refers to the bulk aqueous phase concentration (or more precisely, activity) of a potential-determining ion which gives zero surface charge. Also, since the surface potential exists as a direct result of the presence of surface charge, zero surface charge implies zero surface potential. That is, at the pzc,

$$\sigma_o = 0, \; \psi_o = 0 \tag{29}$$

In the case of oxides or hydroxides, the point of zero charge may be determined experimentally by measuring the net uptake of hydrogen and hydroxide ions as a function of pH. The resulting surface charge versus pH curves obtained for different constant electrolyte concentrations intercept at the pzc.

For an amphoteric surface, it follows from the condition $\sigma_o = \psi_o = 0$ and Equations 23-25, and 28 that the pzc is given by,

$$pH_{pzc} = -(1/2) \log(K_1 K_2) = (1/2) \; (pK_1 + pK_2) \tag{30}$$

When the pH differs from pH_{pzc}, combination of Equations 23-25 with Equation 30 gives:

$$\psi_o = -(2.303kT/e) \; (pH - pH_{pzc}) + (2.303kT/2e)\log([X^-]/[XH_2^+]) \tag{31}$$

Near the pzc, $[X^-] \approx [XH]$, and therefore Equation 31 simplifies to

$$\psi = -(2.303kT/e) \; (pH - pH_{pzc}) \tag{32}$$

Equation 32, which is a Nernst-type expression, implies that when $pH < pH_{pzc}$, the surface acquires a positive potential; on the other hand when $pH > pH_{pzc}$ the surface potential becomes negative. It should be noted that Equation 32 is valid only in the neighborhood of the pzc.

The presence of a charge at the solid/aqueous interface means that an applied electrical field can generate relative motion between the solid and the liquid. This process is termed electrokinetics (1-4), and depending on the particular experimental situation, it may involve the movement of colloidal particles in a stationary fluid (electrophoresis) or the movement of liquid along a stationary surface (electroosmosis). In either case, a thin immobile film of water adheres to the solid surface. The boundary between the immobilized water layer and the bulk water is termed the shear plane and the potential difference between the *shear plane* and the bulk liquid is denoted as the *zeta potential* (ζ); see Figure 2. The activity of potential-determining ions that gives zero zeta potential (and therefore zero electrophoretic mobility) is designated the *isoelectric point* (iep). Table 1 presents pzc and iep data for selected materials (10,14-16).

Multisite Surfaces

In the previous section we considered an amphoteric surface covered by only *one type* of surface group, i.e., the XH group. However, it is likely that in real systems, several different types of groups are present at the solid/aqueous interface.

TABLE 1.

The point of zero charge (pzc) and the isoelectric point (iep) of selected materials (10,14-16).

Type of Material	Materials	pzc/iep
Halide	AgCl	pAg 4.1 - 4.6 (iep)
	AgBr	pAg 5.6 - 5.9 (pzc); pAg 4.2 - 5.4 (iep)
	AgI	pAg 4.8 - 6.05 (pzc); pAg 5.1 -6.2 (iep)
	CaF_2	pCa 2.6 - 7.7 (iep)
Sulfate	$BaSO_4$	pBa 3.9 - 7.0 (iep)
	$PbSO_4$	pPb 5.1 (iep)
Phosphate	$AlPO_4 \cdot 2H_2O$	pH 4 (pzc)
	$FePO_4 \cdot 2H_2O$	pH 2.8 (pzc); pH 3.4 (iep)
	$(Ce, La, Th) PO_4$	pH 3.4 (iep)
	$Ca_5(PO_4)_3 (F, OH)$	pH 6.9 - 8.5 (pzc); pH 4-6 (iep)
Carbonate	$CaCO_3$ (calcite)	pH 9.5 - 10.8 (iep)
	$MgCO_3$ (magnesite)	pH 5.2 (iep)
	$CaMg (CO_3)_2$ (dolomite)	<pH 8.5 (iep)
Tungstate	$CaWO_4$	pCa 4.8 (iep)
Oxide	α-Al_2O_3	pH 9.1 (pzc)
	γ-Al_2O_3	pH 8.5 (pzc)
	α-Fe_2O_3	pH 9.3 (pzc)
	α-FeOOH	pH 7.6 - 8.3 (pzc)
	δ-MnO_2	pH 2.8 - 4 (pzc)
	α-MnO_2	pH 4.5 (iep)
	β-MnO_2	pH 4.6 - 7.3 (iep)
	γ-MnO_2	pH 5.5 (iep)
	SiO_2	pH 3.0 (pzc)
	SnO_2	pH 5.5 (pzc)
	TiO_2	pH 6.0 (pzc)
	ZrO_2	pH 6.0 (pzc)
Silicate	Al_2SiO_5 (kyanite, andalusite, sillimanite)	pH 5.2 - 7.9 (iep)
	$Mg_3Si_2O_5(OH)_8$ (Chrysotile)	pH 10 - 12 (iep)
	Mg_2SiO_4 (Forsterite)	$\leq$pH 8.4 (iep)
Clay	Kaolinite	pH 2 - 4.6 (iep)
	Montmorillonite	$\leq$ pH 2.5 (iep)

Thus, in the case of metal (hydr) oxides, different types of surface oxygens may be distinguished, based on the number of coordinated metal ions per oxygen and on the coordination number of the metal ion (7-10,12,13,17,18).

The ionization equilibria of a multisite surface may be modeled by viewing such a surface as a composite of idealized surface sites, i.e., acidic, amphoteric, zwitterionic, and basic groups. A zwitterionic surface has two types of surface groups, e.g., XH and B, which ionize to give oppositely charged surface species, i.e., X^- and BH^+. A two-site amphoteric surface consists of two different sets of amphoteric sites (e.g., XH, YH), whereas a two-site amphoteric/basic surface contains an amphoteric group (XH) plus a basic group (B).

Silicon nitride (Si_3N_4) is an example of a surface which possesses both amphoteric (i.e., silanol) and basic (i.e., amine) sites (18). The formation of these different surface sites may be visualized as illustrated in Figure 5.

Quantitative treatments of the surface ionization equilibria of multisite surfaces are available in the literature (7,8,10,17,18). Thus, for example, it can be shown that the pzc of a zwitterionic surface is given by:

$$pH_{pzc} = (1/2) \left\{ \log(2f_B) - \log\left[(1-2f_B) K_2 \pm \sqrt{(2f_B-1)^2 + 4f_B (1-f_B)K_1K_2}\right.\right. \tag{33}$$

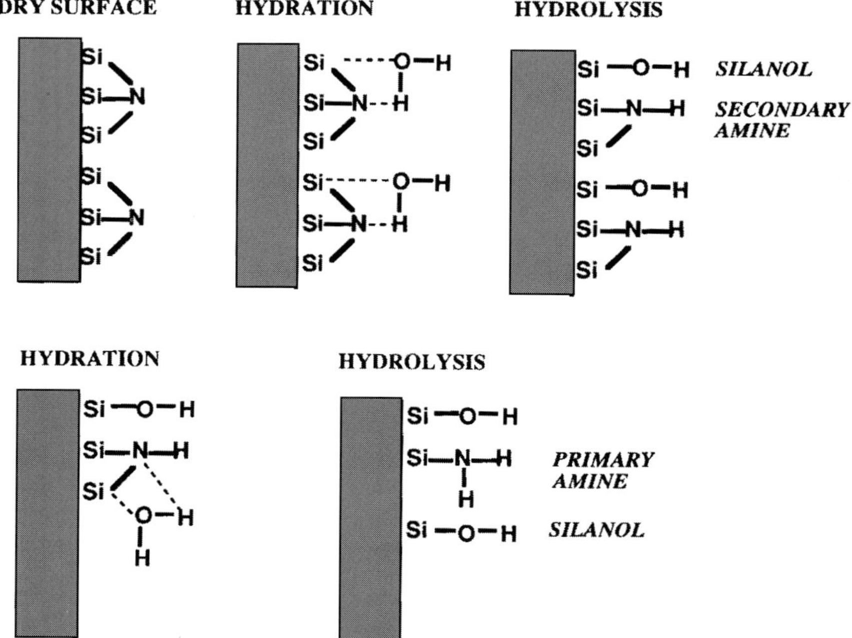

FIGURE 5. Surface chemistry of silicon nitride: formation of silanol and amine sites.

where f_B is the fraction of basic sites, and K_1 and K_2 are the deprotonation constants for the XH and BH$^+$ sites respectively.

SPECIFIC ADSORPTION AND SURFACE COMPLEXATION

Site Occupancy Models

Specific adsorption models may be broadly classified into two main groups, depending on whether they are based on *surface occupancy* or on *site binding*. Surface occupancy models view adsorption primarily in terms of the relationship between unoccupied surface sites and the bulk aqueous phase concentration of the adsorbable species. On the other hand, in the case of site binding models, adsorption is treated as a chemical reaction between specific surface groups and the adsorbable species.

The *Stern isotherm* represents the basic surface occupancy model. According to the Stern model, specifically adsorbed species are located in the compact, i.e., inner layer (see Figure 4). In this adsorption model (1-4), both the solid/liquid interface and the bulk aqueous solution are divided into occupiable sites. Let N_s be the maximum number of surface sites per area, and N the number of occupied sites per area. Further, let N_b be the mole fraction of the adsorbable solute in the bulk solution. Then it is assumed that the ratio of occupied to unoccupied surface sites (i.e. $N/(N_s - N)$) is related to the bulk mole fraction, N_b, by a Boltzmann distribution law:

$$N/(N_s - N) = N_b \exp (-\Delta G_{ads}/kT) \tag{34}$$

where ΔG_{ads} is the free energy of adsorption given by,

$$\Delta G_{ads} = \Delta G_{coul} + \Delta G_{sp} \tag{35}$$

In Equation 35 ΔG_{coul} ($= ze\psi$) is due to the electrostatic interaction between the charge on the adsorbing species and the surface charge, and ΔG_{sp} represents the specific adsorption energy which accounts for the nonelectrostatic interaction of the adsorbate at the solid/liquid interface.

The Stern adsorption model has been exploited extensively in the analysis of surfactant adsorption (19-21). In this connection, the specific energy term, ΔG_{sp}, may be expressed as:

$$\Delta G_{sp} = \Delta G_{cc} + \Delta G_{cs} + \Delta G_{chem} \tag{36}$$

where the terms ΔG_{cc} and ΔG_{cs} represent respectively the free energy change due to hydrophobic interactions of the organic groups of the surfactant molecules with each other, and with the solid surface, and ΔG_{chem} accounts for chemical bond formation between the adsorbing species and the solid surface.

The hydrophobic bonding contributions to ΔG_{sp} depend on the organic chain length of the surfactant molecule. The greater the number of CH_2 groups, the greater the interaction energy. Thus, for straight chain surfactants,

$$\Delta G_{cc} = N_c\ \phi_{cc} \text{ and } \Delta G_{cs} = Nc\ \phi_{cs} \tag{37}$$

where N_c represents the number of CH_2 groups per molecule, and ϕ_{cc} and ϕ_{cs} are respectively the molar cohesive free energies associated with each CH_2 group for chain-chain and chain-solid interactions. The free energy change accompanying the removal of a CH_2 group from water (i.e., ϕ_{cc}) is about - 1RT per mole (19,20). On the other hand, if the adsorbed surfactant consists of horizontally oriented chains, then the corresponding free energy chain (i.e., ϕ_{cs}) is about - 1/2RT. It is reasonable that $\phi_{cs} = 1/2\phi_{cc}$, since with horizontal chain orientation, half the chain surface is next to the solid surface whereas the other half is still exposed to water (21).

The James-Healy model (22) is an application of the Stern model to the adsorption of metal ions. This model has been extended by other investigators (23,24) to the adsorption of heavy metal ions in the presence of complexing ligands. The surface equilibrium reaction can be represented by:

$$(S - S_1) + M_i \underset{}{\overset{K'_i}{\longleftrightarrow}} S_i \tag{38}$$

where: M_i = i-th metal-containing species in the bulk solution, S_i = surface area occupied by the i-th species, S = total surface area available for adsorption, S_1 = total surface area occupied by adsorbed species, K'_i = the equilibrium constant = $\exp(-\Delta G_{ads,i}/RT)$). If $\Gamma_{m,i}$ is the monolayer coverage of the i-th species (moles per unit area), then the total adsorption density, Γ_T, is the sum of contributions by all the surface species and it is given by:

$$\Gamma_T = \frac{\sum_i \Gamma_{m,i} K'_i K_i [M_o][OH^-]^i + \sum_j \Gamma_{m,j} K'_j K_j [M_o][A]^j}{1 + \sum_i K'_i K_i [M_o][OH^-]^i + \sum_j K'_j K_j [M_o][A]^j} \tag{39}$$

where $M_o = M^{2+}$ for i = 0, A refers to the ligand, K_i is the formation constant of the ith metal-hydroxo complex, and K_j is the formation constant of the jth metal-ligand complex.

The free energy of adsorption for each of the solution species, $\Delta G_{ads,i}$, can be considered as the sum of three terms:

$$\Delta G_{ads,i} = \Delta G_{coul,i} + \Delta G_{solv,i} + \Delta G_{chem,i} \tag{40}$$

where $\Delta G_{coul,i}$ is the electrostatic interaction term, $\Delta G_{solv,i}$ is the change in solvation energy of the ion on moving from the bulk solution to the interfacial region, and $\Delta G_{chem,i}$ is a specific energy term which allows for contributions other than the coulombic and solvation terms. The solvation energy change is due to the fact that the free energy required to establish an electric field around an interfacial ion differs from that for the bulk solution. This situation in turn arises because the dielectric constant of the interfacial region is lower than that of the bulk solution.

Figure 6 shows a comparison of the model (Equation 39) with experimental results for copper uptake by TiO_2, Fe_2O_3, Al_2O_3, and SiO_2 in aqueous ammoniacal solution (23). The corresponding computed adsorption densities for the various ammine and hydroxo species indicate that the hydroxo species are largely responsible for the metal uptake (23). Similar conclusions were reached by Crawford (24) in a recent investigation in which aspects of the site dissociation model were incorporated into the James-Healy model.

Site Binding Models

The site binding or surface complexation models view the solid surface as consisting of a fixed number of functional groups which can bind cations and anions (7-10,13,25-27). Consider the surface of a hydrous metal oxide in contact with an aqueous solution. The adsorption of a metal cation (M^{z+}) can be treated as a complex-formation reaction, where the surface OH group acts as a Lewis base:

$$XOH + M^{z+} = XOM^{(z-1)} + H^+ \qquad\qquad \beta_1 \qquad\qquad (41)$$

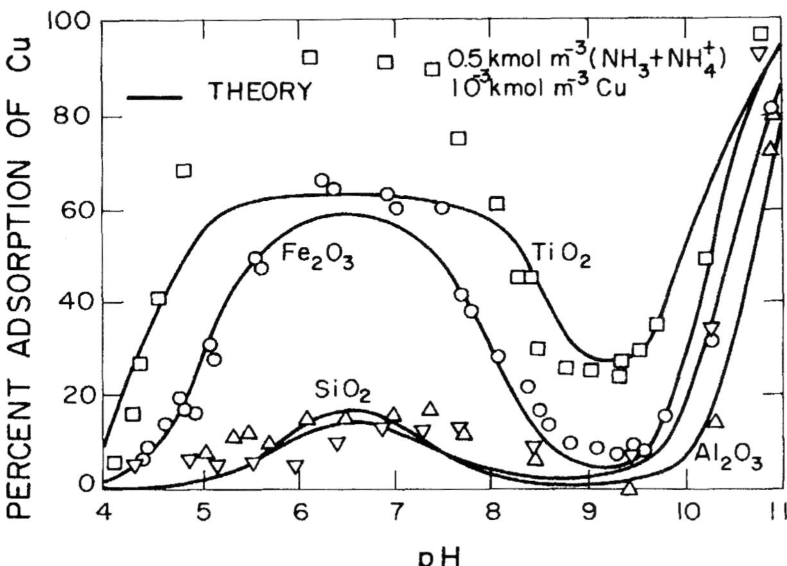

FIGURE 6. Comparison of the experimental and theoretical isotherms for the adsorption of copper on various substrates (160 m^2/L) as a function of pH (23).

$$2XOH + M^{z+} = (XO)_2M^{(z-2)} + 2H^+ \qquad\qquad \beta_2 \qquad\qquad (42)$$

On the other hand, the adsorption of an anion (A^{z-}) can be treated in terms of a ligand-exchange reaction:

$$XOH + A^{z-} = XA^{(1-z)} + OH^- \qquad\qquad \beta_1' \qquad\qquad (43)$$

$$2XOH + A^{z-} = X_2A^{(2-z)} + 2OH^- \qquad\qquad \beta_2' \qquad\qquad (44)$$

Site binding models have been exploited extensively in the aquatic chemistry and geoscience literatures (9,13,25-27). Applications of these models to dissolution processes (4,28-31) are discussed below.

SURFACE CHEMISTRY AND DISSOLUTION PROCESSES

Proton-promoted Dissolution

Dissolution is an interfacial process in which surface species originating from the solid react with surface species provided by the aqueous phase. Thus, the aqueous phase reactants must first adsorb at the solid/aqueous interface before the dissolution proper, i.e., the solid decomposition reaction, can proceed (4,28-31).

The dissolution of a binary compound MA represents a process of ligand exchange. That is, the nonmetal, A, which is the lattice ligand, is replaced by solvent (i.e., H_2O) molecules. Consider the *proton-promoted* dissolution of an M(III) hydrous oxide. The following mechanism may be envisaged:

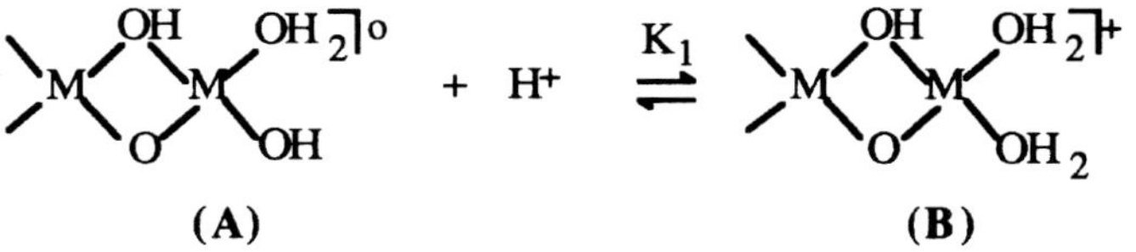

$$(A) + H^+ \;\underset{}{\overset{K_1}{\rightleftharpoons}}\; (B) \qquad\qquad\qquad (45)$$

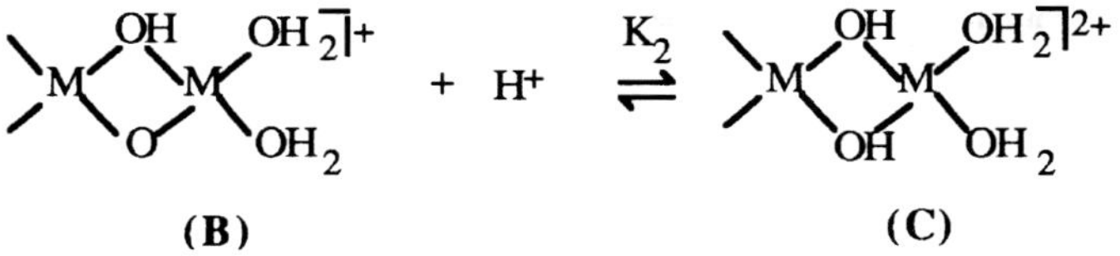

$$(B) + H^+ \;\underset{}{\overset{K_2}{\rightleftharpoons}}\; (C) \qquad\qquad\qquad (46)$$

$$(47)$$

(C)   (D)

$$(48)$$

(D)   (A)

The above equations illustrate the fact that protonation of the M-OH-M group to give M-OH$_2$-M or the M-O-M group to give M-OH-M results in a weakening of the M-O bonds binding the outermost metal ions to the crystal lattice. If the protonation reactions proceed more rapidly than the rate of bond breakage (step 4, Equation 48), then the rate at which metal is released into the bulk aqueous phase is given by:

$$d[M^{3+}]/dt = K_1K_2K_3k_4N_s[H^+]_s^3/(1 + K_1[H^+]_s) \tag{49}$$

where N_s is the total concentration of surface sites.

Hydoxide-promoted Dissolution

The dissolution of a metal oxide in alkaline solution can be viewed as resulting from the attack of the surface atom (M) by hydroxyl ions, leading to the detachment of the metal center.

$$(50)$$

(A)   (B)

$$(51)$$

(B)   (C)

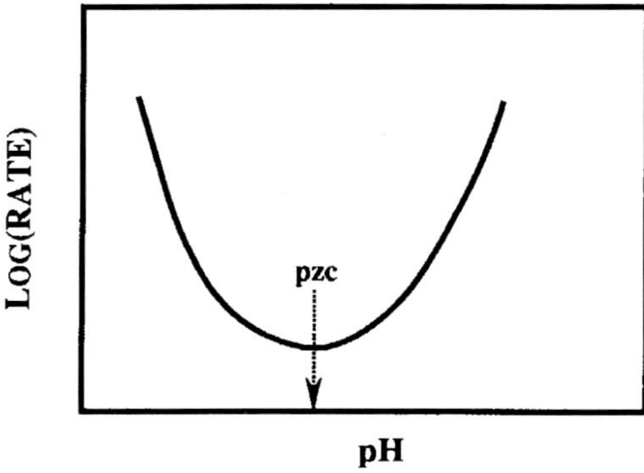

$$\tag{52}$$

(C) ... **(A)**

If the detachment step is rate-limiting, the dissolution rate will be given by

$$d[M]/dt = k_3/K_2K_1[A][H_2O][OH^-]_s, \tag{53}$$

As a result of the ability of protons and hydroxide ions to promote the dissolution of metal oxides, the rate vs. pH curve has a minimum at the point of zero charge, as illustrated schematically in Figure 7 (30,31). Dissolution is lowest at the pzc as a result of the relatively low concentrations of protonated and deprotonated sites. Under these conditions, an important dissolution pathway is the attack of M-O-M bonds by water molecules:

$$-M-O-M + H_2O = -M-OH + -M-OH \tag{54}$$

Ligand-Promoted Dissolution

Ligands other than OH^- and H_2O can also disturb the stability of M-O bonds. As discussed above, anions can adsorb at hydroxylated surfaces via ligand exchange with hydroxyl ions. If the competition between lattice and nonlattice ligands for metal centers favors the latter ligands, dissolution is enhanced above that achievable by protonation alone. Ligand-promoted dissolution is encountered in its simplest form as rapid ligand adsorption (Equation 55) followed by

FIGURE 7. Schematic illustration of the effect of pH on the dissolution rate of a metal oxide.

the slow release of a metal-ligand complex (Equation 56) and a fast surface regeneration step (Equation 57):

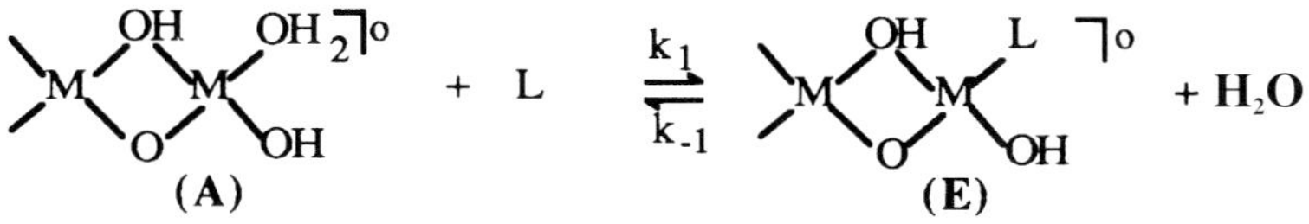

$$\left[\mathrm{M(OH)(O)M(OH_2)(OH)} \right]^{o} + L \;\underset{k_{-1}}{\overset{k_1}{\rightleftharpoons}}\; \left[\mathrm{M(OH)(O)M(L)(OH)} \right]^{o} + H_2O \tag{55}$$

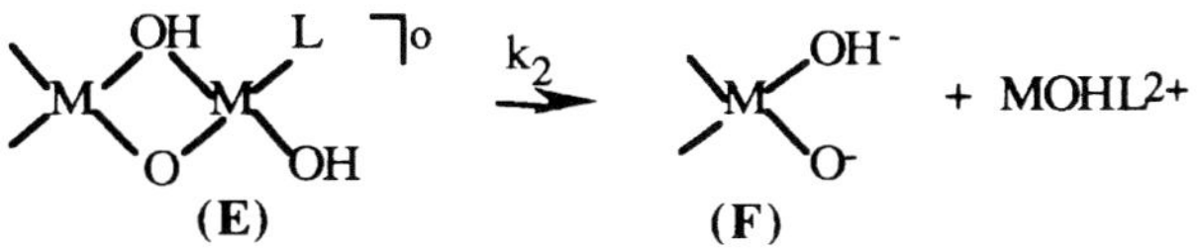

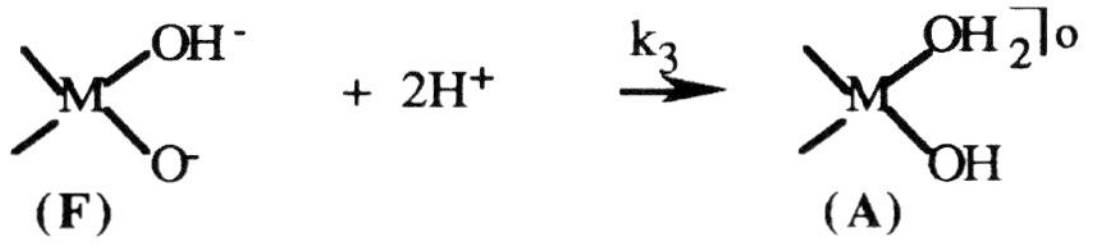

(A) + L ⇌ (E) + H$_2$O

$$\left[\mathrm{M(OH)(O)M(L)(OH)} \right]^{o} \;\overset{k_2}{\longrightarrow}\; \mathrm{M(OH^-)(O^-)} + MOHL^{2+} \tag{56}$$

(E) → (F) + MOHL^{2+}

$$\mathrm{M(OH^-)(O^-)} + 2H^+ \;\overset{k_3}{\longrightarrow}\; \left[\mathrm{M(OH_2)(OH)} \right]^{o} \tag{57}$$

(F) + 2H$^+$ → (A)

The rate equation describing the rate-determining step is:

$$d[M^{3+}] / dt = k_2\,[E] \tag{58}$$

where E represents the surface metal-ligand complex. In recent work from this laboratory (32), this *surface complexation* approach was used to develop a new etching kinetics model for the silicon dioxide-HF system. According to this model, fluoride ions adsorb at the silicon surface centers through a process of inner-sphere surface complexation (Equation 55, L = F⁻). With the detachment step being the slowest step, the dissolution rate is then deduced as:

$$r = kC_{Fs} \tag{59}$$

where C_{Fs} is the surface concentration of adsorbed fluoride.

In Figure 8, the SiF concentrations obtained from the surface complexation model (32) are compared with the etch rate data of Judge (33). It can be seen that both the etch rate and the surface fluoride concentration have essentially the same shape; in particular, each curve exhibits a maximum. The observed correlation between the etch rate and the surface concentration of adsorbed fluoride suggests that the surface detachment-controlled rate law, Equation 59, represents a reasonable description of the etching kinetics.

Chemical-Mechanical Polishing

As a manufacturing process, chemical-mechanical polishing, a process in which aqueous slurries of colloidal particles are used as abrasive agents, has long been used in the optical glass industry (34). It is also an established technology in the preparation of silicon wafers (35). Application of chemical-mechanical polishing as a planarization technique in the large scale manufacture of integrated circuits is also growing (36,37). The important role of surface chemistry in the polishing process is illustrated in Figure 9 for the case of silica glass (34). It can be seen that the surface chemistry of the slurry material, as reflected in the point of zero charge (pzc), has a dramatic effect on the polishing rate.

The observed trends in Figure 9 may be rationalized by considering the nature and concentrations of the surface functional groups on the substrate material and on the particles. An important step in the polishing process is the temporary attachment of the particle to the silica surface via a condensation reaction of the form:

$$Si\text{-}O^{-} + M\text{-}OH \rightarrow Si\text{-}O\text{-}M + OH^{-} \tag{60}$$

where M-OH represents a metal hydroxyl group on the particle surface. Mechanical action provided by the polishing pad subsequently puts a strain on the surface Si-O-M bond. Material removal occurs when a silica tetrahedron bonded to the surface by only one siloxane (Si-O-Si) group is strained sufficiently as to break the underlying Si-O-Si bond, leaving the silica tetrahedron attached to the particle surface. Further reaction with water molecules

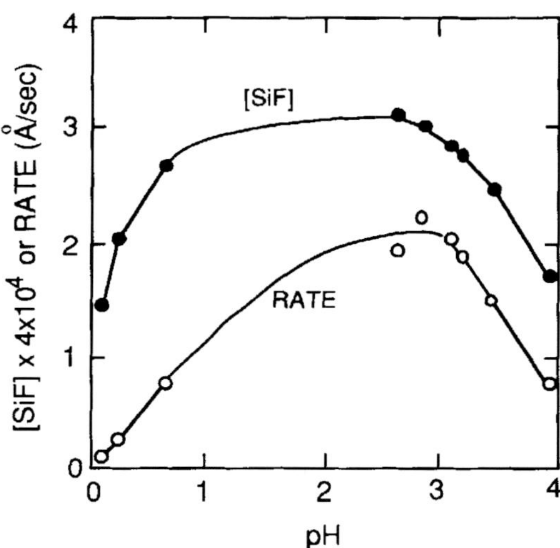

FIGURE 8. Effect of pH on the etching rate of silicon dioxide and on the surface fluoride concentration (32).

releases the chemisorbed silica from the particle surface, producing free silicic acid ($Si(OH)_4{}^\circ$) and regenerating the hydroxyl group on the surface of the colloidal particle.

Glass polishing is typically conducted in the pH 7-9 range. Under these conditions, the silica surface is negatively charged (since the pzc of SiO_2 is at ~pH 2, Table 1). At the same time CeO_2 (pzc = pH 6.8) will possess a nearly neutral surface. Thus, the surface conditions of both the glass and the particle are compatible with the condensation reaction of Equation 60. In the case of a material with a pzc below pH 7, there is a lower concentration of neutral surface sites compared with CeO_2: also the higher negative charge carried by the colloidal particle repels it from the negatively charged silica surface. For materials with relatively high pzc (i.e., pzc > 7), there is again a lower concentration of neutral sites (due to the increased concentration of protonated sites ($MOH_2{}^+$). Here, the main reaction is the transfer of protons to the silica surface:

$$Si\text{-}O^- + M\text{-}OH_2{}^+ \rightarrow Si\text{-}OH + M\text{-}OH \tag{61}$$

It follows from the above discussion that a potentially useful method for selecting slurry particles for a given polishing pH will be to identify a material with a pzc nearest to this operating pH. Apparently this approach has not yet been tested experimentally (34).

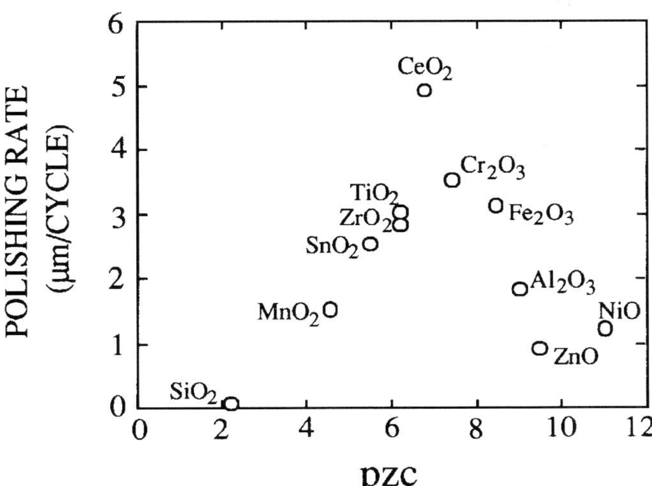

FIGURE 9. Effect of the point of zero charge (pzc) of the polishing compound on the polishing rate of silicate glass (after Cook (34)).

REFERENCES

1. Lyklema, J. 1991. *Fundamentals of Interface and Colloid Science. Volume I: Fundamentals,* Academic, New York.

2. Adamson, A.W. 1990. *Physical Chemistry of Surfaces,* 4th ed., Wiley-Interscience, New York, NY.

3. Hunter, R.J. 1987, 1989. *Foundations of Colloid Science, Volumes I and II,* Oxford University Press, New York.

4. Stumm, W. 1992. *Chemistry of the Solid-Water Interface,* Wiley, New York.

5. Osseo-Asare, K. 1984. "Interfacial Phenomena in Hydrometallurgical Liquid-Liquid Extraction Systems", in *Hydrometallurgical Process Fundamentals,* R. G. Bautista, ed., Plenum, New York, pp. 357-405.

6. Osseo-Asare, K. and F.J. Arriagada. 1990. "Synthesis of Nanosize Particles in Reverse Microemulsions; in *Ceramic Powder Science III,* G.L. Messing, S.Hirano, and H. Hausner, eds., American Ceramic Society, Westerville, OH, pp. 3-16.

7. Healy, T.W. and L.R. White. 1978. "Ionizable Surface Group Models of Aqueous Interfaces", Adv. Colloid Interface Sci., *9,* 303-345.

8. Furlong, D.N., Yates, D.E. and T.W. Healy. 1981. "Fundamental Properties of the Oxide/Aqueous Solution Interface", in *Electrodes of Conductive Metallic Oxides, Part B,* S. Trasatti, ed., Elsevier, New York, NY, pp. 367-432.

9. Schindler, P.W. 1981. "Surface Complexes at Oxide-Water Interfaces," in *Adsorption of Inorganics at Solid-Liquid Interfaces,* M.A. Anderson and A.J. Rubin, eds., Ann Arbor Science, Ann Arbor, MI, pp. 1-49.

10. James, R.O. and G.A. Parks. 1982. "Characterization of Aqueous Colloids by their Electrical Double-Layer and Intrinsic Surface Chemical Properties", Surf. Colloid Sci., *12,* 119-216.

11. Bockris, J. O'M. and A.K.N. Reddy. 1970. *Modern Electrochemistry,* Vols. 1 and 2, Pergamon, New York, NY, pp. 180-202.

12. Westall, J. and H. Hohl. 1980. "A Comparison of Electrostatic Models for the Oxide/Solution Interface", Adv. Colloid Interface Sci., *12,* 265-294.

13. Davis, J.A. and D.B. Kent. 1990. "Surface Complexation Modeling in Aqueous Geochemistry", in *Mineral-Water Interface Geochemistry,* M.F. Hochella, Jr. and A.F. White, eds., Mineralogical Society of America, Washington, D.C.

14. Parks, G.A. 1965. The Isoelectric Points of Solid Oxides, Solid Hydroxides, and Aqueous Hydroxo Complex Systems", Chem. Rev., *65,* 177-198.

15. Parks, G.A. 1967. "Aqueous Surface Chemistry of Oxides and Complex Oxide Minerals", Adv. Chem. Ser., *67,* 121-160.

16. Parks, G.A. 1975. "Adsorption in the Marine Environment", in *Chemical Oceanography,* 2nd ed., J.P. Riley and G. Skirrow, eds., Academic, New York, NY, pp. 241-308.

17. Hiemstra, T., Van Riemsdijk,W.H. and G.H. Bolt. 1989. "Multisite Proton Adsorption Modeling at the Solid/Solution Interface of (Hydr)oxides: A

New Approach. I. Model Description and Evaluation of Intrinsic Reaction Constants", J. Colloid Interface Sci., *133*, 91-104.

18. Harame, D.L., Bousse, L.J., Shott, J.D. and J.D. Meindl. 1987. "Ion-sensing Devices with Silicon Nitride and Borosilicate Glass Insulators," IEEE Trans. Electron Devices, *ED-34*, 1700-1707.

19. Chander, S., Fuerstenau, D.W. and D. Stigter. 1983. "On Hemimicelle Formation at Oxide/Water Interfaces", in *Adsorption from Solution,* R.H. Ottewill, C.H. Rochester, and A.L. Smith, eds., Academic, New York, pp. 197-210.

20. Wakamatsu, T. and D.W. Fuerstenau. 1968. "The Effect of Hydrocarbon Chain Length on the Adsorption of Sulfonates at the Solid/Water Interface", Adv. Chem. Ser., *79*, 161-172.

21. Osseo-Asare, K., Fuerstenau, D.W. and R.H. Ottewill. 1975. "Mechanism of Sulfonate Adsorption at the Silver Iodide-Solution Interface" in *Adsorption at Interfaces,* ACS Symp. Ser. No. 8, pp. 63-78.

22. James, R.O. and T.W. Healy. 1972. "Adsorption of Hydrolyzable Metal Ions at the Oxide-Water Interface. III. A Thermodynamic Model of Adsorption", J. Colloid Interface Sci., *40*, 65-81.

23. Fuerstenau, D.W. and K. Osseo-Asare. 1987. "Adsorption of Copper, Nickel, and Cobalt by Oxide Adsorbents from Aqueous Ammoniacal Solutions", J. Colloid Interface Sci., *118*, 524-542.

24. Crawford, R.J. 1995. *Adsorption and Coprecipitation of Heavy Metals,* Ph.D. Thesis, University of Melbourne, Australia.

25. Stumm, W., Hohl, H. and F. Dalang. 1976. "Interaction of Metal Ions with Hydrous Oxide Surfaces", Croat. Chem. Acta, *48*, 491-504.

26. Stumm, W., Kummert, R. and L. Sigg. 1980. "Ligand Exchange Model for the Adsorption of Inorganic and Organic Ligands at Hydrous Oxide Interfaces", Croat. Chem. Acta, *53*, 291-312.

27. Dzombak, D.A. and F.M.M. Morel. 1990. *Surface Complexation Modeling: Hydrous Ferric Oxide,* Wiley, New York.

28. Furrer, G. and W. Stumm. 1986. "The Coordination Chemistry of Weathering: I. Dissolution Kinetics", Geochim. Cosmochim. Acta, *50*, 1847-1860.

29. Zinder, B., Furrer, G. and W. Stumm. 1986. "The Coordination Chemistry of Weathering: II. Dissolution of Fe(III) Oxides", Geochim. Cosmochim. Acta, *50*, 1861-1869.

30. Guy, C. and J. Schott. 1989. "Multisite Surface Reaction Versus Transport Control During the Hydrolysis of a Complex Oxide", Chem. Geol., *78*, 205-218.

31. Blum, A. and A. Lasaga. 1988. "Role of Surface Speciation in the Low-temperature Dissolution of Minerals", Nature, *331*, 431-433.

32. Osseo-Asare, K. 1996. "Etching Kinetics of Silicon Dioxide in Aqueous Fluoride Solutions: A Surface Complexation Model", J. Electrochem. Soc., *143*, 1339-1347.

33. Judge, J.S. 1971. "A Study of the Dissolution of SiO_2 in Acidic Fluoride Solutions", J. Electrochem. Soc., *118*, 1772-1775.
34. Cook, L.M. 1990. "Chemical Processes in Glass Polishing", J. Non-Cryst. Solids, *120*, 152-171.
35. Mendel, E. 1967. "Polishing of Silicon", SCP and Solid State Technol., Aug., 27-39.
36. Burggraaf, P. 1995, "CMP: Suppliers Integrating, Applications Spreading", Semicond. Int., Nov.,74-82.
37. Kaufman, F.B., Thompson, D.B., Broadie, R.E., Jaso, M.A., Guthrie, W.L., Pearson, D.J. and M.B. Small. 1991. "Chemical-Mechanical Polishing for Fabricating Patterned W Metal Features as Chip Interconnects", J. Electrochem. Soc., *138*, 3460-3465.

Chapter Seven

ELECTROCHEMISTRY

DIGBY MACDONALD

Department of Materials Science and Engineering
Steidle Building
Pennsylvania State University
University Park, PA, 16802

INTRODUCTION

Electrochemistry concerns charge transfer processes at interfaces and, as such, is one of the most pervasive (and complex) disciplines in the physical, chemical, and biological sciences. For example, biological systems convert chemical energy (carbohydrates) into mechanical energy (muscle activation) in a series of complex chemical and charge transfer reactions that employ electrocatalysis. Corrosion of metals and alloys in condensed media is described in terms of electrochemical partial reactions, and the direct conversion of chemical energy into electrical energy (as in batteries and fuel cells) also involves electrochemical processes. The latter can be achieved with very high efficiency which presents humankind with unprecedented opportunities for extending hydrocarbon energy resources. Recently, electrochemistry has been employed in environmental remediation, for micro-machining, for the generation of light, and for the transfer of information. All of these systems and phenomena employ materials that support highly specific charge transfer reactions and processes. Indeed, electrochemistry long ago entered the "materials-by-design" age with the advent of electrocatalysis that enabled the use of fuel cells to provide electrical power for the Apollo missions to the moon.

This chapter provides an overview of the role of electrochemistry in materials science and technology. The goal is not to review the subject in depth, but rather to highlight areas of electrochemistry that have a major impact on the introduction and use of advanced materials in our society. Where appropriate, references are included to direct the reader to more comprehensive reviews of particular topics.

FUNDAMENTAL ASPECTS

Broadly speaking, electrochemistry can be divided into two principal areas: electrochemical thermodynamics and electrochemical kinetics (electrodics). The first of these areas is concerned with specifying the conditions of equilibrium for charge transfer processes and for stipulating the conditions of spontaneity, whereas the second describes the rates at which charge transfer reactions occur. Clearly, in-depth discussions of these topics are beyond the scope of this presentation and the reader is referred to several monographs that provide accounts of varying sophistication on both topics (1-4). However, a number of key fundamental points must be made here in order that the material presented later in this chapter becomes clear.

ELECTROCHEMICAL THERMODYNAMICS

Inclusion of electrical work for charged systems into the First Law of Thermodynamics and the subsequent combination of the First Law with the Second Law of Thermodynamics yields the spontaneity condition as

$$(E - E^e)\, I > O \tag{1}$$

where E^e is the equilibrium potential for the cell, $R'/R \mid P'/P$, E is the applied potential, and I is the current. The Stockholm convention is employed, so that the cell potential is the potential difference between the right hand couple (P'/P) and the left hand couple (R'/R), and a positive current ($E>E^e$) signifies that the right hand couple proceeds in the oxidation sense

$$P \rightarrow P' + pe^- \tag{2}$$

and the left hand couple proceeds in the reduction sense

$$R' + re^- \rightarrow R \tag{3}$$

Electrochemical thermodynamic information is commonly summarized in the form of a potential/pH (Pourbaix) diagram, and such a diagram for iron is shown in Figure 1. In these diagrams, the left side of Cell I is the Standard Hydrogen Electrode (SHE), which is the hypothetical

$$H_2(f_{H_2} = 1)\, /H^+(a_{H^+} = 1) \tag{II}$$

cell in which the hydrogen fugacity and hydrogen ion activity are both unity.

A potential/pH diagram is, in fact, a super-imposition of several sub-diagrams.

The most elementary is that for water, which is represented by the Lines (a) and (b). Line (a) corresponds to the hydrogen electrode reaction

$$H^+ + e^- \Leftrightarrow \tfrac{1}{2}H_2 \tag{4}$$

whence

$$E^e = -\frac{2.303RT}{2F}\log p_{H_2} - \frac{2.303RT}{F}\ pH \tag{5}$$

Thus, the line is plotted for a given partial pressure of hydrogen gas, but Equation (5) predicts that the line moves to more positive potentials as the

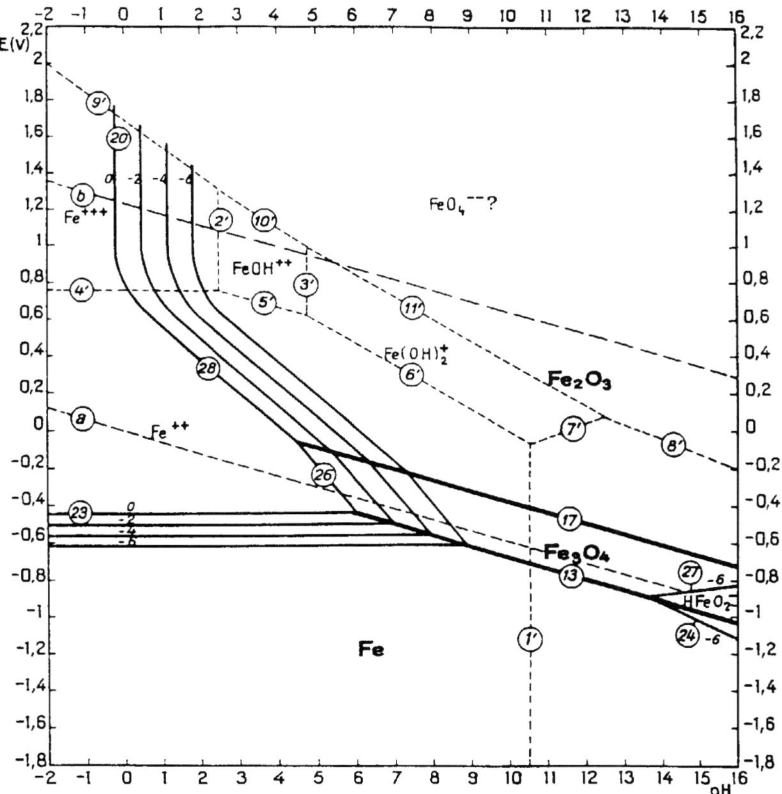

FIGURE 1. Potential-pH equilibrium diagram for the system iron-water, at 25°C (After Pourbaix)

partial pressure of H_2 is lowered. Line (b), on the other hand, corresponds to the reaction

$$O_2 + 4H^+ + 4e^- \Leftrightarrow 2H_2O \tag{6}$$

whence

$$E^e = E^\circ + \frac{2.303RT}{4F} \log p_{O_2} - \frac{2.303RT}{F} pH \tag{7}$$

where E° is the standard potential for the H_2O/O_2 couple and p_{O_2} is the partial pressure of oxygen. In this case, the potential shifts in the positive direction with increasing partial pressure of the gas. However, the rate at which the potential shifts in the positive direction with $\log p_{O_2}$ is only half of that with respect to $\log p_{H_2}$ for Line (a). As both p_{H_2} and p_{O_2} are decreased, the lines move closer together until they coincide at a point where the partial pressures are related by

$$p_{H_2} \cdot p_{O_2}^{1/2} = \exp(-\Delta G^\circ/RT) \tag{8}$$

where ΔG° is the change in standard Gibbs energy for the reaction

$$H_2O \Leftrightarrow H_2 + \tfrac{1}{2}O_2 \tag{9}$$

For any combination of H_2 and O_2 partial pressures, such that $p_{H_2} \cdot p_{O_2}^{1/2}$ exceeds the right side of Equation (8), the difference between $E^e_{O_2}$ and $E^e_{H_2}$ defines the potential range for the thermodynamic stability of water. For $p_{H_2} = p_{O_2} = 1$ atm at 25°C, this range extends 1.23V.

Given that the energies of natural systems range from 3×10^{-4} eV (background temperature of the universe) to $>10^{19}$ eV (cosmic rays), it is remarkable that life, as we know it, can exist only within the very narrow potential range of the thermodynamic stability of water.

The second sub-diagram that exists within Figure 1 is that for iron, and involves solid and ionic species. Because the equilibrium lines define the conditions of co-existence of two (or more) species, the enclosed regions are stability fields, which may be interpreted in a manner that is analogous to those for physical phase diagrams. It is commonly assumed that E/pH diagrams contain only equilibrium information, but that is incorrect. For example, if we extrapolate the equilibrium line for Fe/Fe_3O_4 [i.e. Line (13)] into the stability field for Fe^{2+}, then we define the conditions for the metastable formation of Fe_3O_4 on Fe in acidic solutions. Thus, Fe_3O_4 can exist as a metastable phase only at potentials that are more positive than the extrapolated line, but whether magnetite actually forms is a a question of chemical kinetics. The formation of Fe_3O_4 as a metastable phase accounts for the passivity of iron in acidic media.

Potential/pH diagrams are now available for almost every element in the periodic table except the noble gases and some of the transuranic elements. These diagrams summarize an enormous amount of electrochemical thermodynamic information, and they have found extensive use in corrosion (see also Chapter 22), electrometallurgy, geochemistry, and battery technology, and they have an enormous impact on these branches of chemistry and electrochemistry when they were introduced by Pourbaix in the 1960s.

Electrochemical Kinetics

The theory of the kinetics of elementary charge transfer reactions is now highly developed, and contributions to this area won Marcus a Nobel Prize a few years ago, one of two that have been awarded to electrochemists. The theory has been extensively reviewed in the literature and will not be discussed in detail here. It is sufficient to note that electrochemical processes generally are heterogeneous reactions and that the rate constant has the form

$$k = k_o \exp \left(\pm \alpha nF\eta/RT \right) \tag{10}$$

where k_o is a constant, $\eta = E - E^e$ is the overpotential, and α is the transfer coefficient, which is a measure of the fraction of the potential imposed across the interface that affects the Gibbs energy of activation. The "+" and "-" signs correspond to anodic (oxidation) and cathodic (reduction) processes, respectively.

For a redox reaction of the type

$$R \underset{k_{-1}}{\overset{k_1}{\rightleftarrows}} O + ne^- \tag{11}$$

where R and O are reduced and oxidized species, respectively, the parameter n is the electron number, and k_1 and k_{-1} are the forward and reverse rate constants given by Equation (10). Because the rate can be expressed as a current density through Faraday's law, the rate equation becomes

$$i = nF \left[k_1 C_R(x = 0) - k_{-1} C_O(x=0) \right] \tag{12}$$

where $C(x=0)$ is the concentration at the electrode surface and F is Faraday's constant. The concentration at the surface differs from that in the bulk because of mass transfer effects, which leads to "concentration polarization." By expressing the mass transfer rate in terms of limiting current densities for the forward ($i_{\ell,f}$) and reverse ($i_{\ell,r}$) directions of Reaction (11), we are able to write the observed current (due to the difference between the forward and reverse partial current densities) as

$$i = \frac{e^{\eta/b_a} - e^{-\eta/b_c}}{\dfrac{1}{i_o} + \dfrac{1}{i_{\ell,f}} e^{\eta/b_a} - \dfrac{1}{i_{\ell,r}} e^{-\eta/b_c}} \tag{13}$$

where $\eta = E-E^e$ is the overpotential, b_a and b_c are the Tafel constants, and i_o is the exchange current density. This equation, which is sometimes referred to as the "generalized Butler/Volmer equation" embodies kinetic (i_0,b_a,b_c), thermodynamic (E^e), and mass transfer ($i_{\ell,f}$, $i_{\ell,r}$) effects.

The exchange current density occupies a central place in electrochemistry that is comparable to the position occupied by the rate constant in chemical kinetics. However, the exchange current density can be reduced further by the expression

$$i_o = i_o^o C_O^\gamma C_R^\delta \tag{14}$$

where i_o^o is the standard exchange current density, C_O and C_R are bulk concentrations (actually, these are ratios of the bulk concentrations to the standard state concentrations, and hence are dimensionless), and γ and δ are constants that are derived from the mechanism of the reaction. Both i_o and i_o^o may vary by many orders in magnitude from reaction-to-reaction or even for a single reaction from substrate-to-substrate (1).

The form of Equation (13) for different values of b_a, b_c, and i_o is shown in Figure 2 for the case of a "simple" reaction. One sees that the shape of the current/voltage curve is sensitive to the values chosen for b_a, and b_c, with a symmetrical curve being obtained only if $b_a = b_c$ and $i_{\ell,f} = i_{\ell,r}$. However, additional complications can occur for redox reactions that occur on passive surfaces. An example of this is shown in Figure 3, in which we plot the current/voltage curve for the hydrogen electrode reaction on platinized nickel in high temperature boric acid/borate buffer solution. The sudden drop in the current in the anodic region is possibly accounted for by quantum mechanical tunneling across a passive film that forms on the metal surface.

The treatment of complex, multi-step electrochemical reactions generally follows the methodologies that have been developed in chemical kinetics (1). These methods include the pre-equilibrium hypothesis, in which all steps prior to the (postulated) rate-determining step are regarded as being in equilibrium. The second popular method is the steady-state hypothesis, in which the concentrations of all intermediates are assumed to be independent of time. These two approaches provide a powerful framework for analyzing even the most complicated electrochemical reaction mechanisms.

ANALYSIS OF ELECTROCHEMICAL SYSTEMS

In the remainder of this chapter, a number of electrochemical systems will be reviewed with emphasis on materials issues. The reviews are not exhaustive and, because of limited space, only a few topics could be included.

Anodic Dissolution

Metal dissolution processes have played a central role in the development of electrochemistry and corrosion science, and it seems to be appropriate to begin this section with a brief discussion of this subject. The topic is important from a materials viewpoint, because metal dissolution is involved in electrochemical technologies as diverse as electrochemical machining and high energy density batteries.

In order to appreciate the place that metal dissolution occupies in electrochemical phenomena, it is necessary to explore briefly the polarization curve for an

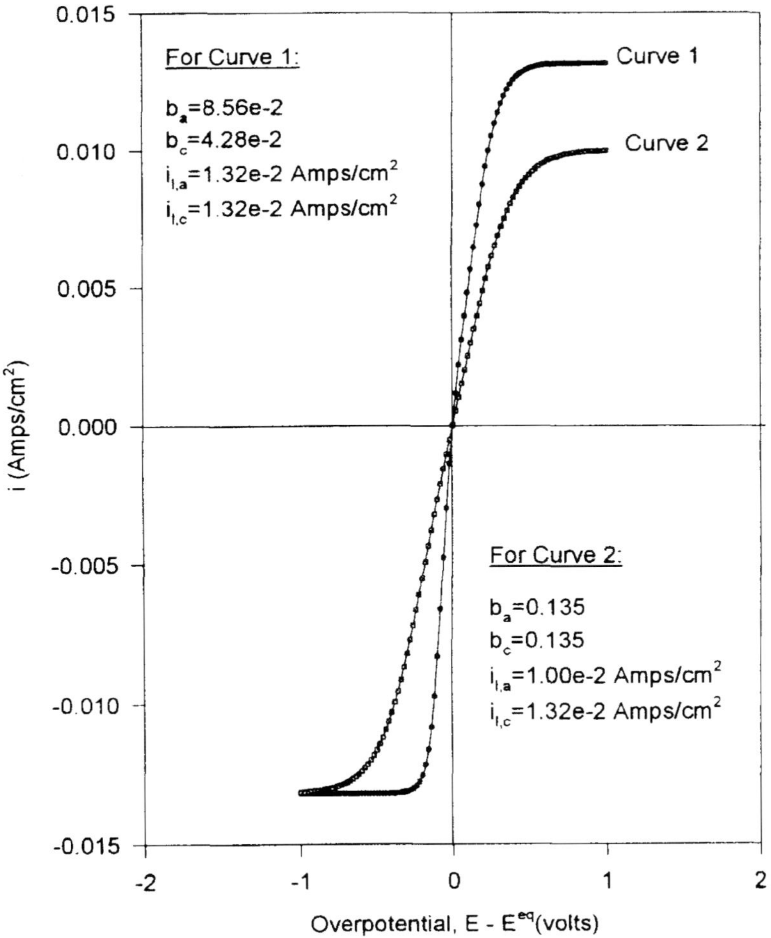

FIGURE 2. Current-voltage curves for the hypothetical redox reaction, O/R, calculated using Equation (13).

active metal, such as iron (Figure 4). Thus, on increasing the potential in the positive direction from the hydrogen evolution region, the current is found to increase exponentially [linearly for ℓn(i)] with voltage over a current range that may extend many orders in magnitude. This finding is consistent with charge transfer theory,

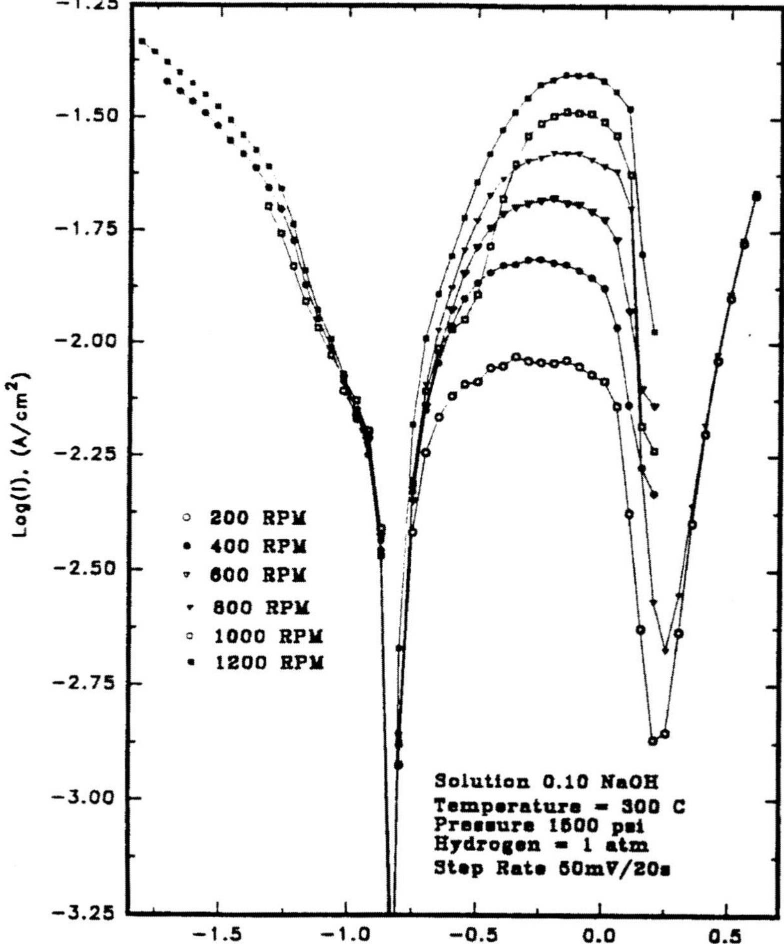

FIGURE 3. Current/voltage curves for the H_2/H^+ electrode reaction in 0.10 m NaOH at 300°C. The rotational velocities are those of a flow-activating impeller such that mass transfer is enhanced by increasing rpm values.

which predicts this "Tafel" relationship for the "simple" transfer of an ion from the metal phase to the solution, unencumbered by mass transfer or solution resistivity effects [see Equ. (10)]. However, few, if any, dissolution reactions are "simple," in that most reactions are thought not to be elementary processes.

Perhaps the simplest dissolution reactions that one can envision is the dissolution of a metal to produce a univalent ion (e.g., $Cu \rightarrow Cu^+$, $Ag \rightarrow Ag^+$). In this case, it might be supposed that ions simply leave the surface from high

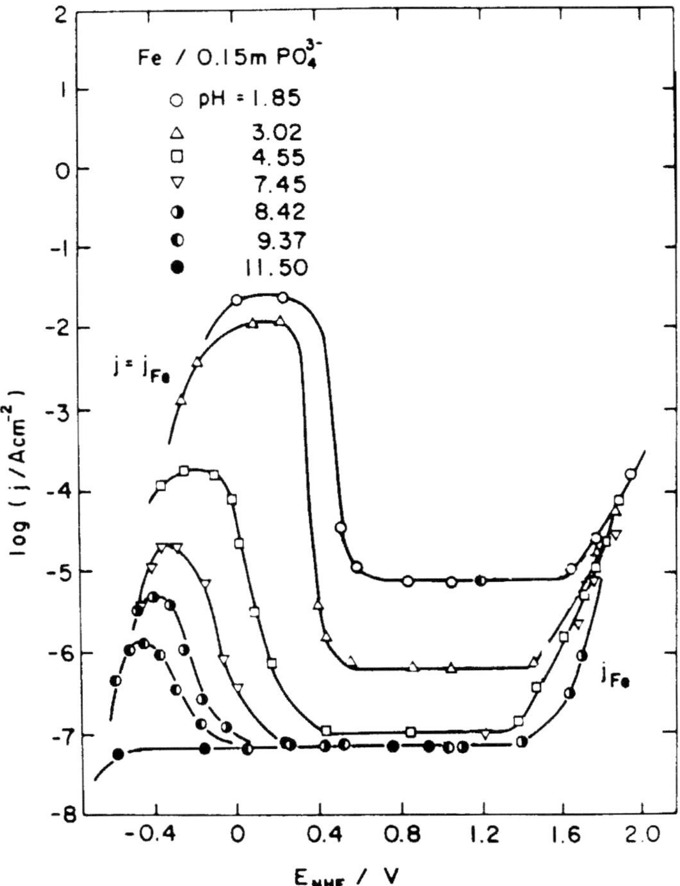

FIGURE 4. Anodic polarization curves for iron in phosphate buffer solutions as a function of pH [After Sato (5)].

energy/low coordination sites (e.g., line sites, ad-atoms, Figure 5) in a single elementary step. However, the (meager) evidence suggests that that is not the case; instead, the dissolution process appears to involve the formation of an adion or adatom that has a finite lifetime on the surface prior to being ejected into the solution (Figure 5). Thus, at the fundamental level, the electrodissolution of such "simple" metals would involve at least three steps

$$M(\text{high energy site}) \rightarrow M_{ad}^{\delta+} + \delta e^- \tag{19}$$

$$M_{ad}^{\delta+} \overset{\text{Surface}}{\underset{\text{Diffusion}}{\rightarrow}} M_{ad}^{\delta\delta+} \tag{20}$$

$$M_{ad}^{\delta+} \rightarrow M^+ + (1 - \delta)e^- \tag{21}$$

where δ is the extent of the ionization of the ad-atom (or ad-ion). For energetic reasons, the most active surface sites are thought to be those with the least coordination. Thus, with reference to Figure 5, the site activity increases in the

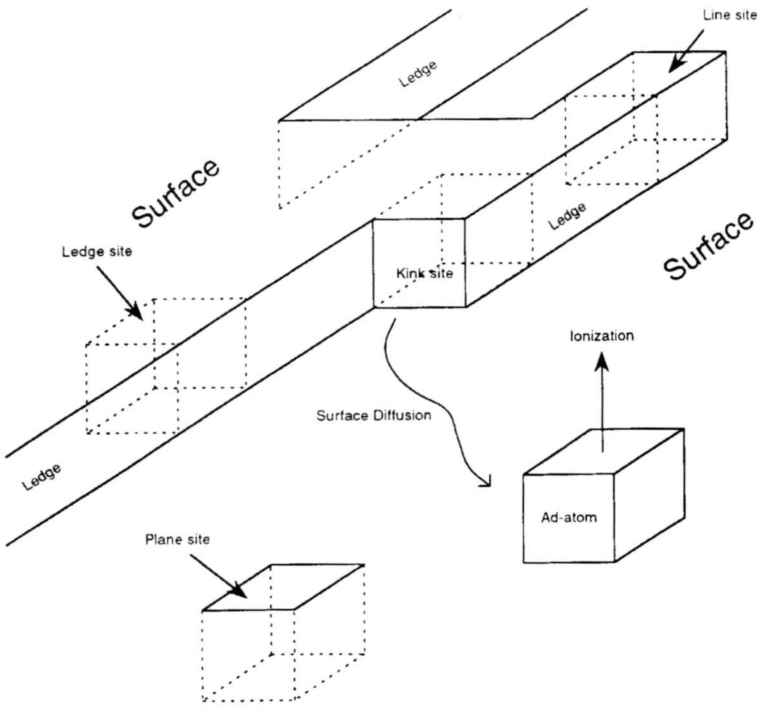

FIGURE 5. Ad-atom mechanism for metal electrodissolution.

sequence: plane [5] < ledge [4] < kink [3] < line [2] < ad-atom [1], where the quantity in parentheses is the coordination number. Although it is possible for direct ionization to occur from a high coordination site, energy arguments suggest that the most likely path is through the ad-atom. At any rate, the surface is evidently a dynamic place with continuous movement occurring in the morphology as the metal dissolves and as the potential is changed. These changes have, over the past five years, been well-documented by in situ scanning tunneling microscopy (STM) studies (6), and a typical example of such a study as reported by Snedden et. al. is shown in Figure 6. No evidence was found in these studies for the specific adsorption of anious, such as ClO_4^-, SO_4^{2-}, and F^-; or if adsorption does occur there was no obvious affect on the morphological dynamics of the interface.

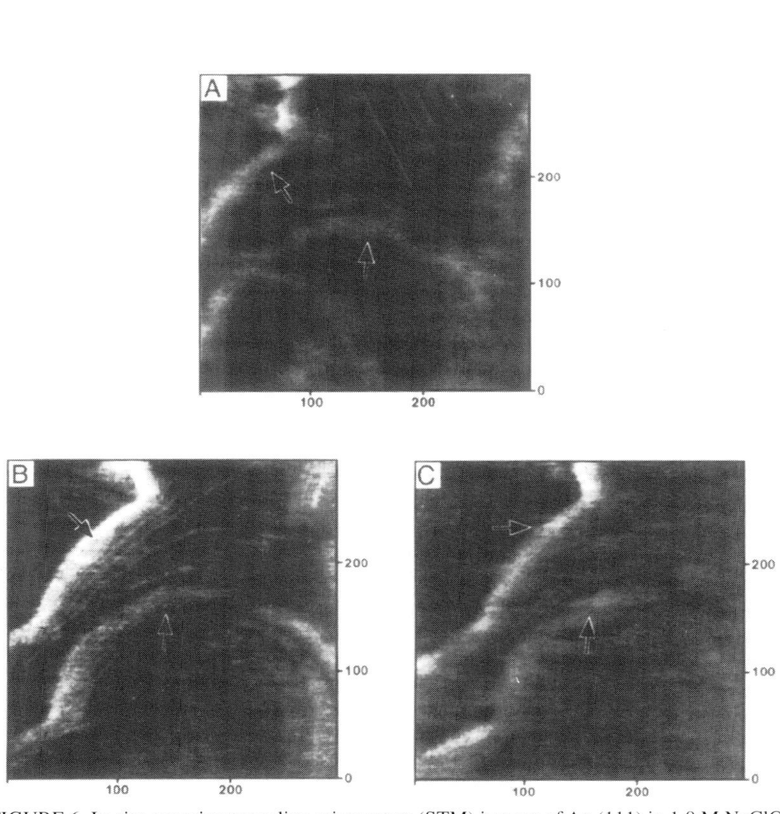

FIGURE 6. In-situ scanning tunneling microscope (STM) images of Ag (111) in 1.0 M $NaClO_4$ under anodic polarization conditions. Note the movement of the ledge with increasing potential. The resolution is such that ad-atoms are not detectable [After Sneddon et. al. (6)].

For more complex metals, such as those that form multi-valent ions (e.g., Fe $\rightarrow$ Fe^{2+}), the highly improbable quantum event of two electrons being transferred simultaneously argues that their dissolution cannot be "simple", even in the sense outlined above for metals such as Ag and Cu. Furthermore, the kinetics of electrodissolution of metals, such as iron, in acidic media are found to depend on pH and the concentrations of anions in the solution, even though these species may not appear in the stoichiometric reaction. For example, the electrodissolution of iron is consistent with the formation of a hydroxy complex

$$Fe + H_2O \rightarrow Fe(OH)_{ads} + H^+ + e^- \tag{22}$$

or even a surface chloro complex

$$Fe + Cl^- \rightarrow FeCl_{ads} + e^- \tag{23}$$

in chloride-containing media. These species undergo further reactions to form even more complex surface species, which eventually leave the surface to form uncomplexed ions (e.g., Fe^{2+}) in the solution. The important point is that these reactions are multi-step, anion-assisted processes that involve a number of sequential and parallel steps. In recent years, with the advent of electrochemical impedance spectroscopy (EIS), the mechanisms of the electrodissolution of metals such as Fe, Ni, Al, and Zn have been explored in depth, and the involvement of surface species of the type shown above is a common theme.

As the potential is increased still further, the current does not continue to follow the Tafel equation, but instead reaches a maximum before plunging to the passive state at the Flade potential (Figure 4). Little is known, in detail, of the processes that are responsible for the active-to-passive transition although several possibilities have been discussed by workers in the field. These include the generation of electrochemically adsorbed species (e.g., oxygen atoms) on the surface that block the high activity sites from which the dissolving metal atoms originate, and the formation of three-dimensional patches of oxide that spread laterally across the surface to block the entire metal surface. A third mechanism exists, in that it is possible to envision the formation of a continuous metal hydroxide film on the surface [e.g., Fe(OH)$_2$] at potentials in the active region, with the active-to-passive transition being triggered by a field-assisted ejection of protons from the layer thereby leading to the conversion of the hydroxide into an oxide. One curious aspect is that the current/voltage curve is seldom reversible in this region, which may be due to the existence of processes with inordinately long relaxation times.

Passivity

The phenomenon of passivity has been studied since the dawn of electrochemistry, and it is one of the most important phenomenon in the physical sciences. Its importance stems from the fact that our metals-based civilization would not be possible without the formation of thin corrosion product films on the surface that effectively isolate the underlying reactive metal from the oxidizing environment. It is perhaps of interest to note that the energy density of aluminum in reaction with oxygen is about the same as that of gasoline, yet we happily fly through the air in aluminum aircraft without appreciating the fact that the structural integrity of the aircraft is maintained entirely by an oxide layer that is too thin to detect with the naked eye.

The initial stages of the formation of anodic oxide films on metal surfaces is the subject of considerable controversy. Two broad classes of mechanisms exist; (a) initial formation of a continuous mono-atomic layer of adsorbed oxygen followed by film thickening, perhaps in the initial stages by place exchange; and (b) the nucleation, growth, and coalescence of oxide patches. The first of these mechanisms appears explain the scanning tunneling microscopy (STM) observation (Figure 7) on the initial formation of anodic oxide films on Au (111), as reported by Vitus and Davenport (7). However, these same authors report that considerable restructuring of the initial "mono-atomic" layer occurs as the potential is changed in the negative direction with the formation of worm-like patches that tended to coalesce. On the other hand, Bockris et. al. (8) report, also on the basis of STM studies, that the anodic oxide film that forms on iron does so initially by the second mechanism. In any event, it is evident that, during the initial stages of passive film formation, the surface is, again, a dynamic place, with significant morphological changes occurring as the potential is varied.

An enormous literature exists on growth of passive films, so that only a brief outline of one approach to describing the growth and breakdown of anodic oxide films will be given. That approach is the Point Defect Model (PDM) that has been extensively developed by the author and his coworkers (9). The essence of the PDM is outlined in Figure 8. Briefly, the theory postulates that the passive film is a defective oxide, with the principal defects being metal vacancies and oxygen vacancies. The vacancies are produced and annihilated at the interfaces, as indicated in Figure 9. The model also postulates that the electric field is independent of the applied voltage, because of buffering by Esaki tunneling (band-to-band quantum mechanical tunneling), which is considered to render the field insensitive to the applied voltage.

The PDM predicts that, regardless of the predominant defect in the barrier layer, the steady-state thickness of the barrier layer should vary linearly with the applied voltage, and this is a ubiquitous observation in experimental studies (see, e.g., Figure 10). Furthermore, the current is predicted to be constant or to depend upon the applied potential, depending upon the predominant defect in the film. By using a diagnostic criteria derived from the PDM, we have been able to show that the barrier layer on nickel is a cation conductor whereas that on tungsten is an oxygen ion (vacancy) conductor. The passive film on zinc is consistent with being an interstitial cation conductor, although the kinetics of the film growth/dissolution appears to be dominated by the oxygen vacancy structure.

The PDM has been extended to describe passivity breakdown and hence the initiation of localized corrosion. The basic mechanism is that the absorption of chloride (or some other aggressive anion) into oxygen vacancies at the barrier layer/environment interface generates an enhanced flux of cation vacancies across the film. If these vacancies cannot be annihilated at a sufficiently high rate at the metal/film interface by Reaction (1), Figure 9, they condense at the metal/film interface, thereby causing the barrier layer to separate from the metal. Dissolution occurs at the film/environment interface, which is not matched by growth of the film into the metal at that local area. Eventually, the film ruptures

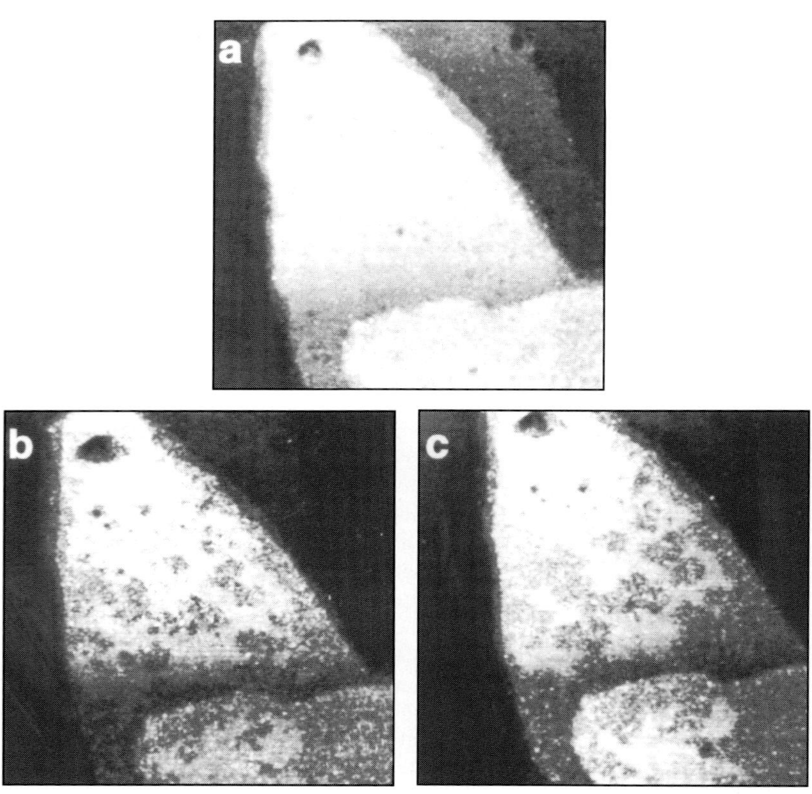

FIGURE 7. A time-sequence showing monolayer oxide formation and propagation at 1.55 V (RHE). The potential was swept from 0.93 V (RHE) at 2 mVs-1 and then held at 1.55 V during image acquisition: (a) time (minutes), t=0:00 (scan direction at t=0:00 is down): (b) t=4.23 (up); (c) t=7:56 (up). Experimental parameters, 200 x 200 nm; bias voltage, -600.0 mV; tunneling current, 2.0 nA. [After Vitus and Davenport (7)]

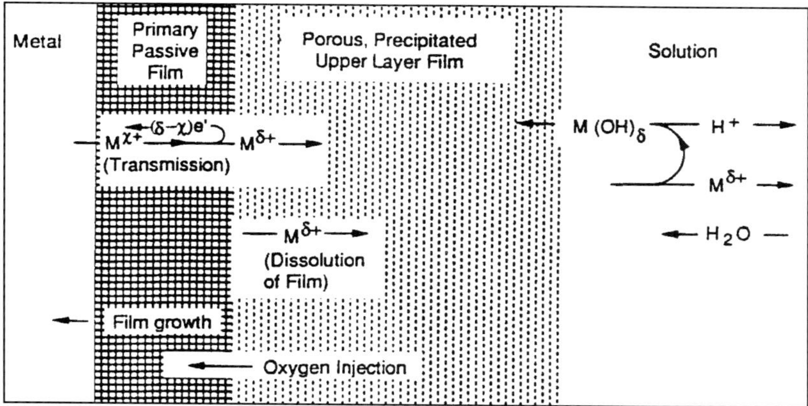

FIGURE 8. Schmatic of processes that lead to the formation of bilayer passive films on metal surfaces.

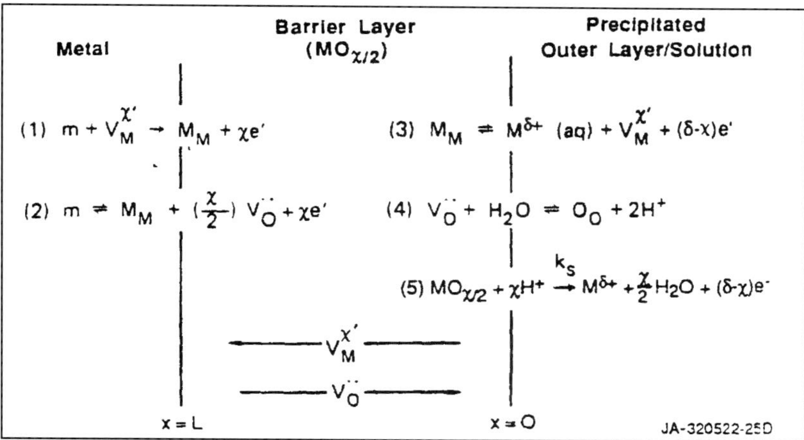

FIGURE 9. Schmatic of physico-chemical processes that occur within a passive film according to the point defect model.

M = metal atom, M_m = metal cation in cation site, O_O = oxygen ion in anion site $V_M^{\chi'}$ = cation vacancy, $V_O^{..}$ = anion vacancy.

During film growth, cation vacancies are produced at the film/solution interface, but are consumed at the metal/film interface. Likewise, anion vacancies are formed at the metal/film interface, but are consumed at the film/solution interface. Consequently, the fluxes of cation vacancies and anion vacancies are in the directions indicated. Note that Reactions (1), (3), and (4) are lattice conservation processes, whereas Reactions (2) and (5) are not.

marking the passivity breakdown event (Figure 11). Blisters (or vacancy condensates), as precursors to passivity breakdown, have been observed by many authors, and this mechanism appears to be quite general.

The PDM has now been developed to yield a deterministic theory for the initiation of damage (13). The key postulate is that the potential breakdown sites are normally-distributed with respect to the cation vacancy diffusivity within the barrier layer. This postulate leads to any analytical function for the distribution in induction time. Coupling of this theory with the Coupled Environment Pitting Model (11) allows one to calculate the damage function (pit density versus depth) as a function of the time of observation (Figure 12). The observation time at which the upper extreme of the damage function exceeds a critical dimension (e.g., thickness of a pipe wall) yields the failure time. To date, we have applied this method to estimate the progression of pitting damage and to estimate failure times for Type 304SS tubes in gas-fired condensing heat exchangers, to estimate failure times for CVD-Ni detector tubes in the neutrino-detection experiment now being initiated

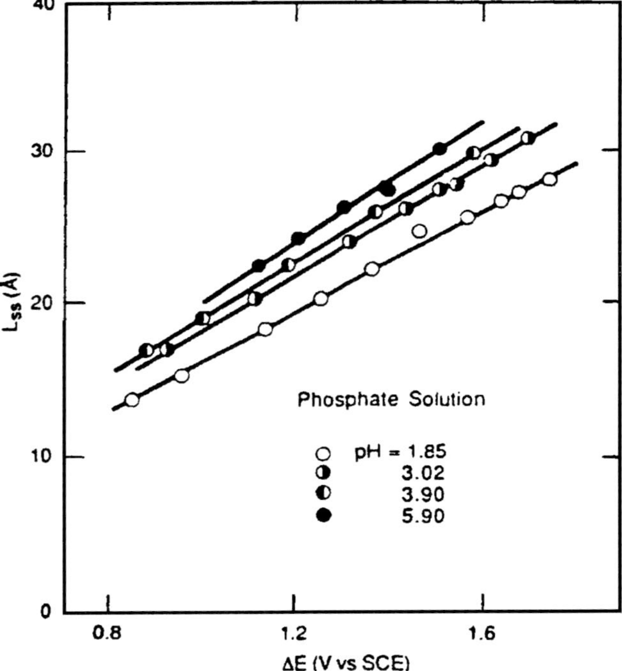

FIGURE 10. Thickness of the barrier oxide layer on iron in acidic phosphate solutions as a function of potential [after Sato and Noda (10)].

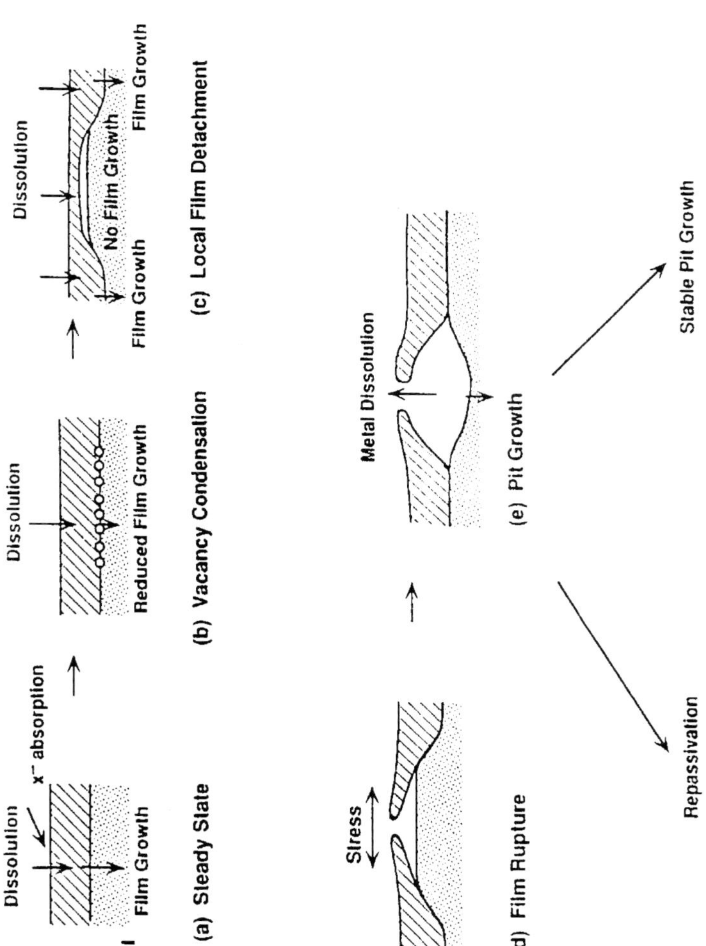

FIGURE 11. Cartoon outlining various stages of pit nucleation according to the Point Defect Model.

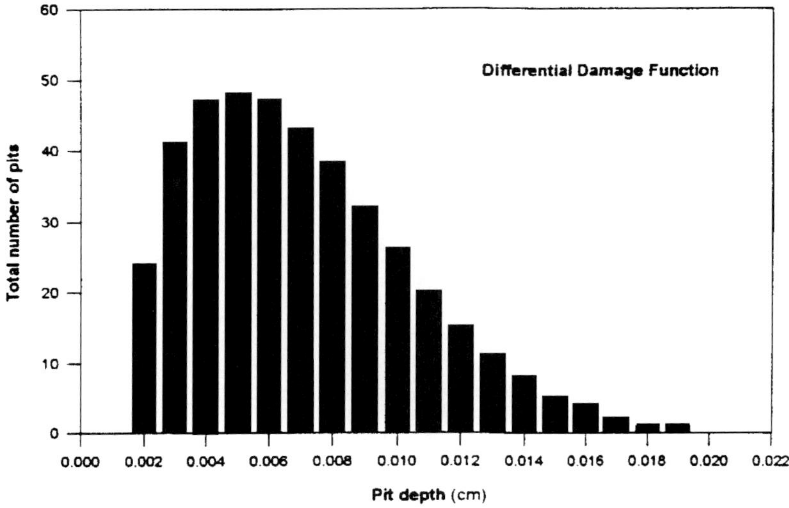

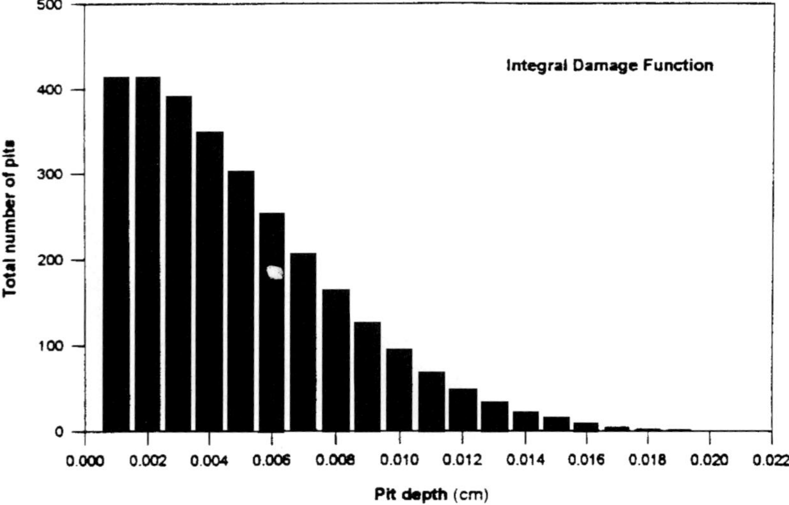

FIGURE 12. Calculated differential and integral damage functions for carbon steel in high temperature pure water. ($T=150°C$, pH=5.82, $[O_2]$=7.5ppm, t_{obs} =1 yr., Kp=100yr^{-1}). The differential damage function corresponds to the distribution in pit depth whereas the integral damage function depicts the distribution in remaining pits as successive layers of material are removed from the surface.

seven miles beneath the Canadian Shield in Ontario, Canada, and to estimate the progression of damage to boiler tubes in a fossil-fueled thermal power station. To our knowledge, the work described above represents the first attempts to predict the distribution of damage due to localized corrosion in a deterministic manner.

Production of Aluminum

As noted elsewhere in this review, aluminum production is a major consumer of electrical energy and, in the absence of a non-electrochemical method for producing this metal, will remain so for the foreseeable future. The economics of aluminum production are such that a (large) source of cheap electricity must be available (thus, the preferred source is hydro), and the recovery process must be very efficient (12).

Metallic aluminum is produced electrolytically in Hall-Heroult cells of the type shown in Figure 13. These cells consist of a molten electrolyte (Al_2O_3/Na_3AlF_6,~10 w/o alumina), carbon anodes, and a molten aluminum cathode. The electrolyte (Al_2O_3 dissolved in fused cryolite, Na_3AlF_6) is an ionic conductor, with the conducting species believed to include oxyfluoroaluminates of the general form $Al_xO_yF_z^{(3,-2,-)}$. The values of x, y, and z are considered to be composition-dependent,

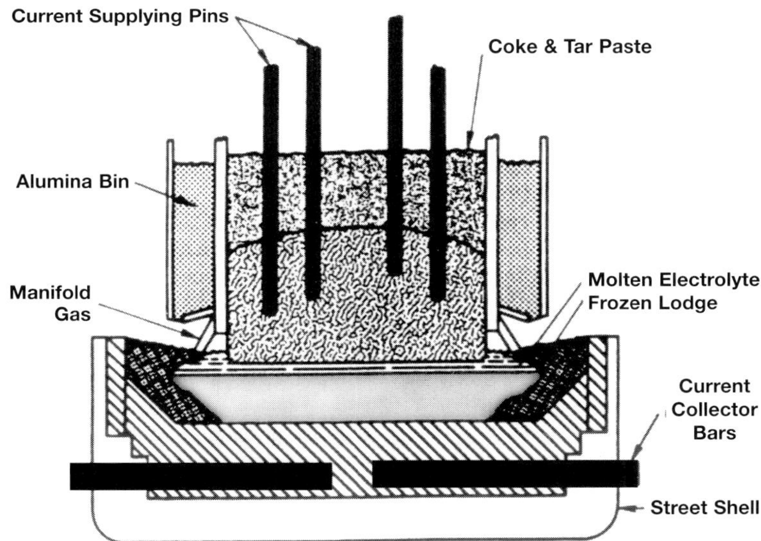

FIGURE 13. Aluminum electrolysis cell with Soderberg anode. [By permission of Kirk-Othmer Encyclopedia of Chemical Technology, Vol. 2, 3rd ed., Wiley, New York (1978).]

such that species complexity increases with Al_2O_3 loading. Although much of the cryolite chemistry remains undefined, and recognizing that the electrolyte chemistry impacts the electrochemistry at the electrodes, it is thought that the cathode reactions are

$$AlF_4^- + 3e^- \rightarrow Al + 4F^- \tag{24}$$

and

$$AlF_6^{3-} + 3e^- \rightarrow Al + 6F^- \tag{25}$$

with the first dominating at low cryolite ratios, $[NaF]/[AlF_3]$, and the second at high ratios.

The anode reaction is thought to be significantly more complicated than the cathode reaction, and various processes have been proposed to account for the consumption of the carbon anode and the production of CO_2 in the cell. These reactions include

$$2Al_2OF_6^{2-} + 2AlF6_6^{3-} + C \rightarrow 6AlF_4^- + CO_2 + 4e^- \tag{26}$$

$$2Al_2OF_6^{2-} + 4F^- + C \rightarrow 4AlF_4^- + CO_2 + 4e^- \tag{27}$$

$$2AlOF_2^- + 2AlF6_6^{3-} + C \rightarrow 4AlF_4^- + CO_2 + 4e^- \tag{28}$$

and

$$2AlOF_2 + 4F^- + C \rightarrow 2AlF_4^- + CO_2 \tag{29}$$

which are considered to occur at different alumina loadings and cryolite ratios, to yield the overall cell reaction of

$$2Al_2O_3 + 3C \rightarrow 4Al + 3CO_2 \tag{30}$$

It is interesting to note that the overall reaction is a carbothermic reaction, and indeed considerabe effort has been expended in trying to develop a chemical carbothermic reaction (i.e. the direct reduction of Al_2O_3 by carbon) for the direct production of aluminum. However, thermodynamics show that the direct reduction process is not possible at temperatures lower than ~1750°C, whereas electrochemically-mediated reaction occurs readily at 940-980°C (the normal operating temperature of the Hall-Heroult cell). Of course, the lower temperature is compensated for by the electrical energy that is fed into the cell.

Cryolite is commonly regarded as being a universal solvent in that few materials are immune to attack. Certainly, common (and low cost) metals, such as iron, are unable to be used in contact with this electrolyte. Even refractory materials, such as Al_2O_3 (of course), SiO_2, TiO_2,..., readily dissolve and hence cannot be used for containment. This problem is addressed in the current technology by maintaining a layer of frozen electrolyte on all boundaries, which usually comprise carbon blocks and linings (Figure 13). While this strategy is effective, it does require careful thermal control in order to maintain the frozen electrolyte layer intact. The development of a material that is stable in contact with the molten electrolyte would be a boon to the aluminum recovery industry.

Given the competitive nature of aluminum production, it is not surprising that producers strive for the highest possible energy efficiency. The losses in the Hall-Heroult cell are principally electrical in nature (IR losses), which may be reduced only by decreasing the resistance between the anode and the cathode. Increasing the conductivity of the electrolyte has generally proved fruitless as a means of reducing the IR losses (although the addition of LiF does significantly reduce the resistivity), so that reducing the anode/cathode separation has been the preferred way of improving the cell efficiency. However, there is a limit to how close the anode and the cathode may be placed without inadvertent shorting. A principal phenomenon limiting the increase in degradation of conductance is the formation of gaseous CO_2 in the gap and adherent gas bubbles on the anode surface (Figure 14). Because of the high magnetic fields and currents that exist in a production cell, electromagnetic forces may establish ripples and waves on the surface of the molten aluminum cathode. The magnitudes of these waves clearly define the lower limit of the interelectrode gap.

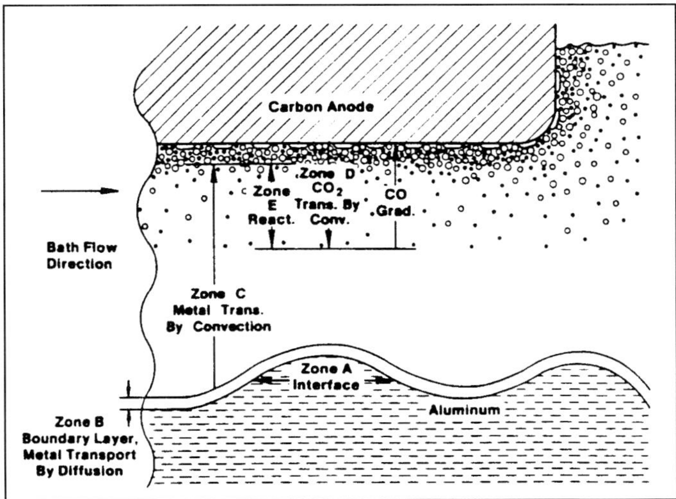

FIGURE 14. Mechanisms of current efficiency losses in Hall-Heroult cells.

Energy Storage and Conversion

Late in the last century, the world had the opportunity to choose between batteries/electric motors and the internal combustion engine (ICE) for automotive propulsion. Indeed, many of the early automobiles were electrically powered and, for a short while, it must have seemed that batteries/electric motors would reign supreme (1). So, why did the world turn to the internal combusion engine for automotive power, in particular, and to heat engines, in general, as the primary sources for the post-industrial revolution? The answer lies in a complex mixture of economic, technical, and sociological factors, and came about in spite of the clear advantages of electrical systems in terms of conversion efficiency and emissions. Much has been written on this subject (13, 14), and we will not attempt to reproduce all of the arguments here. Instead, a couple of key issues will be explored that, perhaps, may help explain the current resurgence of interest in EVs (Electric Vehicles).

Heat engines operate by transferring heat from a source at a high temperature (T_1) to a sink at a lower temperature (T_2) with an intermediate step of converting some of the heat energy into mechanical work. Carnot showed, more than two hundred years ago, that the maximum efficiency of conversion of heat (chemical) energy from the source into mechanical work is given by the simple formula.

$$\varepsilon_c^M = 1 - T_2/T_1 \tag{31}$$

For a thermal power plant (fossil or nuclear) and gasoline-powered ICEs, the efficiency ranges from 25-35%, with 65-75% of the available energy being rejected to the environment in the form of waste heat and entropy. Diesel-powered ICEs are somewhat more efficient (approaching 40%), but still the majority of the input heat energy is rejected. Examination of Equation (31) shows that the only parameter that may be readily manipulated to increase ε_c^M is the temperature of the heat source (T_1), since T_2 is determined by the nature of the environment, immediately down stream of the heat-to-mechanical energy converter (e.g., the turbine in a gas turbine). However, the maximum practical value of T_1 is limited by the materials used in the system. Nevertheless, impressive gains have been made over the past few decades in improving the efficiencies of heat engines. For example, the development of "adiabatic" ICEs, in which the cylinders and pistons are fabricated from toughened ceramics, promises to increase ε_c^M to more than 50%. Such a development would save the US many billions of dollars annually in fuel costs. As a second example, we note the remarkable increase that has occurred in the efficiencies of the gas turbines that power aircraft. This increase has been responsible for a significant decrease in the cost-per-seat-per-mile-of-operation as measured in constant dollars. The improved efficiency is directly attributable to higher turbine operating temperatures, which in turn have been made possible by the development of super-alloys.

In contrast to the above, the efficiency of chemical-to-electrical energy conversion via an electrochemical converter (e.g., a fuel cell) is determined by

$$\varepsilon_C^E = \Delta G_R / \Delta H_R = 1 - T(\Delta S_R / \Delta H_R) \tag{32}$$

where ΔG_R, ΔH_R, and ΔS_R are the changes in the Gibbs energy, enthalpy, and entropy for the cell reaction. Most cell reactions are exothermic and hence ΔH_R is negative. For example, the Gibbs energy, enthalpy, and entropy changes for the hydrogen/oxygen fuel cell, for which the cell reaction is,

$$H_2 + \tfrac{1}{2} O_2 \rightarrow H_2O \tag{33}$$

are -237.18 kJ/mole, -285.83 kJ/mole, and 0.163 kJ/K mole (of H_2), respectively, under standard conditions (T=25°C, $p_{H_2} = P_{O_2} = 1$ bar). Thus, the theoretical efficiency is 83%, which is considerably higher than that of a heat engine under practical conditions. However, close examination of Equation (32) shows that the efficiency is essentially determined by the ratio $\Delta S_R / \Delta H_R$, and for most fuel cell reactions ΔS_R and ΔH_R are of the same sign (<0 for the above case).

Theoretically, it might be possible to find a cell reaction for which ΔS_R and ΔH_R have opposite signs, in which case the theoretical efficiency, ε_C^E, would be greater than one! Of course, this does not mean that we have found electrochemistry's holy grail and that perpetual motion is just around the corner. Instead, it means that the enthalpy content of the surroundings must also be considered a "fuel." Thus, consider the (most likely) case where $\Delta S_R > 0$, corresponding to the case where a liquid fuel (perhaps) is converted within the cell to gaseous products, such that the total number of moles of gas in the products exceeds the number of moles of gaseous reactant (O_2). The greater-than-one efficiency is attributed to a transfer of heat from the surroundings. This additional "fuel" is not included in the definition of the efficiency, because it is tacitly assumed that the boundaries of the cell define the "system", in the thermodynamic sense. However, the intriguing possibility that ε_C^E is real, although the author is unaware of any case where this has been demonstrated experimentally.

It might be argued that, for some systems, the comparison between heat engines and electrochemical converters is unfair, because the outputs are not the same (mechanical energy for an ICE and electrical energy for a fuel cell). In order to correct for this disparity, the efficiency of an ICE should be expressed as

$$\varepsilon_C^E = \varepsilon_C^M \cdot \varepsilon_M^E = (1 - T_2/T_1) \, \varepsilon_M^E \tag{34}$$

and that of an electrochemical converter to

$$\varepsilon_C^M = \varepsilon_C^E \cdot \varepsilon_E^M = [1 - T(\Delta S_R / \Delta H_R)]\varepsilon_E^M \tag{35}$$

when comparing a fuel cell to an ICE. Thus, using the ICE as the basis, we write the ratio of the ideal efficiency for chemical-to-electrical energy conversion via an ICE/generator combination to that for direct conversion via a fuel cell as

$$\phi_C^E = (1 - T_2/T_1)\varepsilon_M^E /[1 - T(\Delta S_R/\Delta H_R)]\varepsilon_M^E \tag{36}$$

On the other hand, if we use an electrochemical energy converter as the basis, then we define the ratio of the ideal efficiency for chemical-to-mechanical energy conversion via a fuel cell relative to that for a heat engine as

$$\phi_C^M = [1 - T(\Delta S_R/\Delta H_R)]\varepsilon_E^M/(1 - T_2/T_1) \tag{37}$$

Typical values for ϕ_C^E and ϕ_C^M based on Reaction (34) are 0.5 and 1.5, respectively. The advantage of electrochemical systems for energy conversion is clear ($\phi_C^E < 1$ and $\phi_C^M > 1$). However, the fact remains that society chose heat engines over electrochemical energy converters, so that other factors must have dominated the "decision" (or, at least, those who made the decision were not electrochemists!). These additional factors, again, are a complex mix of technology, economics, and sociology, but important materials factors came (and still come) into play. However, the most important factor was the ready availability of cheap hydrocarbon fuels derived from petroleum.

Fuel Cells

Because a fuel cell uses an external source of fuel and oxidant, it is commonly defined as an electrochemical energy converted (ECEC) whose energy density is independent of the size of the cell. This definition differentiates a fuel cell from a primary battery, whose reactive materials (or at least one) are contained within the boundaries of the cell and which cannot be recharged. The most common "fuel cells" that fall within this definition include:

- Solid oxide cells
- Phosphoric acid cells
- Alkaline cells
- Polymer electrolyte membrane (PEM) cells
- "Primary" metal/air (e.g., A_1/air) and metal/water (e.g., Li/H_2O cells, and
- Molten carbonate cells

Fuel cells are, in fact, quite old as far as electrochemical energy conversion technologies are concerned, with cells having been devised by Groves during the last century. However, because of the highly attractive theoretical conversion efficiency and low emissions, fuel cells have been extensively explored for both stationary and mobile applications. Perhaps the best-known application was in the Gemini and Apollo space missions, with the explosion of a fuel cell in Apollo-13 being the event that most people remember. The particular fuel cell technology used in these missions (H_2/O_2 PEMs and alkaline cells) were ideal, because of their reliability (notwithstanding the explosion), ease of electrolyte management, and simplicity. Also, cost was no barrier. However, cost is an issue for both stationary and mobile commercial applications, and the cost is defined by the prices of conventional fuels (natural gas, oil, and coal) for existing technologies (e.g., thermal power plants). The currently low costs of these fuels, and the large investments that have been made in these facilities, discourage the introduction of large fuel cells (or any new technology, for that matter), but, even so, a number of multi-megawatt pilot and/or demonstration plants are now operating world-wide. The kW cost of these facilities is currently $3-5,000, but this capital cost must be reduced to < $1,000 in order to compete with conventional systems.

Now for the materials issues. All practical fuel cells that have been devised to date use hydrogen as the fuel (and O_2 as the oxidant) either directly (as in Gemini and Apollo) or indirectly (by reforming hydrocarbons or partially-oxidized hydrocarbons, such as methanol). The problem is that no electrocatalysts are available that will support the direct oxidation of hydrocarbons (e.g., CH_4) or oxy-hydrocarbons (e.g., CH_3OH) at the anode. Even the most "catalytic" materials, such as Pt and other platinum-group metals, are ineffective, at least for the near-ambient temperature systems (i.e. all of those listed above except the solid oxide system). Thus, in the present systems, any hydrocarbon-based fuel needs to be reformed to produce hydrogen, and the reformation facilities add greatly to the complexity (and hence the cost) of the system.

There appear to be two approaches to solve this problem, at least for near-ambient temperature systems: i) finding a better electrocatalyst, or, ii) devising higher temperature systems.

The prospects of finding better electrocatalysts would appear to be small but are nevertheless finite. The problem is that the oxidation of a hydrocarbon fuel (e.g., CH_4 or CH_3OH) is a complex reaction involving multiple adsorbed species. In electrochemical systems, such reactions are normally "slow", with the kinetic limitation residing in a step that is only weakly catalytic. Furthermore, despite a great deal of excellent basic research on the charge transfer and adsorption behaviors of model systems (e.g., CO and H on Pt), the detailed mechanisms of the electrocatalytic oxidation of hydrocarbons and hydrocarbon-derived fuels (e.g., CH_3OH) remain obscured.

The alternative approach of significantly increasing the temperature of an "low-temperature" cell (e.g., PEMs, phosphoric acid, alkaline) has not been extensively explored in a systematic manner, even though the phosphoric acid and alkaline systems commonly operate at "elevated" temperatures (in excess of 150°C). Of particular interest is the PEM cell, because of its inherent simplicity. The principal problem here is that the available membranes (e.g., sulfonated polystyrenes, such as NAFION®) dehydrate at temperatures in excess of 80-120°C. We estimate that temperatures in excess of 200-250°C will be required in order to achieve reasonable oxidation rates or methanol, even for the most electrocatalytic electrode surfaces. Thus, one requires a proton conducting membrane that will remain hydrated at this temperature, that will have acceptable mechanical properties, and that will have minimal carry-over of the fuel to the oxidant side of the membrane. This is a tall order, but it is one that is worth pursuing.

Not all of the materials problems are associated with the fuel side of the membrane. On the oxidant side, a myriad of problems exist, depending upon the type of cell and the materials employed. Common problems include the oxidation of carbon supports for the catalyst, the formation of hydroxyl radicals that oxidize organically-based membranes, and deactivation of the catalyst. These are all pressing problems, in that their solution is required in order for the technologies to become truly practical.

Two particularly attractive high temperature fuel cell technologies for stationary applications are solid oxide cells and molten carbonate cells. Solid oxide fuel cells (SOFCs) employ an oxygen ion conductor as an electrolyte (e.g., yttria-stabilized zirconia) and electronically conducting electrocatalysts on both the fuel side and the oxidant side of the membrane. Although the direct oxidation of a hydrocarbon is feasible, most variants of this technology employ reforming to produce a hydrogen-rich fuel. While this technology is attractive for a number of reasons, it suffers significant disadvantages, particularly when scaled-up to multi-megawatt units. These drawbacks include the use of brittle ceramic electrolytes that must separate otherwise explosive mixtures of fuel and oxidant, the susceptibility of ceramic materials to thermal shock, difficulties in matching coefficients of thermal expansion of electrolyte/electrocatalyst/current collector composites, and cost.

The final technology that will be discussed very briefly is the molten carbonate cell. This system employs a liquid eutectic mixture of Li_2CO_3 and K_2CO_3 as the electrolyte, with the anodic (fuel) reaction and oxidant (oxygen) reaction being formally represented as

$$H_2 + CO_3^{2-} \rightarrow CO_2 + H_2O + 2e^- \tag{38}$$

and

$$\tfrac{1}{2}O_2 + CO_2 + 2e^- \rightarrow CO_3^{2-} \tag{39}$$

respectively. The principal difficulty with this technology is the corrosiveness of the electrolyte; indeed, molten carbonates (like supercritical water and cryolite) are generally regarded as being "universal solvents," and few materials survive over extended exposure periods. Likewise, the erosion/corrosion of electrode materials is severe, and although pilot plants of up to a few megawatts have been operated, the materials problems must be solved if the cost per kilowatt is to be reduced from the current $3-5,000 to less than $1,000, at which the technology becomes economically attractive.

Batteries

"Storage" batteries date back to the earliest times of electrochemistry with the Volta pile possibly being the first "practical" battery. However, it was the invention and development of the lead-acid battery towards the end of the nineteenth century, and its subsequent use for SLI (starting, lighting, and ignition) in ICE-powered vehicles, that brought the secondary battery into the mainstream of commercial acitivity. The other development that catalyzed the development of secondary batteries (i.e. ones that can be electrically-recharged) was the rise of the portable electronics industry. Indeed, the current lack of a secondary battery that will allow the continuous operation of a portable computer on a five-hour transcontinental flight (i.e. across the US from east to west) is seen by many as a major goal to be achieved in the further development of portable electronic devices (e.g., computers and cellular telephones). The literature on the development of primary (non-electrically rechargeable) and secondary batteries is enormous, and no justice can be done to the field here. Instead, this review concentrates on a small number of relatively recent developments, with the view in mind of tracing the likely course of development in the future.

Before discussing some specific batteries, it is worth noting that the eventual application has a profound and determining influence over the likelihood that any given technology will be successful. A listing of these factors, as compiled by McBreen (15), is shown in Table 1. Note that cost is not always the determining

TABLE 1

Electrochemical Energy Converter Requirements for Various Applications [After McBreen (15)]

Civilian consumer market			Industrial market		Noncivilian market		Emerging markets	
Auto SLI	Electronic	Motor driven	Traction	Emergency power	Military	Aerospace	Electric vehicles	Load leveling
Cost	Sealed operation	Sealed operation	Cost	Calendar life	Specific energy	Specific energy	Cost	Cost
Specific power	Compactness	Specific power	Maintenance	Maintenance	Specific power	Specific power	Specific energy	Cycle life
Temperature range	Life	Compactness	Life	Cost	Compactness	Compactness	Specific power	Energy efficiency
Ruggedness	Cost	Life			Ruggedness	Sealed operation	Compactness	Temperature range
Life		Cost			Temperature range	Ruggedness	Ruggedness	Maintenance
					Life	Life	Cycle life	
							Temperature range	
							Energy efficiency	
							Maintenance	

factor, particularly for military and emergency applications, where the cost of downtime of the system being powered may far exceed the cost of any battery.

Batteries can be divided into two broad classes: i) Primary batteries, and ii) Secondary batteries. A classification of primary batteries, which are not electrically rechargeable, has been developed by Barak (16) and a modified version of this classification is reproduced in Table 2. Aqueous, non-aqueous, molten salt, and solid electrolyte systems are included, and the conditions under which the various cells operate are indicated. A similar classification for secondary batteries, which are electrically rechargeable, has been devised by McBreen (15) and an updated version of this classification is given in Table 3. Finally, an updated listing of various individual secondary battery technologies that was originally compiled by McBreen (15) is given in Table 4. The myriad of materials that are employed in these various technologies is evident, but it is to be noted that only a few of these systems will ever penetrate the market place. Thus, few high temperature batteries develop beyond the laboratory, because of the difficulty of consumer acceptance. Likewise, batteries that contain complicated auxiliary systems (e.g., pumps, heat exchangers, precipitators) are at a significant disadvantage compared with their "black-box" counterparts. Finally , many of these batteries have been proposed as power sources for electric vehicles, and hence they must compete with the ubiquitous internal combustion engine operating on inexpensive gasoline (at least in the US).

As noted above, the lead-acid battery was the standard-bearer in electrical energy storage technology for many decades, having found extensive use for SLI in automobiles, for starting and emergency power in aircraft, and in submarines and submersibles, to name only a few of the more common applications. Until the 1970s, lead-acid batteries were heavy and, while they exhibited good power densities (power per unit volume) and reasonable specific powers (power per unit mass), the energy density (energy per unit volume) and specific energy (energy per unit mass) were only marginal. However, the great attribute was their excellent cycling characteristics, in that cycle-lives of many hundreds or even thousands of deep discharge cycles were possible under ideal conditions. Since that time, remarkable developments have occurred in lead-acid battery technology, all of which can be traced to specific advances in materials. The first significant advance was the development of the light-weight, sealed "maintenance-free" battery, in which it was no longer necessary to vent gases on charging or to service the battery with electrolytes. These consumer-enhancing features were made possible by the development of alloyed electrodes and by the extensive use of plastics for the case and internal components. Subsequently, the development of gelled electrolytes, which enable lead-acid batteries to be used in any orientation and prevented acid spills, has led to the penetration of the lead-acid battery into the aircraft market as a competitor to nickel/cadmium. Thus, the "lead-acid" battery, which currently dominates the secondary battery market, has evolved significantly over the past three decades and many of these advances can be traced to the introduction of new materials. This evolution will continue into the future because the advanced lead/acid battery is a leading contender for the near-term electric vehicle market.

TABLE 2
Classification of Primary Batteries [After Barak (16)]

| Designation or application | Electrodes | | | Cell Voltage V |
	Cathode	Anode	Electrolyte	
1. Aqueous electrolyte-solutions of inorganic salts				
"Dry" Lechanché	MnO_2	Zn	$NH_4Cl/ZnCl_2$	1.50
Air depolarized	O_2	Zn	$NH_4Cl/ZnCl_2$	1.40
Torpedo W/A[a]	AgCl	Mg	NaCl; seawater	1.30
High power W/A	Cl_2[b]	Al alloy	$AlCl_3$ + NaCl; seawater	2.10
Radar-Sonde W/A N/D	Cu_2Cl_2	Mg	NaCl or KCl	1.20
Meteorological balloon W/A N/D	PbO_2	Mg	NaCl or KCl	1.80
2. Aqueous electrolyte-solutions of alkali or acid				
Alkaline Manganses "Premium" Leclanché	MnO_2	Zn	KOH	1.50
Military applications	MnO_2	Mg	KOH	1.65
Mercury-zinc, Rubens-Mallory	HgO	Zn	KOH	1.30
Mercury-indium	HgO	In/Bi	KOH	1.16
Air depolarized	O_2	Zn	KOH	1.40
Lalande-Chaperon N/D	CuO	Zn	NaOH	1.10
Torpedo: RD fuze	PbO_2	Pb	$HClO_4$ (reserve)	1.95
Lifeboat radio N/D	PbO_2	Zn	H_2SO_4 (reserve)	2.20
3. Metal/air and metal/water				
Aluminum/air	air	Al	KOH	1.6
Lithium/water	water	Li	LiOH	3.0
4. Nonaqueous electrolytes				
Panasonic high energy	$(CF_x)_n$	Li	Propylene carbonate with $LiClO_4$	2.80
Lithium	$SOCl_2$	Li	Thionyl chloride ($SOCl_2$) with $LiCl$-$AlCl_3$	3.60
Lithium	$SOCl_2$	Li	Acetonitrile or polypropylene carbonate with $LiClO_4$	2.95
Lithium	CuO	Li	Propylene carbonate with $LiClO_4$	2.35
5. Solid electrolyte				
Iodide cell	RbI_3	Ag	Rubidium silver iodide ($RbAg_4I_5$)	0.66
Note: Some of the systems in groups 3 and 4 are still in the research and development stage.				
6. Molten salt electrolyte				
Thermal batteries	Chromates	Ca or	KCl + LiCl	
Military applications	Fe_2O_3	Mg or Li	300 to 600°C	0.80
7. Ammonia vapor activated (AVA)				
Military applications	V_2O_5 or	Mg or	(NH_3)	
	PbO_2 or	Zn	KCNS or LiCNS	1.50
	$mDNB$[d]	Mg		
8. Standard cells-aqueous electrolytes				
Clark cell	Hg	Zn	Sat. aq. $HgSO_4$, plus $ZnSO_4$	1.457
Weston cell	Hg	Cd	Sat. aq. $HgSO_4$ plus $CdSO_4$	1.018

[a] Water activated
[b] Trichlorotriazinetrione.
[c] No longer in demand
[d] Metadinitrobenzene.

Other secondary batteries are rapidly penetrating the market place. These range from the nickel/hydrogen system that is used in low-earth-orbit satellites, nickel/cadmium batteries for use in consumer electronics, aircraft, satellites and a myriad of other applications, and nickel/metal hydride batteries that are now emerging as a competitor for NiCads. All three systems have benefited from significant advances in materials science and engineering with the pay back being higher energy and power densities and longer cycle-life. For example, a few years ago nickel/cadmium batteries had cycle-lives that were measured in tens or hundreds of cycles, depending upon how they were discharged and

TABLE 3

Classification of Secondary Batteries [After McBreen (15)]

Current battery technology

    (a) Lead-acid battery
    (b) Alkaline electrolyte batteries
        Iron-nickel oxide
        Cadmium-nickel oxide
        Zinc-silver oxide
        Cadmium-silver oxide

Emerging battery technologies

Aqueous electrolyte batteries
(1) Zinc batteries
    (a) Alkaline electrolye batteries
        Zinc-nickel oxide
        Zinc-manganese dioxide
    (b) Zinc-halogen batteries
        Zinc-chlorine
        Zinc-bromine
    (c) Aluminum-sulfur
(2) Batteries with one gaseous reactant
    (a) Metal-air (oxygen) batteries
        Zinc-air (oxygen)
        Iron-air (oxygen)
        Cadmium-air (oxygen)
    (b) Hydrogen-metal oxide batteries
        Hydrogen-nickel oxide
        Hydrogen-silver oxide
    (c) Nickel-metal hydride
(3) Batteries with fluid reactants
    (a) Hydrogen-oxygen
    (b) Hydrogen-halogen batteries
        Hydrogen-chlorine
        Hydrogen-bromine
    (c) Redox batteries
        Titanium-iron
        Chromium-iron

Nonaqueous electrolyte batteries
(1) Organic electrolyte batteries
    (a) Lithium-bromine batteries
    (b) Lithium-metal chalcogenide batteries
        Lithium-titanium disulfide
        Lithium-niobium diselenide
(2) High-temperature batteries
    (a) Sodium batteries
        Sodium-sulfur
        Sodium chloroaluminate batteries
    (b) Lithium batteries
        Lithium-iron disulfide
        Lithium-chlorine
        Lithium charge storage batteries
(3) Solid polymer electrolyte batteries

TABLE 4
Characteristics of Various Secondary Batteries [After McBreen(15)]

Battery type and system	Cell voltage, V	Theoretical specific energy, Wh/kg	Operating temperature °C	Specific energy 1-hr rate, Wh/kg	Specific energy 5-hr rate, Wh/kg	Energy efficiency, % approx.	Cycle life, approx.	Calendar life, years
			Current battery technology					
Pb-H$_2$SO$_4$-PbO$_2$	2.1	184	-35-70	20-30	30-40	60	500	3-5
Cd-KOH-NiOOH	1.35	218	-40-65	30-40	40-50	75	>1000	5-15
Cd-KOH-Ag$_2$O$_2$	1.41, 1.15	269	-40-80	50-60	65-75	70	700	3-5
Fe-KOH-NiOOH	1.37	267	-40-80	30-35	35-45	50	>1000	>10
Zn-KOHAg$_2$O$_2$	1.86, 1.60	490	-40-55	100-110	85-95	70	500	3-5
			Emerging battery technology					
Ambient temperature batteries								
Zinc Batteries								
(a) Alkaline zinc batteries								
Zn-KOH-NiOOH	1.7	341	-40-65	50-60	65-75	75	300	3-5
Zn-KOH-MnO$_2$	1.55, 0.95	484	-40-65	50-60	50-60	75	200	3-5
(b) Zinc-halogen batteries								
Zn-ZnCl$_2$-Cl$_2$	2.1	826	15-60	60-80	75-90	55-65	100	N.A.*
Zn-ZnCl$_2$-Cl$_2$•6H$_2$O	2.1	460	15-60	60-80	75-90	55-65	100	N.A.
Zn-ZnBr$_2$-Br$_2$	1.8	428	0-50	50-60	65-75	75	300	>3
Zn-ZnBr$_2$-(CH$_3$)$_4$NClO$_4$Br$_9$	1.8	364	0-50	50-60	65-75	75	300	>3
Batteries with one gaseous reactant								
(a) Metal-air (oxygen) batteries								
Zn-KOH-air (O$_2$)	1.6	1054	25-65	100	110-170	<50	500	N.A.
Fe-KOH-air (O$_2$)	1.2	716	25-60	55-65	80-90	<30	>1000	>3
Cd-KOH-air (O$_2$)	1.18	432	0-80	60-70	80-90	<40	500	N.A.
(b) Hydrogen-metal oxide batteries								
H$_2$-KOH-NiOOH	1.36	393	-20-65	60-70	70-80	65	>1000	>3
H$_2$-KOH-Ag$_2$O$_2$	1.38, 1.12	532	-20-80	70-80	80-90	60	700	3
Ni-metal hydride	1.4	>300	0-65	90	100	-	>1000	>3

TABLE 4 *Continued*

Battery type and system	Cell voltage, V	Theoretical specific energy, Wh/kg	Operating temperature °C	Specific energy		Energy efficiency, % approx.	Cycle life, approx.	Calendar life, years
				1-hr rate, Wh/kg	5-hr rate, Wh/kg			
Batteries with fluid reactants								
H_2-KOH-O_2	1.23	3663	-20-180	-	-	50	500	>5
He-HCl-Cl_2	1.36	1000	0-100	75-90	75-90	80	-	-
H_2-HBr-Br_2	1.07	354	0-100	45-60	45-60	75	-	-
$TiCl_2$-$FeCl_3$	0.61	58	0-100	<25	<25	70	-	-
$CrCl_2$-$FeCl_3$	1.20	122	0-100	<30	<30	70	-	-
Lithium batteries								
Li-PC-TiS_2	2.15	484	0-70	50-60	110-120	60-75	-	>3
Li-Al-PC-TiS_2	1.85	339	0-70	40-50	80-90	60-75	100	>3
Li-PC-$NbSe_3$			0-70	40-50	80-90	60-75	150	>3
Li-$PCOBr_2$	4.05	1249		40-50	80-90	60-75	1000	>3
Li-SPE-TiS_2	3.2-1.8	>400	25-100	<100	120-180	-	100	-
Li-SPE-MnO_2	3.2-1.8	>400	25-100	<100	120-180	-	100	-
Li-SPE-V_6O_{13}	3.2-1.8	>400	25-100	<100	120-180	-	100	-
High temperature batteries								
Li-LiCl-KCl-FeS_2	2.1, 1.6	1342	400-450	-	-	70	100	-
Li-LiCl-KCl-FeS	1.6	842	400-450	-	-	70	-	-
Li-Al-LiCl-KCl-FeS_2	1.8, 1.3	650	400-450	80-100	140-160	70	300	1
Li-Al-LiCl-KCl-FeS	1.3	447	400-450	-	-	-	120	-
Li_4Si-LiCl-KCl-FeS_2	2.05-1.26	944	400-450	-	-	-	-	-
Li_4Si-LiCl-KCl-FeS	1.55-1.26	637	400-450	-	-	-	-	-
Na-β-Al_2O_3-S	2.08-1.75	763	300-400	80-90	100	70	-	N.A.
Na-β-Al_2O_3-S	2.08-1.75	521	300-400	-	-	-	-	-
Na-β-Al_2O_3-S-Na_2S_{s2}	2.08-1.75	308	300-400	-	-	-	-	-
Na-β-Al_2O_3-S-NaAlCl_4-M_xCl_y	2.7-3.2	792-1034	210	-	-	-	-	-
Li-Al-LiCl-KCl-C-$TeCl_4$	N.A.	N.A.	400	-	-	-	-	-
Li-LiCl-KCl-LiF-Cl_2	3.6	2275	450	150	200	75	700	0.1

*N.A. = not available.

charged, but recent systems have cycle-lives of thousands or tens of thousands of cycles. The quest for higher cycle-life is the principal driving force in the development of a secondary battery, particularly towards the end of the development cycle. This is so, because a longer cycle-life relieves the pressure on cost and enhances consumer acceptability.

One other battery system deserves specific mention and that is the secondary lithium battery. Even the most cursory analysis shows that lithium batteries would revolutionize the portable electronics market if a practical technology can be developed. The manner in which this would happen can be judged from the "transcontinental airplane test." Thus, at the current time, no battery is available that will allow the continuous operation of a modern lap-top computer on an airplane flight from New York to Los Angeles without recharging. Even the best NiCads and nickel/metal hydride batteries don't meet this test, much to the chagrin of the busy executive. Similar tests can be formulated for cellular telephones, portable TVs, and other portable electronic devices.

Primary lithium batteries have been available for many years, and are commonly used to power watches and other small electronic devices. They were also used extensively by the military, and despite the occasional explosion, they have served very well. However, consumers demand secondary batteries that can be electrically-recharged from household outlets. This demand has led to extensive efforts to develop rechargeable lithium batteries, with most efforts being centered on the solid polymer electrolyte systems. In these batteries, the anode (lithium metal or lithiated carbon) is separated from the cathode (an intercalation compound with desirable electrochemical characteristics) by a polymer that conducts lithium ions (Figure 15). Early batteries employed polyethylene oxide (PEO)/LiO_3SCF_3 (lithium trifluoromethane sulfonate, or "triflate") as the electrolyte in structures of the type

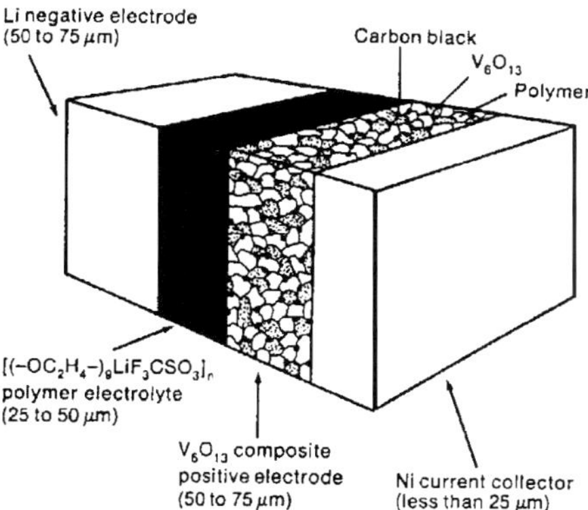

FIGURE 15. Schematic structure of solid polymer electrolyte battery. Adapted from D.F. Shriver and G.C. Farrington, *Chem. Eng. News,* May 20, p.53 (1985).

$$\text{Li/PEO(Li Triflate)IC} \tag{41}$$

where IC is an intercalation cathode (e.g., TiS_2, V_6O_{13}, MnO_2), but the high resistivity of the electrolyte dictated that useful power could only be obtained at elevated temperatures (80+.°C). This temperature is too high for the consumer electronics market. More recently (past decade), other polymer electrolytes have been devised (e.g., MEEP, polymethyl methacrylates), and batteries that provide useful power at ambient temperature have been developed. One great advantage of these batteries is the great flexibility in packaging that is afforded by the thin film, solid polymer electrolyte structure. Two such configurations are shown in Figure 16. The two remaining issues appear to be the safety of lithium and the poor cycle-life. Although lithium is probably no more dangerous than gasoline, a few lithium fires have been sufficient to scare liability- conscious developers away from lithium metal and towards graphitized carbon as an intercalation anode. The price paid is a reduction in the energy density of the anode (although this is less than it may seem because of the need to use excess lithium in the case of the metal anode). Interestingly, a fire has also occurred in a battery with an intercalated carbon anode, under circumstances where the battery was electrically "abused." The impact of that event on the future of the lithium ion/SPE (solid polymer electrolyte) battery is yet unknown.

As noted earlier in this review, the cycle-life tends to be enhanced towards the end of the development cycle of a secondary battery, and we expect the lithium/SPE batteries not to be an exception. Currently, cycle-lives of the order of 100 to 150 cycles have been achieved in the laboratory, but lives of > 1000 are required for most consumer-oriented applications. The cause(s) of the limited

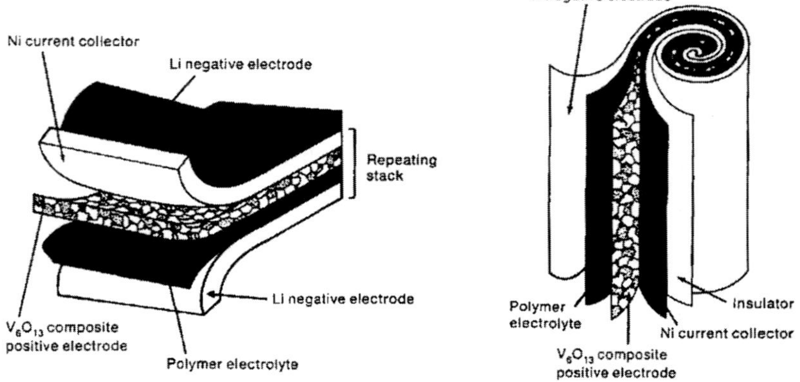

FIGURE 16. Two typical configurations of solid polymer electrolyte battery. (a) Bi-polar structure (b) "Swiss role" configuration. Adapted from D.F. Shriver and G.C. Farrington, *Chem. Eng. News*, May 20, p. 53 (1985).

cycle lives is (are) obscure, but it (they) appears to be related to the powdering of the lithium at the anode and to the irreversible restructuring of the intercalation cathode. Either or both could lead to the sudden increase in resistance, which is the normal mode of failure.

THE FUTURE

Because of space limitations, only a few of the exciting areas of electrochemistry have been addressed and even these were done so in a most inadequate manner. The topics chosen for discussion above illustrate the very wide penetration of electrochemistry into our everyday lives and the impact that this discipline has on the economic structure of our society. Looking ahead, it is possible to identify areas where electrochemical science and technology will likely have a profound impact on the way we live. Some of these areas are as follows:

- Electric vehicles (EVs) — their initial introduction will be legislated, but in the long term (100+ years), waning petroleum resources being replaced by nuclear (possibly fusion) electrical power generation will make EVS commonplace.
- Plastic batteries — light-weight and flexible, they will operate with proton-conducting electrolytes and protolytic anodes and cathodes.
- Molecular/electrochemical computers — data storage and manipulation will occur at nonometer dimensions using charge transfer processes in highly conjugated systems (50+ years).
- Artificial nerves — nerves are ionically-conducting channels that transmit electrical impulses from sensory points to the central processing unit (the "brain") of a living organism. In humans, nerves are non-regenerative and are impossible to transplant. However, ionically conducting polymer fibers, in which ions move through longitudinal channels, are on the far horizon (50+ years) and may be suitable for artificial nerve applications. A major problem will be impedance matching.
- Displays — electrochemical "flat panel" displays are already under active development, but they lack longevity, switching speed, and definition (20+ years). Major advances are likely in this area as the mechanisms of switching and degradation are understood. For example, it is generally accepted that the switching speed is determined by the rate at which a counter ion (e.g., an anion) can move to compensate the local charge, and there is little that can be done to speed up this process. However, it may be possible to synthesize electrochemically-activated chromophores in which charge compensation is achieved by concerted proton displacement along a conjugated chain. This would greatly increase the switching speed.
- New electronic devices will be synthesized by the direct electrochemical oxidation of designed substrate alloys (10+ years). Controlled segregation of alloying elements into passive films will permit specific electronic properties

to be achieved, through a balance between defects and alloying elements as dopants. At this time, passive films have been synthesized that exhibit remarkable electronic properties, including "super-slow" electronic relaxation times (>100s) for films as thin as 10 nm, and whose electronic properties may be altered by electromagnetic radiation.

- New, optically active molecules (particularly biologically-active systems) can be expected from electrochemical synthesis in the future. Unlike homogeneous chemical reactions, which allow equal probability of a reacting entity to either side of an asymetric center, heterogeneous electrochemical reactions at surfaces can be arranged to allow the entering group to approach from one side only, thereby producing a chiral center. Although this technology has been known for a long time, the full potential of electrochemistry as a synthetic route for biologically-active compounds has yet to be fully realized.
- Electronics/biological systems interface (100+ years). In these systems, an electrochemical interface with the nervous system will allow the "reprogramming" of the brain, the correction of nervous system disorders, and the direct control of machines through "thought" processes.

The above are but a few of the opportunities that one can identify in electrochemistry of the future, but these are surely only a few of those that are obvious at this moment. As in all areas of science, revolution is always just around the corner.

REFERENCES

1. Bockris, J.O'M. and A.K.N. Reddy. 1973. *Modern Electrochemistry,* Plenum Press, New York, N. Y.

2. Bard, A.J. and L. Faulkner. 1980. *Electrochemical Methods,* John Wiley & Sons, New York, N. Y.

3. Macdonald, D.D. 1977. *Transient Techniques in Electrochemistry,* Plenum Press, N.Y.

4. Vetter, K.J. 1967. *Electrochemical Kinetics,* Academic Press, New York, N.Y.

5. Sato, N. 1978. in *Passivity of Metals,* (Ed. R.P. Frankenthal and J. Kruger), The Electrochem. Soc., Princeton, N. J.

6. Sneddon, D.N., Sabel, D.M. and A.A. Gewirth. 1995. *J. Electrochem. Soc., 142,* 3027.

7. Vitus, C.M. and A.J. Davenport. 1994. *J. Electrochem. Soc., 141,* 1291

8. Bhardwaj, R.C., Gonzalez-Martin, A. and J.O'M. Bockris. 1991. *J. Electrochem. Soc., 138,* 1901.

9. Macdonald, D.D. 1992. *J. Electrochem. Soc., 139,* 3434.

10. Sato, N. and Noda. 1973. *Japan Inst. Met., 37,* 951.

11. Macdonald, D.D., Liu, C., Urquidi-Macdonald, M., Stickford, G., Hindiu, B. and A.K. Agrawa. 1994. *Corrosion, 50,* 761.

12. Haupin, W.E. and W.B. Frank. 1981. "Electrometallurgy of Aluminum", in Compr. Treat. Electrochem. (Ed. J. O'M. Bockris, B.E. Conway, E. Yeager,

and R.E. White), Plenum Press, N.Y., *2*, 301.

13. Cairns, E.J. and E.H. Hietbrink. 1981."Electrochemical Power for Transportation", in *Compr. Treat. Electrochem* (Ed. J. O'M. Bockris, B.E. Conway, E. Yeager, and R.E. White), Plenum Press, N. Y., *3*, 421.

14. Kordesch, K. 1981. "Electrochemical Energy Storage", in *Compr. Treat. Electrochem.* (Ed. J. O'M. Bockris, B.E. Conway, E. Yeager, and R.E. White), Plenum Press, N. Y., *3*, 123.

15. McBreen, J. 1981. "Secondary Batteries-Introduction", in *Compr. Treat. Electrochem.* (Ed. J. O'M. Bockris, B.E. Conway, E. Yeager, and R.E. White), Plenum Press, N. Y., *3*, 303.

16. Barak., M. 1981. "Primary Batteries-Introduction", in *Compr. Treat. Electrochem.* (Ed. J. O'M. Bockris, B.E. Conway, E. Yeager and R.E. White), Plenum Press, N.Y., *3*, 191.

Chapter Eight

DIFFUSION IN OXIDE SYSTEMS

V. S. STUBICAN

Department of Materials Science and Engineering
Steidle Building
Pennsylvania State University
University Park, PA 16802

INTRODUCTION

Solid state diffusion is the migration of atoms, ions and point defects in a solid. Migration of these species can be caused by a concentration gradient, a thermal gradient or an electrical gradient. Diffusion is a kinetic process which can be described quantitatively by the diffusion coefficient, D, or diffusivity. In the solid state the making and breaking of the chemical bond almost always requires some diffusive motion by the ions or atoms. As a result the energetics and kinetics of diffusion controls the rate of nearly all solid state reactions.

Mass transport of solids is of considerable practical and economical importance. Furthermore, the field of study of diffusion brings a very interesting scientific challenge. Diffusion in the inorganic materials was extensively studied in metals as well as in non-metallic materials. In general it can be stated that the results obtained with metals are more reliable than with the non-metallic materials.

The purpose of this review article is to critically describe some results of diffusion studies on stoichiometric and nonstoichiometric oxides.

MECHANISMS OF DIFFUSION IN OXIDES

Several recent books describe extensively mechanisms of diffusion in solids(1). In oxides there are three basic mechanisms by which diffusion takes place: (i) vacancy; (ii) interstitial and (iii) interstitialcy mechanism. If the predominant point defects in an oxide are vacancies they provide a path for diffusion and ions jump to vacant lattice positions. If the predominant defects are interstitial ions diffusion will occur along interstices of a crystal. A variant of this mechanism is if an interstitial ion "pushes" an ion on the regular lattice site into an interstitial position, taking its place.

If we consider diffusion of atoms as an uncorrelated process then the mean square displacement of atoms is

$$\overline{R^2} = n\alpha^2 \tag{1}$$

where n is the number of jumps which occur during certain times and α is the jump distance. According to Einstein the mean square displacement is related to the random diffusion coefficient by the equation

$$\overline{R^2} = 6D_r t \tag{2}$$

However, if diffusion coefficients are measured by the tracer technique diffusion of the tracer is no longer completely uncorrelated. The diffusion coefficient of tracer atom/ion D* is given by

$$D^* = fD_r \tag{3}$$

where f is the correlation factor and depends on the diffusion mechanism and the crystal structure. By determining the value of the correlation factor it is possible to distinguish between different mechanisms (vacancy, interstitial, interstitialcy). For the vacancy mechanism f=0.3333-0.7815 for the interstitialcy f=0.6666-0.9855 and for the interstitial mechanism f=1. There are two methods of determining the value of f: (i) to study the isotope effect, i.e. by comparing diffusion rates of different isotopes with different masses; ii) by comparing measured tracer diffusion coefficients with that evaluated from the ionic conductivity.

Diffusion is a thermally activated process and temperature dependence of Dr for vacancy diffusion is given by the expression

$$D_r = \alpha\lambda^2\upsilon\exp[(\Delta S_d + \Delta S_m) / k]\exp[-(\Delta H_d + \Delta H_m) / kT] \tag{4}$$

where α is a geometrical factor, λ is the jump distance, υ is the Debye frequency, ΔS_d and ΔS_m are entropies of formation and of motion of vacancies respectively, ΔH_d and ΔH_m are enthalpies of formation and motion of vacancies and k is the Boltzmann constant. Empirically it was found that regardless of the diffusion mechanism temperature the dependence of D can be described by the Arrhenius type equation

$$D = D_o \exp(-Q/kT) \text{ in } cm^2/s \tag{5}$$

where D_o is called pre-exponential factor and Q activation energy.

The activation energy Q is equal to $\Delta H_d + \Delta H_m$ if diffusion is controlled by the thermally created defects (intrinsic region). However, if experiments are carried out under such conditions that aliovalent impurities determine the vacancy concentration then $Q=\Delta H_m$ (extrinsic region). For interstitial and interstitialcy diffusion $Q=\Delta H_m$.

LATTICE DIFFUSION IN OXIDE SYSTEMS

The field of diffusion in oxides is very broad and many data have been collected in the last forty years(2). Most of these data will not be repeated here, rather some of the typical examples will be discussed.

The most common method for studying selfdiffusion in oxides are those employing isotopes. For the determination of diffusion profiles (concentration-distance curves) several methods are used such as sectioning technique, ion probe and microprobe, SIMS, electron probe microanalysis, photoelectron spectroscopy, etc. Using solutions of the second Fick's law for different boundary conditions diffusion coefficients can be calculated. For diffusion of impurities radioactive tracers or nonradioactive ions can be used.

Diffusion in high melting stoichiometric oxides was measured and many discrepancies in the data were observed. Typical example is MgO. Crystals of MgO contain impurities such as Ca, Fe, Na, Al and Si. The MgO crystals are not available in very pure form i.e. with the concentration of aliovalent impurities less then ~50-80 ppm by weight. Harding and Price(3) who studied selfdiffusion of an Mg-isotope in MgO observed a break in the Arrhenius plot at $\approx 2170K$. They claimed that below this temperature extrinsic diffusion occurs with $D_o=7.48 \times 10^{-6}$ cm^2/s and Q=1.57eV. Above this temperature intrinsic diffusion is dominant with $D_o=7.43 \times 10^{-2}$ cm^2/s, and Q=3.47eV. These results could not be repeated by Wuensch et al(4). According to these authors there is no break in the Arrhenius plot, and only extrinsic (impurity controlled) diffusion is observed in the temperature region 1273-2673K with $D_o=4.10^{-4}$ cm²/s and Q=2.76eV.

Earlier results of Paladino and Kinngery(5) on selfdiffusion of an Al-tracer in Al_2O_3 also indicate that diffusion is impurity controlled. With high melting stoichiometric oxides such as MgO, Al_2O_3, ZrO_2, $MgAl_2O_4$ etc. a few parts per million of aliovalent impurities is sufficient to cause mass transport to be impurity controlled (extrinsic) up to the melting point. This is due to the relatively small amount of thermally generated point defects. Diffusion rates of oxygen ions in the high melting stoichiometric oxides are usually ~3-5 orders of magnitude smaller at 1500K than the selfdiffusion rates of cations(6).

There have been numerous studies of impurity diffusion in oxides but few systematic investigations have been carried on. Wuensch and Vasilos(7) found that the activation energy for diffusion of divalent cations in MgO increases exponentially with the ratio ionic radius/ionic electronic polarizability. But the trend is not universal as the results for Be do not fit this picture. According to

Peterson(8) in NiO and CoO impurities with small ionic radii diffuse more slowly with a larger activation energy than impurities with large ionic radii.

It seems, however, that relationships between various parameters for impurity diffusion is not simple. It is difficult to compare diffusivities of impurity ions in a lattice without taking into consideration bonding energies between impurity ions and vacancies(9,10). The effect of impurity charge has not been systematically investigated. This effect can be quite strong as shown on Figure 1. Trivalent ion Cr^{3+} in a divalent matrix, MgO, has a rather large impurity-vacancy bonding energy which seems to reduce Mg ion-vacancy exchange rate in the neighborhood of the Cr- ion. Large activation energy and low diffusivity rates for Cr^{3+} -ion are observed(11).

Nonstoichiometric phases are found in many oxide systems, especially in oxides in which the cations can exist in several oxidation states. Non-stoichrometric systems are thus found in the transition metal, rare earth, and actinide oxides. With the large concentration of defects present even at moderate deviation from stoichiometry appreciable interaction between the defects occurs. Thus clustering, long-range ordering and crystallographic shearing mechanism must be taken into consideration to understand diffusion behavior of these

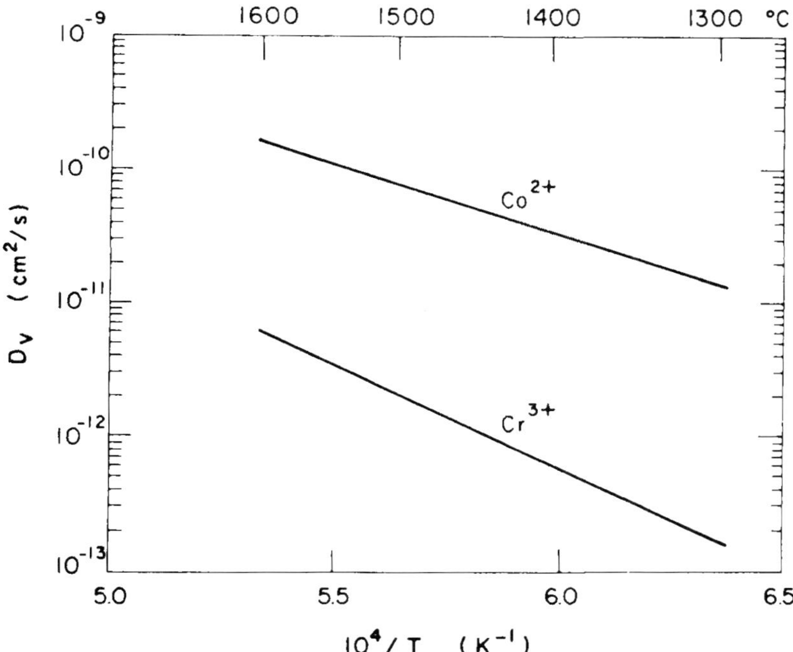

FIGURE 1. Arrhenius plots for volume diffusion of Co^{2+} and Cr_{3+} ions in MgO. Ref. 11.

oxides. Diffusion in nonstoichiometric oxides described quite extensively by Matzke(12). Here we will show one example where the diffusion mechanism can be explained in terms of noninteracting point defects and one example where appreciable interaction of defects strongly influences diffusion behavior.

Dieckmann and Schmalzried(13) measured D^*_{Fe} in Fe_3O_4. The plot of log D^*_{Fe} vs. log P_{o_2} shows a minimum as seen in Figure 2. These results can be explained in terms of noninteracting point defects. At low P_{o_2} the intrinsic defects are Frenkel pairs, thus iron diffusion is dominated by the more mobile interstitial ions. As P_{o_2} increases and vacancies form on the iron sublattice the iron interstitial ions concentration decreases, causing D^*_{Fe} to decrease. However, as P_{o_2} continues to increase, a point is reached beyond which the vacancy component completely dominates the diffusion process; from that point on, D^*_{Fe} increases with increasing P_{o_2}. The results on Figure 2 show a slope ∂ log D^*_{Fe}/∂ log P_{o_2} = -2/3 at low P_{o_2}, and at high P_{o_2} a slope near + 2/3 is observed which is entirely consistent with the defect model.

Extensive diffusion data exist for wustite $Fe_{1-x}O$ where x can be as large as 0.15. The data of Chen and Peterson (14) are considered the most reliable because of the technique used and the high quality of their single crystals. Their results of

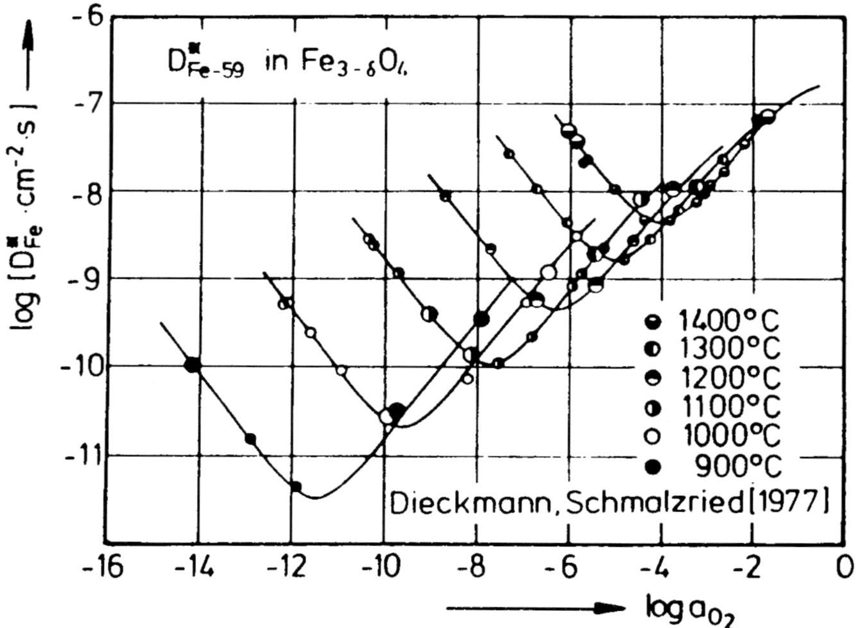

FIGURE 2. Tracer diffusion coefficients of iron in magnetite as a function of oxygen activity and temperature. Ref. 13.

selfdiffusion are shown on Figure 3 and were explained as follows: At low temperatures (e.g. 802°C) D^*_{Fe} decreases with x because an increased number of vacancies form immobile clusters, whereas at high temperatures (e.g. 1200°C) D^*_{Fe} tends to increase with x since the thermal energy kT decreases the number of vacancies which form immobile clusters. Small and large type clusters were detected in $Fe_{1-x}O$ using neutron defraction and x-ray techniques(15,16).

Existing diffusion data on nonstoichiometric oxides (e.g. TiO_2, UO_2, PuO_2, etc.) where ordering, formation of clusters or shear phases occur are often inconsistent. This is due to the difficulty in obtaining well defined specimens and achieving equilibrium structures. Similar difficulties are present with highly doped fluorite type materials such as ZrO_2, HfO_2 and CeO_2.

DIFFUSION ALONG GRAIN BOUNDARIES

The transport of ions along grain boundaries and dislocations was studied intensively because of the scientific and technical importance. Atkinson and

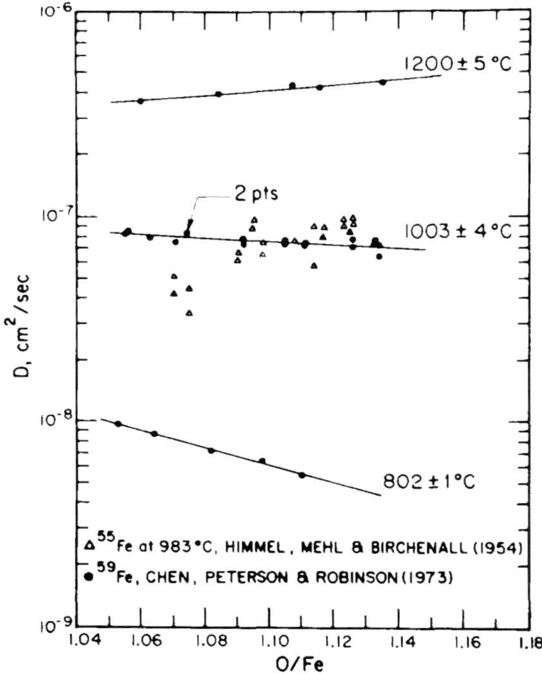

FIGURE 3. LogDv versus O/Fe for cation self-diffusion in $Fe_{1-x}O$. Ref. 14.

Monty(17) gave an appraisal of the field from early days to the year 1988. The purpose of this contribution is to draw some general conclusions from existing data. There are some difficulties to obtain a general picture of the mechanisms of diffusion in grain boundaries of oxides because of impurity effects.

Techniques for the measurement of diffusion at grain boundaries are direct or indirect, (i.e. sintering or creep measurements). Direct method involves monitoring the motion of tracer of lattice or impurity ions along the interfaces in question. For the direct measurements of the grain boundary diffusion bicrystals or dense polycrystalline specimens are used. Concentration profiles are determined by mechanical serial sectioning or more modern techniques which are also used for determination of lattice diffusion.

The solution of the diffusion equations for the analysis of grain boundary diffusion data have been discussed in detail by LeClair(18). In almost all phenomenological models the interface is treated as a slab of different materials having a width, δ , and the grain boundary diffusion coefficient D' larger than the volume (lattice) diffusion coefficient D_v. Two solutions for the grain boundary diffusion are mostly used. The thin film Suzuoka's solution and Whipple's thick film solution under most experimental conditions yield results within 10% of each other. Both solutions give values for D'. δ where δ is the effective grain boundary thickness which is of the order of ≈ 1nm(17). For solute diffusion (impurity diffusion) a solute segregation factor α should be considered so that in this case both solutions give values for $\alpha D'\delta$ which is taken as the grain boundary diffusion parameter. Although there are inconsistencies in the experimental data on grain boundary diffusion in oxides results obtained show some similarity to grain boundary diffusion in metals. Generally it can be concluded: (i) In most of the systems studied the D'/D_v ratio is in the range 10^3-10^5. (ii) The activation energy for grain boundary diffusion is 0.5 to 0.8 times that for volume diffusion. (iii) Diffusion in single grain boundary is anisotropic. (iv) The mechanism of grain boundary diffusion is a vacancy or an interstitial one. (v) Grain boundaries fabricated in different ways may display different diffusivity, which may be due to different impurity concentrations or to different boundary structure.

The enhanced diffusion of Ni and oxygen ions in grain boundaries and dislocations of NiO is clearly evident in Figure 4. Furthermore, grain boundary diffusivity of Ni-ion is several orders of magnitude higher than the diffusivity of O-ions. Experimental values for activation energies for grain boundary diffusion and for lattice diffusion for Ni-ion are $Q'=1.78$eV and $Q_v=2.56$eV($Q'/Q_v=0.70$) and for O-ion $Q'=2.56$eV and $Q_v=5.6$eV ($Q'/Q_v=0.45$). Theoretical calculations of these parameters are in reasonable agreement with the experimental results (19).

Chen and Peterson(20) have studied the diffusion of impurities Co^{2+} amd Cr^{3+} along grain boundaries in bicrystals and large grain size (>500μm) polycrystalline NiO. Diffusion of both tracers was significantly enhanced as compared with bulk diffusion, however, the grain boundary activation energy was almost the same as that for volume diffusion (~2.5eV). Chen and Peterson suggested that this high activation energy may result from a negative binding enthalpy between the cobalt or chromium ions and the grain boundary.

Barbier et al (21) studied a wide range of bicrystals of NiO and could not detect rapid diffusion of Co in grain boundaries. According to these authors D' is less than $10^3 D_v$ and not of the order of $10^5 D_v$ as previously reported(22). Lack of reproducibility between various studies mentioned above may be due to various factors such as impurities effects or different boundary structure.

The influence of point defects (cation vacancies) was first studied by using MgO bicrystals doped with the increasing concentration of Cr^{3+} ions(23,24). It was found that diffusivity product for the grain boundary diffusion of Cr-isotope

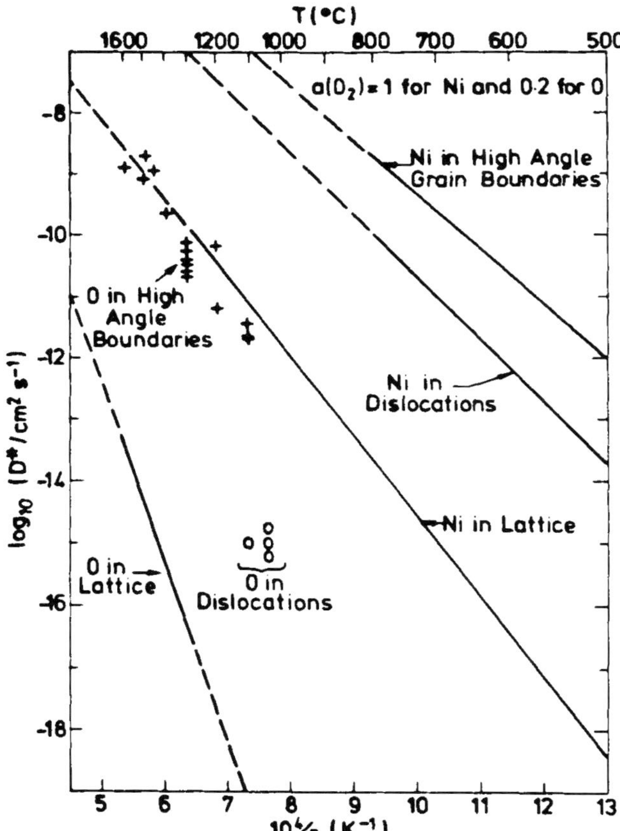

FIGURE 4. Diffusion of Ni and O in NiO. The grain boundary diffusion coefficients are given assuming $\delta=1$nm. Ref. 17.

increased with the increase in the dopant concentration. Further proof of the influence of point defects on the grain boundary diffusivity was given by Atkinson and Taylor(22). They showed significant increases in the grain boundary diffusivity of Ni and Co isotopes in NiO when the partial pressure of oxygen was increased (increase in the metal vacancy concentration). Perinet et al(25) have measured oxygen selfdiffusion in the lattice and in the grain boundaries of $Cu_{2-x}O$. Their results indicate the main defects responsible for the oxygen diffusion are neutral oxygen interstitials. A very clear example of the influence of point defects on the grain boundary diffusion in oxides was obtained by Carinci and Stubican(26) who measured grain boundary diffusivity of Co-isotope in Fe_3O_4 as a function of the partial pressure of oxygen. Two distinct regions, shown in Figure 5 were observed as in the lattice diffusion(13). It could be concluded that similar diffusion mechanisms operate during grain boundary and lattice diffusion.

The enhancement of diffusion in pure grain boundaries can be explained by the higher concentration of point defects and/or higher mobility of defects in the grain boundaries than in the lattice. To understand the mechanism of grain

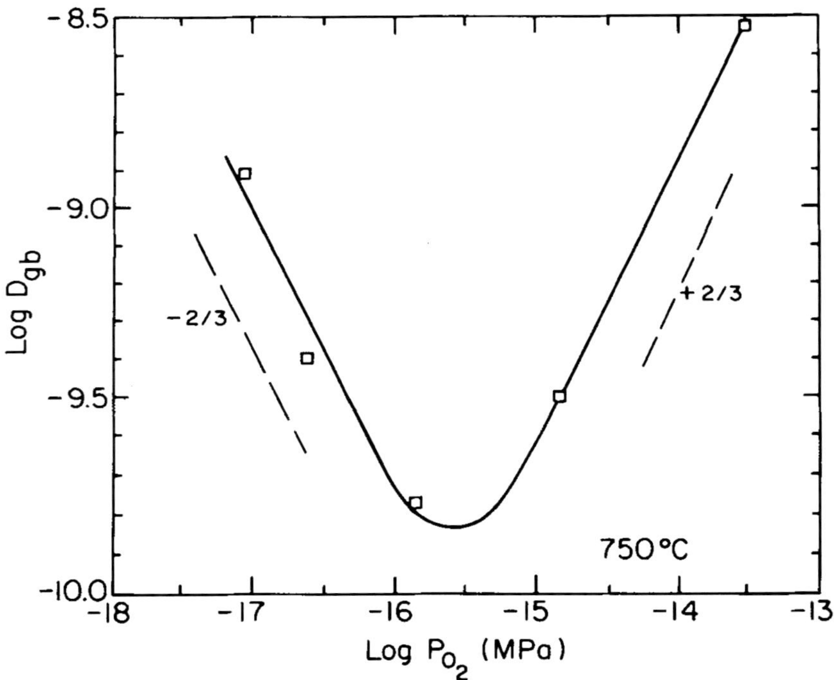

FIGURE 5. Grain boundary diffusion coefficients of Co-ions for polycrystalline Fe_3O_4 as a function of the partial pressure of oxygen ($\alpha=1$, $\delta=1nm$). ref. 26.

boundary diffusion in oxides Stubican and Osenbach(23) have studied the anisotropy of grain boundary diffusion. The grain boundary diffusion parameters $\alpha D'\delta$ as a function of orientation both parallel and perpendicular to the growth direction of bicrystals of MgO, Cr-doped MgO, Al_2O_3 and $MgAl_2O_4$ were determined. Cr-isotope and tilt angles $\approx 5\text{-}25°$ were used in their study. The results obtained could be combined on one plot, Figure 6, which is plot of $D'_{\shortparallel}/D'_{\perp}$ versus tilt angle for all investigated oxides. It is evident that the ratio $D'_{\shortparallel}/D'_{\perp}$ which reflects the anisotropy of grain boundary diffusion, is constant for tilt angles above $\approx 10°$, and increases sharply as tilt angles become smaller than $\approx 10°$. These results are in accord with the dislocation model for the low angle grain boundaries postulated by Turnbull and Hoffman(27).

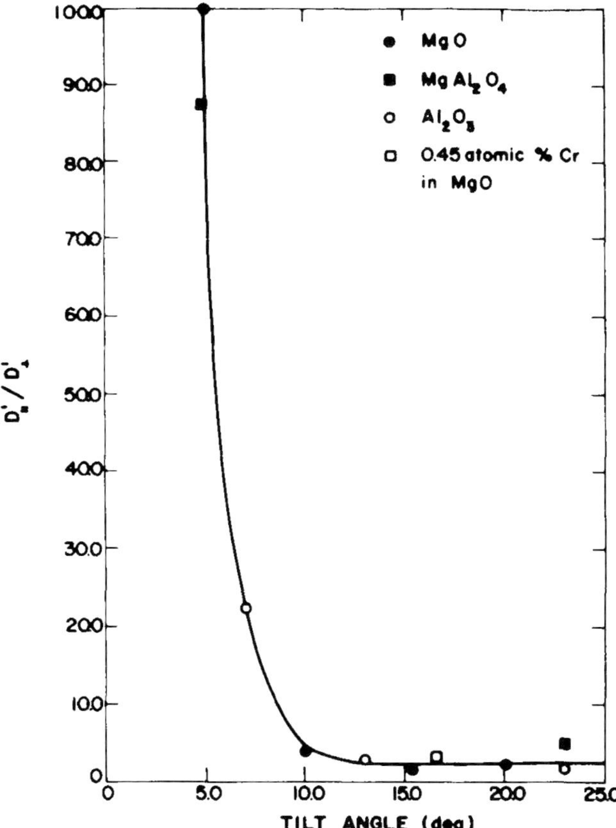

FIGURE 6 Dependence of the anisotropy of the grain boundary diffusion on tilt angle for Cr diffusion in magnesia, ^{51}Cr-doped magnesia, alumina and spinel. Ref. 23.

There is now a considerable amount of data on grain boundary diffusion in oxides, however, many inconsistencies still exist. Several factors which contribute to discrepancies of data such as segregation of impurities on grain boundaries, influence of different fabrication routes in the boundary structure, influence of the precipitated phases etc. Further research in this area is necessary with well defined materials.

SURFACE DIFFUSION IN OXIDES

Our knowledge of surface diffusion in oxides is much less advanced than it is for the lattice diffusion. Recently Matzke(28) has reviewed the available data on surface diffusion in ceramic systems. It is unfortunate that the direct methods such as field electron emission and field ion emission which are used for the surface diffusion studies in metals can not be used for oxide materials. Studies of the surface diffusion in metals have shown that the mechanism can be vacancy or adatom type.

Two different types of techniques exist to measure surface diffusion coefficient D_s or surface diffusivity product $\alpha D_s \delta$ (δ is the thickness of the high diffusivity layer). Mass transport techniques are based on investigating changes in the geometry of surfaces caused by the net flow of material along crystal plane. Since for mass transport all components have to be transported, for multiatomic ceramics rate determining species are undefined. Furthermore, results obtained by the mass transport techniques are often influenced by volume and vapor transport of material.

Most of studies of surface diffusion by the mass transport techniques involved MgO, Al_2O_3, UO_2 and NiO. The results obtained gave relatively high values for $D_o(>10^{-1}cm^2/s)$ and $Q_s(>2.5eV)$.

The first reliable data on surface diffusion in oxides measured by tracer technique were obtained by Zhou and Orlander(29). They measured the spreading of an U-tracer on UO_2 single crystals in temperature regions 1760-2110°C. Their values for the surface diffusivity are 3-4 orders of magnitude larger than those obtained by the mass transport methods. They interpreted the large D_o (5.10^6 cm^2/s) by the Bonzel theory(30) of "nonlocalized" surface migration of UO_2 molecules at high temperature. In the nonlocalized diffusion molecular species travel in a manner similar to gaseous motion. These species are only partially desorbed from the surface and still react with the surface atoms. The surface diffusion of Co- and Cr- isotopes on MgO was studied recently by Stubican and Lin(31). As evident from Figure 7 the slope of Arrhenius plot changes markedly at ≈1050-1100°C for Co and Cr-isotopes. Stubican et al(32) have shown that change in the Arrhenius slope is due to the overriding influence of a vapour-transport mechanism at high temperatures, which can be described as nonlocalized motion of the isotope. Figure 7 also shows that the divalent Co-ion diffuses more rapidly than the trivalent Cr-ion. This is also true for volume diffusion (Figure 1).

The influence of the point defects on near surface diffusion of ions is very similar to the influence on volume diffusion(33). Figure 8 shows the influence of the partial pressure of oxygen on surface diffusion of Co-isotope on NiO surface. Due to the presence of aliovalent impurities surface diffusivity of Co-isotope is constant at low oxygen pressures (extrinsic region). At high oxygen pressures ($>10^{-7}$ MPa) surface diffusivity becomes increasingly dependent on the intrinsic cation vacancy concentration. It is reasonable to assume that α and δ do not change with P_{o_2} than changes in $\alpha D_s \delta$ shown in Figure 8 reflect the changes in D_s.

Surface diffusion of Co-isotope on Fe_3O_4 (110) surface was also investigated at different P_{o_2} as shown on Figure 9 (33). It was found that the influence of point defects on surface diffusion in this case is very similar to the influence of point defects on volume diffusion (Figure 2).

The activation energy for the localized surface diffusion of Co-isotope in the surface layer of NiO is 1.5eV(34) and the activation energies for grain boundary diffusion and lattice diffusion are 1.9eV and 2.4eV respectively (22). It seems feasible in this case to compare surface, grain boundary and volume diffusivities at least semi-quantitatively. This is done in Figure 10 where two lines for the

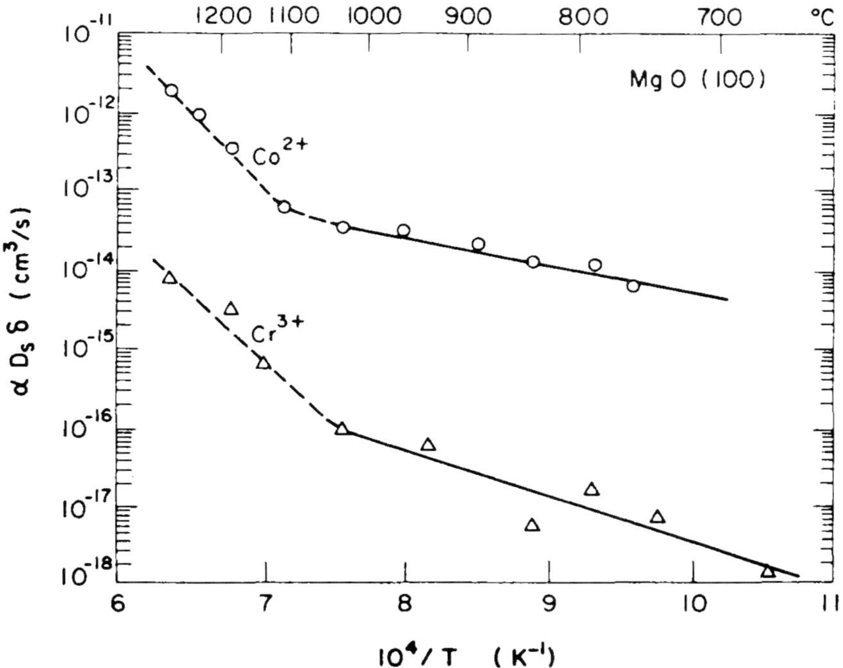

FIGURE 7. Ahrrenius plots for surface diffusion of ^{57}Co, and ^{51}Cr on MgO (100) plane. Ref. 31.

surface diffusion are given. Line $\underline{a}$ was calculated by assuming the value of $\delta = 1.7$ Å which is the diameter of Ni^{2+} ion, and for the line $\underline{b}$, δ, was assumed to be 4.18 Å which is the unit cell dimention of NiO. According to the obtained data surface diffusion of Co-ion on NiO (110) surface is ~6-7 orders of magnitude faster than the volume diffusion at 800°C, and ~5 orders of magnitude faster at 1200°C (35).

CONCLUSIONS

It is now generally accepted that selfdiffusion in the presently available crystals of high melting stoichiometric oxides such as MgO, Cr_2O_3, Al_2O_3, ZrO_2, HfO_2, etc. is impurity controlled (extrinsic). In these crystals the number of thermally created point defects is less than the number of impurity created point defects at all temperatures below the melting point. Diffusion rates of oxygen in these oxides are 3-5 orders of magnitude lower than diffusion rates of cations. Diffusion processes in nonstoichiometric oxides such as $Ni_{1-y}O$, $Co_{1-y}O$, $Fe_{3-y}O_4$ where the degree of nonstoichiometry is not large can be explained by using simple point defect models. On the other hand the diffusion results obtained with

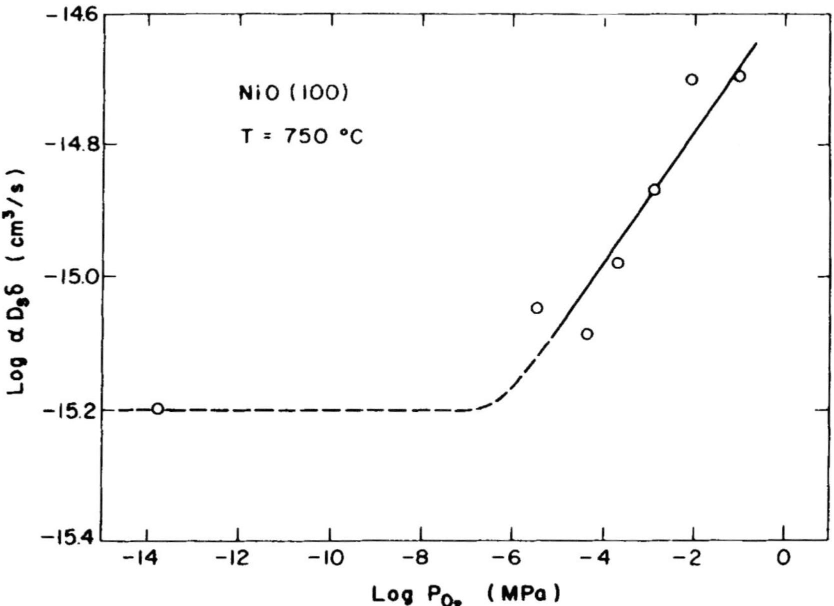

FIGURE 8. Surface diffusion product $\alpha D_s \delta$ of ^{57}Co in NiO surface layer as function of the partial pressure of oxygen. Ref. 33.

oxides such as $Fe_{1-y}O$, $Mn_{1-y}O$, TiO_{2-y}, CeO_{2-y} UO_{2-y}, etc. which display wide range of nonstoichiometry interaction of defects (clustering) must be considered in the interpretation of results. Diffusion mechanism in grossly defective oxides phases such as Ti_mO_{2n-1} Magneli phases) or in crystallographic shear phases of Wo_{3-y}, Nb_2O_{5-y}, etc. are not well understood. More research in this area is necessary.

In the fluorite-type oxides (doped and/or nonstoichiometric) oxygen ions diffuse faster than cations.

Diffusion of numerous impurities with various charge state has been studied in oxides. In the motion of homeovalent impurities size and polarizability dominate diffusion. For aliovalent impurities the other important term is the impurity-vacancy binding energy .

Grain boundary diffusion data in oxides show considerable inconsistencies. This can be attributed to the strong influence of impurities, segregation effects and methods of fabrication of specimens. However, it appears that in most of the systems studied $D'/D_v=10^3-10^5$, $Q'/Q_v=0.5-0.7$ which indicates considerable enhancement of diffusion along the grain boundaries. Diffusion in a single grain boundary is anisotropic and the mechanism of diffusion is a vacancy or intersti-

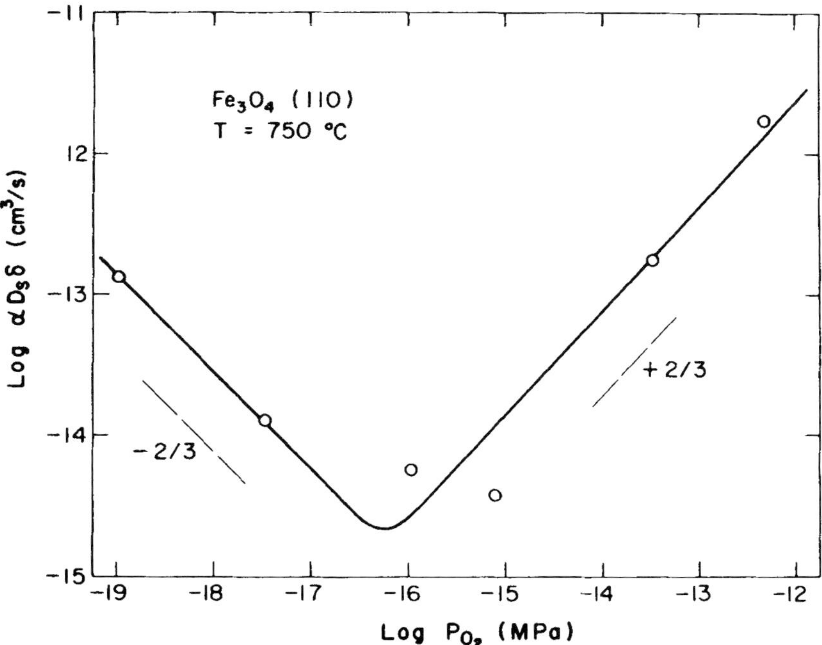

FIGURE 9. Surface diffusion product $\alpha D_s \delta$ of ^{57}Co in Fe_3O_4 surface layer as a function of partial pressure of oxygen. Ref. 33.

tial (interstitialcy) mechanism. Further studies should concentrate on high purity materials and structurally well defined grain boundaries.

Surface diffusion data obtained by the mass transport techniques and by the tracer techniques are often in disagreement. In mass transport experiments all components are transported and in tracer techniques only tracer ion diffusion is measured.

Recent results obtained by the tracer technique show that the surface diffusion of isotopes in many systems proceed by localized ionic motion at low temperature (<1300K) and at high temperature by delocalized motion of ionic or molecular species. Surface diffusion of tracers at ~1100K is $\approx$6-7 orders of magnitude faster than volume diffusion and the activation energy is $\approx$0.3-0.4 times that for lattice diffusion. The mechanism of surface diffusion in oxides seems to be a point defect mechanism.

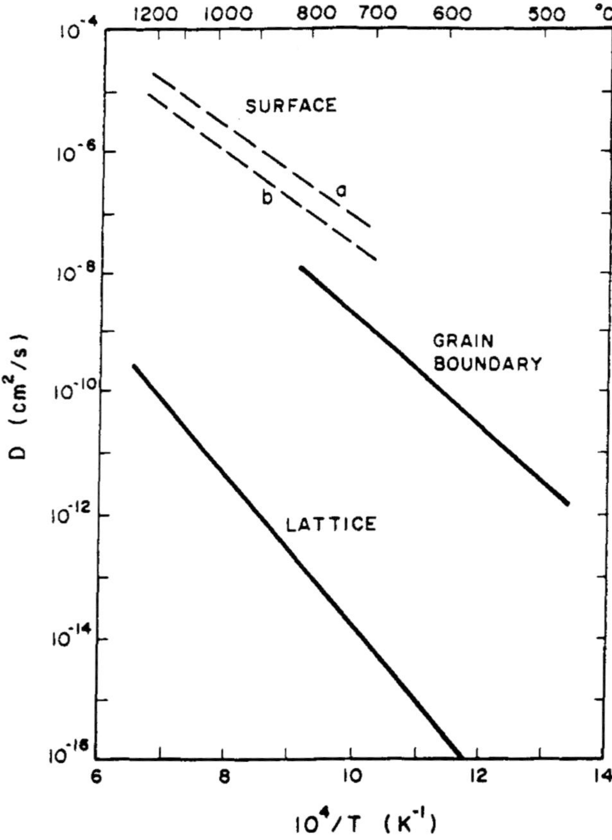

FIGURE 10. Arrhenius plots for surface, grain boundary and volume diffusion of Co ion in NiO. Line $\underline{a}$ was calculated using δ=1.7Å and line $\underline{b}$ by using δ=4.18Å. Ref. 35.

Further research in this area is necessary to obtain data on anisotropy of surface diffusion and on the influence of temperature on the thickness of the high diffusivity layer. Instrumentation for direct observation of diffusion of ions on surfaces of the high dielectric materials should be developed.

ACKNOWLEDGEMENT

The author thanks Dr. G. Simkovich for correcting the manuscript.

REFERENCES

1. See e.g. (a.) A.R. Allnat and A.B. Lidiard. 1993. *Atomic Transport in Solids.* Cambridge University Press, Cambridge, England
    (b) J. Philibert. 1991. *Atom Movements, Diffusion and Mass Transport in Solids,* Les Editions de Physique, Les Ulis, France.
    (c) Shewmon P. 1989. *Diffusion in Solids,* the Minerals, Metals, and Materials Soc., Warrendale, Pennsylvania.
    (d) H. Schmalzried. 1995. *Chemical Kinetics of Solids,* VCH, Weinheim, Germany.
2. R. Free. 1980. "Bibliography. Selfdiffusion and Impurity Diffusion in Oxides", J. Mater. Sci., *15,* 803.
3. B.C. Harding and D.M. Price. 1972. "Cation Self-Diffusion in MgO up to 2250°C", Phil. Mag., *26,* 253.
4. B.J. Wuensch, Steele, W.C. and T. Vasilos. 1973. "Cation Self-Diffusion in Single-Crystal MgO", J. Chem. Phys., *58,* 5258.
5. A.E. Paladino and W.D. Kingery. 1962. "Aluminium Ion Diffusion in Aluminium Oxides", J. Chem. Phys., *37,* 957.
6. J.D. Cawley, Halloran, I.W. and H.R. Cooper. 1991. "Oxygen Tracer Diffusion in Single-Crystal Alumina", J. Am. Ceram. Soc., *74,* 2086.
7. B.J. Wuensch and T. Vasilos. 1962. "Diffusion of Transition Metals Ions in Single Crystals of MgO", J. Chem. Phys., *36,* 2917.
8. N.L. Peterson. 1984. "Impurity Diffusion in Transition-Metal Oxides", Solid State Ionics, *12,* 201.
9. A.B. Lidiard. 1957. "Ionic Conductivity" in *Encyclopedia of Physics,* Ed. S. Flügge, Springer Verlag, Berlin, Germany. pg. 246.
10. G.W. Weber, Bitler, W.R. and V.S. Stubican. 1980. "Diffusion of ^{51}Cr in Cr-Doped MgO", J. Phys. Chem. Solids, *41,* 1355.; W.J. Minford and V.S. Stubican. 1974. "Interdiffusion and Association Phenomena in the System NiO-Al$_2$O$_3$", J. Am. Ceram. Soc., *57,* 8.
11. V.S. Stubican, Lin, C. and E. Macey. 1987. *Study of Surface Diffusion in Oxides by using Isotopes, in Nonstoichiometric Compounds,* Eds, C.R.A. Catlow and W.C. Mackrodt, in Advances in Ceramics, Vol. 23, Publ., The

Am. Ceram. Soc., Westerville, Ohio, pg. 97.

12. Hj. Matzke. 1981. "Diffusion in *Nonstoichiometric Oxides*", in *Non-stoichiometric Oxides,* Ed. O.T. Sorensen, Academic Press, New York, New York, pg. 156.

13. R. Dieckmann and H. Schmalzried. 1977. "Defects and Cation Diffusion in Magnetite", Ber. Bunsenges. Phys. Chem., *81,* 344; *81,* 414.

14. N.L. Peterson and W.K. Chen. 1973. "Mass Transport in Solids", Ann. Rev. Mat. Sci., *3,* 75.

15. W.L. Roth. 1960. "Defects in the Crystal and Magnetic on Ferrous Oxide", Acta Crystallogr. *13,* 140.

16. F. Koch and J.B. Cohen. 1969. "The Defect Structure of $Fe_{1-x}O$", Acta Crystallogr., *25* B, 275.

17. A. Atkinson and C. Monty. 1988 "Grain Boundary Diffusion in Ceramics", in *Surfaces and Interfaces of Ceramic Materials,* Eds. L.C. Dufour, C. Monty and G. Petot-Ervas, Kluwer Acad. Publ., Dodrecht, pg.273.

18. A.D. LeClair. 1963. "The Analysis of Grain Boundary Diffusion Measurements", J. Appl. Phys., *14,* 351.

19. A. Atkinson and R.I. Taylor. 1981. "The Diffusion of ^{63}Ni along Grain Boundaries in Nickel Oxide", Phil. Mag., *43A,* 979.

20. W.K. Chen and V.L. Peterson. 1980. "Grain Boundary Diffusion of ^{60}Co and ^{51}Cr in NiO", J. Am. Ceram. Soc., *63,* 566.

21. F. Barbier, Monty, C. and M. Déchamps. 1988. "On the Grain Boundary Diffusion of Co in NiO", Phil. Mag. *58A,* 475.

22. A. Atkinson and R.I. Taylor. 1982. "Diffusion of Co-57 along Grain Boundaries in NiO", Phil. Mag. *45A,* 583.

23. V.S. Stubican and J.W. Osenbach. 1984. "Influence of Anisotropy and Doping on Grain Boundary Diffusion in Oxides", Solid State Ionics, *12,* 375.

24. V.S. Stubican and J.W. Osenbach. 1983. "Grain Boundary and Lattice Diffusion of ^{51}Cr in Alumina and Spinel", in *Structures and Properties of MgO and Al₂O₃ Ceramics,* Ed. W.D. Kingery in Advances in Ceramics Vol. 10, Publ. The Am. Ceram. Soc., Columbus Ohio, pg. 394.

25. F. Perinet, LeDuigou, J. and C. Monty. 1989. "Oxygen Self-Diffusion in Volume and in Grain Boundaries of $Cu_{2-x}O$" in *Non-Stoichiometric Compounds, Surfaces, Grain Boundaries and Structural Defects,* Eds. J. Nowotny and W. Wepner, Kluwer Acad. Publ., Dordrecht, Netherland, pg. 387.

26. L.R. Carinci and V.S. Stubican. 1995. "Grain Boundary Diffusion in NiO and Fe_3O_3", submitted to J.Am. Ceram. Soc.

27. D. Turnbull and R.H. Hoffmann. 1954. "The Effect of Relative Crystal and Boundary Orientation on Grain Boundary Diffusion Rates", Acta Metal., *2,* 419.

28. Hj. Matzke. 1988. "Surface Diffusion and Surface Energies of Ceramics" in *Surfaces and Interfaces of Ceramic Materials,* Eds. L.C. Dufour, Claude Monty and G. Petot-Ervas, Kluwer Acad. Publ., Dordrecht, Netherland, 241.

29. S.Y. Zhou and D.R. Orlander. 1984. "Tracer Surface Diffusion on Uranium Oxide", Surf. Sci., *136,* 82.

30. H.P. Bonzel. "Mass Transport by Self-Diffusion", in *Surface Mobilities on Solid Materials,* Ed. Vu Thieu Binh, Plenum Press, New York, 1983, pg. 195.
31. C.M. Lin and V.S. Stubican. 1987. "Influence of Ionic Charge on Diffusion in the Surface Layer of Magnesia", J. Am. Ceram. Soc., *70,* C73.
32. V.S. Stubican, Huzinec, G. and D. Damjanovic. 1985. "Diffusion of ^{51}Cr in Surface Layers of Magnesia, Alumina, and Spinel", J. Am. Ceram. Soc., *68,* 181
33. C.M. Lin and V.S. Stubican. 1990. "Point Defects and Surface Diffusion in Magnetite and Nickel Oxide", J. Am. Ceram. Soc., *73,* 587.
34. E.P. Macey and V.S. Stubican. 1993. "Diffusion of ^{57}Co in Surface Layers of Nickel Oxide and Alumina", J. Am. Ceram. Soc. *76,* 557.
35. V.S. Stubican. 1993. "Diffusion of Isotopes in Surface Layers of Some Oxides", Phil. Mag. *68A,* 809.

Chapter Nine

INTERDIFFUSION IN METALS

WILLIAM R. BITLER and SUZANNE MOHNEY

Department of Materials Science and Engineering
Steidle Building
Pennsylvania State University
University Park, PA 16802

INTRODUCTION

Interdiffusion is the mixing of atoms which takes place when dissimilar elements are brought into contact with each other. Whenever any two different elements are brought into contact with each other at a finite temperature there is a thermodynamic driving force present acting to cause interdiffusion arising from the entropy of mixing. That is, a system composed of two different elements in contact with each other is inherently thermodynamically unstable. It is recognized of course that for systems composed of A and B atoms in which there is a large size mismatch or in which the A-A and B-B interactions are much stronger than the A-B interactions, the solubility limits of A in B or B in A can be vanishingly small at low temperatures as typified for example by systems with wide miscibility gaps, eutectic, or eutectoid systems. Still the effort required to remove residual impurities from an intrinsic semiconductor is a measure of the forces tending to cause mixing. Finally it is important to keep in mind that even low levels of diffusion induced by the entropy of mixing can have significant effects on the mechanical properties of a metal if the interdiffusing species are segregated at the grain boundaries and of course low impurity levels can significantly affect the electrical behavior of semiconductors and the nature of junctions in semiconductor devices. At ambient temperatures diffusion is typically a slow process in solids and the mixing may occur only over very short distances which for many situations can be ignored. Exceptions occur if the integrity of the system is interface sensitive or if the dimensions of the system become comparable to the distance over which the mixing occurs. For example, in the case of galvanized steel, the current material of choice for automobile panels, reaction between the zinc coating and the steel substrate may occur during the application of the zinc by the hot dipping process or, in the case of electrodeposited Zn, during the curing of the overlying paint, resulting in the production of undesirable intermediate phases between the zinc and the steel. As another example, the electrical connector industry typically uses gold coatings for applications requiring low electrical contact resistance. Because of the high price of gold there is an incentive to keep the gold coatings as thin as possible, but not so thin that the substrate will diffuse through the gold coating to the surface during the service life of the connector. Finally in the design of semiconductor devices for the computer industry there is continual pressure to miniaturize the devices so as to speed up their response time and result in higher speed computers. The technology has already so miniaturized computer hardware that the ability to suppress interdiffusion is of concern.

Alternatively, thermodynamic instability can arise from an abrupt change in temperature, again giving rise to a driving force for matter transport but in this case the transport may be such as to mix or unmix the system. An example of the latter would be the quenching of a solid solution alloy into a miscibility gap and then letting the alloy react within the miscibility gap. As a consequence a second topic of significant import to the materials scientist is that of solid-solid reactions such as those that take place during the thermal processing (heat treatment) of metallic alloys. Many of these reactions are diffusion controlled.

BACKGROUND

The study of interdiffusion, the mixing of materials on a microscopic, i.e. on an atomistic scale (as distinct from convective or mechanical mixing considered in this context to be macroscopic processes), has a long scientific and technological history. Diffusion was first given a quantitative description with the publication of Fick's Laws of Diffusion in 1855 (1). These were phenomenological equations which proposed a macroscopic mathematical description of the diffusion process. Specifically Fick's first and second laws are:

First Law
$$J_x = -\frac{\partial c}{\partial x} \tag{1}$$

Second Law
$$\frac{\partial c}{\partial t} = D\frac{\partial^2 c}{\partial x^2} \tag{2}$$

(here it has been assumed that D is a constant and is independent of time and position, which may or may not be true in real physical systems - more to be said about this later) in which J is the flux, the amount of material in units of moles, kilograms, or number of atoms for example, passing through a unit area per unit of time, D is the proportionality constant called the diffusivity and is still tabulated in most references in units of cm²/sec, c is the concentration in units consistent with those chosen for the flux and x is the position, again in units consistent with those used for the flux and the concentration. Fick's first law is mathematically equivalent to Fourier's law for heat flow and to Ohm's law for electric current flow, permitting their solutions under the same boundary conditions to be used interchangeably. For example, a good source for solutions to Fick's equations is Carslaw and Jaeger's book on heat flow (2). It is worth noting that diffusion in solids is a thermally activated process and as a result the diffusivity, D, is well represented in most cases by the Arrhenius relation:

$$D = D_o exp(-\frac{Q}{RT}) \qquad (3)$$

where D_o is a material constant, usually tabulated in units of cm²/sec, and has a value of the order of one cm²/sec for many metals, Q is the activation energy for the process and is typically of the order of 50,000 calories/mol for lattice diffusion, R is the molar gas constant, and T is the absolute temperature, usually in kelvins. It is because of this exponential dependence of the diffusivity on the inverse temperature that most materials interdiffuse at high rates only at elevated temperatures. Subsequent discussion of diffusion in this article will be restricted to the interdiffusion of metals unless otherwise noted.

Following the publication of Fick's Laws, the experimental determination of D for various metals continually improved and it became clear that D was not a single material constant but varied not only from one binary system to another but could vary by orders of magnitude across a binary system. The source of a major part of this variation became apparent when it was recognized that the appropriate driving force for matter transport by diffusion was the gradient of the chemical potential rather than the gradient of the concentration or composition. Thus the variation of D, as defined by Fick's first law, with composition is a reflection of the deviation of the alloy system from ideality, i.e. the variation of γ, the activity coefficient, with composition. It should also be noted that in this context, for binary systems with intermediate phases, the variation of the enthalpy across a given phase may be the dominant cause for mass transport rather than the entropy of mixing. Darkin (3) reformulated the equations for mass transport by diffusion in terms of the gradient of the chemical potential and these equations are now commonly referred to as Darken's equations:

$$D_{chem} = X_1 D_2 + X_2 D_1 \qquad (6)$$

$$= (X_1 D_2^* + X_2 D_1^*)(1 + \frac{\partial \ln\gamma_1}{\partial X_1}) \qquad (7)$$

that is

$$D_i = D_i^* (1 + \frac{\partial \ln\gamma_i}{\partial \ln X_i}) \qquad (8)$$

where D_1 and D_2 are the intrinsic diffusion coefficients, D_{chem} is the chemical or interdiffusion coefficient, X_1 and X_2 are the mol fractions of components 1 and 2, i.e. $X_1 + X_2 = 1$, D_1^*, D_2^*, and D_i^* are the tracer coefficients measured in a chemically homogeneous sample.

In parallel with the evolution of the macroscopic description of diffusion culminating in Darken's equations, studies were in progress to determine the atomistic mechanism(s) for diffusion in solids. Landmark investigations by Kirkendall (4) and Smigelskas and Kirkendall (5) showed that markers placed at the interface of a copper-brass binary diffusion couple moved during the diffusion anneal. Darken's equations which assign an intrinsic diffusivity for each component in a binary diffusion couple incorporate the marker motion into the analysis (3). An atomistic analysis of diffusion mechanisms in metals, see for example Huntington (3) and Bardeen and Herring (3), related this motion to a vacancy mechanism for diffusion. Subsequently Mehl and da Silva (6) systematically investigated a large number of binary metal diffusion couples and found marker motion in all of the couples they investigated. It is now generally accepted that diffusion in solid solution binary metal couples occurs by a vacancy mechanism with the possible exception of some bcc metals whose diffusivities are considered anomalous and are characterized by a D_o that is orders of magnitude smaller than the 1cm²-sec typical of most metals, e.g. uranium with a D_o of 1.2 x 10⁻³ cm²/sec. (7,8). This chapter is entitled interdiffusion of metals and restricts its discussion to diffusion of uncharged particles as distinct from ionic diffusion in ionic solids where charge neutrality constraints play a significant role.

Not only was the variation of diffusivity with composition determined experimentally as the experimental techniques improved but measurements at lower temperatures became feasible and it soon became apparent that the mechanism for diffusion could vary with temperature, i.e. the Arrhenius plot of ln D verses 1/T for polycrystalline samples did not extrapolate linearly to low temperatures. An example typical of this phenomena resulting from investigations of the interdiffusion of palladium with copper (9,10) carried out at Penn State is shown below, see Figures 1 and 2. This immediately suggests that more than one mechanism is contributing to the diffusion process. In the late 40's and early 50's Achter and Smoluchowski (11), Hoffman and Turnbull (12) and Turnbull (13) established that grain boundaries contributed to the observed volume diffusion. Fisher (14), Suzuoka (15), and Whipple (16) provided mathematical analyses of the phenomena which permitted the evaluation of grain boundary diffusivities from volume concentration profiles. A review article on the mechanisms of grain boundary diffusion by Balluffi (17) is a good reference for this topic. As indicated in the figure, grain boundary diffusion dominates the diffusion process at low temperatures and can be orders of magnitude larger at ambient or near ambient temperatures than that predicted from an extrapolation of high temperature data to low temperatures. It should be noted that the smaller the grain size the larger is the contribution of grain boundary diffusion to the total diffusion flux. This has significant technological implications as coatings become thinner or devices become smaller such that diffusion distances at ambient temperatures become of concern.

More recently, in the 1980's, a new mechanism for transport was proposed which was first recognized and reported by den Broeder (18) in 1972. In 1978 Hillert and Purdy (19) reported similar observations of the interaction of solid iron with zinc vapor. Both of these examples involved moving grain boundaries whereas previous studies of grain boundary diffusion had been restricted to that occurring in stationary boundaries. An extensive literature concerned with this phenomenon (known as diffusion induced grain boundary motion, DIGM) has since been published as well as some good review articles, see e.g. A.H. King (20). DIGM can result in enhanced volume transport of matter over that resulting from direct lattice diffusion or from grain boundary diffusion through static grain boundaries. This has been observed in the tin-copper system (21) important to the electronics industry because of the wide spread use of solder connections.

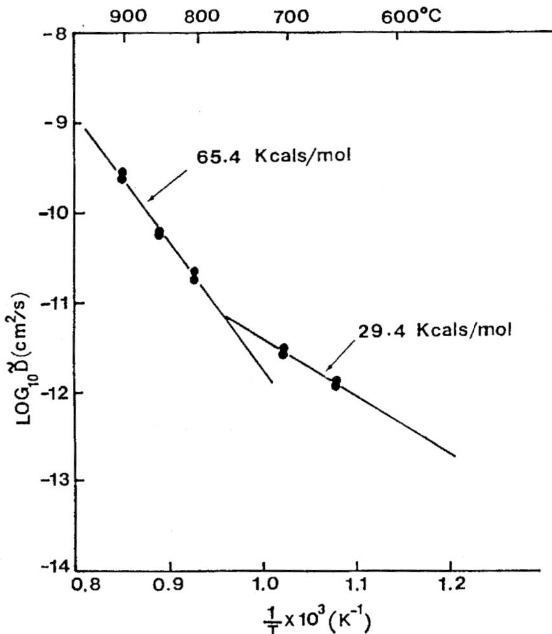

FIGURE 1. Copper-wrought palladium system. Arrhenius plot of the interdiffusivities at 50 atom% copper.

It is worth while pointing out that unlike the situation in ionic solids in which the diffusivity can be varied by orders of magnitude with minute chemical additions of aliovalent species to the system of interest, in metals small, e.g. less than 5%, variations in the composition in metals cause correspondingly small changes in the lattice diffusivity, although, as noted previously, cumulatively the changes may be large from one end of a binary system to the other end. There is, however, evidence in the literature, e.g. see Gleiter and Chalmers (22), that grain boundary diffusion may be sensitive to small changes in chemical composition.

To illustrate some areas of current interest in interdiffusion, the three examples already mentioned previously, which have significant current economic importance, i.e. reactions between zinc and steel of concern to the automobile industry which currently use this material as the material of choice for automobile panels, diffusion barriers between gold electrodeposits and their substrates used extensively as electrical contacts, and interdiffusion between metals and compound semiconductors of import for optoelectronics will be discussed.

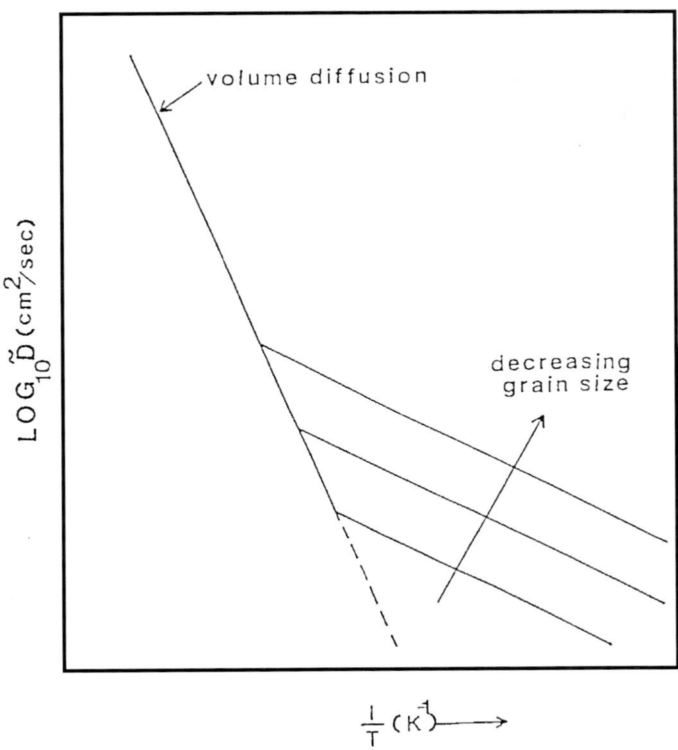

FIGURE 2. Schematic of an Arrhenius plot of interdiffusivity for samples of varying grain sizes.

GALVANIZED STEEL

Analysis of the interdiffusion of steels coated with Zn-based alloys, i.e., galvanized steel, has become increasingly important, particularly in the automotive industry. According to Peter Wray (23) the automobile industry used galvanized steel produced by electrodeposition at the rate of 6.6 x 10^6 tons per year and hot dipped galvanized steel at the rate of 2.3 x 10^6 tons per year as of 1995. The automobile industry is attempting to overcome the type of body panel corrosion exhibited in the car shown in Figure 3. The primary function of the Zn based coating is to provide cathodic protection for the underlying steel sheet by functioning as a sacrificial anode. However, economic considerations may limit the coating thickness (currently at ~10 μm) which may also influence other coating properties such as formability, wearability, and paintability. Keep in mind that galvanized steel used for automobile panels is produced as flat sheet or coils that subsequently must be formed in dies into fenders, side panels, etc., then spot welded during car assembly and must at the end have a finish suitable for painting so that the resultant painted car has the mirror like finish that the customer now expects. It is surprising at first glance that more isn't known about the iron-zinc system.

Over the past 25 years the Fe-Zn phase diagram has undergone numerous changes. Differences in published phase diagrams have made it clear that uncertainties existed concerning the phase boundaries of the solution phases as well as the number of intermediate compound phases and their stoichiometries. These discrepancies in the literature can be attributed to the difficulty in preparing homogeneous alloys of known composition due to the high vapor pressure of Zn relative to Fe. Therefore, phase boundary determinations and thermodynamic

FIGURE 3. Body panel corrosion on an automobile.

measurements are difficult to perform accurately. Reproducible results have been obtained by producing alloys by isothermic diffusion (24) and by a technique known as liquidus-hot-pressing (25).

The Zn-rich portion of the Fe-Zn phase diagram was drastically changed in the 1970's as a result of several publications on the topic at that time. Kubaschewski compiled the most up-to-date Fe-Zn phase diagram which includes experimental observations of the Fe-Zn system through 1980 (26). A recent thermodynamic evaluation (27) of the Fe-Zn system supports the diagram by Kubaschewski. The Fe-Zn phase diagram was redrawn by ASM Int (28) and an adapted version is shown in Figure 4. In addition to the terminal solid solution αFe and ηZn phases, the stable intermetallic phases located in the Zn rich portion of the Fe-Zn phase diagram are named, gamma, Γ, gamma sub one, Γ_1, delta, δ, and zeta, ζ.

An ideal way to study the reaction kinetics between solid Zn and Fe would be to prepare Fe-Zn diffusion couples, react them for various lengths of times at various temperatures and then to perform a transmission electron microscope, (TEM), study of TEM samples prepared in turn from these diffusion couples. Ideally, for such a study, the thin TEM samples would be oriented so that a thin section of Zn on one side of the sample would adjoin a thin section of Fe on the other side and the electron beam would be directed down the interface between

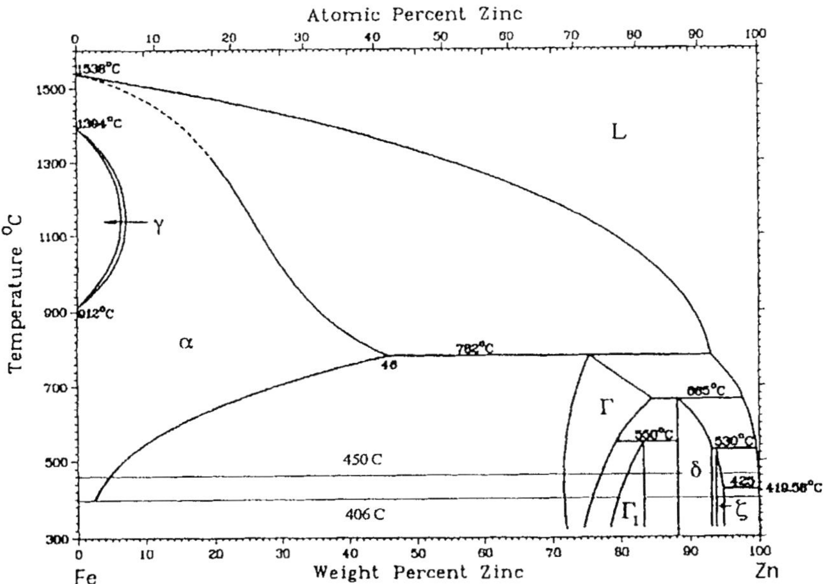

FIGURE 4. The Fe-Zn phase diagram.

the Zn and Fe regions, see Figure 5. Again because of the very different physical and chemical properties of Zn and Fe no such samples had been fabricated until the late 1980's. L.A. Giannuzzi (29) developed a workable TEM sample preparation technique (30,31) see Figures 6 and 7 and subsequently studied the reactions between electrodeposited Zn and Fe. See Ph. D. Thesis by Giannuzzi (29) and resultant publications (32-36). Still requiring study are the kinetics and reaction products produced when commercial steel is reacted with Zn alloy coatings.

GOLD COATED CONNECTORS

In making electrical connections, if the electrical signals are low level, e.g. those connected by the tuner switch in your TV set or by the cable connectors of your PC computer, and the mechanical force available to make the connection is limited, then it is essential that the connector have a low electrical contact resistance. The ideal material for making such an electrical connection by mechanical contact is gold. This is because gold itself has a low electrical resistance and more significantly forms no stable compounds with relatively high electrical resistivities, such as an oxide, under ambient conditions. As a result gold for many connector applications has become the material of choice. Of course to fulfill this application it is essential that the substrate material does not diffuse to the surface of the gold coating during the service lifetime of the connector. For applications requiring long service lifetimes e.g. switching systems in telephone

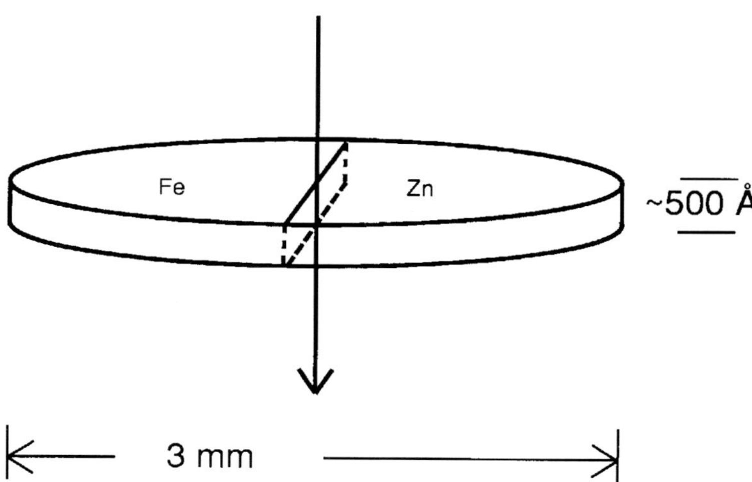

FIGURE 5. Ideal sample geometry for studying the interfacial reactions between Fe and Zn.

networks, this required relatively thick coatings. When President Nixon took the U.S. off the gold standard in the late 60's the price of gold rose ten fold and there was great pressure on the connector industry to drastically reduce its use of gold. This it did by relaxing standards when they were overconservative but more

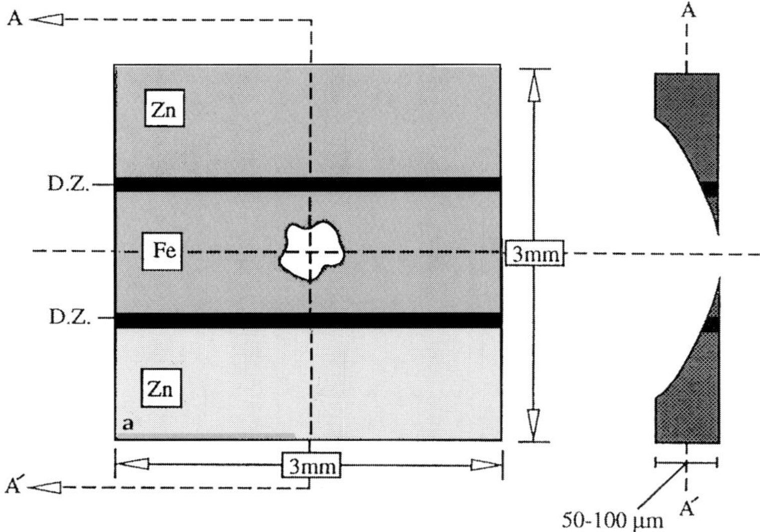

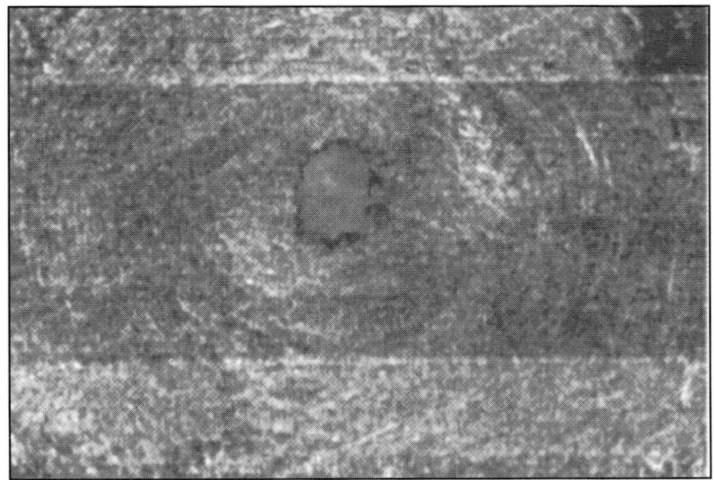

FIGURE 6. a) A schematic of an Fe-Zn couple which was dimpled to perforation. b) An image of an Fe-Zn couple dimpled to perforation.

importantly by developing diffusion barriers between the gold and the metal, typically copper or copper alloy, substrates. Even gold is not exempt from the laws of supply and demand. Figures 8 and 9 show the consumption of gold by the plating industry and the price of gold during the late 70's and the early 80's and

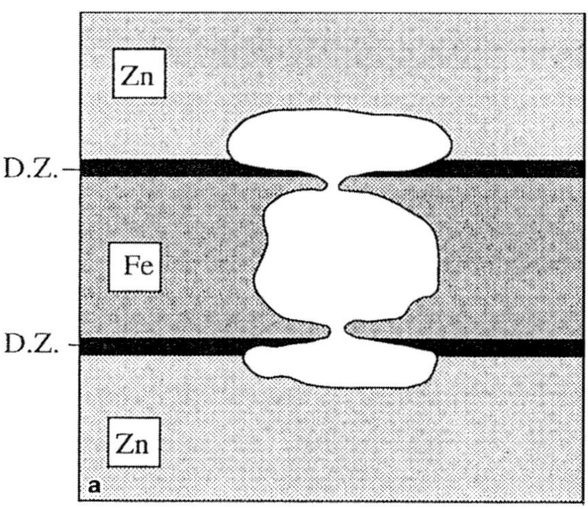

FIGURE 7. a) A schematic of an Fe-Zn couple after liquid nitrogen ion milling. b) An image of an Fe-Zn couple after liquid nitrogen ion milling.

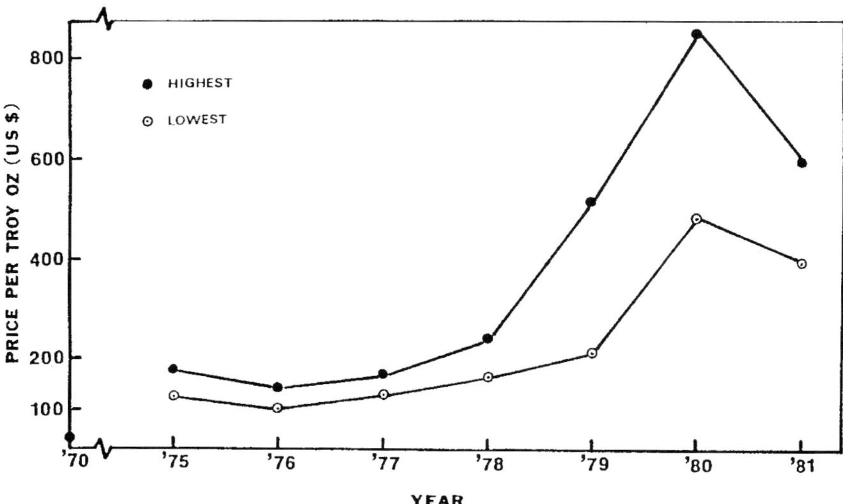

FIGURE 8. Consumption of gold in the plating industry in the USA during the 1970s.

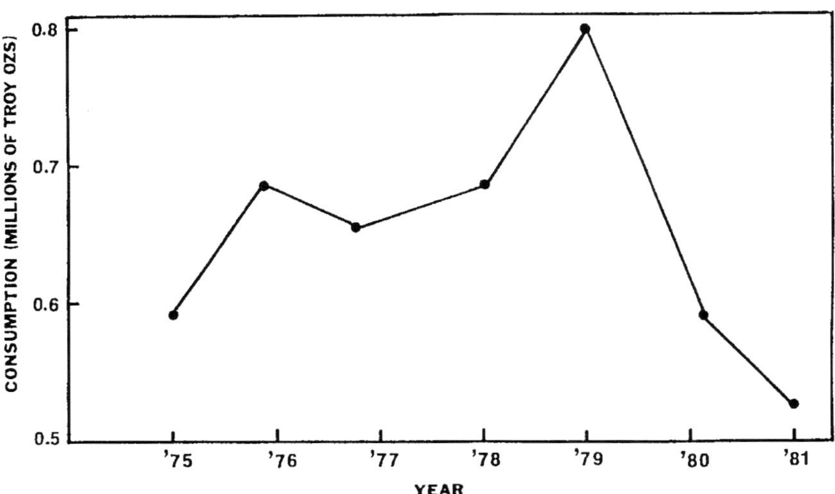

FIGURE 9. Variation in the price of gold in the 1970s.

Figure 10 the price of gold up through the 90's. It is obvious during the latter half of the seventies the use of gold rose continuously and this was largely a result of its increased use by the electronics and connector industries. As technology improved and the gold required for these applications was reduced, reducing the overall consumption of gold the price in turn was forced downwards. Thus technology in this case had a double effect, 1) it reduced the amount of gold required for a given application and 2) as a result of supply and demand pressures it reduced the cost of the gold being consumed. Of course the above remarks should be considered with much reservation, the price of gold is a very complex subject in its own right, e.g. it is also sensitive for example to the value of the dollar, see e.g. the discussion in the Wall Street Journal (59,60).

The problem of interdiffusion of gold with its substrate at ambient temperatures is further compounded by the effects of grain boundary diffusion. For many connector and electronic applications, gold is applied by electrodeposition. Electrodeposits, particularly as applied under production conditions, are deposited far from thermodynamic equilibrium resulting in high nucleation rates and very small grain sizes. The analysis of their low temperature diffusion behavior is then further complicated by the fact that electrodeposits are readily recrystallized resulting in grain boundary diffusivities whose contribution to the volume diffusion is changing with time if the diffusion anneal induces recrystallization concurrently with the diffusion anneal (39-41). Ideally, a suitable diffusion barrier should have a low electrical resistivity and interdiffuse at a slow rate both with the substrate and the gold deposit. The material of choice and used extensively by industry is Ni. It is fortunate that the commercial Ni plating conditions, i.e. bath chemistry, temperature and current density are such that a relatively large (for electrodeposits) grain size results. Studies at Penn State (9,39-42) found more effective diffusion barriers for the Au - Cu system, e.g.

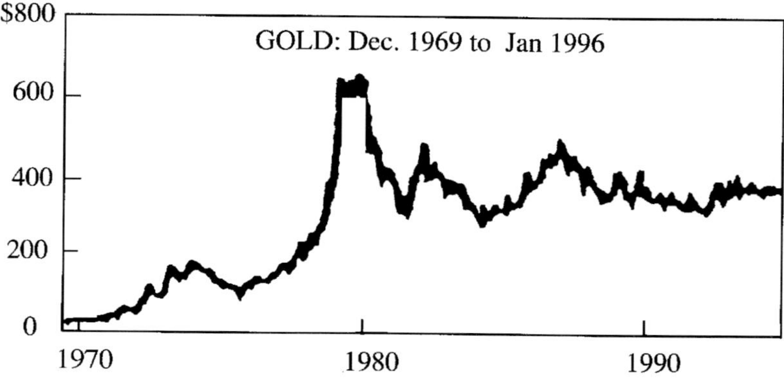

FIGURE 10. Price of gold in more recent years.

Cr or Co (9) but they have not been utilized commercially because of economic constraints, toxicity problems of the affluents from the plating process and/or ease of deposition. The search continues for more effective diffusion barriers and for a more inexpensive material to substitute for gold.

One such substitute which has been studied extensively (9,10) at Penn State and that is finding limited commercial use is Pd. Pd has the drawback that under plating conditions so far developed for the electrodeposition of Pd, the grain size is very small and copper diffuses through it rapidly at ambient or near ambient temperatures. As a result, it, too, requires an intermediate Ni layer when it is plated on copper. Because of the complexity of the electrodeposition process, there is a strong reluctance in industry to change from the use of gold with a nickel diffusion barrier whose properties and use are well documented.

INTERDIFFUSION BETWEEN METALS AND COMPOUND SEMICONDUCTORS

Interdiffusion between metals and compound semiconductors can impact the electrical properties, morphology, and thermal stability of ohmic contacts and Schottky barriers used in a variety of transistors and optoelectronic devices. Although there are a few exceptions, the intermediate reaction products that form between most metals and compound semiconductors are metallic phases; hence, it is entirely appropriate to consider these systems in a chapter on interdiffusion in metals.

As first discussed by Lin et al. (44), it can be difficult to predict the sequence of phases that will form when a metal and compound semiconductor interdiffuse. According to Kirkaldy and Brown (45), a unique diffusion path, or sequence of single and two phase regions, should evolve between the end members of a ternary diffusion couple at a given temperature, provided thermodynamic equilibrium is attained locally at the interfaces between phases. However, sufficient thermodynamic and diffusional data are rarely available to actually predict this sequence.

For metal contacts on compound semiconductors, the situation can be even more complicated. Since the contacts are only on the order of 100 nm thick, a diffusion path for which local equilibrium is attained at the interfaces may not have time to evolve before the metal is completely consumed in reaction. Non-equilibrium interfaces or even metastable phases can play an important role in the initial stage of the reaction of thin film contacts, even in cases for which the diffusion path of bulk Metal/GaAs couples eventually evolves to attain local equilibrium at the interfaces between phases (44).

One technologically useful feature of the reaction between many metals and compound semiconductors is the formation of M_xIIIV phases, where M represents a metal, III represents a group III element (Al, Ga, or In), and V represents a group V element (P, As, or Sb). The ratio of the group III to group V element is close to one in these phases, and when they are formed in the initial stage of reaction, they may be useful in the formation of ohmic contacts through a solid phase regrowth mechanism (46). Phases of nominal composition M_xIIIV have been observed in crystalline or amorphous form by transmission electron microscopy following limited reaction in Ni/GaAs (47), Pd/GaAs (47), Co/GaAs (48), Pt/GaAs (49), Ni/InP (50), Pd/InP (51), Pt/InP (52), Ni/AlAs (53), Pd/GaP (54), and Ni/GaSb (55) contacts. Except for the amorphous Pt_xGaAs phase (49), which grows only a few nanometers before it decomposes, corresponding ternary phases or binary phases with extensive solubility have been observed in the equilibrium phase diagrams of these Metal-III-V systems. (The actual composition of the phases found in the thin film contacts, however, are sometimes outside the equilibrium ranges of homogeneity.)

Because of the technological application of the ternary phases, it is important to understand under what circumstances they are favored to form in an annealed contact. While their existence in the equilibrium phase diagram appears to be important, kinetic factors must also play a significant role. Lin et al. (44) have suggested that the M_xGaAs phases appear initially in the reaction of many metals with GaAs because nucleation of the M_xGaAs phases does not require redistribution of Ga and As; diffusion of the metal alone is sufficient for the phase to form. Sands et al. (56) have also correlated the heats of vaporization of the late transition metals with the kinetics of these Metal/GaAs reactions. They suggest that Pd, Ni, and Co, which exhibit the weakest bonding (as indicated by the lowest heats of vaporization), are available for reaction at temperatures so low that Ga and As are essentially immobile. Thus, the formation of the M_xGaAs phases is favored. These reactions can begin at temperatures below 200°C.

The crystal structure of the M_xIIIV phase may also be an important factor in its sustained growth. For example, a phase of nominal composition Ni_3GaAs has been observed to grow hundreds of microns thick in bulk Ni/GaAs diffusion couples, even at temperatures as high as 700°C (57), despite the non-equilibrium interfaces in the couple. The crystal structures of the Ni_3GaAs phase and its related superlattices are derivatives of the NiAs structure, and it is believed that Ni is able to diffuse much more rapidly than Ga and As through this structure due to the availability of suitable interstitial sites (57). Thus, the stoichiometry of the growing ternary phase can be maintained during prolonged growth. This situation is in contrast to systems such as Ni/InP, for which the kinetic stability of the ternary phases is very limited (58). Unfortunately, diffusion through ternary and solution phases of interest for M/IIIV contacts have not been examined from an atomistic point of view for crystal structures other than the NiAs derivatives.

As the reactions between metals and a wider variety of compound semiconductors are studied, it should be possible to extend our understanding of the kinetic factors that influence the diffusion path in metal contacts. Ultimately, such an understanding may aid in the design of improved contacts with desirable electrical and morphological characteristics for a greater number of semiconductors.

REFERENCES

1. Fick, A. 1855. *Pogg. Ann.* 94.
2. Carslaw, A.S. and J.C. Jaegar. 1959. Conduction of Heat in Solids. Oxford Univ. Press.
3. Darken, L.S. 1951. Formal Basis of Diffusion Theory, pp. 1-25. *Atom Movements.* American Society for Metals. (This book is a veritable Who's Who of the leaders in the study of solid state diffusion and includes articles such as "Mechanisms of Diffusion," by H.B. Huntington, "Diffusion in Alloys and the Kirkendall Effect," by J. Bardeen and C. Herring, "Grain Boundary and Surface Diffusion," by D. Turnbull, etc.)
4. Kirkendall, E.O. 1942. Diffusion of Zinc in Alpha Brass. *Trans. AIME* 147:104.
5. Smigelskas, A.D. and E.O. Kirkendall. 1947. Zinc Diffusion in Alpha Brass. *Trans. AIME* 171:130.
6. da Silva, L.C.C. and R.F. Mehl. 1951. *Trans AIME* 155:73.
7. Pound, G.M., Bitler, W.R. and H.W. Paxton. 1961. Theory of the Diffusion Coefficient in Cubic Metals. *Phil. Mag.* 6(64):473.
8. *Diffusion in Body-Centered Cubic Metals.* American Society for Metals. 1965.
9. Kaja, S. Interdiffusion of Electrodeposits with their Substrates and Diffusion Induced Grain Boundary Migration. Ph.D. Thesis, Penn State, 1985.
10. Bitler, W.R., Pickering, H.W. and S. Kaja. 1985. Interdiffusion Kinetics of Copper with Palladium. *Plat. and Surf. Fin.* 72:60.
11. Achter, M.R. and R. Smoluchowski. 1949. Grain Boundary Diffusion. *Phys. Rev.* 76:470.
12. Turnbull, D. and R. Hoffman. 1954. The Effect of Relative Crystal and Boundary Orientations on Grain Boundary Diffusion Rates. *Acta Met.* 2:419.
13. Turnbull, D. 1949. Rate of Self-Diffusion in Single and Polycrystals of Silver. *Phys. Rev.* 76:471.
14. Fisher, J.C. 1951. *J. Appl. Phys.* 22:74.
15. Suzuoka, T. 1961. *Trans. Jap. Inst. Metals* 2:25.
16. Whipple, R.T.P. 1954. *Phil. Mag.* 45:1225.
17. Balluffi, R.W. 1982. Grain Boundary Diffusion Mechanisms in Metals. *Met. Trans. A.* 13A-2069.
18. den Broeder, F.J.A. 1972. Interface Reactions and a Special Form of Grain Boundary Diffusion in the Cr-W System. *Acta Met.* 20:319.
19. Hillert, M. and G.R. Purdy. 1978. Chemically Induced Grain Boundary Migration. *Acta Met.* 26:333.
20. King, A.H. 1987. Diffusion Induced Grain Boundary Migration. *Int. Mat. Revs.* 32(4):173-189.
21. McGaughey, M.K. Interdiffusion Between Tin Coatings and Copper Nickel Alloy Substrates. M.S. Thesis, Penn State, 1993.
22. Gleiter, H. and B. Chalmers. 1975. *Progress in Materials Science.* Vol. 16. Pergamon Press.
23. Wray, Peter. Inland Steel (ret.), private communication.

24. Bastin, G.F., van Loo, F.J.J. and G.D. Rieck. 1974. *Z. Metallkde.* 65:656.

25. Gellings, P.J., DeBree, E.W. and G. Gierman. 1979. *Z. Metallkde.* 70:312.

26. Kubaschewski, O. 1982. *Iron-Binary Phase Diagrams.* Springer Verlag, Berlin, p. 173.

27. Petersen, S., Spencer, P.J. and K. Hack. 1988. *Thermochima Acta.* 129:77.

28. Binary Alloy Phases Diagrams. 2nd Edition. T.B. Massalski, editor-in-chief. ASM International, p. 1797. 1990.

29. Giannuzzi, L.A. A Transmission Electron Microscopy Study of Twinning in Electrodeposits, Diffusion Induced Recrystallization During Ion Beam Milling, and the Interdiffusion Regions of Iron-Zinc Couples. Ph.D. Thesis, Penn State, 1992.

30. Giannuzzi, L.A., Howell, P.R., Pickering, H.W. and W.R. Bitler. 1992. Preparation of Cross-Sectional TEM Samples of Fe-Zn Couples, In Ron Anderson, Bryan Tracy, and John Bravman (Ed) *Specimen Preparation for Transmission Electron Microscopy of Materials-III.* Materials Research Society, Pittsburgh, PA, p. 159.

31. Gianuzzi, L.A., Howell, P.R., Pickering, H.W. and W.R. Bitler. 1993. Techniques for the Production of Thin Foils From the Interfacial Regions of Iron-Zinc Couples. *Materials Characterization.* 30(1):55-60.

32. Giannuzzi, L.A., Howell, P.R., Pickering, H.W. and W.R. Bitler. Interfacial Characterization of Zinc-Coated Steels by Transmission Electron Microscopy-A Preliminary Study, p. 121. In G. Krauss and D.K. Matlock (Ed.) *Zinc-Based Steel Coating Systems: Metallurgy and Performance,* TMS, Warrendale, PA.

33. Giannuzzi, L.A., Howell, P.R., Pickering, H.W. and W.R. Bitler. 1991. Transmission Electron Microscopy of Iron-Zinc Couples. SUR/FIN '91 Conference Proceedings, Toronto, Canada, AESF, p. 927.

34. Giannuzzi, L.A., Howell, P.R., Pickering, H.W. and W.R. Bitler. 1991. A Preliminary Characterization of the Defect Structure of the Zeta Phase in the Interfacial Region of Fe-Zn Couples, p. 233. In A.D. Romig Jr., D.E. Fowler, P.D. Bristowe (Ed.) *Structure/Property Relationships for Metal/Metal Interfaces,* Materials Research Society, Pittsburgh, PA.

35. Giannuzzi, L.A., Ramani, A.S., Howell, P.R., Pickering, H.W. and W.R. Bitler. 1992. Observation of Pseudo 10-Fold Symmetry in the Ordered Iron-Zinc Delta Phase, p. 36. In G.W. Bailey, J. Bently, and J.A. Small (Ed.) *Proc 50th Annual Meeting of the Electron Microscopy Society of America,* San Francisco Press, San Francisco, CA.

36. Giannuzzi, L.A., Howell, P.R., Pickering, H.W. and W.R. Bitler. 1992. A Microstructural Evaluation of Fe-Zn Couples by Cross-Section Transmission Electron Microscopy, pp. 43-51. SUR/FIN '92 Conference Proceedings, Atlanta, GA, AESF.

37. "The Characterization of Intermediate Phases of Electrogalvanized-Iron Couples by Cross-Section Transmission Electron Microscopy - A Brief Overview," in *Galvatech* '92, Proc. 2nd International Conference on Zinc

and Zinc Alloy Coated Steel Sheet, Amsterdam, Netherlands, 8-10 Sept. 1992, 461-467.

38. Gianuzzi, L.A., Howell, P.R., Pickering, H.W. and W.R. Bitler. 1993. Transmission Electron Microscopy of the Interdiffusion Regions of Iron-Zinc Couples, pp. 212-223. In G.F. Vander Voort, F.J. Warmuth, S.M. Purdy, and A. Szirmae (Ed.) *Metallography: Past, Present, and Future (75th Anniversary Volume)*. American Society for Testing and Materials, Philadelphia, PA.

39. Marx, D.R. Interdiffusion of Electrodeposits with Basis Metals. M.S. Thesis, Penn State, 1976.

40. Katz, J.D., Pickering, H.W. and W.R. Bitler. 1980. Low Temperature Recrystallization Kinetics in Nickel Electrodeposits. *Plat. and Surf. Fin.* 67:45.

41. Katz, J.D., Pickering, H.W. and W.R. Bitler. 1980. Low Temperature Diffusion Kinetics in Nickel Electrodeposits. *Plat. and Surf. Fin.* 68:55.

42. Turn, J.C., Jr. Interdiffusion of Electro-deposits with Basis Metals. M.S. Thesis, Penn State, 1973.

43. Hyres, J.W., II. Low-Temperature Interdiffusion and a Transmission Electron Microscope Grain Size Study of Palladium Electrodeposits. M.S. Thesis, Penn State, 1987.

44. Lin, J.-C., Schulz, K.J., Hsieh, K.-C. and Y.A. Chang. 1989. *J. Electrochem Soc.* 136:3006.

45. Kirkaldy, J.S. and L.C. Brown. 1963. *Can. Metal* Q. 2:89.

46. Sands, T. 1989. *Mater. Sci. Eng.* B1:289 and references therein.

47. Sands, T., Keramidas, V.G., Yu, A.J., Gronsky, R. and J. Washburn 1987. *J. Mater. Res.* 2:262.

48. Shiau F.-Y., Chang, Y.A. and J.-C. Lin. 1992. *Mater. Chem. Phys.* 32:30.

49. Ko, D.-H., and R. Sinclair. 1994. *Ultramicroscopy,* 54:166.

50. Sands, T., Chang, C.C., Kaplan, A.S., Keramidas, V.G., Krishnan, K.M. and J. Washburn. 1987. *Appl. Phys. Lett.* 50:1346.

51. Caron-Popowich, R., Washburn, J., Sands, T. and A.S. Kaplan. 1988. *J. Appl. Phys.* 64:4909.

52. Mohney, S.E. and Y.A. Chang, 1993. *J. Appl. Phys.* 74:4403.

53. Poudoulec, A., Guenais, B., Guivarc'h, A., Caulet, J. and R. Guerin. 1991. *J. Appl. Phys.* 70:7613.

54. Mohney, S.E., Lin, C.F. and Y.A. Chang. 1993. *Appl. Phys. Lett.* 63:1225.

55. Guivarc'h, A., Caulet, J., Minier, M., Le Clanche, M.C., Deputier, S. and R. Guerin. 1994. *J. Appl. Phys.* 75:5061.

56. Sands, T., Keramidas, V.G., Yu, K.M., Washburn, J. and K. Krishnan. 1987. *J. Appl. Phys.* 62:2070.

57. Jan, C.-H., Swenson, D.. Zheng, X.-Y., Lin, J.-C. and Y.A. Chang. 1991. *Acta Metall. Mater.* 39:303.

58. Mohney, S.E. and Y.A. Chang. 1995. *J. Appl. Phys.* 78:1342.

59. *The Wall Street Journal* 29 Jan 1996 pg A14.

60. *The Wall Street Journal* 6 Feb 1996 pg C1.

Chapter Ten

AN INTRODUCTION TO POLYMER BLENDS

MICHAEL M. COLEMAN and PAUL C. PAINTER

Department of Materials Science and Engineering
Steidle Building
Pennsylvania State University
University Park, PA 16802

PREAMBLE

Polymer blends are simply physical mixtures of two or more different polymers and are analogous to metal alloys, materials that date back to the dawn of civilization. Just as ancient metallurgists (some of whom are still lurking about our department!) sought to improve the chemical, physical and mechanical properties of metallic materials by mixing different metals, so polymer scientists over the past half century have used the same approach and mixed together various polymers in an attempt to produce useful materials having specific desired properties. As we will see below, we must not push this analogy too far, however, because there are fundamental differences in the mixing of small molecules or metals compared to those of chain-like macromolecules or polymers.

Commercial and academic interest in polymer blends is at an all time high and it is unlikely that this will abate in the foreseeable future. In large part the reasons are economic. The cost of research, development, scale up and introduction of a new polymer into the marketplace is enormous and it has been argued by many that all the large scale economically viable (i.e. cheap) polymers have now been discovered. Accordingly, it follows logically that if desired properties can be attained by simply mixing two or more existing commercial polymers, surely this will be more cost effective than developing a new single polymeric material from scratch. While this is a persuasive argument, those of us who are approaching senior citizen status well remember similar arguments being expressed immediately prior to the discovery of Ziegler-Natta and metal oxide catalysts, which resulted in the introduction of isotactic polypropylene (PP), high density polyethylene

(HDPE), ethylene-co-propylene elastomers (EPDM), etc.! And there appears to be no limit to the ingenuity of the polymer chemist. New metallocene and other catalysts promise "designer" polymers where the desired structural properties of the polymer are "dialed in" and produced from the appropriate catalyst. Notwithstanding these advances, polymer blends are here to stay. It has been estimated (1) that by the end of the decade fully 90% of all polymers synthesized will be incorporated into blends and composites.

SOME DEFINITIONS: MISCIBLE/ IMMISCIBLE; COMPATIBLE/ INCOMPATIBLE; COMPATIBILIZER.

Polymer blends are conveniently subdivided into two major categories, based upon thermodynamic behavior. Polymer mixtures that are mixed at the molecular level and exist as a single phase are termed *miscible*. These polymer mixtures, depicted in the cartoon shown on the left hand side of Figure 1, are in the distinct minority of all polymer blends. Miscible polymer blends are true solutions of one polymer in the other and exhibit physical properties typical of single phase materials i.e. a single glass transition temperature (T_g), optical transparency etc. Examples of miscible polymer blends include poly(phenylene oxide) (PPO) – polystyrene (PS), which is a commercial engineering plastic produced by General Electric under the trade name Noryl®, poly(vinyl chloride) (PVC) – ethylene-co-vinyl acetate (EVA), which is essentially PVC containing a miscible macromolecular plasticizer, and poly(2,6-diisopropyl-4-vinyl phenol) (PDIPVPh) – poly(tetrahydrofuran) (PTHF), an example of a macromolecular antioxidant that is miscible with PTHF (2).

Single Phase

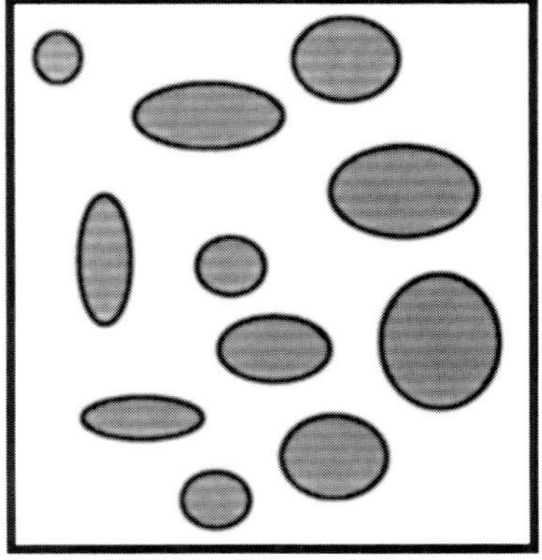

Two Phase

FIGURE 1. Cartoon of single and two phase polymer systems.

Immiscible or two phase binary polymer blend systems are the norm. In fact, if one were to take two polymers from the shelf at random and mix them in a blender of some type, the most likely outcome would be a grossly phase separated material, with one polymer poorly dispersed in a matrix of the other, as depicted on the right hand side of Figure 1. These blends are typically opaque and generally exhibit poor physical and mechanical properties.

The reluctance of most polymers to mix with one another has important ramifications in the plastics recycling industry. While some believe it is simply an industrial conspiracy that prevents us from recycling all plastics materials, it is actually a formidable scientific problem. Consider, for example, the ubiquitous soda bottle, illustrated schematically in Figure 2.

This is quite a complex piece of engineering, involving many different plastics materials, which are necessary because of the different chemical, physical and mechanical requirements of the various different parts of the bottle. Soda bottles are recycled in the USA, but not before the different plastics materials are separated from one another, and then only the poly(ethylene terephthalate) (PET) blown bottle and the HDPE base cup are actually recycled; the remaining plastics are committed to landfill.

Why can't we just grind up the whole bottle and put the mixture into an extruder or injection molding machine and make something useful? It just doesn't work. PET is completely immiscible with HDPE. For that matter, HDPE is immiscible with PP; two polymers that are chemically quite similar. Why is it that the vast majority of binary polymer mixtures are immiscible? We will attempt to answer that question in the next section of this article, but before we

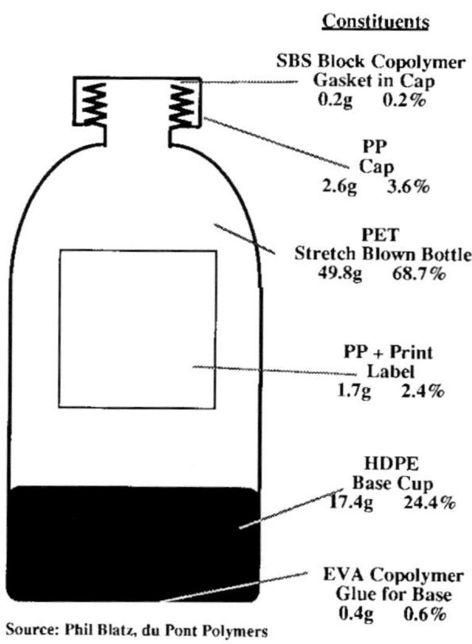

FIGURE 2. A standard 2 litre soda bottle

PET

HDPE

PP

get to that we need to consider the meaning of compatible and incompatible, two words that are commonly used (or misused) in the world of polymer blends.

Compatible is defined in Webster's Dictionary as "Capable of coexisting in harmony". This is a broad definition lacking scientific precision and herein lies a problem that pervades the scientific and patent literature. In scientific terms, the word compatible is almost meaningless. Compatible has been variously used to describe very disparate polymer blend systems from the truly miscible (single phase) polymer blends (here compatible is synonymous with miscible), to immiscible blend systems where the phase size of the minor component is relatively small and, at the other extreme, it has even been used to describe grossly phase separated materials where the component polymers simply have similar melt viscosities. The antonym to compatible, *incompatible,* has suffered less abuse, even though it has been used by some interchangeably with immiscible (remember, an immiscible polymer blend is two phase, but the two phases could be very small and contain significant concentrations of one polymer in the other). Incompatible is usually reserved for grossly phased separated materials that generally have poor mechanical properties. As the reader may have deduced, the authors dislike the use of the words compatible and incompatible to describe the phase behavior of polymer blends and usually avoid their usage like the plague!

Paradoxically, the terms *compatibilizer* and *compatibilization,* which for purists of the English language are as jarring as a fingernail down a blackboard, do have reasonably well defined meanings. This is best illustrated by a description, albeit rather simplistic, of rubber toughened or high impact polymeric materials. These materials are exemplified by the commercially successful high impact polystyrene (HIPS) and acrylonitrile/butadiene/styrene (ABS) materials that find application in such articles as car dashboards, car bumpers, refrigerator panels etc. Polystyrene is a glass at room temperature ($Tg \approx 100°C$) and as such is a brittle material unsuited for the applications just mentioned. However, it has the distinct advantages of being a strong, stiff material that is easily processed and yields articles exhibiting high gloss. It is just that polystyrene has very poor impact strength, as anyone who has squeezed a transparent plastic beer "glass" quickly finds out. Elastomers like natural rubber, polybutadiene or styrene/butadiene (SBR), on the other hand, with Tgs well below -40°C, are naturally resilient and inherently impact resistant, but they lack strength, stiffness and are dull in appearance. What if we blended together polystyrene with, say, 15-30% of some crosslinked rubber latex? Wouldn't we have the best of both worlds, where the matrix was the strong, stiff, glossy polystyrene into which was dispersed a minority component of energy absorbing rubber balls? Shouldn't this give us a high impact material? The answer is no. As depicted schematically in Figure 3, a propagating crack in such a material progresses as though the rubber particles were not there. Brittle fracture occurs and the effect of the rubber particles on the impact strength of the blend is marginal.

This is, of course, a disappointing result, but in hindsight not a surprising one. Polystyrene and rubber "hate" each other and form grossly phase separated immiscible blends with insignificant interpenetration at the interface, which implies minimal adhesion between the two phases. How then can one overcome this problem? – the answer, use a compatibilizer!

Compatibilizers are typically block or graft copolymers as illustrated below.

A—A—A—A—A—A—A—A—A—A—B—B—B—B—B—B—B—B—B—B—B—B
Block Copolymers (Diblock)
A—A—A—A—A—A—A—A—A—A—R—A—A—A—A—A—A—A—A—A—A—A
|
B
|
B
|
B
|
B
|
B
|
B

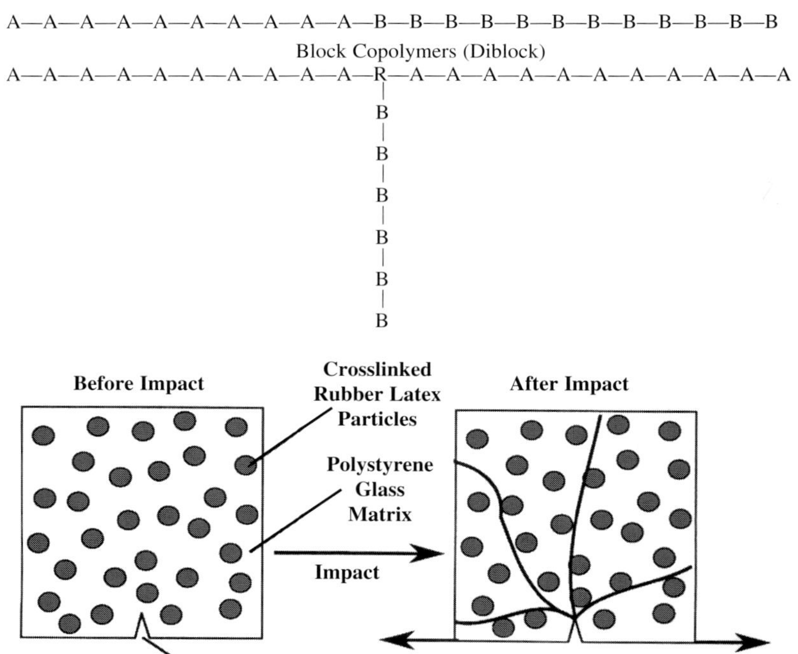

FIGURE 3. Schematic diagram of a blend of polystyrene and crosslinked rubber latex particles.

where A might be, for example, butadiene and B styrene.

Such block and graft copolymers are unhappy molecules. As we mentioned before, the styrene block or graft hates the rubber part of the molecule and *vice versa,* but they are covalently bonded together and have to make the best of it. What is interesting is that if a small amount of the styrene/rubber block or graft copolymer is introduced, (by deliberate addition or by formation *in situ* during synthesis), into the blend of polystyrene and crosslinked rubber latex, the effect on the impact strength of the blend is dramatic. As illustrated in Figure 4, the block or graft copolymer acts as a surface active agent (similar to a soap or surfactant), with the polystyrene part of the molecule residing in the polystyrene matrix and the rubber part residing in the rubber particle. In other words, the block or graft copolymer serves to "bind" the rubber particles to the polystyrene matrix.

The mechanism of rubber toughening in HIPS and ABS materials is complex, but it is known that it involves processes called sheer banding and, perhaps of even greater importance, craze formation. Figure 5 depicts the formation of a craze following an impact directed onto a HIPS or ABS type material. A craze may be thought of as a crack filled with a foam-like structure containing fibrils and many voids. The material maintains most of its inherent strength, however. In large part, the enhanced adhesion between the polystyrene matrix and the rubber particles, derived from the presence of the graft or block compatibilizer,

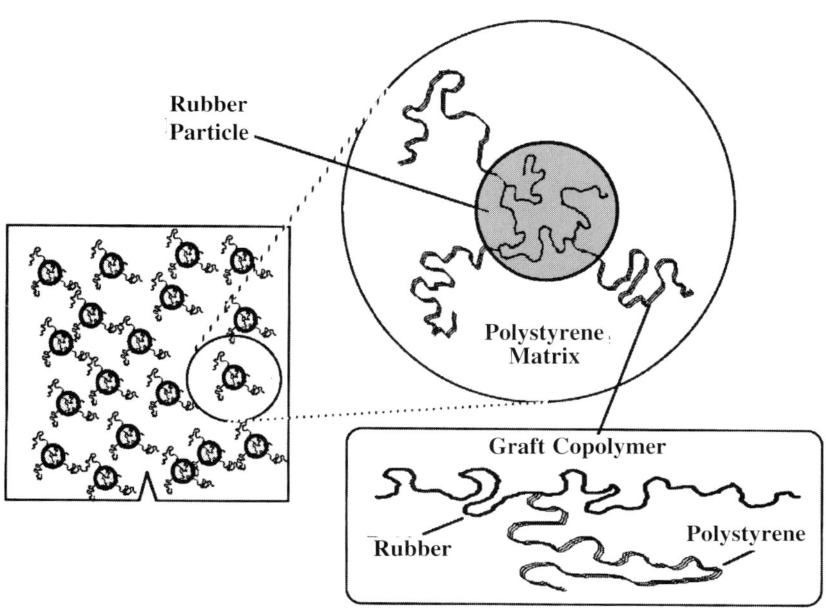

FIGURE 4. Schematic diagram of a polystyrene/rubber blend containing a graft copolymer.

is responsible for the absorption of the energy upon impact through the formation of crazes. This results in a high impact material which does not fail in a brittle manner.

Finally, we should note that compatibilizers (block, graft and certain random copolymers), are added to many other immiscible polymer blend systems in order to reduce the phase size of the "droplets" of the minor component and increase dispersion. Compatibilizers serve to modify the surface tension and wetting properties between the two blend components, and if they are used in conjunction with a careful matching of melt viscosity parameters, very fine dispersions can be achieved which can result in significantly enhanced impact strengths. Compatibilization in this context thus means the formation of a fine stable dispersion.

WHY ARE THE MAJORITY OF POLYMER MIXTURES IMMISCIBLE?

We all have the intuitive knowledge from everyday life experiences that if we mix two low molecular weight materials together the chances are that they will mix and form a single phase. So, if we were to take say benzene and toluene off the shelf and mix them in equal proportions, we would not be surprised if we were to obtain a miscible (single phase) mixture. Yes, we know that oil and water phase separate, but such examples are in the minority and there are extraordinary differences involved in this case – we will come back to this later. Organic chemists, or for that matter anyone who has taken a freshman organic laboratory course, intuitively knows the rule of thumb, "like likes like". This is a commonsense expres-

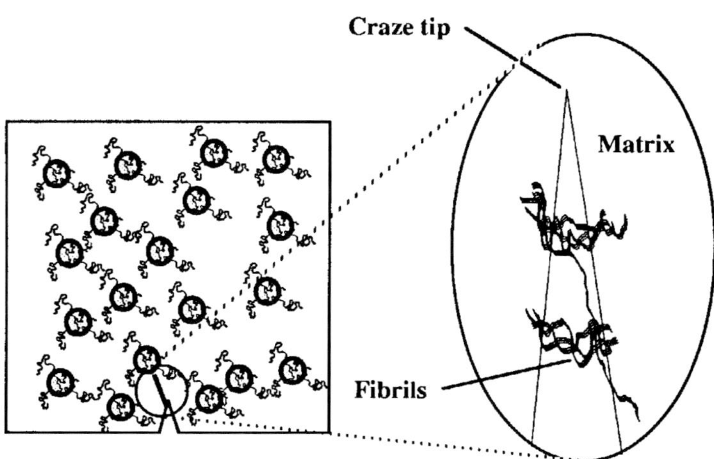

FIGURE 5. Schematic diagram of a craze.

sion that reflects the general observation that two low molecular weight materials are most likely to be miscible if they have similar chemical structures. So we would expect chloroform to mix in all proportions with carbon tetrachloride. And they do.

To understand why this rule of thumb works it is informative to consider the Gibbs free energy of mixing equation:

$$\Delta G_m = \Delta H_m - T\Delta S_m \tag{1}$$

For a binary mixture to be miscible (single phase), at a particular temperature across the entire composition range, the free energy of mixing, ΔG_m, must be negative (and additionally the second derivative of the free energy with respect to composition, $\partial^2\Delta G_m/\partial\Phi^2$, must be positive). Figure 6 depicts the mixing of two chemically similar low molecular weight molecules, benzene and toluene. It is important to recognize that in this case the change in enthalpy upon mixing, ΔH_m, is endothermic (albeit rather weakly) and unfavorable to mixing because it is a positive contribution to the free energy of mixing. On the other hand, the change in entropy upon mixing, ΔS_m, is strongly positive, which yields a large negative component to the free energy and completely overwhelms the

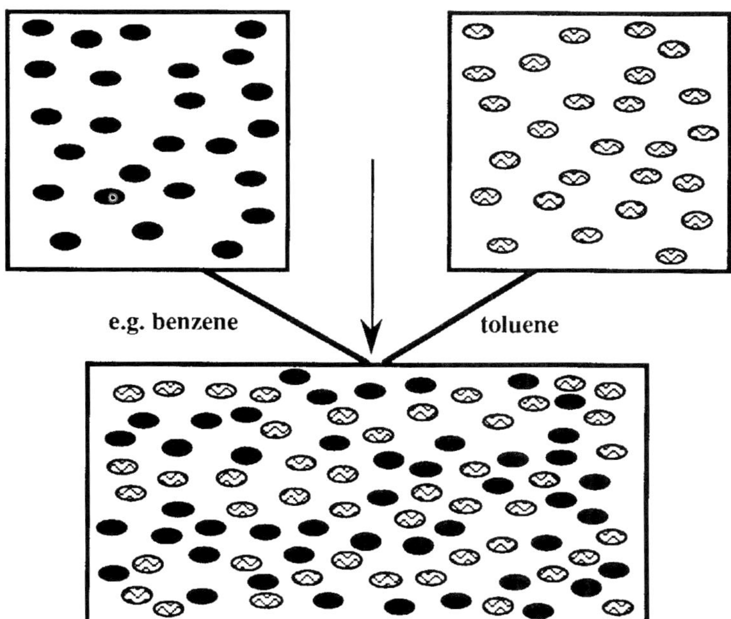

FIGURE 6. Schematic diagram of the mixing of two low molecular weight molecules.

unfavorable ΔH_m contribution. Accordingly, "like likes like" works because mixing is driven by combinatorial entropy#.

When we come to mixing of polymers, the adage "like likes like" fails miserable. Figure 7 depicts the potential mixing of two chemically similar polymers, polystyrene (PS) and poly(α-methyl styrene) (PαMS) that are analogous to the low molecular molecules considered above, benzene and toluene. PS and PαMS form immiscible blends. We have mentioned before when describing the soda bottle, HDPE and PP also do not mix in the melt (amorphous state), even though they are chemically quite similar. This is in fact the rule for the vast majority of all polymer blends. Again, the change in enthalpy upon mixing, ΔH_m, is weakly endothermic and unfavorable to mixing. However, unlike the low molecular weight case, the change in combinatorial entropy arising from the mixing of two high polymers is negligibly small (see later for an explanation). Combinatorial entropy is no longer the driving force for mixing, and the unfavorable enthalpy term dominates, resulting in immiscibility.

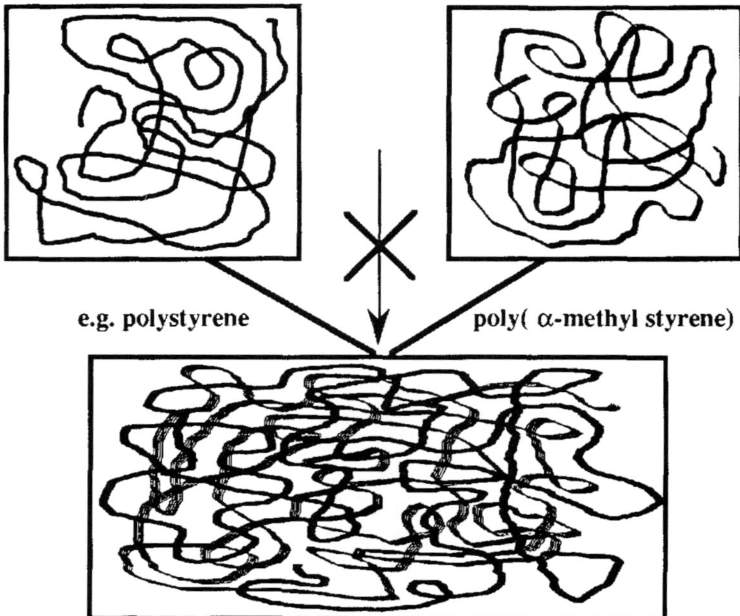

FIGURE 7. Schematic diagram of the mixing of two high molecular weight polymers.

Incidentally, in the case of oil and water, ΔH_m is very strongly endothermic, partly due to the extensive hydrogen bonding present in water (self-association), and even the large favorable contribution from combinatorial entropy is not sufficient to overcome the unfavorable enthalpy term.

Since the very small change in combinatorial entropy from the mixing of two polymers is inevitable, the only way to "drive" the mixing of two polymers is through the manipulation of the enthalpy term. In fact, Paul Flory wrote in his classic 1952 textbook, *Principles of Polymer Chemistry (3)*;

"The critical value of the interaction free energy is so small for any pair of polymers of high molecular weight that it is permissible to state as a principle of broad generality that two high polymers are mutually compatible with one another only if their free energy of interaction is favorable, i.e. negative. Since the mixing of a pair of polymers, like the mixing of simple liquids, in the great majority of cases is endothermic, incompatibility of chemically dissimilar polymers is observed to be the rule and compatibility the exception. The principal exceptions occur among pairs possessing polar substituents which interact favorably with one another."

Similarly, Sonya Kraus (4), in a 1978 review of polymer – polymer miscibility wrote, "It is best to expect odd instances of compatibility between polymers, – – whenever hydrogen bonding is possible". In other words, if one can find chemically dissimilar polymers containing functional groups that attract one another, the free energy contribution from specific interactions can be favorable and the primary driving force for miscibility. The authors of this article have spent a considerable portion of their careers searching for such miscible hydrogen bonded polymer blends and attempting to predict their phase behavior (5). One example should suffice; poly(4-vinyl phenol) (PVPh) is miscible with poly(ethyl methacrylate) (PEMA) for all compositions in the experimental accessible temperature range from room temperature to the onset of significant thermal degradation ($\approx$ 220°C). PVPh strongly self-associates through the formation of intra- and intermolecular hydrogen bonds between the hydroxyl groups of the 4-vinyl phenol segments in the same and other polymer chains, as illustrated schematically in Figure 8. PEMA, on the other hand, does not self-associate to any measurable extent, but does interassociate through hydrogen bonding to PVPh (Figure 8). This is an equilibrium process, and if upon blending PVPh with PEMA there is a favorable contribution to the free energy of mixing due to the changing pattern of hydrogen bonds in the blend vis-a-vis the pure components, this can be the driving force towards polymer / polymer miscibility. To put this on a more rigorous basis we should briefly discuss some of the simple models that have been developed to describe the mixing of two polymers.

SIMPLE MODELS FOR THE MIXING OF POLYMERS – FLORY/HUGGINS AND THE ASSOCIATION MODEL

Flory (3,6) and Huggins (7) independently and concurrently extended the ideas of Fowler and Guggenheim (8) and used a lattice model to develop a free energy of mixing equation, ΔG_m, that is appropriate for *non-polar* polymer solutions and blends. The resulting Flory/Huggins (F/H) equation (equ.2) is considered one of the major achievements of polymer theory, and while the model has many limitations, it is simple to understand and, most importantly, quantitatively rationalizes why most polymer blends are immiscible.

$$\frac{\Delta G_m}{RT} = \left\{ \frac{\Phi_A}{N_A} \ln \Phi_A + \frac{\Phi_B}{N_B} \ln \Phi_B \right\} + \chi_{AB} \Phi_A \Phi_B \qquad (2)$$

In this equation (which is written in terms of a mole of lattice sites), Φ_A, Φ_B and N_A, N_B are the volume fractions and degrees of polymerization of polymer A and polymer B, respectively, and χ is the well known Flory interaction parameter. The first two terms contained in the bracket of equation 2 are the contribution from combinatorial entropy and it is informative to roughly calculate the magnitude of the combinatorial entropy contribution for a typical low molecular weight mixture of A and B, a solution of a polymer polyA in a solvent B and finally a polymer blend consisting of polyA and polyB. Let us assume that A and B each have a molar volume of 100 cm³ mole^{-1} and that polyA and polyB have the same degree of polymerization $N_A = N_B = 100$ (this would be equivalent to a moderate molecular weight of 10,000 if the repeat units have a molecular weight $M_0 = 100$ g mole$_{-1}$). If we consider a 1:1 mixture by volume (i.e. $\Phi_A = \Phi_B = 0.5$), then the magnitude of the combinatorial entropy contribution to the free energy is -0.69, -0.35 and -0.001 for the low molecular weight mixture, polymer solution and polymer blend, respectively. While there is about a 50% decrease in going from the low molecular weight mixture to the polymer solution, there is a greater than

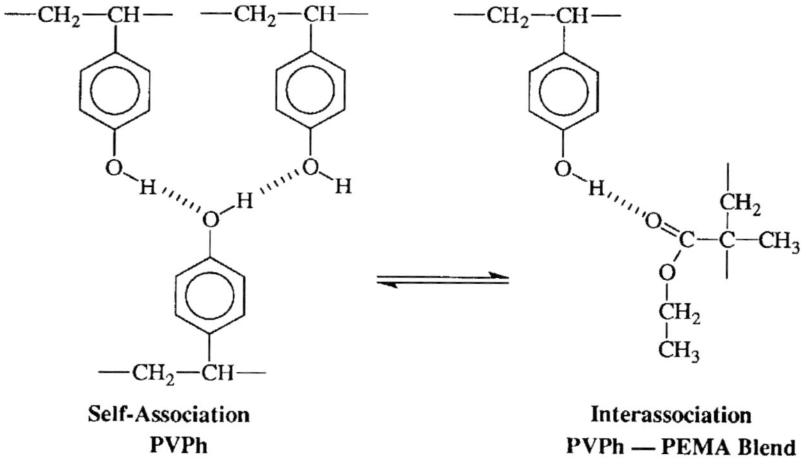

FIGURE 8. An example of a miscible hydrogen bonded polymer blend

a two orders of magnitude decrease to the polymer blend. Since equation 2 is essentially a balance between the favorable to mixing (negative) entropic contribution and the unfavorable (positive) enthalpic contribution (the $\chi_{AB}\Phi_A\Phi_B$ term), the upshot of these dramatic differences in the contribution from combinatorial entropy is that only a very small value of χ_{AB} can be tolerated if a miscible polymer blend is required. A critical value of χ_{AB} can be determined by equating the second and third derivatives of equation 2 with respect to composition (Φ_A) and setting them to zero, which gives the following result:

$$\chi_{Crit} = \frac{1}{2}\left[\frac{1}{N_A^{0.5}} + \frac{1}{N_B^{0.5}}\right]^2 \tag{3}$$

Accordingly, using the same parameters as those mentioned above, χ_{crit} values of 2, 0.6 and 0.02 are calculated for the low molecular weight mixture, polymer solution and polymer blend, respectively. If the molecular weight of the two polymers is increased to 10^3 g mole^{-1} (i.e., $N_A = N_B = 1000$) then χ_{crit} equals 0.002. This is a very small value of χ_{crit} and there are very few examples of non-polar polymer blends where the value of χ is less than χ_{crit}. Like the English language, there are always exceptions to the general rule, however, and Trask and Roland (9) report a very low value of $\chi = 0.00017$ for the miscible *cis*-1,4-polyisoprene blends with poly(vinyl ethylene).

Let us assume that we have two essentially non-polar polymers A and B, that have a value of χ at room temperature less than that of the critical value, which infers a miscible polymer blend. In its simplest form Flory[3] has shown that χ varies with the inverse of temperature, so that as we decrease temperature the value of χ increases. At some temperature, below room temperature (always assuming that we are above the T_g) the value of χ will equal that of χ_{crit} and upon further cooling χ will exceed χ_{crit} and the blend will phase separate.

We can calculate a spinodal phase diagram by simply taking the second derivative of the F/H free energy equation (equ.2) with respect to composition as a function of temperature assuming χ varies as T^{-1}. The spinodal line is the locus of the points where the second derivative of the free energy equals zero. A typical example of such a spinodal phase diagram is shown in Figure 9. The very top of the two phase region is called the upper critical solution temperature (UCST), which mathematically may be calculated by equating the third derivative of the free energy to zero. While the UCST is not usually observed experimentally in most of the known miscible polymer blend systems because of limitations caused by the T_g, there are many examples of polymer solutions that exhibit UCST behavior. The existence of a UCST is pleasing because it confirms the commonsense adage that mixing is assisted by heating.

So far so good. But now we come to an interesting experimental observation, which the F/H theory completely fails to predict and is somewhat counterintuitive. Some blends, for example, polystyrene blends with either poly(vinyl methyl ether) or poly(phenylene oxide), are miscible at say room temperature, but phase separate upon heating. Conversely, the blend may be immiscible at a relatively high processing temperature, but form a miscible single phase material upon cooling (assuming appropriate diffusion kinetics). This type of phase diagram is illustrated schematically in Figure 10 and appears very strange. There is nothing in the F/H treatment to explain such a phenomenon. Recall, that the Flory χ parameter deals with relatively weak intermolecular interactions that are present in essentially non-polar polymers, so we are not observing the effects of strong interactions such as hydrogen bonds, discussed later. The presence of a lower critical solution temperature (LCST) has been ascribed to a so-called free volume effect, which is a small unfavorable contribution to the free energy of mixing and increases with increasing temperature (10). The free volume effect has its origin in the motion and vibrations of molecules that lead to density fluctuations. This can be modeled by introducing holes on the lattice (11). A mismatch in the free volumes of the two polymers in the blend, even though they are only quite small differences, is enough to yield a correspondingly small unfavorable contribution to the overall free energy of mixing. To reiterate, for non-polar polymer blends the magnitude of the individual contributions from combinatorial entropy and the $\chi_{AB}\Phi_A\Phi_B$ term are relatively small and small changes in these values can have an enormous effect on the phase diagram. The free volume effect is likewise a small contribution in absolute terms, but since all the contributions to the free energy involve rather small numbers, the contribution from free volume tips the balance at temperatures exceeding the LCST resulting in phase separation.

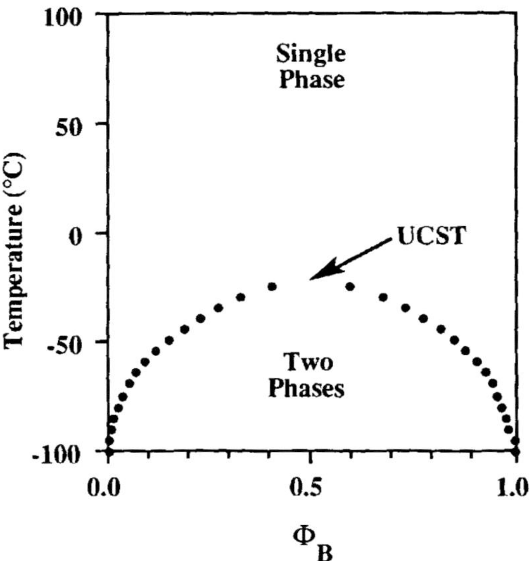

FIGURE 9. Phase diagram calculated from the F/H equation (equ.2)

The preceding discussion in this section has been limited to polymer blend systems involving relatively weak interactions (dispersive and perhaps weak polar forces). What happens if there are stronger dipolar or even weak hydrogen bonds present in the blend? For example, poly(vinyl chloride) (PVC) is miscible with poly (ϵ-caprolactone) (PCL) in the amorphous state, and it is known from infrared spectroscopic studies that there is a significant interaction (in the range of $\approx$ 1-2 kcal mole^{-1}) between PVC and itself (self-association) and also between the PVC methine hydrogen and the PCL carbonyl group.

The potential exists for a favorable contribution from the presence of these interactions and one way of dealing with this has been to permit the value of χ in the F/H equation (equ. 2) to assume negative values, which, of course, guaran-

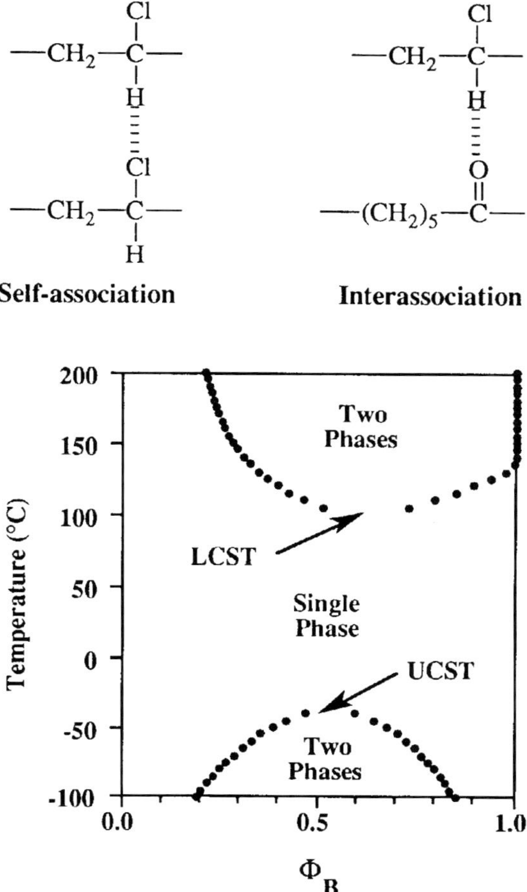

FIGURE 10. Phase diagram illustrating UCST and LCST behavior.

tees a negative free energy of mixing. However, in the opinion of the authors of this article, simply permitting the value of χ to be negative, especially as the interactions become stronger and stronger, is a poor approximation that violates theoretical principles and is tantamount to lumping together two disparate factors that have very different compositional and temperature dependencies. This leads us to the free energy equation that we have developed for hydrogen bonded polymer blends, where we have effectively separated interactions into contributions from unfavorable "physical" (essentially dispersive and weak polar forces) and favorable "chemical" (strong polar and hydrogen bonding forces) (5).

A Flory type lattice approach, applied in conjunction with an association model to describe hydrogen bonding interactions, was used to derive equation 4.

$$\frac{\Delta G_m}{RT} = \left\{ \frac{\Phi_A}{N_A} \ln \Phi_A + \frac{\Phi_B}{N_B} \ln \Phi_B \right\} + \chi_{AB} \, \Phi_A \Phi_B + \frac{\Delta G_H}{RT} \tag{4}$$

This equation has a similar form to that of the Flory-Huggins equation, but with an added contribution, $\Delta G_H/RT$. Equation 4 can be thought of as consisting of three major contributions. Combinatorial entropy; a very small but nonetheless favorable contribution to the free energy of mixing that is contained in the first two logarithm terms. Thus the free energy of mixing is dominated by the balance of the last two terms, $\chi\Phi_A\Phi_B$, which is an unfavorable contribution from what we term "physical" (i.e. van der Waals) forces and the $\Delta G_H/RT$, a favorable contribution derived from hydrogen bonding or so-called "chemical" forces. The positive contribution from "physical" forces is determined using a Flory type χ parameter that is, in turn, estimated from solubility parameters calculated from non hydrogen bonded group molar attraction and molar volume constants. The negative contribution from "chemical" forces is determined from equilibrium constants and enthalpies of hydrogen bond formation that are derived from infrared spectroscopic data and describe self- and inter-association and the distribution of hydrogen bonded species in the polymer blend.

A major assumption is that we can separate the contributions to the free energy of mixing from "strong" and "weak" forces. It should be underscored that separation of strong and weak forces is not just a matter of convenience. The contributions to the free energy of mixing from "weak" and "strong" forces have very different temperature and compositional dependencies, which markedly affect the shape and form of phase diagrams (a factor often ignored by those attempting to describe blends involving strong interactions by a single negative χ parameter). Miscibility thus depends primarily upon the balance between the contributions from the $\chi\Phi_A\Phi_B$ and $\Delta G_H/RT$ terms. If there is no contribution to the free energy of mixing (and its derivatives) from "chemical" forces, we revert to the Flory-Huggins case. However, if there is a significant contribution from "chemical" forces, i.e. $\Delta G_H/RT < 0$, then χ_{crit} will be >0.002 (for $N_A = N_B = 1000$) and the actual value of χ_{crit} will depend upon the magnitude of the $\Delta G_H/RT$ term.

It is not unusual in hydrogen bonded polymer blends, where there is a large favorable contribution from "chemical" forces, to have χ_{crit} values in excess of 1.0.

So how do we calculate the $\Delta G_H/RT$ contribution and how well does equation 4 predict the phase behavior of hydrogen bonded polymer blends? Infrared spectroscopy is uniquely sensitive to hydrogen bonding and in the spectra of appropriate polymer blends, (e.g. PVPh blends with (co) polymers containing carbonyl groups such as polyacrylates, polymethacrylates, polyesters, etc.), one can observe two well resolved infrared bands attributed to carbonyl groups that are hydrogen bonded and to those that are not (commonly called "free" carbonyl groups) (5,12). Typical infrared spectra are shown in Figure 11. The band at $\approx$ 1740 cm^{-1} corresponds to the C=O stretching vibration of "free" carbonyls, while that observed at $\approx$ 1708 cm^{-1} is due to the C=O stretching vibration of hydrogen bonded carbonyls. A quantitative analysis of the fraction of hydrogen bonded carbonyl groups, as a function of composition and temperature, can be readily determined by curve resolving the spectra into the two components (areas) and taking into account their differing absorptivity coefficients (5,12). From the fraction of hydrogen bonded carbonyl group data a set of equilibrium constants and enthalpies of hydrogen bond formation can then be determined that describe the distribution of hydrogen bonds in blends containing similar

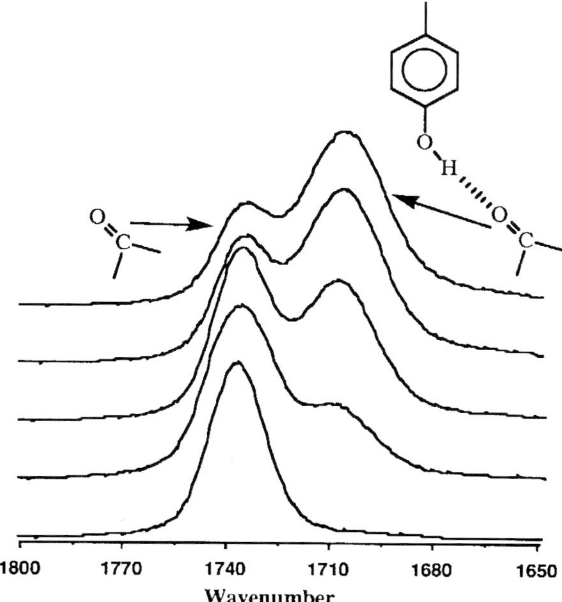

FIGURE 11. Typical infrared spectra of PVPh blends with an ester containing polymer. From bottom to top: 0, 20, 40, 60 and 70% PVPh.

functional groups. Armed with these equilibrium constant values and enthalpies of hydrogen bond formation, together with the molar volumes of the polymer repeat units, we can now calculate the contribution from the $\Delta G_H/RT$ term in equation 4, without any adjustable parameters. Since the rest of equation 4 is simply the F/H equation discussed previously, we are now in a position to calculate the free energy of mixing and phase diagrams for hydrogen bonded polymer blends.

Figure 11 shows typical spinodal phase diagrams calculated for PVPh blends with a series of poly(n-alkyl methacrylates); poly(ethyl methacrylate) (PEMA), poly(n-butyl methacrylate) (PBMA) and poly(n-hexyl methacrylate) (PHMA).

Although the chemical structure of the poly(n-alkyl methacrylates) is much the same, differing only in the number of methylene groups in the side chain

PEMA   **PBMA**   **PHMA**

(1, 3 and 5, respectively, for PEMA, PBMA and PHMA), their predicted phase diagrams in blends with PVPh runs the gambit from the essentially miscible (PEMA), to one containing a closed loop type (PBMA) and finally to an immiscible one shaped like an hour glass (PHMA). PVPh blends are predicted to be

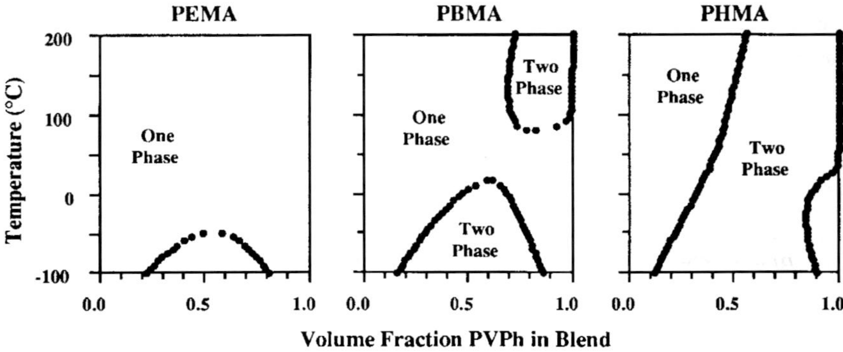

FIGURE 12. Predicted phase diagrams for PVPh blends with PEMA, PBMA and PHMA.

miscible with PEMA from the UCST at about -50°C to above the onset of degradation ($\approx$ 220°C). Experimentally, we cannot test the UCST because the T_g of the blends is above 80°C, but we have determined that PVPh/PEMA blends are miscible (single phase across the whole composition range) above 100°C. At the other extreme, PVPh/PHMA blends are predicted to be immiscible over the entire experimentally accessible temperature range, which in fact they are. The most interesting case, however, is that of the intermediate PVPh/PBMA blends. Here the predicted spinodal phase diagram has two regions of calculated two phase areas. The first is that associated with the UCST, while the second is a high temperature closed loop region (theoretical calculations above 200°C show that a closed loop is indeed formed and at even higher temperatures, inaccessible experimentally , complete miscibility is again predicted). This is characteristic of hydrogen bonded low molecular weight mixtures such as nicotine and water and is a consequence of the complex free energy changes (enthalpic and entropic) involved with the forces of self-association and interassociation. The essential features of the PVPh/PBMA phase diagram have been verified experimentally (5). Blends of PVPh and PBMA are miscible at room temperature, and those compositionally rich in PVPh do phase separate above about 80°C. All in all, the prediction of the phase diagrams from equation 4 has been remarkably successful for PVPh blends with poly(n-alkyl methacrylates). We should also mention that free volume effects are not required to explain the existence of LSCTs in the case of hydrogen bonded blends. In fact, free volume effects can usually be ignored in such systems as phase behavior is dominated by the balance of the "physical" $\chi\Phi_{AB}\Phi_{B}$ and "chemical" $\Delta G_H/RT$ terms.

LAST WORDS

To summarize, most non-polar polymers do not mix with one another. However, by introducing the appropriate types of functional groups into a polymer chain mixing can be driven by strong, specific interactions (i.e. hydrogen bonds). The resulting phase diagrams are rich and there are many complications that we have not mentioned here. We expect these to consume the rest of our working lives. God help us and those poor students who work with us!.

REFERENCES

1. Utracki, L.A. 1995. *Encyclopaedic Dictionary of Commercial Polymer Blends.* ChemTech Publishing, Toronto, Ontario, Canada.
2. Coleman, M.M., Yang, X., Stallman, J.B. and P.C. Painter, 1995. *Makromol. Symp., 94,*1.
3. Flory, P.J. 1953. *Principles of Polymer Chemistry.* Cornell University Press, Utica, N.Y.

4. Krause, S. 1978. Polymer-Polymer Compatibility. Chapter 1 in: D.R. Paul and S. Newman (Ed.), *Polymer Blends*, Academic Press, NY.

5. Coleman, M.M. and P.C. Painter 1991. *Specific Interactions and the Miscibility of Polymer Blends*. Technomic Publishing Co. Inc., Lancaster, Pennsylvania, USA.

6. Flory, P.J. 1941. *J. Chem. Phys. 9*, 660.

7. Huggins, M.L. 1941. *J. Chem. Phys., 9*, 440

8. Guggenheim, E.A. 1952. *Mixtures*. Clarendon Press, Oxford.

9. Trask, C.A. and C.M. Roland 1989. *Macromolecules, 22*, 256.

10. Patterson, D. and A. Robard 1978. *Macromolecules, 11*, 690.

11. Sanchez, I.C. and R.H. Lacombe 1978. *Macromolecules, 11*, 1145.

12. Coleman, M.M. and P.C. Painter 1984. Appl. *Spectrosc. Revs., 20*, 255.

Chapter Eleven

FUNCTIONAL POLYOLEFINS:
Synthesis and Applications

T. C. CHUNG

Department of Materials Science and Engineering
Steidle Building
Pennsylvania State University
University Park, PA 16802

INTRODUCTION

Polyolefins, especially polyethylene (PE), polypropylene (PP), poly(1-butene) and ethylene-propylene copolymers (EP rubber), are a very important class of materials accounting for over 25 billion pounds (1) of polymer per year in US, close to 60% of the total polymer produced. They greatly influence our day-by-day life, from bread bags, milk jugs, hoses, tires to bullet-proven jackets. They possess an excellent combination of mechanical, chemical and electronic properties along with low cost, light weight (low density) and superior processability. Most of the polyolefins can be processed by all available methods, including extrusion and injection molding, compression molding, injection blow molding and calendering, to various sized and shaped (films, sheets, pipes, fibers) products. Their properties can be further widened by copolymerization, orientation, and other techniques. By controlling molecular weight and crystallinity, polyolefins with a wide range of thermal and mechanical properties, from plastic to rubber to wax, have been produced for many commercial applications. Because olefin monomers are oil refining by-products and extremely high polymer/catalyst turnovers are possible, polyolefins represent one of the least expensive high performance polymers. It is their excellent performance versus cost that make polyolefins so useful in today's society. In 1955 the first low pressure (Ziegler process) polyethylene production plant went on stream and in 1957, only three years after its initial discovery, the first isotactic polypropylene plant went into operation. Since that beginning the market and the production of PP, PE, higher polyolefins, and various olefin copolymers has continually expanded. The principal commercial polyolefins are high density polyethylene (HDPE), isotactic polypropylene (i-PP), polybutene, polyoctene, poly(4-methy-1-pentene) and their copolymers, such as linear low density polyethylene (LLDPE), ethylene-propylene copolymer (EP rubber).

The world market for polyolefins has steadily increased. It is projected that U.S. demand for polypropylene alone will grow from 6.4 billion lb in 1990 to 11.5 billion lb by the year 2000, posting a 5.8% annual growth rate (1).Western Europe and Japan are expected to see an even higher growth rate in the near future. With increasing foreign competition, especially from the oil-producing and the pacific rim countries, polyolefin research continues with the aim of improving production processes, lowering operating costs, increasing catalyst efficiency, and developing new grades of products. With huge world-wide capacity, producers are continually searching for new markets and new applications for these ubiquitous polymers. One of the strategies for increasing applications of polyolefins has been to transform them from commodity to specialty materials through copolymerization, blending, rubber toughening, fiber reinforcing, introducing fillers and additives.

The increasing awareness of environmental issues and the impact upon human safety of the use of chemicals means that significant pressure is placed on manufacturers to choose the environment-friendly materials in their products. Many polymers have been facing increasing resistance to their use in many areas of the world. Since polyolefins are the polymeric form of petroleum products, they naturally behave as solid fuel with similar burning characteristics as the regular liquid fuels. Polyolefins are readily recyclable due to their thermoplastic properties with melting points below 250°C. Usually, plastizers and lubricants are not needed in their processing. As landfill spaces are becoming less available, with no chance of improving in the foreseeable future, the choice of material is clearly favorable to easy recycling and clean incineration materials. Polyolefins are becoming favorable choice of materials.

ZIEGLER-NATTA POLYMERIZATION

Olefins are only polymerized to linear high molecular weight polymers by transition metal catalysts, usually Ziegler-Natta catalysts (2). In addition, Ziegler-Natta catalysis is one of the few methods capable of stereoregularly polymerizing α-olefins. Isotactic polyolefins with repeating stereoregular structure as shown in Figure 1 leads to helical chains which can efficiently pack together in crystals. The catalyst controlled isotactic structure is responsible for the excellent thermal and mechanical properties of stereoregular polyolefins.

CATALYST DEVELOPMENT

The discovery of Ziegler-Natta (Z-N) catalysts started from Professor Karl Ziegler at the Max Planck Institute in Germany. While studying the reactions and oligimerization of ethylene by aluminum alkyls, known as "Aufbau" reaction (2). He found by accident that contamination from the nickel reaction vessel altered the reaction. With this clue, Ziegler began investigating reactions of a series of transition metal salts with $AlEt_3$ and ethylene. In 1953 the first actual Ziegler polymerization, as we know it, used zirconium acetylacetonate and $AlEt_3$ to form high molecular weight linear polyethylene. The process was soon improved by the use of $TiCl_4$ and $AlEt_3$. Giulio Natta, a consultant with Montecatini Co. under an agreement with Ziegler, soon undertook further studies of these catalysts. It was in 1954 that the second major breakthrough took place. Natta realized that the Ziegler catalysts could polymerize propylene to crystalline, stereoregular high polymer. Immediately after these initial experiments larger research groups were set up in the U.S., Italy, and Germany to study these compounds and their resulting polymers. Ziegler and Natta were jointly awarded the Nobel Prize for Chemistry in 1963 for their roles in the start of the "Golden Age" of polymer science (2).

In the broadest definition Z-N catalysts are a mixture of a metal alkyl of base metal alkyls of group I to III and a transition metal salt of metals of group IV to VIII. The base metal component, most often an aluminum alkyl, serves as an alkylating agent for the transition metal salt. The most common one, $TiCl_3$ is a crystal of alternating layers of Ti^{+3} and Cl^{-1} ions. At the edges, cracks or defects the alkylation takes place on titanium atoms with an unsaturated coordination sphere. As shown in Figure 2, two

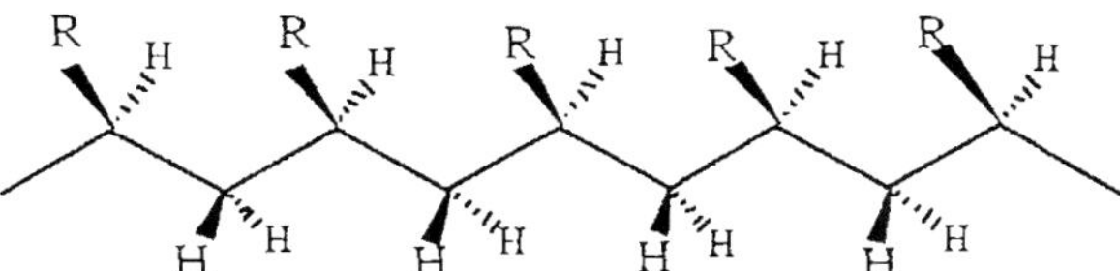

FIGURE 1. Molecular structure of isotactic polyolefin.

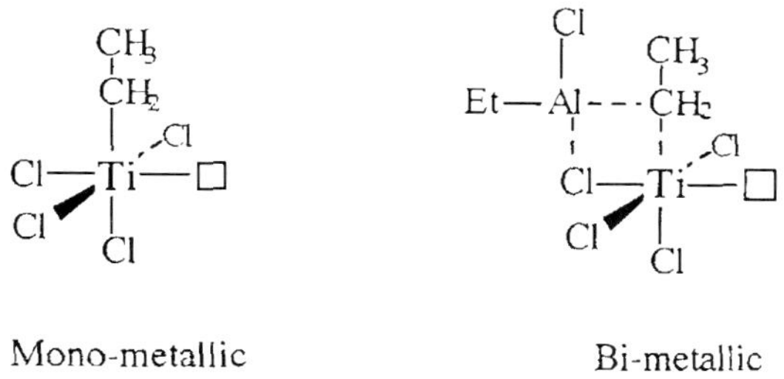

Mono-metallic   Bi-metallic

FIGURE 2. Two active species in Ziegler-Natta catalyst.

different active species are proposed, the mono-metallic and bi-metallic active centers. There is evidence that both species might be present in the heterogeneous catalysts.

Both homogeneous and heterogeneous catalysts are useful in Z-N processes. In the past, the heterogeneous systems offered better stereospecific addition for the preparation of isotactic structure. The soluble systems are generally used for poly-ethylene and EP copolymers where tacticity is irrelevant. However, a new generation soluble metallocene catalysts with high isospecific control will be discussed later.

POLYMERIZATION MECHANISM

To this day there is still controversy over the exact mechanism of Z-N polymer-ization. Even the propagating active center of the reaction is disputed. The most rea-sonable theory states that the polymer growth takes place at a transition metal-carbon bond. "Cossie was the first to propose a mechanism in which the polymer chain grows by successive olefin insertion" (3) as shown in Figure 3. Cossie's mechanism was later proven by deuterium labeling experiments done by Grubbs (4).

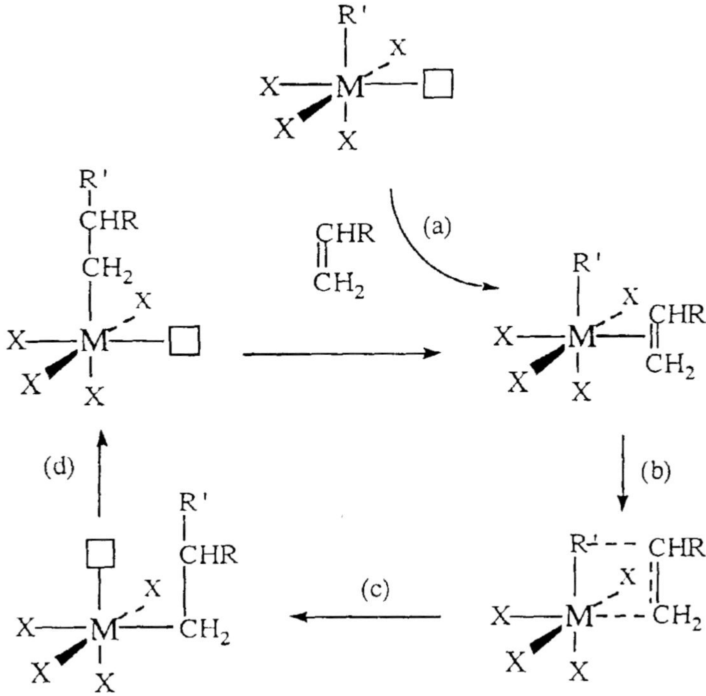

FIGURE 3. Cossie's mechanism of Ziegler-Natta polymerization.

The first step of the reaction is the coordination of the C=C bond's π electrons to the vacant d orbitals of the octahedral Ti (a). The olefin complex then inserts into the Ti-C bond as shown (b and c). The empty orbital and the alkyl group can then switch octahedral positions (d) to generate the correct stereochemistry for the next insertion. These reactions (a-d) are repeated thousands of times to successively add to the growing polymer chain. As long as the empty orbital migration step takes place faster than the next insertion an isotactic structure is produced.

RECENT ADVANCES IN CATALYSIS

The discovery of Z-N catalysts sparked a wave of research in organometallic catalysis and polymer science that is still growing. The most important advances in Z-N catalysis are in the area of soluble metallocene systems. One example is 1,1'-ethylenedi-η^5-indey lzirconium dichloride in which rotation of the η^5-indenyl about the metal center is prevented by the presence of a bridge (ethylene or silane) between the complexed aromatic rings (5-7). The racemic mixture with the presence of methaluminoxane yields PP with an isotactic index of 98% with reactivities as high as 43,000 kg of PP per gram of Zr, an order of magnitude greater than heterogeneous catalysts (8).

Usually, the soluble catalysts have a single active site unlike the numerous different active sites found on the heterogeneous transition metal catalysts. The resulting polymers have a much narrower molecular weight distributions. Often reactivity differences between co-monomers are lowered, which allows for more control of co-monomer composition.

TECHNOLOGICAL CHALLENGES

Despite the new manufacture technology and the excellent combination of polyolefin properties, including low price, low density, easy processing, favorable properties, safe, recyclable and able to be incinerated, they also have certain shortcomings that must be addressed, especially involving other materials, such as pigments, paints, glass fibers and carbon black. Due to their low surface energy, lack of chemical functionalities, and semi-crystallinity pure hydrocarbon polyolefins possess almost no chemical and poor physical interactions with other materials. Polyolefins exhibit inadequate compatibility with other synthetic polymers and virtually no adhesion to metal or glass. Traditionally, polyolefins are known to be inadequate for applications where adhesion, compatibility, wetability, printability, or reactivity (i.e., functional chemical groups) are required instead they are favored in applications requiring low cost, chemical and thermal stability, easy processing, and excellent mechanical properties. For example, polyolefins are most commonly used for films and molded articles where single polyolefin is required. If the afore mentioned problems could be overcome it would phenomenally expand the available market for polyolefin applications.

The improvements of compatibility and adhesion of polyolefins are essential in broadening their end uses. Most of these difficulties would be resolved by the introduction of chemical functionalities or by grafting with a suitable polar monomer, which could enhance interactions between polyolefin and other materials in polymer blends and composites. Accordingly, the chemical modification of polyolefins, especially PE, PP and EP, has been an area of intense interest as a route to improve these commodity polymers. This is not as easy as it might sound.

APPROACHES IN THE FUNTIONALIZATION OF POLYOLEFINS

Theoretically, there are two processes in the functionalization chemistry. One is the direct process by introducing functional groups containing monomers during copolymerization reactions. The other involves chemical modification of pre-formed polymers. As discussed that the Ziegler-Natta and metallocene catalysts, using early transition metals, are the most important methods for preparing poly-olefins (2). But the direct polymerization of functional monomers by these methods are normally very difficult, because of catalyst poisoning and other reactions (9,10). The Lewis acid components (Ti, V, Zr and Al) of the catalyst will tend to complex with nonbonded electron pairs on N, O, and X (halides) of functional monomers, in preference to complexation with the π-electrons of the double bonds. The net result is the deactivation of the active polymerization sites by formation of stable complexes between catalysts and functional groups, thus inhibiting poly-merization.

Some reports (11,12) described the use of protecting groups to sterically protect heteroatom during the copolymerization reactions. However, the resulting copoly-mers are as inert as pure hydrocarbon polymers. The other protection chemistry (13-15) involved the pretreatment of functional group (Lewis base) with an organo-aluminum (Lewis acid) to prevent catalyst poisoning. Ester copolymers are especially useful because they are relatively stable to such polymerizations and easily converted to an acid functionality after polymerization. However, there are several drawbacks, mainly the use of large excess of Lewis acid protecting agent, the low levels of ester functional group incorporated to polymer, the deprotecting reaction and the high viscosity of polymer solution in the polymerization process, due to ionic association of polar groups in hydrocarbon solvent. Moreover, the ester poly-mer is not a versatile intermediate. It is difficult to convert the ester functional poly-mer to other functional polymers by simple mild chemical conditions.

On the other hand, the post-polymerization processes have also faced many dif-ficulties. Polyolefins with saturated hydrocarbon bonds are inert to most of reagents under normal reaction conditions. Much attention has been focused on the free radical reactions involving various free radical production mechanisms, such as thermal (16), flame, shock waves (17), UV(18) and γ (19-20) radiations. Many attempts were carried out by mixing polyolefin with a free radical sensitive reagent, such as meleic anhidride (21-22), maleic acid (23) and methacrylate (24), with the presence of peroxide initiator in melt or solution at high temperature. Despite the extensive research efforts to tune the reaction conditions and reagents, the free radical chemistry suffer from many undesirable side reactions, such as the crosslinking (25) and degradation (26) of the polyolefin backbone and homopolymerization of monomers. Overall, the composition and structure of functionalized polyolefins were difficult to control.

It is clear that there is a fundamental need to develop a new chemistry, which can address the challenge of preparing functional polyolefins with wide varieties and a wide concentration range of functional groups and preserving the original physical properties, such as crystallinity and thermal transition temperatures, of polyolefins.

THE BORANE APPROACH IN FUNCTIONAL POLYOLEFINS

In the past few years, we have been investigating a new approach (27-29) for functionalizing polyolefins using the borane (boron alkys) intermediates. The initial idea was based on a logical thinking, mainly the unique location of boron in the Periodic Table as illustrated in Figure 4.

Line A in the Periodic Table separates the metallic region (left) and non-metallic region (right). Most of the elements in the non-metallic region are used in organic functional groups. On the other hand, line B separates the electron-deficiency elements (Lewis acids) in the left side and electron-rich elements (Lewis bases) in the right side. Boron is the only element which is located in the non-metallic region (right side of line A) and is a Lewis acid with electon deficient (left side of line B). The Lewis acid nature of borane offers itself a very good chance to coexist with transition metals (Lewis acids). In addition, the size of the boron atom is relatively small, steric protection can be effectively applied if it needed. Therefore, an α-olefin containing borane group shall be able to be incorporated into the polymer using transition metal catalysts. Boron is situated next to carbon in the Table. Both elements are similar in the atomic size and the B-C bond is covalent in nature, as the regular C-C bond. Therefore, the borane-containing polymers will behave like regular hydrocarbon polymers with similar solution properties (solubility, viscosity) and solid state properties (glass transition temperature, mechanic strength). More importantly, the same reaction conditions and processes for the polymerization of α-olefins can be directly applied to the borane monomer (borane containing α-olefin) cases. High molecular weight and high yield of borane containing polymers were achieved as those of pure polyolefins. Borane groups can be effectively transformed to a remarkably fruitful variety of functionalities under mild reaction conditions. On the basis of stability, solubility and versatility of the borane group, we have synthesized a broad range of new functionalized polyolefins (27-32), containing various functional groups. It is also possible to control the locations of functional groups in polyolefins as illustrated in Figure 5.

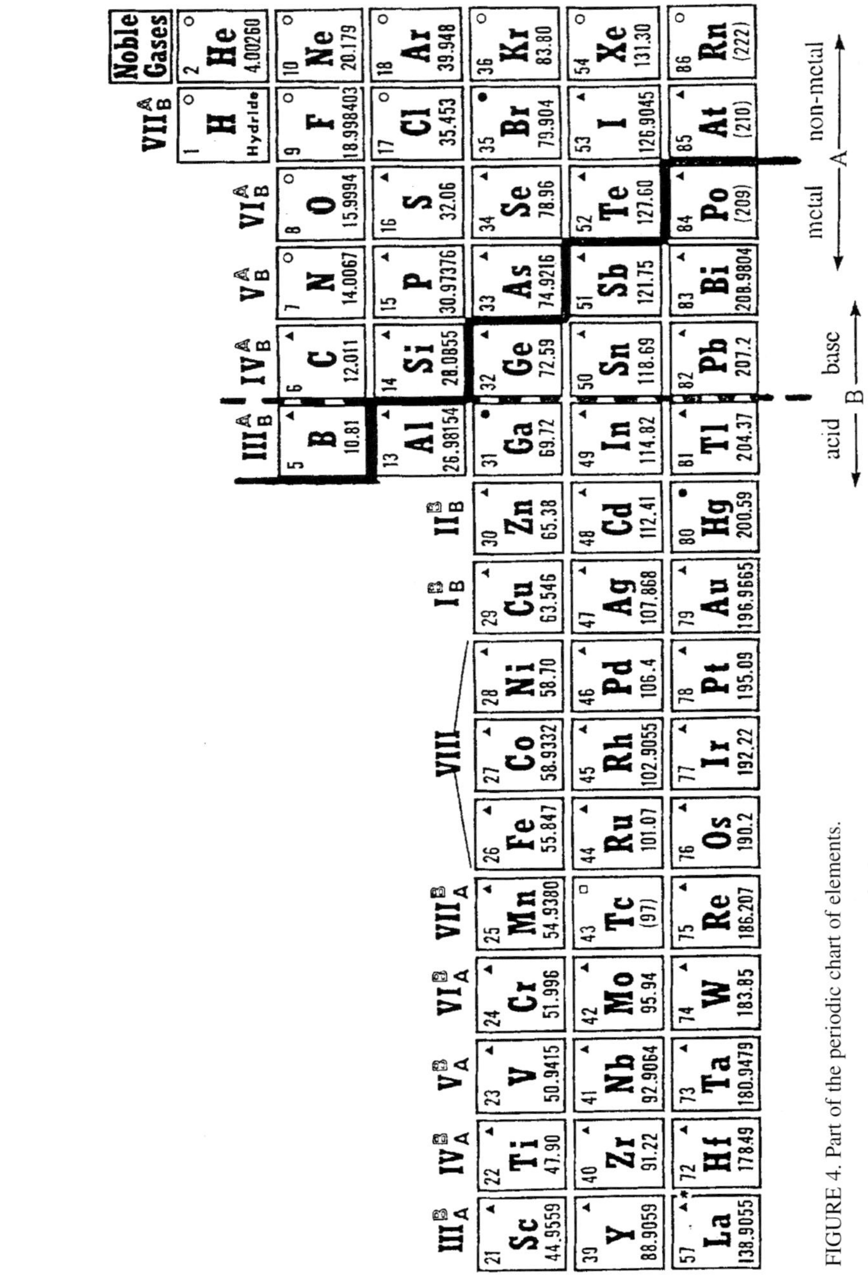

	III B/A	IV B/A	V B/A	VI B/A	VII B/A	VIII			I B	II B	III A/B	IV A/B	V A/B	VI A/B	VII A/B	Noble Gases
															1 **H** Hydride	2 **He** 4.00260
											5 **B** 10.81	6 **C** 12.011	7 **N** 14.0067	8 **O** 15.9994	9 **F** 18.998403	10 **Ne** 20.179
											13 **Al** 26.98154	14 **Si** 28.0855	15 **P** 30.97376	16 **S** 32.06	17 **Cl** 35.453	18 **Ar** 39.948
	21 **Sc** 44.9559	22 **Ti** 47.90	23 **V** 50.9415	24 **Cr** 51.996	25 **Mn** 54.9380	26 **Fe** 55.847	27 **Co** 58.9332	28 **Ni** 58.70	29 **Cu** 63.546	30 **Zn** 65.38	31 **Ga** 69.72	32 **Ge** 72.59	33 **As** 74.9216	34 **Se** 78.96	35 **Br** 79.904	36 **Kr** 83.80
	39 **Y** 88.9059	40 **Zr** 91.22	41 **Nb** 92.9064	42 **Mo** 95.94	43 **Tc** (97)	44 **Ru** 101.07	45 **Rh** 102.9055	46 **Pd** 106.4	47 **Ag** 107.868	48 **Cd** 112.41	49 **In** 114.82	50 **Sn** 118.69	51 **Sb** 121.75	52 **Te** 127.60	53 **I** 126.9045	54 **Xe** 131.30
	57 **La** 138.9055	72 **Hf** 178.49	73 **Ta** 180.9479	74 **W** 183.85	75 **Re** 186.207	76 **Os** 190.2	77 **Ir** 192.22	78 **Pt** 195.09	79 **Au** 196.9665	80 **Hg** 200.59	81 **Tl** 204.37	82 **Pb** 207.2	83 **Bi** 208.9804	84 **Po** (209)	85 **At** (210)	86 **Rn** (222)

FIGURE 4. Part of the periodic chart of elements.

Most of polymers would be very difficult to prepare by other existing methods. The new functionalized polyolefins were used to improve the adhesion and compatibility of polyolefins in coating and composites. Some evaluations will also be discussed in this article.

DIRECT FUNCTIONALIZATION PROCESS

The general route to incorporate borane to polyolefins is by using borane monomers, the α-olefin containing ω-borane group (33-35). As shown in Equation 1, one of borane monomers, 5-hexenyl-9-BBN (9-BBN: 9-borabicyclo [3.3.1]nonane), is copolymerized with α-olefins using a commercial Ziegler-Natta catalyst. This chemistry is very general and can be applied to various α-olefins, such as ethylene, propylene, 1-butene, 1-octene and their mixtures.

The copolymerization between borane monomer (5-hexenyl-9-BBN) and various α-olefins, such as ethylene, propylene, 1-butene and 1-octene have been carried out in an inert gas atmosphere using both heterogeneous catalysts, such as $TiCl_3.AA$ (AA:aluminum reduced and activated)/Et_2AlCl and homogeneous metallocene catalysts (36), such as Cp_2ZrCl_2 (Cp:cyclopentadiene) and $Et(Ind)_2ZrCl_2$ (Ind: indene) with methylaluminoxane (MAO). In general, the homogeneous catalysts were very effective for polyethylene copolymer syntheses, and isospecific heterogeneous catalysts were used for propylene polymerization to obtain PP copolymers with high melting point. The polymer solution was very dependent on the nature of the α-olefin. In 1-octene case, a homogeneous solution was observed through the whole copolymerization reaction. On the other hand,

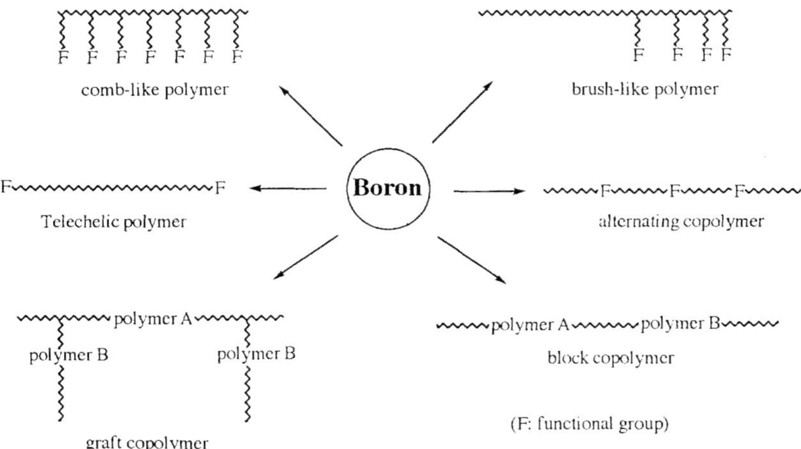

FIGURE 5. Various functionalized polymer structures have been prepared by borane containing monomers and polymers.

$$CH_2{=}CH\!-\!R$$

$$+$$

$$CH_2{=}CH\!-\!(CH_2)_n\!-\!B$$

$$\xrightarrow[\text{Z-N Catalyst}^{*}]{}\ -(CH_2\text{-}CH)_x(CH_2\text{-}CH)_y-$$

with R and $(CH_2)_n\!-\!B$ substituents

$\xrightarrow{\text{NaOH / H}_2\text{O}_2}\ -(CH_2\text{-}CH)_x(CH_2\text{-}CH)_y-$ with R and $(CH_2)_n\!-\!OH$

$\xrightarrow[\text{CH}_3\text{COONa}]{\text{NaI / Chloramine-T-hydrate}}\ -(CH_2\text{-}CH)_x(CH_2\text{-}CH)_y-$ with R and $(CH_2)_n\!-\!I$

$R = H,\ CH_3,\ C_2H_5$ and C_6H_{13}

* Z-N Cat. = $TiCl_3/EtAlCl_2$, Cp_2ZrCl_2/MAO and $C_2H_4[Ind]_2ZrCl_2/MAO$

$n > 2$

EQUATION 1. Copolymerization of borane monomer and α-olefin in Ziegler-Natta Process.

almost immediately white precipitate could be seen in the PE and PP cases, which are due to the formation of semi-crystalline structures. The copolymerization was terminated after a certain reaction time by addition of isopropanol to destroy the active metal species. These borane containing copolymers were isolated from solution by simple filtration and then washed repeatedly with isopropanol. The borane containing copolymers are stable for long periods of time (6 months in dry-box) or at elevated temperatures as long as O_2 is excluded. Usually, the borane groups in polymers were completely converted to the stable functional groups, such as OH, NH_2 and halides, by ionic processes.

Table 1 summarizes the experimental results of the functionalized polyethylene copolymers. Overall, the homogeneous zircorocene/MAO catalysts, especially, $Et(Ind)_2ZRCl_2/MAO$ with stained ligand geometry (32) (due to ethylene bridge between two indene ligands) which results in an opened active site (empty d-orbital) for accommodating relatively large size of borane monomer, show satisfactory copolymerization at ambient temperature with stained ligand geometry and an opened active site for accommodating relatively large size of borane monomer. About 50-60% of borane monomers was incorporated into the PE copolymers after about half hour reaction time.Comparing runs A-1 to A-4, the concentration of borane groups in polyethylene is basically proportional to the concentration of borane monomer feed. It is very unexpected that the catalyst activity systematically increases with the concentration of borane monomer in $Et(Ind)_2ZrCl_2$/MAO catalyst system. Obviously, no retardation due to the borane groups is shown in these cases. The copolymerization of borane monomers in Cp_2ZrCl_2/MAO system (shown in run B-2) is significantly more difficult, only 1.22 mole % of borane monomer incorporated in PE copolymer even high concentration of borane monomer used. On the other hand, the heterogeneous $TiCl_3 \cdot AA/Et_2AlCl$ catalyst shows no detectable amount of borane group in the copolymer as shown in run C-1.

The borane groups in polymers were converted to the corresponding hydroxy groups by reacting with $NaOH/H_2O_2$ reagents at 40°C for 3 hours. The effective interconversion in heterogeneous reaction condition is due to the high surface area of borane groups in the semicrystalline microstructure of the copolymers. The borane groups located in the flexible side chains may migrate to the surfaces or amorphous phases where the chemical reagents can be easily reached. It is very interesting to note that the resulting functionalized polyethylene (LLDPE-OH), containing a hydroxy group located at the end of the side chain, is structurally similar to that of linear low density polyethylene (LLDPE).

The isospecific heterogeneous $TiCl_3 \cdot AA/Et_2AlCl$ catalyst was employed in the copolymerization reactions of 5-hexenyl-9-BBN and high α-olefins, such as 1-propene, 1-butene and 1-octene. In general, the borane monomer behaves like a high α-olefin in heterogeneous catalyst, the bigger the size of monomer the lower the reactivity. Reactivity ratios of propylene or 1-butene (M_1) ($r1=k_{11}/k_{12}$) and 5-hexenyl-9-BBN (M_2) ($r_2=k_{22}/k_{21}$) are estimated by Kelen-Tudos method. The calculation is based on the Equation 2.

$$\eta = r_1 \xi - r_2/\alpha (1 - \xi)$$
$$\eta = G/\alpha + F \text{ and } \xi = F/\alpha + F \qquad \text{Equation 2.}$$

TABLE 1.

A Summary of Copolymerization Reactions Between Ethylene (m_1) and 5-Hexenyl-9-BBN (m_2)

Run No.	Cat. Type*	Comonomers m_1/m_2 (psi)**/(g)	Reaction Temp./Time (°C/Min.)	Cat. Activity (Kg/mol·hr)	Borane in Copolymer (mole %)
A-1	I	40/0	30/70	350	0
A-2	I	45/0.22	30/30	480	1.25
A-3	I	45/0.61	30/30	660	2.15
A-4	I	45/0.82	30/30	850	2.30
B-1	II	45/0	30/70	110	0
B-2	II	45/5	30/70	210	1.22
C-1	III	80/10	60/110	1.3	0

*Catalysts: $Et(Ind)_2ZrCl_2$/MAO(I), Cp_2ZrCl_2/MAO(II) and $TiCl_3 \cdot AA/Et_2AlCl$ (III)
 Solvent: 100 ml toluene
**Corresponding to 0.38 mol/L

where $x = [M_1] / [M_2]$ in feed and $y = d[M_1] / d[M_2]$ in copolymer, $G = x(y-1)/y$, $F = x^2/y$, $\alpha = (F_m \times F_M)^{1/2}$. F_m and F_M are the lowest and highest values of F. Figure 6 shows the plot η versus ξ and the least squares best fit line.

The extrapolation to $\xi = 0$ and $\xi = 1$ gives $-r_2/\alpha$ and r_1. We obtain $r_1 = 70.476$, $r_2 = 0.028$, and $r_1 \times r_2 = 1.973$ for propylene/5-hexenyl-9-BBN and $r_1 = 7.13$, $r_2 = 0.41$, and $r_1 \times r_2 = 2.92$ for 1-butene/5-hexenyl-9-BBN respectively. It is clear that both copolymerization reactions are not ideal cases. The values of $r_1 \times r_2$ are far from unity, and the reaction is favorable for α-olefin incorporation, especially in the copolymerization of propylene and 5-hexenyl-9-BBN. In the batch reaction

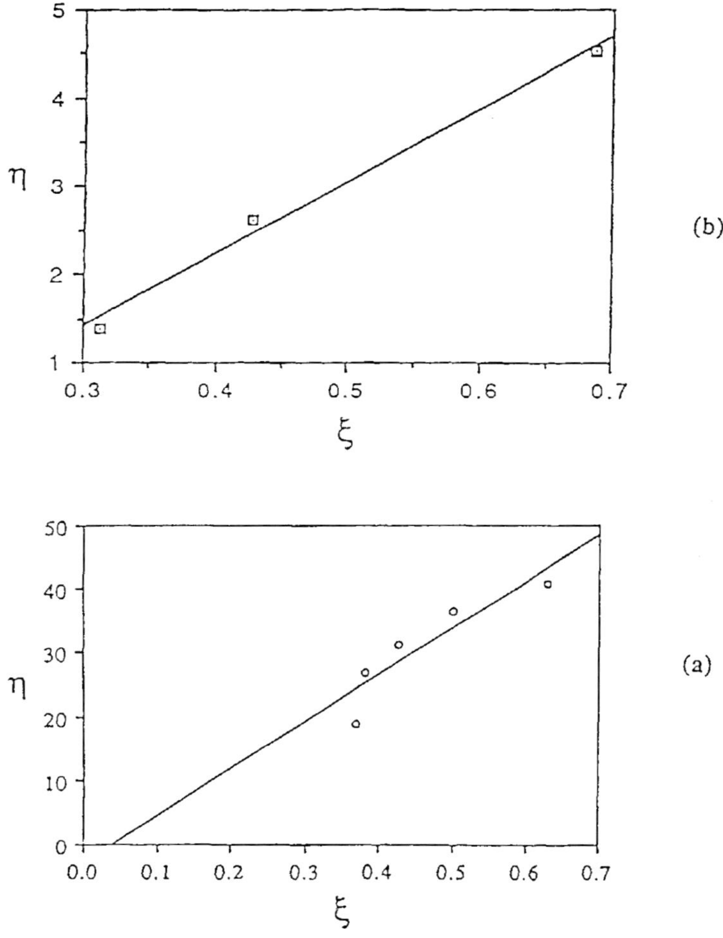

FIGURE 6. Kelen-Tudos plot of η verse ξ for (a) propylene and (B) 1-butene with hexenyl-9-BBN.

with the fixed monomer ratio of propylene/5-hexenyl-9-BBN, either a narrow compositional distribution was obtained at low conversion (at extremely low yield) or a broad distribution of copolymer composition for high conversions.

CONTINUOUS PROCESS

Due to the significant differences in comonomer reactivity ratios, a continuous monomer feeding process was carried out during the copolymerization. Usually, the more reactive α-olefin monomer was added periodically in order to level the ratio of α-olefin and borane monomer. The α-olefin was added in decreasing amounts to account for the consumption of borane monomer during the reaction. Table 2 summarized the experimental results (34) of a copolymerization between 1-propene and 5-hexenyl-9-BBN, the 1-propene monomer was added at 30-min intervals for 6 times with the consecutive reduced amounts.

In general, the copolymerization reactions by continuous addition of 1-propene were very effective to produce functionalized PP copolymers with high molecular weight, high yield and controllable concentration of functional groups. In addition, this process can produce PP copolymer with "brush-like" structure (34) as shown Figure 7.

Due to the fluctuation of 1-propene concentration during the copolymerization process, the side chains (with functional groups) in polymer are concentrated at the chain end of polyolefin backbone. The functionalized PP with "brush-like" structure has many interesting physical properties. The polymer not only contains functional groups but also preserves original physical properties, such as crystallinity, melting point and thermal stability, of pure polypropylene. The polypropylene sequence in copolymer can co-crystallize with pure polypropylene. The functional group located at the end of side chain preferably moves away from hydrophobic PP crystalline matrix to the surfaces, which can then interact with other materials.

TABLE 2.
The Summary of Copolymerization of 1-Propene (m_1 and 5-Hexenyl-9-BBN (m_2) by Continuous Reaction and $TiCl_3 \cdot AA/Et_2AlCl$ Catalyst.

Comonomers m_1/m_2	Reaction Temp./Time (°C/hr.)	Yield (%)	Borane in Copolymer (mole%)	Intrinsic Viscosity (η)	Mv (g/mole)
100/0	25/2	93	0	2.07	230,000
95/5	25/3	87	1.7	1.95	210,000
90/10	25/4	72	3	1.78	183,000
85/15	25/5	55	6	1.71	174,000

POST-POLYMERIZATION REACTION

Another approach to prepare borane containing polyolefins involves hydroboration reaction (37) of unsaturated polymers, e.g. poly(ethylene-co-1,4-hexadiene), and the subsequent oxidation reaction to convert borane group quantitatively to various functional groups as shown in Equation 3.

Both hydroboration and oxidation reactions (37) were not inhibited by the insolubility of polyethylene. The effective reaction must be due to the high surface area of reacting sites. While the polyethylene segments are crystallized, the side chains containing the double bonds are expelled out to the amorphous phase which is swellable by the appropriate solvent, e.g. THF and toluene, during the reaction. In addition, the high reactivities of hydroboration reaction and the oxidation reaction certainly enhance the efficiency of functionalization.

The concentration of functional groups can be controlled by the quantity of borane reagents as well as the percentage of 1,4-hexadiene units in pre-formed copolymer. The functional group is a secondary one due to the internal double bonds in polyethylene. In the cases of direct copolymerization using borane monomer, the primary functional group located at the end of the side chain was obtained in the functionalized PE and PP copolymers. Comparing poly(ethylene-co-1,4-hexadiene) before and after functionalization, there is no appreciable change in intrinsic viscosity and molecular weight after functionalization. The results are consistent with

FIGURE 7. Pictorial illustration of hydroxylated polypropylene with "brush-like" structure.

$$CH_2=CH_2 \quad + \quad CH_2=CH\text{-}CH_2\text{-}CH=CH\text{-}CH_3$$

Z-N Catalysts: Et(Ind)$_2$ZrCl$_2$/MAO
Cp$_2$ZrCl$_2$/MAO
TiCl$_3$·AA/(Et)$_2$AlCl

EQUATION 3. Chemical modifications of unsaturated polyethylene.

those observed with homogeneous systems, such as 1,2-polybutadiene (38), without detectable side reactions (such as crosslinking and degradation).

POLYOLEFIN GRAFT COPOLYMERS

It is very interesting to expand the scope of the application of borane-containing polymers, especially, to prepare graft and block copolymers (39-43) which contain both polyolefin and functional polymer segments. Both segments in the polymer are chemically bound, but are phase separated into different domains. A combination of relevant physical properties can be achieved through control of polymer morphology. In other words, a high concentration of hydrophilic functional groups exist in the polyolefin without disturbing the hydrophobic properties of the polyolefin.

In chemistry, it is very desirable to increase the efficiency of borane groups in polyolefins. Instead of converting one borane to one functional group, it is very useful to use borane as inititator to create a functional polymer segment containing hundreds of functional groups. The general scheme in the preparation of polyolefin graft copolymers is shown in Equation 4.

$$
\begin{array}{c}
\text{R} \\
| \\
\text{-(CH}_2\text{-CH)}_x\text{-(CH}_2\text{-CH)}_y^- \\
\text{(CH}_2)_4 \\
| \\
\text{B}
\end{array}
\xrightarrow{\;\text{NaOH/H}_2\text{O}_2\;}
\begin{array}{c}
\text{R} \\
| \\
\text{-(CH}_2\text{-CH)}_x\text{-(CH}_2\text{-CH)}_y^- \\
\text{(CH}_2)_4 \\
| \\
\text{O} \\
| \\
\text{H}
\end{array}
$$

(I)   (II)

$\downarrow$ O$_2$ / MMA

i) n-BuLi
ii) Et$_2$AlCl
iii) ε-CL

$$
\begin{array}{c}
\text{R} \\
| \\
\text{-(CH}_2\text{-CH)}_x\text{-(CH}_2\text{-CH)}_y^- \\
\text{(CH}_2)_4 \\
| \\
\text{PMMA}
\end{array}
\qquad
\begin{array}{c}
\text{R} \\
| \\
\text{-(CH}_2\text{-CH)}_x\text{-(CH}_2\text{-CH)}_y^- \\
\text{(CH}_2)_4 \\
| \\
\text{PCL}
\end{array}
$$

(III)   (IV)

EQUATION 4. Graft from copolymerization reactions by using borane-containing poloyolefins.

By simply exposing the borane-containing polyolefin (I) to air, the borane group become a suitable reactive site (41,42) for free radical polymerization and copolymerization. The autoxidation of borane group takes place at room temperature. The reaction proceeds to produce boron peroxide as shown in Equation 5.

$$
\text{R-CH}_2\text{-B}\langle \;+\; \text{O}_2 \longrightarrow \text{R-CH}_2\text{-O-O-B}\langle \longrightarrow \bullet\text{O-B}\langle \;+\; \text{R-CH}_2\text{-O}\bullet
$$

$$
\downarrow \text{R-CH}_2\text{-B}\langle
$$

$$
\text{R-CH}_2\text{-O-B}\langle \;+\; \bullet\text{O-B}\langle \;+\; \text{R-CH}_2\bullet
$$

EQUATION 5. The oxidation mechanism of alkyl-9-BBM

The boron peroxides will homolytically cleave to generate an alkoxy radical and a B-O• radical or react further with an alkyl borane to yield an alkyl radical and a borinate B-O• radical which is relatively stable and is inactive for polymerization.

Both alkyl and alkoxy radicals can then initiate the polymerization of methacrylates, styrene, acrylamide, vinyl acetate, acrylonitrile, etc. at room temperature. In the copolymer cases, the borane-containing polyolefin (I) acts as a polymeric free radical source to initiate graft-from polymerization. The free radical polymerized polymers are chemically bonded to the side chains of polyolefin. Some interesting polymers (39-42), such as PP-g-PMMA, PP-g-PVA, PE-g-EVOH and EP-g-PMMA, have been synthesized with controlled composition and molecular microstructures. Most of them would be otherwise very difficult to prepare by other existing methods.

The other type polyolefin graft copolymers (43), containing condensation polymer segments (polyester, nylon), can also be prepared by the subsequent functional polyolefins obtained from borane-containing polymers. For example, shown in Equation 4, the hydroxylated polyolefin (II) can initiate ring opening polymerization of ε-caprolactone (ε-CL). The anionic ring opening reaction was carried out at room temperature by the insertion and ring opening of e-CL into the Al-O-bond. Overall, the weight percent PCL in the resulting graft copolymer (IV) increased linearly with increasing ε-CL in the feed. On the other hand, polycaprolactone (PCL) forms truly miscible blends with numerous polymers and possesses considerable interaction with many more. For example, PCL forms miscible blends with PC, PET, PVC, $PVCl_2$, SAN, nitrocelluose, and chlorinated PE to name a only few (44). Therefore, PP grafted with PCL would potentially be an extremely useful compatibilizer for many different polypropylene blend combinations.

APPLICATIONS OF FUNCTIONALIZED POLYOLEFIN

Polyolefin Coating

The poor interaction between polyolefin and a substrate has been the major hurtle to successful coating applications. The successful bondings between polyolefin to glass or aluminum are both scientific and commercial interests. The hydroxylated polypropylene was used as surface modifier (45,46) to improve PP adhesion. The flexibility of hydroxy groups located at the ends of side chains in PP-OH may enhance the interaction of PP-OH to substrates. On the other hand, the high crystallinity of PP-OH (similar Tm as PP) may allow co-crystallization with PP, which provides strong adhesion at PP-OH/PP interface. Both drawn and undrawn PP films (commercial products) were laminated with PP-OH treated substrates. As shown in Table 3, the peel test results are compared with those obtained from standard acid etched samples which involve surface modification of both PP and Al by a dichromate-sulfuric solution and then following the same lamination procedure.

PP/Al laminates (45) bonded by PP-OH were found to exhibit an extraordinary 7-10 fold increase in peel strength over acid etched samples. Contact angles of peeled Al and PP surfaces reveal typical hydrophobic PP surfaces with contact angel 130 to 140° measured by using the water drop method. The same results were revealed in SEM studies. As shown in Figure 8, both peeled Al and PP surfaces show similar morphologies with the PP fibers stretching across the two surfaces.

The failure clearly does not occur at the interface. When peeling Al from well bonded PP the failure path appeared to propagate within the PP layers. The adhesive energy of laminated structure is higher than the cohesive energy in polypropylene matrix. The high adhesive energy gives rise to the high peel strengths observed for these PP/Al laminates. It also helps to explain the peel strength of drawn PP/Al laminates are greater than those of undrawn PP sample, because of higher cohesive energy in the drawn PP film.

The same results were observed in the PP/Glass laminates (46). Most of the PP/Glass laminates, including both undrawn and drawn PP samples with various glass surfaces, showed high peel strength. In particular, the drawn PP/acid etched E-

TABLE 3.
Peel Strength of PP/Al Laminates

Sample	Peel Strength (N/m)
Acid etched	
- Undrawn PP/Al	126 ± 26
- Drawn PP/Al	130 ± 34
PP-OH Solution Cast	
- Undrawn PP/Al	675 ± 44
- Drawn PP/Al	1155 ± 52

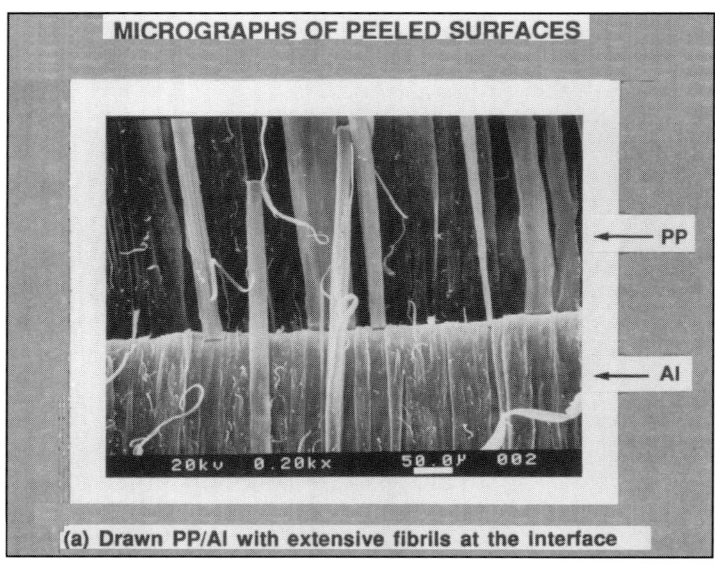

FIGURE 8. SEM micrographs of the peeled surfaces of drawn PP/Al laminate bonded by PP-OH.

glass laminates, with PP-OH interfacial agent and hot press process, have considerably high peel strength (between 950 to 1400 N/M). For comparison, a control sample prepared by laminating acid etched PP film with acid etched E-glass, without the use of PP-OH, only shows very low adhesion (too low to get any significant number in the 90° peel test). This result indicated the important role of the PP-OH layer as the interfacial agent between glass and PP film. The results of SEM and the contact angle values of water drops on the peeled surfaces all indicate cohesive failures in PP/PP-OH matrix. The strong adhesion between PP-OH and glass surface may be primary due to the chemical bonding (Si-O-C) at the interface. Reflection IR studies (46) shown in Figure 9 provide the experimental evidence of the chemical reaction between free Si-OH groups on glass surface and hydroxy groups in PP-OH.

Polyolefin Blends and Composites

In polymer blends and composites, it would be extremely advantageous if the inexpensive, commodity PP could be effectively compatibilized in blends. The PP graft copolymers have been found to be very effective compatilizer (41-43) in poly-

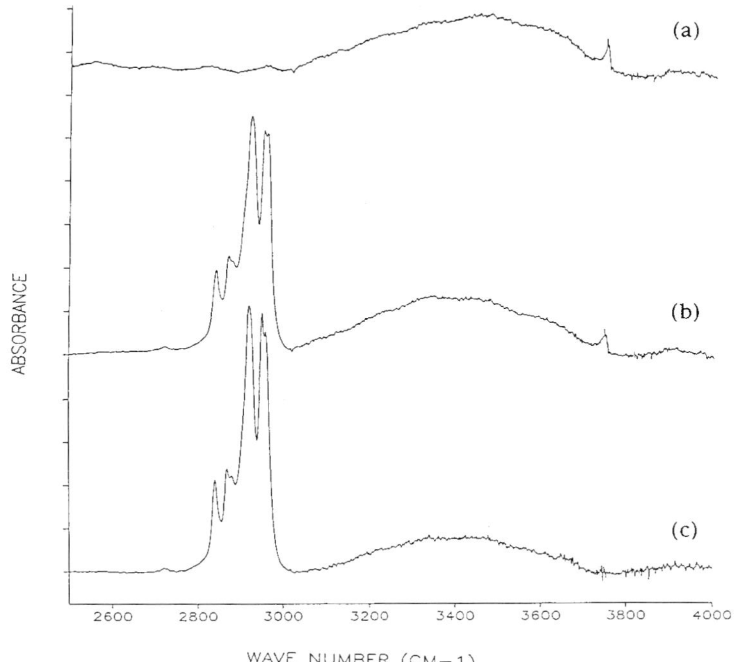

FIGURE 9. Reflection IR spectra of (a) glass surface, (b) PP-OH (thickness about 1000 A°) coated on glass and (c) PP-OH/Glass after thermal treatment.

mer blends. One example of using PP-g-PCL for PP and bisphenol-A polycarbonate (PC) blend was examined by polarized optical microscopy as shown in Figure 10.

Top picture shows the simple mixing of PP and polycarbonate in a 70/30 blend. As expected, a gross phase separation of the spherulitic PP and the featureless amorphous PC phases. The combination of gross phase separation and poor interaction between two domains results in poor mechanical properties. In the same polymer blend, a small quantity of PP-g-PCL was added to obtain the final composition of PP/PC/PP-g-PCL=70/30/10. The micrograph (bottom) shows only small distorted spherulites and a few very small distinct PC phases. The PP-g-PCL is clearly proven to be an effective compatibilizer (43) for PP and PC blends. The similar results of PP/PVC/PP-g-PCL and PP/PMMA/PP-g-PMMA blends were also observed by optical microscopy.

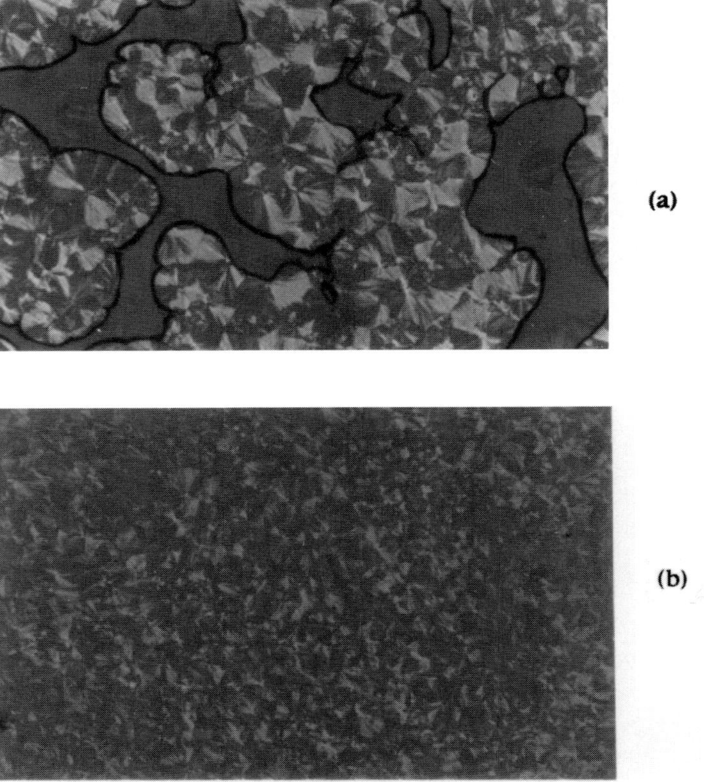

FIGURE 10. Optical micrographs of (a) two homopolymer blend with i-PP/PC = 70/30, (b) two homopolymers with PP-g-PCL, i-PP/PP-g-PCL/PC = 70/10/30 (100x).

Immobilized Catalysts On Polyolefins

Another application of functionalized polyolefins is in the chemical immobilization of soluble catalysts, so that catalysts can be recovered and reused for many reaction cycles. This process presents several important advantages in chemical production processes. These include waste reduction, lower cost and simplier purification. All of these save energy and are environmentally friendly. These considerations become even more important when the reactions require a large quantity of catalyst, such as in the oligomerization of olefins.

The semicrystalline functionalized polyolefins, such as the hydroxylated polypropylene or poly (1-butene), in various forms (film, fiber and particle) were reacted with Lewis acids, such as $EtAlCl_2$ and BF_3, as shown in Equation 6.

The immobilization reaction (47,48) was very effective at room temperature to produce chemically bonded-$OAlCl_2$, -OBF_2 groups on polypropylene. The polymer supported catalyst was then used as the Lewis acid catalyst in the carbocationic polymerization of isobutylene as shown in Equation 6. After the reaction, polyisobutylene (PIB) was obtained by filtering out the supported catalyst, then removing solvent and unreacted monomers under vacuum. The recovered catalyst was mixed with another isobutylene/hexane solution, followed by the same separation and recovery processes. This reaction cycle was repeated a number of times. Table 4 shows a typical result of PIB obtained by using a supported $PP\text{-}OAlCl_2$ catalyst. Typically, the monomer (isobutylene) to catalyst($\text{-}OAlCl_2$) ratio was about 500 to 1.

In most reaction cycles, the conversion from monomer to polymer was complete within 15 minutes. The same catalyst activity was maintained for several reaction cycles. This unusually high and stable catalyst activity in the

EQUATION 6. Polymerization of isobutylene by an immobilized $PP\text{-}O\text{-}AlCl_2$ catalyst.

Table 4.
A Summary of PIB Prepared by PP-O-AlCl₂ Catalysts in hexane

Run #	Temp (°C)	Time (Min)	Mn (g/mole)	PDI	yield (%)
A-1	25	90	1,050	2.0	100
A-2	25	60	1,150	1.6	100
A-3	25	20	1,150	1.8	100
A-8	25	15	1,180	1.5	100
A-10	0	15	4,450	2.6	100
A'-1	25	180	1,370	3.0	100
A'-2	25	120	1,660	2.6	100
A'-3	25	60	1,230	2.4	65
A'-14	25	180	1,320	2.6	100
A'-15	0	300	5,450	2.6	76
*	25	15	1,180	2.3	100
*	0	15	5,450	2.6	100

*C_5-O-AlCl₂ catalyst ** GPC measurement

polyolefin-supported system must be due to a high and stable surface area, which could be a consequence of small particle size, the crystallinity of polyolefin and the flexibility of the side chain between the catalyst and polymer backbone.

CONCLUSION

The difficulties and desire in the functionalization of polyolefins facing to the polymer research community present not only scientific challenge but also commercial importance. The novel borane approach has opened up a useful route to prepare functionalized polyolefins, with a broad range of compositions and controlled molecular structure. The success of this chemistry is *lied* on the combination of advantages, (a) the stability of borane moiety to transition metal catalysts, (b) the solubility of borane compounds in hydrocarbon solvents (hexane and toluene) used in transition metal polymerizations, (c) the reactivity of boronhydride to unsaturated polyolefins and (d) the versatility of borane groups, which can be transformed to a remarkable variety of functionalities, including radical initiator for the preparation of graft and block copolymers.

The resulting functionalized polyolefins have been demonstrated to be very effective interfacial modifiers to improve the adhesion between polyolefin and substrates and the compatibility in polyolefin blends and composites. The desirable chemical and physical properties of polyolefin products now can be achieved through blending and fiber-reinforcement. With the combination of this bottleneck breakthrough in the functionalization chemistry and the existing advantages, such as low price, low density, easy processing, safe, recyclable and able to be incinerated, polyolefin presents a very attractive material of choice in many applications. The significantly wider use of polyolefin in both commodity and specialty products could be expected in the near future.

REFERENCES

1. Ainsworth, S. 1990. *Chemical & Engineering News,* June 15, 9.
2. Boor, J. Jr., 1979. *Ziegler-Natta Catalysts and Polymerizations,* Academic Press.
3. Crabtree, R., 1988. "The Organometallic Chemistry of the Transition Metals", John Wiley and Sons, New York.
4. Grubbs, R.H.; J. Am. Chem. Soc. 1985. 107, 3377.
5. Cheng, H.; , J. Ewen. Makrol. Chem. 1989. 190, 1931.
6. Zambelli, L.; Grassi, A.; Resconi, L.; Albizzati, E.; Mazzocchi, R. 1988. Macromolecules. 21, 617.
7. Kaminsky, W.; Bark, A.; R. Spiehl. 1988. "Isotactic Polymerization of Olefins with Catalyst for Olefin Polymerization", Ed. Kaminsky, W; Sinn, H.; Springer-Verlag.
8. Odian, G. 1991. "Principals of Polymerization", 3rd Ed., Wiley-Interscience, New York.
9. Schulz, D.N., Kitano, K., Burkhardt, T.J. and A.W. Langer. 1984. US Patent 4,518,757.
10. Giannini, U., Bruckner, G., Pellino, E. and A. Cassata 1967. *Polymer Lett.,* 5, 527.
11. Langer, Jr. A.W. 1973. US Patent 3, 755, 279.
12. Clark, K.J. and W.G City. 1970. US Patent 3, 492, 277.
13. Purgett, M.D. and O. Vogl. 1988. *J. of Polymer Sci., A. Polymer. Chem.* 26, 677.
14. Landoll, L.M. and D.S. Breslow. 1989. *J. of Polymer Sci., A, Polymer. Chem.* 27, 2189.
15. Padwa, A.R. 1989. *Prog. Polym. Sci.* 14,811.
16. Gabara, W. and S Porejko. 1967. *J. of Polymer Sci., A-1. 5,* 1539.
17. Hamann, S.D. 1967. *J. of Polymer Sci., A-1.* 5, 2939.
18. Tazuke, S. and H. Kimura.. 1978. *J. of Polymer Sci., Polym. Lett.* 16, 497.
19. Lang, J.L., Pavelich, W.A. and H.D. Clarey. 1963. *J. of Polymer Sci., A-1.* 1, 1123.
20. Decker, C. and F.R Mayo. 1973. *J. of Polymer Sci., Polymer. Chem. Ed.* 11,2847.
21. Braun, D. and U. Eisenlohr. 1976. *Die Angew. Makromol. Chemie* 55, 43.
22. Gaylord, N.G. and M. Metha. 1982. *J. of Polymer Sci., Polym. Lett.* 20, 481.
23. Boelle, E. and M. Mazoyen. 1960. *Franch Patent* 1,229,660.
24. Song, Z. and W.E. Baker. 1990. *J. of Appl. Polymer Sci.* 41, 1299.
25. Borsig, E., Capla, M., Fiedlerova, A. and M. Lazar. 1990. *Polym. Commun.* 31, 293.
26. Ruggeri, G., Aglietto, M., Petragnani, A. and F. Ciardelli. (1983) *Eur. Polymer J.* 19, 863.
27. Chung, T.C. 1988. US Patent 4,734,472 and 1989. US Patent 4, 812, 529.
28. Chung, T.C. 1988. *Macromolecules* 21, 865.

29. Chung, T.C. Ramakrishnan, S. and M.W. Kim. 1991. *Macromolecules* 24,2675.
30. Chung, T.C. 1991. *Polymer* 32, 1336.
31. Chung, T.C. and M. Chasmawala. 1991. *Macromolecules* 24, 3718.
32. Chung, T.C. and M. Chasmawala. 1992. *Macromolecules* 25, 5137.
33. Chung, T.C. and D. Rhubright. 1991. *Macromolecules* 24, 970.
34. Chung, T.C. and D. Rhubright. 1993. *Macromolecules* 26, 3019.
35. Chung, T.C. 1994. *Macromol. Chem. and Phys.* (in print).
36. Kaminsky, W., Hahnsen, H., Kulper, K. and R. Woldt. 1985. U.S. Patent 4, 542,199.
37. Chung, T.C. and D. Rhubright. 1993. *J. of Polymer Sci., Polymer. Chem. Ed.,* 31, 2759.
38. Chung, T.C., Raate, M., Berluche, E. and D.N. Schulz.. 1988. *Macromolecules* 21, 1903.
39. Chung, T.C., Jiang, G.J. and D. Rhubright.. 1994. U.S. Patent 5,286,800.
40. Chung, T.C. and G.J. Jiang. 1992. *Macromolecules* 25, 4816.
41. Chung, T.C., Rhubright, D. and G.I. Jiang. 1993. *Macromolecules* 26, 3467.
42. Chung, T.C., Janvikul, W., Bernard, R. and G.I. Jiang. 1994. *Macromolecules,* 27, 26.
43. Chung, T.C. and D. Rhubright. 1994. *Macromolecules,* 27, 1313.
44. Koleske, J. Polycaprolactone Blends, Chapter 22 in Polymer Blends Vol. 2, Academic Press: New York,1978.
45. Chinsirikul, W., Chung, T.C. and I. Harrison. 1993. J. *Thermoplastic Composite Materials* 6, 18.
46. Lee, S.H., Li, C.L. and T.C. Chung. 1994. *Polymer,* 35, 2980.
47. Chung, T.C. and A. Kumar. 1992. *Polymer Bulletin,* 28, 123.
48. Chung, T.C., Rhubright, D. and A. Kumar. 1993. *Polymer Bulletin,* 30, 385.

Chapter Twelve

NEW SEMICONDUCTOR MATERIALS

JERZY RUZYLLO

Department of Electrical Engineering
111H Electrical Engineering West
Pennsylvania State University
University Park, PA 16802

INTRODUCTION

The importance of semiconductors is difficult to overestimate as in one form or another they find applications in all types of industries and areas of daily life. This overwhelming demand is responsible for the unprecedented growth of semiconductor technology over the last four decades. Such growth was possible not only through the continuous advancements in device design and manufacturing technologies, but also through the continuous progress in semiconductor materials engineering. Fundamental properties of semiconductor materials determine the performance of devices built into these materials. As the demands regarding performance of devices are continuously growing, a continuous, ever accelerating search for new, or at least improved semiconductor materials follows.

The goal of this Chapter is to provide a brief overview of semiconductor materials that are currently emerging as candidates for wide range of either new applications, or applications which result in the improved performance of the already existing technologies. It should be noted that the scope and objectives of this chapter will limit its content to the rather brief treatment of this very vast topic. Also in agreement with its purposes, this chapter will not include discussion of well established semiconductors such as silicon and gallium arsenide. On the other hand, semiconductors which are not yet mature enough for industrial device applications, most notably diamond, or are of marginal practical importance also will not be included in this overview.

To introduce and discuss the most important emerging semiconductors in an orderly fashion we must adopt criteria to distinguish various types of semiconductors. Semiconductors can be classified in several ways. First, semiconductors can be classified as elemental, comprising elements which belong to the group IV of the periodic table, and compound which include binary and ternary compounds of elements from either groups III and V (III-V semiconductors), or groups II and VI (II-VI semiconductors) of the periodic table. Second, semiconductors may be subdivided into various groups based on their physical properties. For instance, the width of the energy bandgap structure may be used as a criterion. From this point of view, a rather arbitrary subdivision of semiconductors into narrow-bandgap and wide-badgap materials may be applied. The bandgap structure is a better defined criterion based on which semiconductors can be subdivided into direct-gap and indirect-gap materials. Yet another way of classifying semiconductors relies on their structure. Accordingly, a semiconductor may be amorphous, polycrystalline, or single-crystal, and the crystalline structure will have a very strong effect on the properties of semiconductors.

A subdivision may also be based on the device consequences of basic physical properties of specific semiconductor material. This is because the inherent physical properties of each semiconductor material determine applications for which it is being considered. For instance, materials with a narrow bandgap will be excluded from applications in high power devices in which the material is required to preserve its operational characteristics at elevated temperatures. Moreover, semiconductors with indirect bandgap, in which radiative recombination is inefficient, will not be considered for photonic applications. Therefore, these materials can only be used to make devices using the electric charge of electrons as an information carrier, or in other words electronic devices. On the other hand, semiconductors with a direct bandgap which feature radiative recombination, are inherently compatible with systems using photons as information carriers. These materials are suitable for photonic device fabrication.

Considering the scope and objectives of this chapter, in the following discussion we shall review emerging semiconductor materials based on their perceived device applications. This approach appears to be the most appropriate as developments and progress in semiconductor materials engineering is driven primarily by the needs of semiconductor devices. For the sake of clarity of this brief overview, we shall focus our attention only on semiconductors for electronic and photonic devices which represent by far the largest segments of the semiconductor industry leaving other applications, such as for instance in galvanomagnetic devices to the more extensive elaborations.

In this Chapter, we shall at first define key distinguishing characteristics of semiconductors upon which their applications are determined and show examples of semiconductor materials engineering geared toward modifications of characteristics of semiconductors. Then, semiconductors for electronic applications will be considered. The subsequent section will discuss materials, for photonic applications, and the chapter will be concluded with a brief summary.

CHARACTERISTICS OF SEMICONDUCTORS

The definition of fundamental material characteristics of semiconductors is needed to better understand trends and driving forces in semiconductor materials engineering. Table 1 lists key characteristics of semiconductors and outlines device implications of each of them.

It is important to note that inherent physical properties of a material define its potential but do not decide on its final usefulness in semiconductor device manufacturing. For instance, material that is theoretically superior from the physical properties point of view will remain useless if its properties will be difficult to control and reproduce, or defect density will be too high, or its doping type and level cannot be precisely controlled. All these, as well as other characteristics listed in Table 1 must be taken into account while defining semiconductor material usefulness for industrial device fabrication.

TABLE 1
Characteristics of Semiconductors and Their Importance in Device Applications

MATERIAL CHARACTERISTICS	DEVICE IMPLICATIONS
Energy gap width	*Electronic Devices* - wide band gap desired for better power, temperature handling. *Photonic Devices* - energy gap width determines wavelength of emitted radiation and wavelength of absorbed (detected) radiation.
Type of energy gap	*Indirect* - inefficient radiative recombination *Direct* - efficient radiative recombination (suitable for light emitters).
Electron mobility	Determines speed of device operation.
Saturation electron velocity	Determines speed of device operation at high internal electric fields.
Oxidation characteristics	Determine ability to form high quality native oxide.
Susceptibility to doping	Both n-type and p-type doping must be possible.
Defect density	Defects deteriorate device performance.
Breakdown strength	Determines resistance to high electric field.
Thermal conductivity	Determines ability to dissipate heat generated during device operation.
Thermal stability/Melting point	Determine resistance to high temperature - important during device processing.
Radiation hardness	Determines degree of undesired sensitivity of semiconductor to high energy radiation.
Mechanical stability	Needed to prevent damage of the semiconductor substrate during device fabrication.

In this context, it should be stressed that at this time only very few materials displaying semiconductor properties can be labeled as "new" semiconductors. The vast majority of semiconductors, including those which have never been used to make practical devices, have been known and synthesized, or their properties have been predicted many years ago. The fact that they are only now emerging as practically useful materials results from our mastering of techniques of their fabrication in the form compatible with device manufacturing requirements on one side, and the need for devices displaying specific operational characteristics on the other. These considerations were adopted as guidelines while preparing the overview of semiconductors contained in this chapter.

SEMICONDUCTOR MATERIALS ENGINEERING

The elemental semiconductor silicon is a workhorse of microelectronic applications and as such is the most widely used, and hence, the most important semiconductor. Also, elemental carbon in the form of diamond, which theoretically is the best semiconductor from the device characteristics point of view, has potential for a wide range of applications. In addition, germanium, although practically not used at present is yet another elemental semiconductor. Besides these three elements however, all other semiconductors that are either used, or show promise in device applications are compounds. Therefore, a brief overview of directions in semiconductor materials engineering appears to be an adequate introduction to the discussion of new and emerging semiconductors.

The key semiconductor materials, either elemental or compound comprise elements from groups II through VI of the periodic table. Part of the periodic table that contains group IV elemental semiconductors, and elements that can be

TABLE 2
The Semiconductor Series in the Periodic Table

II	III	IV	V	VI
	5	6	7	8
	B	**C**	**N**	**O**
	Boron	Carbon	Nitrogen	Oxygen
	13	14	15	16
	Al	**Si**	**P**	**S**
	Aluminum	Silicon	Phosphorous	Sulfur
30	31	32	33	34
Zn	**Ga**	**Ge**	**As**	**Se**
Zinc	Gallium	Germanium	Arsenic	Selenium
48	49	50	51	52
Cd	**In**	**Sn**	**Sb**	**Te**
Cadmium	Indium	Tin	Antimony	Tellurium

used to make III-V and II-VI compound semiconductors is shown in Table 2. Depending on the selection of elements semiconductor materials with a vast variety of properties can be obtained. This is because each element contributes to the properties of the compound semiconductor in the fashion determined by its valence electron structure which may vary significantly . An example is a gradual change in properties of the group IV elemental semiconductors with the increase of the atomic number. As the atomic number increases, the inter-atomic cohesive forces are weakened, and this is reflected by an increase in the size of the atom and a decrease in the melting point and the bandgap width (1).

Following the above reasoning, most of the characteristics of semiconductors listed in Table 1 can be modified in a controlled fashion by adequate materials engineering. This concerns in particular two key characteristics of semiconductors which are type and width of the energy gap. By proper selection of starting components both can be altered to yield semiconductors displaying drastically different characteristics. For example, some of the III-V semiconductors are direct-gap semiconductors, for instance GaAs, and some are indirect-gap semiconductors, for instance AlAs. The GaAs-AlAs system, in which components exhibit complete solid solubility in each other, shows a transition from direct gap GaAs to indirect gap AlAs as the amount of Al in $Al_xGa_{1-x}As$ is increased. Resulting materials display optoelectronic properties very different from the starting components. In general, psudo-binary compounds of the form $A^{III}_xB^{III}_{1-x}C^V$ or $A^{III}B^V_yC^V_{1-y}$ where x and y can range continuously from 0 to 1, offer materials with a broad range of properties, and thus, applications. Notably, these compositional changes result in gradual changes of the bandgap of respective semiconductors which permits either the formation of superlattices, or the targeting of desired wavelengths of emitted radiation. An example is a ternary III-V compound AlGaN in which depending on the molar fraction of Al, the energy gap can be varied from 3.5 eV to 6.2 eV (2).

The II-VI compounds, formed by combining the elements in groups II and VI in the periodic table are another important group of highly engineered semiconductors. Like III-Vcompounds, ternary semiconductor compounds of the form $A^{II}_xB^{II}_{1-x}C^{VI}$, or $A^{II}B^{VI}_yC^{VI}_{1-y}$ can also be obtained from many elements from groups II and VI. Here again the properties of the resulting compound vary gradually with the fraction x, or y. For instance, it has been possible to grow crystals of ZnS_xSe_{1-x} with x ranging from 0 to 1.

Mixing and matching of elements geared toward semiconductors displaying superior characteristics concerns also group IV elemental semiconductors (Table 2). An example is silicon carbide, SiC, which exhibits properties superior to silicon on one hand, and is easier to obtain in device compatible form than diamond on the other. Also, silicon germanium, SiGe, offers characteristics that make it very attractive in several electronic device applications.

The purpose of the above considerations was to demonstrate that semiconductors encompass a very large group of materials with a broad range of possibilities for material engineering. Further in this chapter, in agreement with its scope, only those will be considered which over the period of the last few years have proven their maturity by contributing to the breakthroughs in device engineering.

SEMICONDUCTORS FOR ELECTRONIC DEVICE APPLICATIONS

For the purpose of this discussion we shall define electronic devices as those which are designed to control the electric current in larger electronic systems. In other words, in these devices carriers of electric charge, electrons and holes, are acting as information carriers. Referring to Table 1, usefulness of semiconductors for electronic device applications will be determined by the characteristics such as bandgap width and electron mobility, and also device manufacturability related features of general importance.

Silicon is, and will remain for the foreseeable future the most widely used semiconductor in electronic applications. In fact, about 95% of all semiconductor devices are made using silicon. The dominant role of this material is especially visible in microelectronics which represents by far the largest segment of the semiconductor industry. The most advanced integrated circuits, including memories and microprocessors, are made almost uniquely in single-crystal silicon. Also, the use of polycrystalline and amorphous Si is gaining momentum mainly due to their expected wide use in thin-film transistor structures for active matrix liquid crystal displays.

The success of silicon in electronic device manufacturing is not due to the outstanding physical characteristics of this material. It is rather due to our ability to form large wafers, up to 300 mm, of nearly perfect adequately doped single-crystal silicon at relatively low cost, advanced state of silicon device processing, and the fact that the high quality native oxide of silicon, SiO_2, can be very easily formed on the silicon substrates. These features led to the unprecedented growth of silicon device technology which has reached a degree of maturity unmatched by any other semiconductor material.

Even the most advanced device technology, allowing fabrication of sub-0.5 micron devices in silicon, cannot entirely compensate, however, for the inherent shortcomings of silicon. As shown in Table 3, both electron mobility and energy gap width are lower in silicon than in gallium arsenide, GaAs. Consequently, this

TABLE 3
Key Parameters for Si, GaAs and SiC.

	Si	GaAs	6H-SiC
Bandgap (eV)	1.12	1.43	2.9
Electron Mobility (cm²/V sec)	1500	8500	500
Breakdown field ($\times 10^5$ V/cm)	2.5	3	40
Thermal conductivity (W/cm K)	1.48	0.46	4.9
Saturated Electron Drift Velocity ($\times 10^7$ cm/s)	1	1	2

last compound was targeted as a potential replacement of silicon particularly in high-speed, high-power device technology. There are many types of discrete and integrated devices successfully implemented in GaAs. On the other hand, however, some difficult to overcome GaAs device manufacturability problems prevented GaAs from contributing to the overall microelectronics technology at the anticipated level. Among those problems lack of large-diameter and defect-free substrate wafers, poor thermal stability of GaAs, inferior oxidation characteristics, and fragility complicating wafer handling during device manufacturing appear to be the most troublesome. In addition, high cost of GaAs wafers and potential shortages of gallium and arsenic which are likely to drive prices even higher were preventing mass manufacturers of digital microchips from a stronger commitment to GaAs. Overall, the performance advantages of GaAs in these applications are at this time too small to justify investments needed to overcome the advanced state of processing in silicon. Yet, the GaAs devices are bound to grow due to their increasingly broad use in emerging areas such as wireless communication systems.

As considerations above indicate, each for different reasons, either Si or GaAs may not be able to meet satisfactorily all the long range goals of semiconductor electronics. Consequently, the search for alternative semiconductors for electronic applications continues, and covers a wide range of materials. Reserving the right to subjective interpretation of emerging trends, we introduce below semiconductors which appear to exemplify main directions of this search. Interestingly, most of these materials are in one form or another derivatives of either Si, or GaAs which shows that the most successful are solutions based on the established technologies.

SILICON CARBIDE

As seen in Table 3, silicon carbide, SiC, offers several characteristics superior to both Si and GaAs. Specifically, wider bandgap and higher thermal conductivity on one hand, and higher saturated electron drift velocity on the other lead to superior performance of SiC devices in high temperature/high power power devices, as well as in very high speed operation under high electric stress, respectively. In addition, among three wide-bandgap semiconductors with potential in microelectronics applications, i.e. diamond, boron nitride, and SiC, the last one is certainly the most mature in terms of material quality, and advancements in device processing (3). Introduced a few years ago commercial SiC substrate wafers (4), brought this material even closer to wider practical implementation. Besides wider bandgap allowing superior high power, high temperature performance, another important advantage of SiC over Si and GaAs is the higher breakdown voltage of this material. Device modeling and experiments carried out on Si, GaAs, and SiC MESFETs clearly demonstrated that for high voltage applications (drain voltage above 10 volts in Figure 1) the SiC MESFET has the highest absolute power density (5).

Silicon carbide exists in well over 100 different polytypes which vary in the details of long-range stacking order within the crystal, and which feature some-what different properties. Several aspects of silicon carbide's use to make electronic devices have been studied using both alpha-SiC (hexagonal) and beta-SiC (cubic) families. In terms of device performance, overall superior results are usually reported for hexagonal SiC films in the form of either 4H or 6H poly-types (6). This is particularly evident in high power applications as the critical electric field for breakdown for 6H-SiC is about two times larger than in 3C-SiC (7). Significant advances have been made in both crystal growth and epitaxial growth of 6H-SiC. P-type and n-type SiC can be obtained by doping with aluminum or nitrogen, respectively. Dopant atoms can be introduced either *in situ* during crystal growth, or by ion implantation (8). Thermal oxidation of SiC was studied fairly extensively, e.g. (9-11), and key mechanisms controlling kinetics of this process for various surface orientation have been identified. Also, silicon carbide's gas-phase etching characteristics have been investigated, e.g. (12). Moreover, contact properties of both alpha-SiC, e.g. (6), and beta-SiC, e.g. (13,14), with various metals have been studied. These accomplishments, in con-junction with applications of other processing methods commonly used in silicon device technology were sufficient to demonstrate feasibility of the high power 6H-SiC based devices such as MOSFETs for elevated temperature operation (7.15) including 6H-SiC p-channel MOSFETs operable at temperatures above 500°C (4). Moreover, 6H-SiC MOSFETs with saturation drain current three

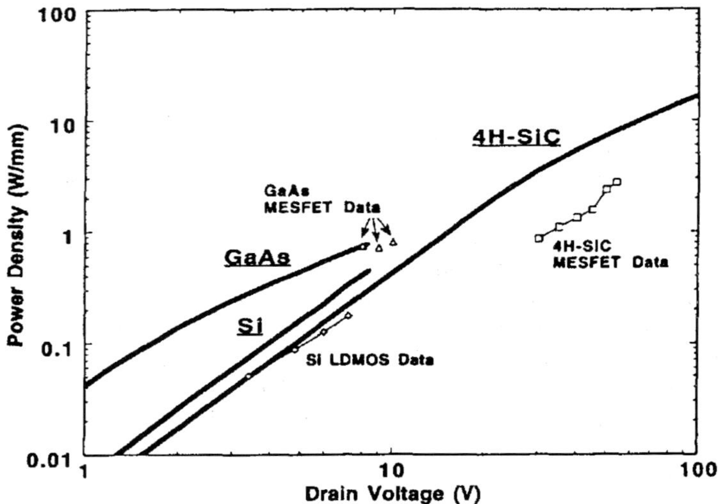

FIGURE 1. Modeled power density of GaAs, Si, and 4H-SiC MESFETs versus drain voltage. For these calculations the channel length was assumed to be 3 μm (5).

times higher than the best reported to date have been demonstrated (16). Also, Schottky diodes, p-n junctions, and high-power, high-frequency MESFETs and HBTs have been fabricated several years ago with encouraging results for commercial applications (17, 18). An example of SiC potential in high-power, high-temperature devices is a 6H-SiC rectifier which exhibits a record voltage blocking capability of 4.5 kV (19).

An important factor in the rapid growth of SiC technology is the availability of commercial 6H-SiC substrate wafers. The SiC wafers, up to 30 mm in diameter, are processed out of single-crystal SiC obtained by a high temperature (2400°C) sublimation of pure SiC powder. Undoped SiC wafers have n-type background doping.

Silicon carbide for electronic devices can also be formed by CVD hetero-epitaxial deposition of 3C-SiC on single-crystal Si while homoepitaxial layers of 6H-SiC can be successfully grown on 6H-SiC substrates. Typically, silane, SiH_4, is used as a source of silicon, while propane C_2H_4, is delivering the carbon needed for SiC formation. In X-ray mask research, a CVD process using $SiCl_4$ and propane at 1100°C was used to grow SiC on silicon (20). Plasma enhancement of the CVD process allows significant reduction of the deposition temperature, but resulting SiC films are non-crystalline.

The use of single-crystal SiC in microelectronics is limited due to the small diameters of SiC wafers and to the presence of a relatively large number of defects. The full potential of SiC in this key application can be attained when SiC wafers comparable in size and quality to Si wafers would become available. At present, this appears to be a significant barrier. However, this limitation may potentially be overcome by mastering the technology of SiC epitaxy on large diameter Si wafers. If available, the SiC-Si substrates may significantly alter future developments in Dynamic Random Access Memories (DRAM) considered to be a technology driver in advanced microelectronics.

In summary, SiC has physical properties which give this material a strong advantage for use in high-temperature, high-power, and high-frequency electronic applications (21). Figure 2 shows that other than diamond, development of which for electronic device applications is still at the very early stage, SiC offers the best performance. In addition, as discussed later in this chapter, its emission spectrum in the blue range makes SiC suitable for blue light emitters. As a result, SiC emerges as a key semiconductor for the future with SiC device market expected to grow over 20 times from 1995 to 2000 (23).

SILICON GERAMANIUM (SiGe)

As indicated earlier, there is a strong link between progress in semiconductor materials and semiconductor device technologies. Very often semiconductor materials which were around for a long time are brought to a new life due to the needs of the specific device structure. Even more often, however, the device structures that were introduced many years ago are becoming of great practical interest due to the advancements in materials engineering and device fabrication methods. A good example of this symbiosis is a Heterojunction Bipolar Transistor (HBT) which requires a wide bandgap emitter and a narrower bandgap base. This combination opened up a wide range of possibilities for mixing and matching of various semiconductors in order to implement the HBT structure, and to take advantage of the potential of this device in high speed applications. One of the most actively pursued approaches to HBT implementation uses a silicon/silicon germanium (Si/SiGe) heterostructure. The addition of Ge reduces the bandgap of Si and results in an alloy that can be used to form the base region of HBTs. This approach, which is a main driving force behind the rapidly growing interest in SiGe, is increasingly popular because it not only produces devices with advantageous characteristics, but also because it relies on the very mature silicon bipolar technology.

The very thin layers of $Si_{1-x}Ge_x$, typically with average Ge mole fraction of about 20% are grown on silicon by either Molecular Beam Epitaxy (MBE), or low-temperature CVD epitaxy. In the latter case, SiH_4 and GeH_4 are typically used as sources of silicon and germanium. In general, the goal is to keep the thickness of epitaxial layers below critical values so that the mismatch between the alloy and the Si substrate is accommodated elastically and no misfit dislocations form (24). The resulting SiGe pseudomorphic layers are considerably

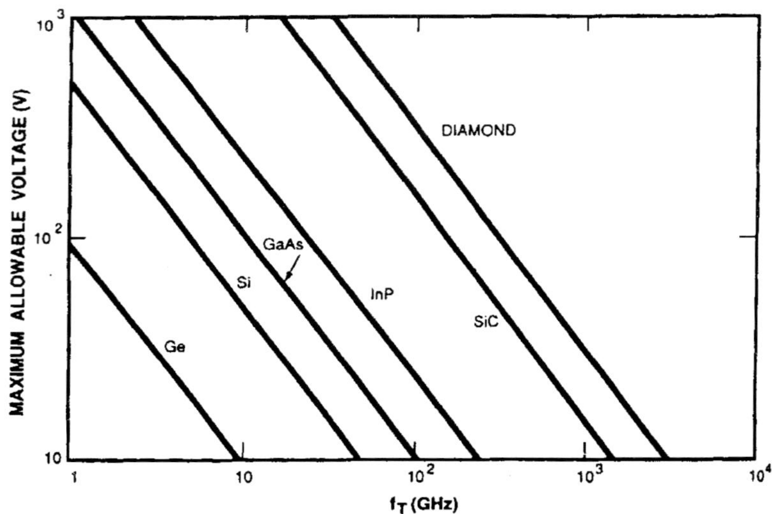

FIGURE 2. Plot of the maximum operating voltage for transistors made of selected semiconductors as a function of estimated cut-off frequency fT. The fT estimates are based on the steady-state velocity-field curve for each material (22).

strained. If the layer is too thick, or the Ge concentration is too high, then strain energy builds up and eventually misfit dislocations are injected to relieve the strain with catastrophic effects on HBT performance due to shunts through the base (25). If, however, the strain is maintained in the layer, than due to the details of the band structure of Si this strain results in major benefits for HBT operation. This is because the overall bandgap shrinkage of the strained SiGe layer is increased over that of unstrained material with the same Ge content. Moreover, for the direction of transport perpendicular to the heterojunction the effective mass of the electron is significantly reduced as compared to the unstrained case (24). This leads to the superior performance of Si/SiGe HBTs in the frequency range well above 100 Ghz. It is very likely that in the future main application of SiGe technology will be in the BiCMOS integrated circuits in which high frequency SiGe HBTs for RF applications will be merged with Si CMOS for digital applications.

The Heterostructure Field Effect Transistor (HFET) is another device in which the use of SiGe may result in performance breakthroughs. Although less developed than HBT, the HFET structure offers a wide range of advantages in a variety of FETs. In particular, striking performance predictions can be made for SiGe Hetero Complimentary Metal Oxide Semiconductor (HCMOS) FETs. The attractiveness of this approach lies in the potential of SiGe HCMOS devices to accomplish characteristics exceeding those exhibited by standard Si CMOS devices at significantly relaxed device geometry. For instance, while the standard Si CMOS accomplishes transconductances around 400 mS/mm at a gate length below 0.2 µm. the SiGe HFET can yield transconductances twice as high at a gate length of 1.2 - 1.4 um (26). This means that the future demands of digital microelectronics may be accomplished using SiGe based structures without huge capital investments required to drive standard Si devices to deep sub-micron geometries.

III-V COMPOUNDS

As discussed earlier, GaAs is the most important III-V semiconductor compound in electronic device applications. Compared to other III-V compounds, GaAs offers a good balance of physical properties making it attractive for a wide range of device applications. In addition, GaAs device technology relative to other III-V compounds is significantly more mature. This last point is exemplified by the availability of commercial single-crystal GaAs wafers of up to 100 mm in diameter. Overall, GaAs is the second most important semiconductor after Si in electronic device applications, although, the manufacturability problems and high cost have slowed progress in GaAs technology down to levels lower than anticipated. Still, GaAs devices find increased applications particularly in microwave systems, wireless communication, and others.

At present, GaAs is particularly important in n-channel Metal-Semiconductor Field Effect Transistors (MESFET) which dominates the transistor markets for many microwave and high-speed digital applications. However, the most dynamic growth of GaAs based technology in electronic devices is expected to occur in the highly engineered heterostructure material systems using GaAs and its derivatives, as well as GaAs in conjunction with other III-V compounds. The driving force behind this anticipated growth is related to the substantial gains in the performance of both bipolar and FET type transistors that can be accomplished by intentionally changing the semiconductor material composition within the device. This leads to a gradually changing bandgap across the device, or its part. As a result, forces on electrons and holes in the device can be engineered separately providing a powerful new degree of freedom in the design of very high performance devices (24).

Due to their fundamental features, compound semiconductors are particularly suitable for devices requiring a very high degree of bandgap engineering. Desired modulation of the bandgap is accomplished by stacking up extremely thin layers, often of thicknesses smaller than 10 nm, of materials of different composition. The resulting series of alternating epitaxial layers of two mismatched materials, each layer having a thickness below the critical thickness, form a strained-layer superlattice which is acting as a thick layer of dislocation free material with unusual electronic and optical properties. It is clear that the success of superlattice devices is determined almost entirely by the accuracy of advanced epitaxial growth methods such as molecular beam epitaxy (MBE), metal-organic chemical vapor deposition (MOCVD).

The AlGaAs/GaAs heterostructure, in which AlGaAs represents larger bandgap material is the most common heterostructure material system based on III-V compounds used in HBTs and other heterostructure devices. Its attractiveness results from an excellent lattice match in this material system and a very good control over the bandgap difference providing abrupt heterojunctions with significant energy barriers. Moreover, ultra-thin layers of AlGaAs of various compositions can be grown with excellent control using MBE and MOCVD methods. Finally, AlGaAs/GaAs is used to fabricate a variety of optoelectronic devices including light emitting diodes, lasers, modulators and detectors of various types. These devices can be monolithically integrated with circuits based on AlGaAs/GaAs HBTs to combine electronic and photonic functions on the same chip (24).

While the AlGaAs/GaAs HBTs are the most common, there is intense interest in the use of InGaP emitters to avoid the oxidation and deep level problems in AlGaAs and also to take advantage of the favorable energy band lineup of InGaP/GaAs and the high wet and dry etching selectivities (27).

The HFET is another device in which III-V compounds are used to accomplish ultra-high speed microwave and digital functions. Here again the most effective is the Al_xGaAs_{1-x}/GaAs material system in which x may be varied from 0 to 1. For instance, at x=0.3 AlGaAs reaches a bandgap of 1.65 eV as opposed to 1.42 eV for GaAs. In HFETs, a larger bandgap doped AlGaAs is grown epitaxially on an undoped lower bandgap GaAs. Such structure is often referred to as a modulation-doped FET, or MODFET, and its physics is such that the charge transport within an electronic device can be optimized by tailoring the layer thickness and doping levels (28). The MODFET is an example of a quantum well with a triangle potential well. Among alternative names for the devices based on the modulation doping principle, high electron mobility transistor, or HEMT, is the most common. All these devices are based primarily on the AlGaAs/GaAs materials system, although, heterostructures of interest for HFET applications include other systems such as InAlAs/GaAs and InGaAs/GaAs. Superior frequency response of these heterostructure devices over more conventional GaAs and Si devices is demonstrated in Figure 3.

In III-V electronic device technology heterostructures based on substrates other than GaAs are also used. For instance, GaP and InP are used to form heterostructure material systems consisting of $GaAs_xP_{1-x}/GaP$ and $Ga_xIn_{1-x}As/InP$, respectively. There are numerous other combinations of lattice-matched and strained lattice heterostructure material systems based on III-V compounds. In general, the goal is to optimize device performance by selecting a materials combination allowing the highest electron transfer velocities in the device. The selection of materials for these structures will largely depend on the degree of control of composition, thickness, and doping provided by the epitaxial growth process of each material involved.

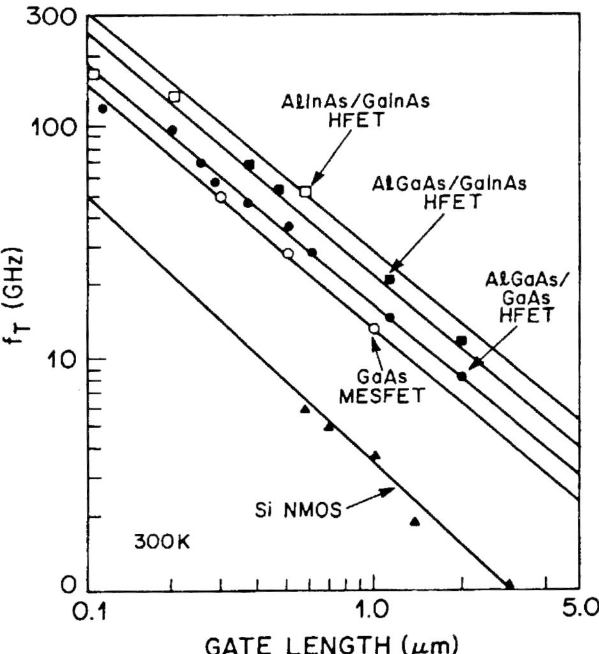

FIGURE 3. Cutoff frequency vs. gate length of various high speed devices (28).

The role of III-V compounds in electronic device applications is expected to grow at an accelerated pace mainly due to the growth of the GaAs based integrated circuits. The exploding market for communication equipment is a driving force behind GaAs growth. In particular, expected high volume use of analog GaAs ICs in 2 GHz and higher wireless communication systems will contribute significantly to this growth.

SEMICONDUCTORS FOR PHOTONIC DEVICE APPLICATIONS

The term "photonic" applies to devices in which photons rather than electric charges are used as information carriers. Since the photonic interactions are either driven by, or converted into an electrical signal, term "optoelectronic" is also common in defining these types of devices.

Based on their operational principles photonic devices can be subdivided into groups among which light emission devices and light detection devices are the most prominent. In the following discussion we shall first give a brief overview of semiconductors for these applications, and then focus our attention on materials which show the best promise for light emission devices in the short wavelength range, namely, silicon carbide, SiC, gallium nitride, GaN, and wide-bandgap II-VI compounds. The focus on green and blue light emitters reflects current trends in the search for new semiconductor materials for photonic devices.

OVERVIEW

In order to facilitate further discussion, Figure 4 summarizes basic information regarding bandgaps and cutoff wavelengths for photonic semiconductor materials. For light detection purposes the bandgap of the semiconductor should be smaller then the energy of radiation to be detected. This means for instance that to detect low-energy infrared radiation only semiconductors with a narrow bandgap from about 0.2 eV to about 1.6 eV are suitable. On the other hand, the energy of radiation emitted by the device will be determined by the bandgap of the semiconductor used. For instance, shortwavelength radiation in the blue and ultraviolet regime can only be generated by devices made of semiconductors featuring a bandgap larger than about 2.6 eV. Additional requirement for the material for light emitting devices is that it must feature a direct bandgap for the efficient radiative recombination between conduction and valence bands to occur.

As far as light detection is concerned, the semiconductor bandgaps range from 0 to above 3.5 eV, providing versatile detection systems. Not all of them are equally suitable for light detection, however, as in the direct bandgap materials absorption of light is more effective than in indirect bandgap semiconductors. In this latter case, lattice vibrations accompany electron transfer from the edge of the valence band to the edge of the conduction band and weakens the absorption of light.

In light detection devices of particular importance is detection of invisible radiation in the infrared range used in night vision and thermal imaging systems. As seen in Figure 4, there are several semiconductors featuring a bandgap compatible with the infrared wavelength range.Very useful is the ternary II-VI compound HgCdTe in which, depending on composition, the bandgap can be varied from 0 to about 1.4 eV. Among III-V compounds InGaAs and InGaAsP which allow modulation of the bandgap and are lattice compatible with InP are useful in light detection devices.

In the case of light emitting devices, the requirements and constraints regarding materials are somewhat different than in the case of light detectors. At present, in terms of material availability the best accommodated are the needs of light emitting diodes (LED) and laser diodes operating in the range of radiation from infrared to yellow. The main efforts in this radiation regime appear to be directed now toward development of new applications based on existing high performance device technologies. For instance, AlInGaP red, orange and yellow devices introduced in recent years have luminous efficiencies higher than incandescent lamps and are finding applications such as exterior vehicle lighting and highway signs (29).

In general, the most important materials in light emitting devices are based on III-V compounds. An important factor, other than strictly physical properties of the material, which predetermines the usefulness of a given material in large scale industrial manufacturing of light emitters is the commercial availability of high quality substrate wafers preferentially in diameters of at least 50 mm. The

BANDGAPS AND CUTOFF WAVELENGTHS FOR SOME OPTICAL MATERIALS

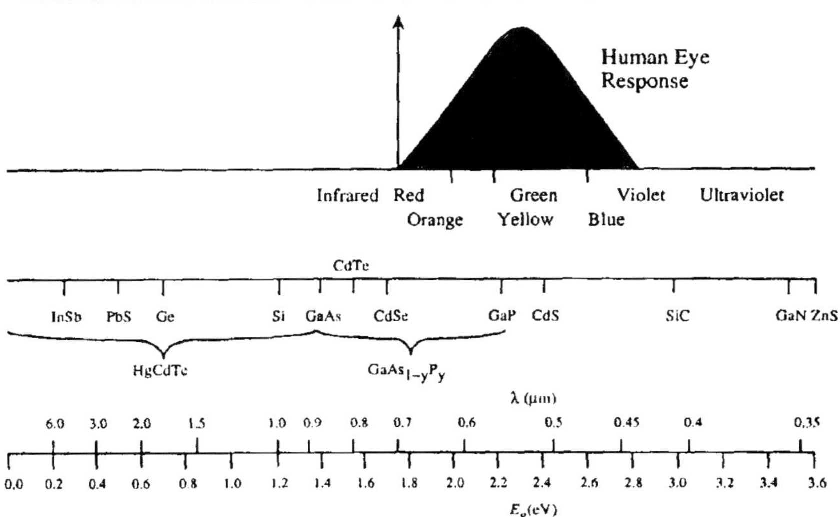

FIGURE 4. The bandgap and cutoff wavelengths for several semiconductors (30).

availability of high quality substrates is extremely important in epitaxial technology. The most mature substrates available for LEDs are GaAs and InP. Materials used for active layers will consequently have to be selected from among semiconductors and their alloys which can lattice match with these substrates. Important semiconductor materials exploited in LEDs are the alloy $Ga_xAl_{1-x}As$ which is very well lattice matched to GaAs substrates: $In_{0.53}Ga_{0.47}As$ and $In_{0.52}Al_{0.48}As$ which are lattice matched to InP: InGaAsP which is a quaternary material whose composition can be tailored to match with InP, and GaAsP which has a wide range of bandgaps available (30).

Although structures of advanced laser diodes are more complex than structures of LEDs, basically the same families of III-V compounds are used in both types of devices. Also, in the case of lasers GaAs and InP are the most common substrates while AlGaAs, InGaAs and GaInAsP are typically used for active layers respectively. It is interesting to note, that the importance of the basic AlGaAs/GaAs material system goes beyond its use in photonic devices. As discussed in the previous section, the same heterostructure is a key building block in several ultra-high speed electronic devices including HBTs and HFETs.

The greatest challenge facing state-of-the-art photonic device technology is to improve the manufacturability and performance of devices emitting blue, blue-green and green radiation. The availability of light emitters in these radiation regimes is highly inadequate to current and future needs. The potential areas of applications for shortwavelength light emitting devices covers a broad range of commercial and military uses mainly in data storage systems and full color displays, but also in traffic lights, interiors and dashboards of cars and airplanes, as well as in battlefield lasers to mention just a few possible applications. Full color displays require red (longer than 630 nm), green (490-560 nm) and blue (450-490 nm) light as with these primary colors any other color of light can be created. Lack of high brightness blue LEDs was a weak link in the chain for full color displays particularly for outdoor uses. And while substantial progress in blue LED technology has been accomplished, more structurally complex lasers emitting blue light have been demonstrated only very recently.

As seen in Figure 4, wide-bandgap semiconductors suitable for shortwavelength emitters include SiC, GaN, and selected II-VI compounds. These materials are briefly considered below.

SILICON CARBIDE (SiC)

Earlier in this chapter silicon carbide was introduced as a very promising semiconductor material for a range of electronic applications particularly in high power and high temperature devices. Also, as indicated earlier, SiC features a direct bandgap of 2.9 eV which makes this material also a candidate for the manufacture of blue light emitters. Blue LEDs are made in SiC by forming p-n juction in the n-type SiC substrates. A relative weakness of SiC based LEDs is their low brightness, typically limited to 10-20 mcd, making them best suitable for indoor applications. As the significantly higher brightness can be obtained in blue LEDs formed in GaN, the SiC based LED technology is not expected to grow substantially beyond its current status. This does not mean, however, that SiC substrates will not continue to play a key role in the blue LEDs arena as they are successfully serving as substrates in the GaN-SiC heterostructure high efficiency blue LEDs (4).

GALLIUM NITRIDE (GaN)

Due to the direct bandgap of 3.5 eV, gallium nitride, GaN, is rapidly becoming a main target in the search for high efficiency green, blue, and violet light emitters. This material was long recognized for its potential in this application, but until effective p-type doping of GaN was demonstrated, this material was not suitable for widespread applications. At present, while high brightness, long lifetime GaN based LEDs are commercially available in large quantities development efforts are geared toward the next target, a blue GaN laser. As indicated later, a major breakthrough in this last area has been accomplished recently.

As far as GaN LEDs are concerned, a majority of them are built in MOCVD layers grown on sapphire which has a 14% lattice mismatch to GaN and results in high defect densities in the layer. Because of this, it has been found necessary to first grow a low temperature buffer (500-600°C) which acts as a nucleation layer, before the initiation of high temperature (1000-1100°C) GaN growth. Also, alternative substrates such as ZnO (2.2% lattice mismatch to GaN), $LiGaO_2$ (0.2%), $LiAlO_2$ (1.4%), and spinel (10.9%) are being investigated. In addition, SiC is an attractive alternative to sapphire because it is more closely lattice matched to GaN, it exhibits a closer thermal expansion coefficient, a higher thermal conductivity, and it is available in both conductive and resistive substrates. A high efficiency LED consisting of GaN on a SiC substrate has recently become commercially available. It features 500 µW output power and a 430 nm peak wavelength (4). The proof of maturity of existing GaN LEDs technology is that these diodes are rated at lifetimes in excess of 10 years of continuous operation. Still, it appears that an ultimate solution to the epitaxial GaN - substrate lattice mismatch problem can only be provided by the success of an intensive R&D effort on single-crystal GaN substrate technology currently underway.

In the meantime, besides GaN-on-SiC structures, high efficiency blue and blue-green LEDs are manufactured using MOCVD from GaN with strained single quantum well $In_xGa_{1-x}N$ active layers. Green (520nm), blue (450 nm), and violet (405nm) emission is produced by devices with indium composition x of 45%, 20%, and 9% respectively (31,32). An important aspect of this development is that the addition of In increases the operational wavelength of GaN LEDs from 500 nm to 520 nm making them closer to the true green region. This shift is required for green GaN LEDs to potentially penetrate the huge world-wide green signal traffic light market.

A major breakthrough in the light emitting device area was a recently announced development of GaN-based multi-quantum well laser diode operating at 417 nm under pulsed conditions at room temperature (33,34). It consist of multilayer structure of GaN, AlGaN, and InGaN grown by atmospheric pressure MOCVD method on (0001) sapphire substrate. The device needs to be improved to achieve adequate operational characteristics, but the demonstration of the blue laser in GaN by itself is an event which starts a new era in the development of semiconductor photonic devices.

In addition to applications in shortwavelength light emitting devices, the wide bandgap of GaN can also be explored to make UV detectors that do not respond to visible light. The visible-blind photodetectors have important potential applications for missile guidance systems and UV radiometry.

Besides the obvious importance in photonic devices, GaN is potentially also very attractive for electronic device applications. Other than device manufacturability problems that will have to be resolved, GaN features excellent thermal conductivity, low intrinsic carrier concentration, large saturation velocity, large breakdown field, and resistance to caustic chemicals all of which make this material a perfect candidate for high temperature electronic devices operating in chemically hostile environments.

II-VI COMPOUNDS

Several years ago II-VI compounds featuring wide, direct bandgaps were investigated as prime candidates for light emitting devices covering the entire range of the visible spectrum including its green and blue portions. Unfortunately, the structural and electronic quality of these materials was too low to warrant further development. It was only after advanced nonequilibrium epitaxial growth methods, developed initially for III-V compounds, such as MBE and MOCVD were applied to the growth of II-VI compounds that efficient light emitters based on these materials were implemented (35). Currently, even though GaN LEDs are dominating the blue LED arena there is a renewed interest in developing green and blue laser diodes and LED technologies based on II-VI compounds.

Among II-VI materials ZnSe is currently a compound on which efforts to develop blue light emitters are focused. Accomplishment of stable p-type doping of ZnSe using nitrogen combined with advances with MBE deposition of this material formed a foundation for further developments. Because of a lattice mismatch to ZnSe of only 0.25%, the logical substrate choice for ZnSe was GaAs. In spite of using ZnMgSSe cladding layers which can be perfectly lattice matched to GaAs, resulting heterostructures still exhibit high stacking fault densities. Consequently, a key to the continued progress in ZnSe-based device will likely depend on much wider availability of high quality single-crystal ZnSe substrates.

The ZnSe-based lasers have been demonstrated several years before GaN lasers mentioned earlier. The potential advantages of the latter are related to the shorter emitted wavelengths and to the fact that GaN-based devices are capable of preserving their operational characteristics at the relatively high defect densities in GaN. Still, the work on ZnSe blue emitters is bound to continue with possible challenges to the GaN leadership in this arena still to come.

SUMMARY

An overview of new semiconductor materials given in this chapter leads to the following conclusions regarding present and future status of semiconductor materials science and engineering. The silicon will remain the most broadly used semiconductor primarily because of its dominant role in digital microelectronics. At present, growth of single-crystal silicon technology is geared toward introduction of 300 mm wafers to mass production within the next few years. Also, the use of polycrystalline and amorphous Si is gaining momentum mainly due to their anticipated wide use in thin-film transistor structures for active matrix liquid crystal displays. Besides its key role in photonic devices applications, substantial expansion of GaAs based technologies is anticipated through the increasingly important role GaAs devices will play in wireless communication systems.

Among the myriad of other semiconductor materials considered for the variety of applications, the complex compounds based on III-V elements will continue their dominance of the photonic device sector. Seemingly endless possibilities regarding bandgap engineering in III-V superlattice device fabrication, allowing implementation of innovative device structures, will be responsible for their growth in the specialized high-speed electronic devices manufacturing.

An analysis of current trends indicate that besides those listed above, three other semiconductor materials emerge as those that are the most likely to find a widespread use. Among them, SiC offers advantageous characteristics for high-power, high-temperature electronic devices, and will also play a supporting role in photonic device manufacturing. The SiGe in turn will help to bring Si-based technology to the new levels in terms of device performance. Finally, particularly after the invention of the GaN laser, GaN has secured its position as one among key semiconductors in photonic applications. According to some opinions, GaN is the final semiconductor whose exploitation was known to be of profound economic importance. Consequently, future endeavors in semiconductor research and development may stress mastering of existing technologies instead of devoting large portions of its resources to the search for new semiconductor materials.

ACKNOWLEDGEMENTS

Author would like to express his gratitude to Dr. David L. Miller for his useful comments and suggestions regarding this text and to Dr. Richard E. Tressler for taking his time to review the manuscript.

REFERENCES

1. Tyagi, M.S. 1991. Semiconductor Materials and Devices, J. Wiley and Sons, Inc., New York.

2. Shur, M. 1996. Introduction to Electronic Devices, J. Wiley and Sons, Inc., New York.

3. Koba, R. and W. Russell. 1990. Status of beta-SiC, Diamond, and c-BN Semiconductors; Comparison of a Si Power FET to a Hypothetical Diament FET, Proc. Symp. Diamond,Silicon Carbide and Related Wide Bandgap Semiconductors, Eds. J.T. Glass, R.Messier, and N. Fujimori, Material Research Soc., 162: 43-48.

4. Technical Information, Cree Research, Inc., Durham, North Carolina.

5. Weitzel, C.E. 1995. Comparison of SiC, GaAs, and Si RF MESFET Power Densities, IEEE El. Dev. Lett., 16(10): 451-453.

6. Davis, R.F., Palmour, J.W. and J.A. Edmond. 1990. Epitaxial Thin Film Growth and Device Development in Microcrystalline Alpha and Beta Silicon Carbide, Proc. Symp. Diamond, Silicon Carbide and Related Wide Bandgap Semiconductors, Eds. J.T. Glass, R. Messier, and N. Fujimori, Material Research Soc., 162: 43-48.

7. Baliga, B.J. 1995. Power Semiconductor Devices, PWS Publishing Company, Boston.

8. Edmond, J.A., Das, K. and R.F. Davis. 1988. Electrical Properties of Ion-Implanted p-n Junction Diodes in Beta-SiC, J. Appl. Phys., 63(3): 922-929.

9. J.A. Costello, J.A. and R.E. Tressler. 1986. Oxidation Kinetics of Silicon Carbide Crystals and Ceramics, J. American Cer. Soc., 69: 674.

10. Zheng, Z., Tressler, R.E. and K.E. Spear. 1990. Oxidation of Single-Crystal Silicon Carbide - Part I. Experimental Studies, Part II. Kinetic Model, J. Electrochem. Soc., 137: 854 and 2812.

11. Muehlhoff, L., Bozack, M.J., Choyke, W.J. and J.T. Yates. 1986. Comparative Oxidation Studies of SiC(0001) Surfaces, J. Appl. Phys., 60(7):2558.

12. Luther, B., Ruzyllo, J. and D.L. Miller. 1993. Nearly Isotropic Etching of 6H-SiC in NF_3 and O_2 Using a Remote Plasma, Appl. Phys. Lett., 63(2): 171-173.

13. Yoshida, S., Sasaki, K., Sakamura, E., Misawa, S. and S. Gonda. 1985. Schottky Barrier Diodes on 3C-SiC, Appl. Phys. Lett., 46(8): 766-768.

14. Chaudhry, M.I., Berry, W.B. and V. Zeller. 1990. Ohmic Contacts on Beta-SiC, Proc. Symp.Diamond, Silicon Carbide and Related Wide Bandgap Semiconductors, Eds. J.T. Glass, R. Messier, and N. Fujimori, Material Research Coc., 162: 507-512.

15. Kong, H.S., Glass, J.T. and R.F. Davis. 1989. Growth Rate, Surface Morfology, and Defect Microstructures of Beta-SiC Films Chemically Vapor Deposited on 6H-SiC Substrates, J. Mat. Res., 4(1): 204-214.

16. Pan, J.N., Cooper, J.A. Jr. and M.R. Melloch. 1995. Self-Aligned 6H-

SiC MOSFETs with Improved Current Drive, Electronics Lett., 31(14): 1200-1202.

17. Sugii, T., Ito, T., Furumura, Y., Doki, M., Mieno, F. and T. Eshita. 1987. Si Heterojunction Bipolar Transistor with Single-Crystalline Beta SiC Emitters, J. Electrochem. Soc., 134: 2545-2549.

18. Sugii, T., Ito, T., Furumura, Y., Doki, M., Mieno, F. and M. Maeda. 1988. Beta - SiC/Si Heterojunction Bipolar Transistors with High Current Gain, IEEE El. Dev. Lett., 9(2): 87-89.

19. Kordina, O., Bergman, J.P., Henry, A., Janzin, E, Savoge, S., Andre, J., Rarnberg, L.P., Lendefelt, U., Hevmansson, W. and K. Bergman. 1995. A 4.5 kV 6H Silicon Carbide Rectifier, Appl. Phys. Lett., 67(11): 1561-63.

20. Wells, C.M., Palmer, S., Cerrina, F., Purdes, A. and B. Gnade. 1990. Radiation Stability of SiC and Diamond Membranes as Potential X-ray Lithography Mask Carriers, J. Vac. Sci. Technol., B8(6): 1575-1578.

21. Morkoc, H., Strite, S., Gao,G.B., Lin, M.E., Sverdlov, B. and M. Burns. 1994. Large-Band-Gap SiC, III-V Nitride, and II-VI ZnSe Based Semiconductor Device Technologies. J. Appl. Phys., 73(3): 628-633.

22. Geis, M.W., Efremow, N.N. and D.D. Rathman. 1988. Summary Abstract: Device Applications of Diamonds, J. Vac, Sci. Technol., A6(3): 1953-1954.

23. Steele, R. 1995. The Forecast for Silicon Carbide, Compound Semicon., 1(3): 17.

24. Asbeck, P.M. 1990. Bipolar Transistors, in High-Speed Semiconductor Devices, Ed. S.M. Sze, J. Wiley and Sons, Inc., New York.

25. Houghton, D.C. 1995. High Frequency Performance of SiGe HBTs, Compound Semicon., 1(3): 31.

26. Konig, U. and H. Dambkes. 1995. SiGe HBTs and HFETs, Solid St. Electronics, 38(9): 1595-1602.

27. Ren, F., Abernathy, C.R.,Pearton, S.J., Yang, L.W. and S.T. Fu. 1995. Novel Fabrication of Self-Aligned GaAs/AlGaAs and GaAs/InGaP Microwave Power Heterojunction Bipolar Transistors, Solid St. Electronics, 38(9): 1635-1639.

28. Pearton, S.J. and N.J. Shah. 1990. Heterostructure Field-Effect Transistors, in High-Speed Semiconductor Devices, Ed. S.M. Sze, J. Wiley and Sons, Inc., New York.

29. Craford, M.G. 1995. Wide Bandgap LED Overview, Compound Semiconductor, 1(1): 48.

30. Singh, J. 1996. Optoelectronics - An Introduction to Materials and Devices, McGraw-Hill Co., Inc., New York.

31. Nakamura, S., Senoh, M., Iwasa, N. and S. Nagahama. 1995. High-Brightness InGaN Blue, Green and Yellow Light Emitting Diodes with Quantum Well Structures, Japan. J. Appl. Phys., 34: L797-L799.

32. Nakamura, S., Senoh, M., Iwasa, N. and S. Nagahama 1995. High-Power InGaN Single-Quantum-Well Structures for Blue and Violet Light Emitting Diodes, Appl. Phys. Lett., 67(13): 1866-1868.

33. Nakamura, S., Senoh, M., Nagahama, S., Iwasa, N. and T. Yamada. 1996. InGaN-Based Multi-Quantum-Well-Structure Laser Diodes, Japan. J. Appl. Phys., 35: L74-L76.
34. Special Feature. 1996. Nichia Demonstrates the First Nitride Laser, Compound Semicond., 2(1): 7-8.
35. Gunshor, R.L.and A.V. Nurmikko. 1995. II-VI Blue-Green Laser Diodes: A Frontier of Materials Research, MRS Bull., XX(7): 15-19.

Chapter Thirteen

FUNCTIONAL COMPOSITES FOR SENSORS AND ACTUATORS: Smart Materials

ROBERT E. NEWNHAM

Materials Research Laboratory
Pennsylvania State University
University Park, PA 16802

ABSTRACT

"Smart" materials have the ability to perform both sensing and actuating functions. Passively smart materials respond to external change in a useful manner without assistance, while actively smart materials have a feedback loop which allows them to both recognize the change and initiate an appropriate response through an actuator circuit.

One of the techniques used to impart intelligence into materials is "Biomimetics," the imitation of biological functions in engineering materials. Composite ferroelectrics fashioned after the lateral line and swim bladders of fish are used to illustrate the idea. "Very Smart" materials, in addition to sensing and actuating, have the ability to "learn" by altering their property coefficients in response to the environment. Field-induced changes in the nonlinear properties of relaxor ferroelectrics and soft rubber are utilized to construct tunable transducers. Integration of multifunctional ferroic ceramics into compact, robust packages is a major goal in the development of smart materials.

INTRODUCTION

It has been said that life itself is motion, from the single cell to the most complex organism: the human body. This motion, in the form of mobility, change, and adaptation, is what elevates living beings above the lifeless forms (Rogers, 1990). This concept of creating a higher form of materials and structures by providing the necessary life functions of sensing, actuating, control, and intelligence to those materials is the motivation for studying smart materials.

Smart materials are part of smart systems - functional materials for a variety of engineering applications. Smart medical systems for the treatment of diabetes with blood sugar sensors and insulin delivery pumps. Smart airplane wings that achieve greater fuel efficiency by altering their shape in response to air pressure and flying speed. Smart toilets that analyze urine as an early warning system for health problems. Smart structures in outer space incorporating vibration cancellation systems that compensate for the absence of gravity and prevent metal fatigue. Smart houses with electrochromic windows that control the flow of heat and light in response to weather changes and human activity. Smart tennis rackets with rapid internal adjustments for overhead smashes and delicate drop shots. Smart muscle implants made from rubbery gels that respond to electric fields, and smart dental braces made from shape memory alloys. Smart hulls and propulsion systems for navy ships and submarines that detect flow noise, remove turbulence, and prevent detection. Smart water purification systems that sense and remove noxious pollutants. As indicated later, modern automobiles already contain a wide variety of sensors and actuators, but there are many more to come. In a recent newspaper cartoon, Blondie and Dagwood encountered a smart automobile that drives itself back to the finance company when the owner misses a payment!

In this chapter the idea of "smartness" in a material is discussed, along with a number of examples involving ferroelectric components. Some of these smart ceramics are in production, while others have great potential but are thus far limited to laboratory investigations.

To begin our discussion, we first define "smart" to set the limits for classifying smart materials.

PASSIVE SMARTNESS

A passively smart material has the ability to respond to environmental conditions in a useful manner. It differs from an actively smart material in that there are no external fields or forces or feedback systems used to enhance its behavior. Many passively smart materials incorporate self-repair mechanisms or stand-by phenomena which enable the material to withstand sudden changes in the surroundings. The crack-arresting mechanisms in partially stabilized zirconia are a good example. Here the tetragonal-monoclinic phase change accompanied by ferroelastic twin wall motion are the stand-by phenomena capable of generating compressive stresses at the crack tip. In a similar way, toughness can be improved by fiber pull-out or by multiple crack-branching as in the structural composites used in aircraft, or in machinable glass-ceramics.

Ceramic varistors and PTC thermistors are also passively smart materials. When struck by high-voltage lightning, a zinc oxide varistor loses most of its electrical resistance and the current is by-passed to ground. The resistance change is reversible and acts as a stand-by protection phenomenon. Varistors also have a self-repair mechanism in which its highly nonlinear I-V relationship can be restored by repeated application of voltage pulses. Barium titanate PTC thermistors show a very large increase in electrical resistance at the ferroelectric phase transformation near 130°C. The jump in resistance enables the thermistor to arrest current surges, again acting as a protection element. The R(V) behavior of the varistor and the R(T) behavior of the PTC thermistor are both highly nonlinear effects which act as standby protection phenomena, and make the ceramics smart in a passive mode.

ACTIVE SMARTNESS

A smart ceramic can also be defined with reference to sensing and actuating functions, in analogy to the human body. A smart ceramic senses a change in the environment, and using a feedback system, makes a useful response. It is both a sensor and an actuator. Examples include vibration damping systems for outer space platforms and electrically-controlled automobile suspension systems using piezoelectric ceramic sensors and actuators. Other sensor-actuator combinations are illustrated in Figure 1.

A video tape head positioner developed by Piezoelectric Products, Inc., operates on this principle. A bilaminate bender made from tape-cast PZT ceramic has a segmented electrode pattern dividing the sensing and actuating functions of the positioner. The voltage across the sensing electrode is processed through the feedback system resulting in a voltage across the positioning electrodes. This causes the cantilevered bimorph to bend, following the video tape track path. Articulated sensing and positioning electrodes operating at 450Hz near the tape head help keep the head perpendicular to the track.

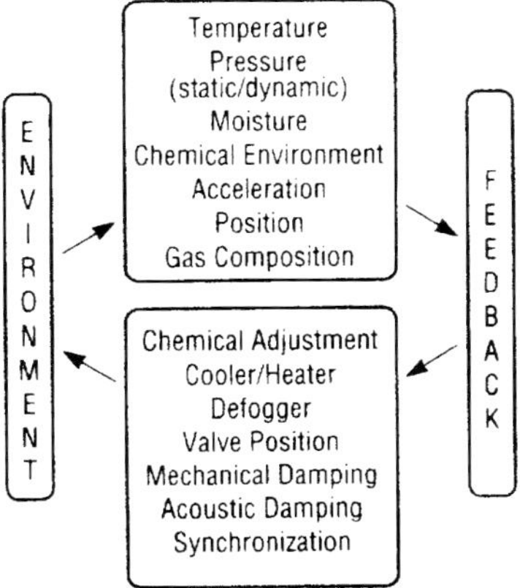

FIGURE 1. Examples of sensing and actuating functions in actively smart system.

RUBBER-LIKE CERAMICS

Every baseball and cricket player knows the importance of "soft hands." In catching a baseball, it is important to withdraw the hands slightly on making contact with the ball. This reduces the momentum of the ball gradually and creates a soft landing. Soft landings are achieved on ceramics in the same way, making them feel as soft as rubber.

To test the concept, controlled compliance experiments have been carried out using PZT sensors and actuators (Newnham et al., 1989). In the test set-up, (Figure 2) one actuator is used as the external driver, and the other as the responder. Sandwiched between the two actuator stacks are two sensors and a layer of rubber.

The upper actuator is driven at a frequency of 100 Hz and the vibrations are monitored with the upper sensor. The pressure wave emanating from the driver passes through the upper sensor and the rubber separator and impinges on the lower sensor. The resulting signal is amplified using a low noise amplifier and fed back through a phase shifter to the lower actuator to control the compliance.

A smart sensor-actuator system can mimic a very stiff solid or a very compliant rubber. This can be done while retaining great strength under static loading, making the smart material especially attractive for vibration control.

If the phase of the feedback voltage is adjusted to cause the responder to contract in length rather than expand, the smart material mimics a very soft, compliant substance. This reduces the force on the sensors and partially eliminates the reflected signal. The reduction in output signal of the upper sensor is a measure of the effectiveness of the feedback system. In our experiments, the compliance of the actuator-sensor composite was reduced by a factor of six compared to rubber (Newnham et al., 1989).

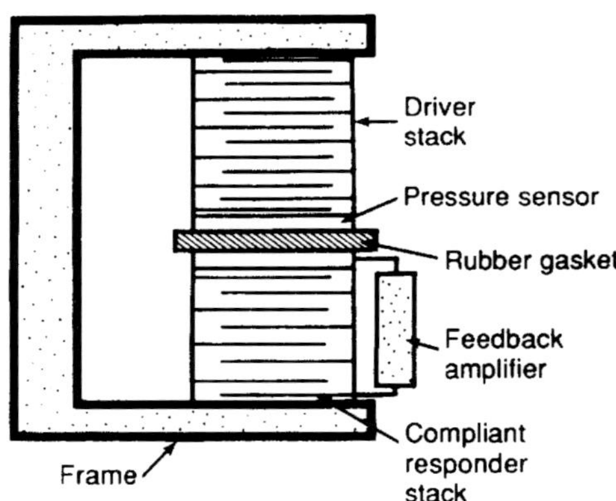

FIGURE 2. Fixture for experimental evaluation of smart materials with controlled compliance.

MODULATED SUSPENSION SYSTEMS

The automobile industry is a very large market in which smart composites and sensors are already widely used. More than fifty electroceramic components can be found in today's high-tech autos, ranging from the air-fuel oxygen sensors used in most autos to the more exotic piezoelectric raindrop sensor, which automatically senses the amount of rain falling, and adjusts the windshield wipers to the optimum speed (Taguchi, 1987).

Controlled compliance with piezoelectric ceramics is utilized in Toyota's piezoTEMS (Toyota Electronic Modulated Suspension), a system which has been developed to improve the drivability and stability of the automobile, and at the same time enhance passenger comfort (Fukami et al., 1994). The TEMS is basically a road stability sensor and shock adjustor, which detects bumps, dips, rough pavement, and sudden lurches by the vehicle, then rapidly adjusts the shock absorbers to apply a softer or firmer damping force, depending on what is necessary to minimize discomfort while maintaining control of the vehicle. The shock absorbers are continuously readjusted as the road conditions change so that rocking or wobbling on soft shocks is eliminated.

The TEMS road surface sensor consists of a five-layer piezoelectric ceramic sensor mounted on the piston rod of the shock absorber. When a bump in the road is encountered, the resulting stress applied to the sensor produces a voltage which is fed into an electronic control unit that amplifies the signal and supplies a high voltage to the piezoelectric actuator. The 88-layer PZT actuator produces a 50 µm displacement on the oil system which is hydraulically enlarged to two millimeters, enough to change the damping force from firm to soft; the entire process takes only about twenty milliseconds (not even enough time to slam on the brakes!). Also figured into the actuator output are the vehicle speed and the driver's preference for a generally softer (American) or firmer (European) ride.

Alternatively, it is possible to damp stresses and vibrations without the need for a sensor-actuator feedback loop; materials which can perform this function are called passive damping materials. In a piezoelectric passive damper, a piezoelectric ceramic is connected in parallel with a properly matched resistor. The external stress creates a polarization in the piezoelectric, which induces a current in the resistor, leading to energy dissipation. A high piezoelectric coupling coefficient is required to induce the maximum voltage and energy dissipation (Suzuki, et al., 1991) (Figure 3).

ACTUATOR MATERIALS

Two of the most popular actuators are the multilayer and bimorph designs shown in Figure 4. The multilayer has fast response, low drive voltage, and high generative force, but only micron-size motions, whereas bimorphs are capable of millimeter movements, but are slow, weak, and unreliable. The moonie is an

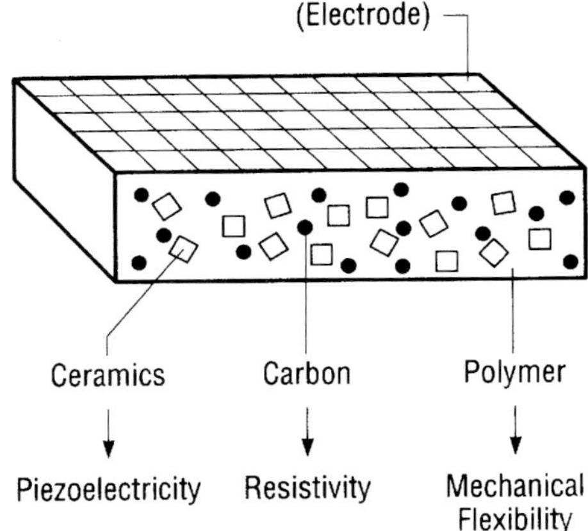

FIGURE 3. Composite damper consisting of piezoelectric PZT grains and conducting carbon grains embedded in a flexible polymer matrix.

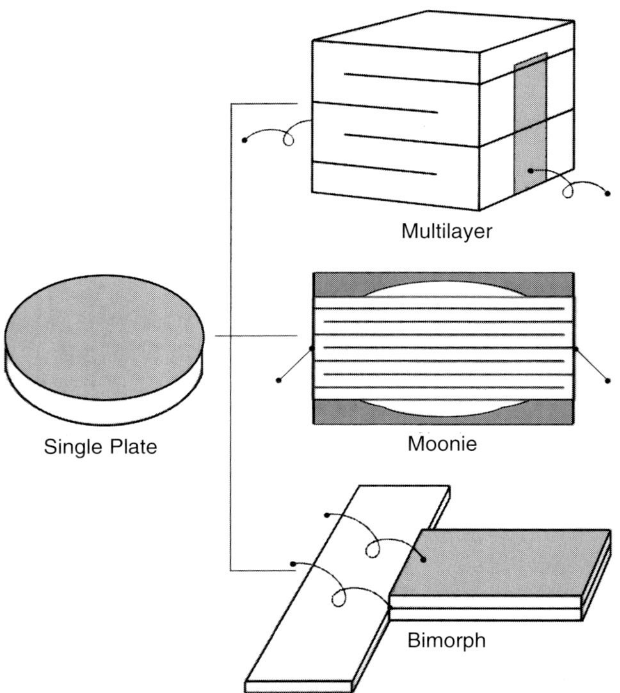

FIGURE 4. Typical designs for ceramic actuators: multilayer, moonie, and bimorph.

excellent compromise which retains the low drive voltage and amplifies the motion with bonded metal caps operating in a flextensional mode (Uchino, 1993).

There are many approaches to controlling vibration and structural deformation. Actuation strain can be controlled by piezoelectric materials (Burke and Hubbard, 1987), electrostrictive materials, (Uchino, 1986), magnetostrictive materials (Hathaway and Clark, 1993), shape-memory metal alloys, (Schetky, 1979), and thermally-controllable materials (Edberg, 1987). Utilizing a system with distributed actuators, it is possible to design structures with intrinsic vibration and shape control capabilities. Among the most important actuator materials are shape memory metals and ceramics. The shape memory effect is exhibited by alloys which undergo thermoelastic martensite transformations. This is a first order displacive transformation in which a body centered cubic metal transforms by shear on cooling to a martensitic phase. When deformed in the martensitic low temperature phase, shape memory alloys will recover this deformation and return to the original shape when heated to a temperature where the martensite reverts back to the parent body-centered cubic structure. Unlike most ferroelectric and ferromagnetic transitions, the shape memory transformation has a large hysteresis which can be troublesome in practice. Non-ferrous shape memory alloys of nickel-titanium alloy (Nitinol) have been developed by Goodyear Aerospace Corporation for spacecraft antennae (Schetky, 1979). A wire hemisphere of the material is crumpled into a tight ball, less than five centimeters across. When heated above 77°C, the ball opens up into its original shape—a fully formed antenna. Although it has seldom been used in service, this antenna demonstrates the magnitude of deformation and reformation possible in shape memory alloys.

While shape memory alloys are more like a solution looking for a problem, it has been suggested that transient and steady state vibration control can be accomplished with hybrid structures in which the shape memory alloy is embedded inside the material (Rogers et al., 1989)

Some ceramic materials also possess a sizeable shape-memory effect; of particular interest are materials which are simultaneously ferroelectric and ferroelastic. Their ferroelasticity ensures that recoverable spontaneous strain is available for contributing to the shape memory effect, and the ferroelectricity implies that their spontaneous strain can be manipulated not only by mechanical forces but also by electric fields (Wadhawan et al., 1981).

ELECTRORHEOLOGICAL FLUIDS

One of the criteria which separates "smart" materials from "very smart" or "intelligent" materials is the ability of the material to not only sense a change and actuate a response, but to automatically modify one or more of its property coefficients during the sensing/actuating process. In effect, this type of material not only warns the user of a change in its environmental conditions and responds to it, but can in addition adjust itself to compensate for future change.

Electrorheological (ER) fluids (Ghandi and Thompson, 1989), and their magnetic analog, ferrofluids (Rosenweg, 1985), are an example of materials that have great potential for use in smart materials and systems. ER fluids are typically suspensions of fine particles in a liquid medium; the viscosity of the suspension can be changed dramatically by applying an electrical field. The electric field causes alignment of the particles in fibril-like branches in the direction of the applied field. The alignment disappears when the electric field is removed, thus creating the desired property of complete cyclic reproducibility.

ER fluids represent an advanced class of composite materials with self-tuning properties, that will find considerable use in vibration control applications. In addition, the compatibility of this technology with modern solid state electronics makes it an attractive component for integration into multifunction, self-contained smart material packages.

BIOMIMETICS - FISH EARS

The word "biomimetic" is not found in most dictionaries so it needs to be defined. It comes from the Greek words "bios," meaning "life," and "mimetikos," meaning "to imitate." Biomimetic means to imitate life, or to use the biological world as a source of ideas for device concepts.

Fish and the other inhabitants of the underwater world have some interesting ways of talking and listening which have been copied using piezoelectric ceramics. For most fish, the principal sensors are the *lateral line* and the *inner ear* coupled to the *swim bladder*. The pulsating swim bladder also acts as a voice, as do chattering teeth in certain fish species.

The lateral line runs from the head to the tail of the fish and resembles a towed array with sensing organs (stitches) spaced at intervals along the nerve fiber. Each stitch contains several neuromasts made up of gelatinous cupulae resembling pimples in shape, and within each cupula are several fibers which vibrate as the fish swims through water and act as sensors for flow noise. The hair-like fibers are extremely thin in diameter ranging from 0.5 to 10 μm. When stimulated by turbulence, the motion of the hairs produces changes in the synapses which are in turn connected to the nerve fiber. The electric signal originates from impedance changes in cell walls which modulate the flow of K^+ ions. The lateral line is especially sensitive to low frequency fluid motion parallel to the length of the fish. In the 50 Hz range, threshold signals are observed for displacements as small as 30 nm! (Bond, 1979; Moyle and Cech, 1988).

The 1-3 composite hydrophones described later are patterned after the hair-filled cupulae of the lateral line. Thin PZT fibers embedded in polymer provide excellent electromechanical coupling to a liquid medium and can be used as both sensors and actuators.

The primary function of the gas-filled swim bladder is to provide buoyancy, but it is also used for sound and pressure reception and in some species is equipped with drumming muscles for sound production. The flexible swim bladder responds to hydrostatic pressure waves by changing volume. Fish with swim bladders can perceive relative pressure changes equivalent to less than 0.5% of the ambient hydrostatic pressure. Direct or indirect linkages from the swim bladder to the inner ear promote the hearing sensation. Fish with no connections perceive low frequency sound (less than 500 Hz), while those with good connections have an upper frequency response of 5000 Hz. As might be expected, the swim bladder is reduced in size with depth, and loses much of its sensitivity as a sensor. The flextensional Moonie transducer (Figure 4) mimics the swim bladder of a fish.

HYDROPHONE MATERIALS

The knowledge which comes from the understanding of "fish talk" can be directly applied to research in materials destined to someday "sleep with the fishes". Hydrophones are underwater listening devices made from piezoelectric materials which respond to hydrostatic pressure waves. Among the applications for hydrophones are sonar systems for submarines, off-shore oil platforms, geophysical prospecting equipment, fish finders, and earthquake monitors.

As the earth's population continues to increase, mankind must continue to search for new and efficient sources of food and nutrition. The world's oceans may provide a solution to this problem, not only through fish farming but through the use of new and varied salt water vegetation that could provide an abundant source of food, especially for third world countries in which poor soil and harsh climates prohibit conventional farming. Smart hydrophone transceivers will receive and transmit fish talk and monitor the growth of underwater vegetation.

The figure of merit for hydrophone materials is the product of the hydrostatic piezoelectric charge coefficient (d_h) and the piezoelectric voltage coefficient (g_h). While good piezoelectric materials such as PZT have high d_{33} and d_{31} piezoelectric coupling coefficients, the d_h value is only about 45 pC/N because d_{33} and d_{31} are opposite in sign, and $d_h = d_{33} + 2d_{31}$. $d_h g_h$ is also inversely related to the dielectric permittivity, ε_{33}, so that low dielectric constants are desirable as well.

Rather than abandon PZT in search of the ultimate hydrophone material, avoiding this problem is often a matter of appropriate engineering of existing materials. Too often in the field of materials research we put too much emphasis on the synthesis of new materials and too little emphasis on new and unique designs for old materials.

A composite design with 1-3 connectivity is similar in design to the hair-filled gelatinous cupula with thin PZT rods embedded in a polymer matrix. The 1-3 piezocomposites have excellent sensitivity to pressure waves in water (Klicker et al., 1981; Guruaraja et al., 1984). The large d_{33} is maintained because the parallel connection results in stress transfer from polymer to piezoceramic, while the d_{31} is destroyed because of series connection in the lateral dimension where the mechanical load is absorbed by the polymer and not transferred to the PZT rods. Finally, ε_{33} is minimized due to the large volume of low ε_{33} polymer present. The $d_h g_h$ values are improved by more than an order of magnitude when very thin rods are used.

Another piezoelectric hydrophone composite maximizes d_h by simply redirecting the applied stresses using specially shaped electrodes (Xu et al., 1991). These are flextensional transducers which mimic the motions of the swim bladder. Shallow air spaces are positioned under the metal electrodes while the PZT ceramic plays the role of the muscle lining the swim bladder. When subjected to a hydrostatic stress, the thick metallic electrodes convert a portion of the z-direction stress into large radial and tangential stresses of opposite signs. The result is that d_{31} changes from negative to positive, so that its contribution now adds to d_{33} rather than subtracting from it. The $d_h g_h$ of these composites is approximately 250 times that of pure PZT.

VERY SMART COMPOSITES: THE TUNABLE TRANSDUCER

By building in a learning function, the definition of a smart material can be extended to a higher level of intelligence: *A very smart material senses a change in the environment and responds by changing one or more of its property coefficients. Such a material can tune its sensor and actuator functions in time and space to optimize future behavior.* With the help of a feedback system, a very smart material becomes smarter with age, something even human beings strive for. The distinction between smart and very smart materials is essentially one between linear and nonlinear properties. The physical properties of nonlinear materials can be adjusted by bias fields or forces to control response.

To illustrate the concept of a very smart material, we describe the tunable transducer recently developed in our laboratory. Five important properties of a transducer are the resonant frequency f_r, the acoustic impedance Z_A, the mechanical damping coefficient Q, the electromechanical coupling factor k, and the electrical impedance Z_E. The resonant frequency and acoustic impedance are controlled by the elastic constants and density, as discussed in the next section. The mechanical Q is governed by the damping coefficient (α) and is important because it controls "ringing" in the transducer. Definitions of the coefficients are given in Table 1.

Electromechanical coupling coefficients are controlled by the piezoelectric coefficient which, in turn, can be controlled and fine-tuned using relaxor ferroelectrics with large electrostrictive effects. The dielectric "constant" of relaxor ferroelectrics depends markedly on DC bias fields, allowing the electrical impedance to be tuned over a wide range as well. In the following sections we describe the nature of nonlinearity and how it controls the properties of a tunable transducer.

TABLE 1.
Important characteristics of an electromechanical transducer.

Fundamental resonant	$f = \dfrac{1}{2t} \sqrt{\dfrac{c}{\rho}}$	where
frequency, f, of the thickness mode		t = thickness dimension, c = elastic stiffness, ρ = density
Acoustic Impedance Z_A	$|Z_A| = \sqrt{\rho \cdot c}$	where c = elastic stiffness,s ρ = density
Mechanical Q	$Q = \dfrac{\pi}{\lambda_a \cdot \alpha}$	where λ_a = acoustic wavelength α = damping coefficient
Electromechanical coupling coefficient k	$k = d \sqrt{\dfrac{c}{\varepsilon}}$	d = piezoelectric charge coefficient, ε = electric permittivity
Electrical impedance Z_e	$|Z_e| = \dfrac{t}{\omega \cdot \varepsilon \cdot A}$	where ω = angular frequency, A = electrode area

ELASTIC NONLINEARITY: TUNING THE RESONANT FREQUENCY

Rubber is a highly nonlinear elastic medium. In the unstressed compliant state, the molecules are coiled and tangled, but under stress the molecules align and the material stiffens noticeably. Experiments carried out on rubber-metal laminates demonstrate the size of the nonlinearity. Young's modulus was measured for a multilayer laminate consisting of alternating steel shim and soft rubber layers each 0.1 mm thick. Under compressive stresses of 200 Mpa, the stiffness is quadrupled from about 600 to 2400 Mpa (Rivin, 1983). The resonant frequency f is therefore doubled, and can be modulated by applied stress.

Rubber, like most elastomers, is not piezoelectric. To take advantage of the elastic nonlinearity, it is therefore necessary to construct a composite transducer consisting of a piezoelectric ceramic (PZT) transducer, thin rubber layers, and metal head and tail masses, all held together by a stress bolt.

The resonant frequency and mechanical Q of such a sandwich structure has been measured as a function of stress bias. Stresses ranged from 20 to 100 Mpa in the experiments. Under these conditions the radial resonant frequency changed from 19 to 37 kHz, approximately doubling in frequency as predicted from the elastic nonlinearity. At the same time the mechanical Q increases from about 11 to 34 as the rubber stiffens under stress.

The changes in resonant frequency and Q can be modeled with an equivalent circuit in which the compliance of the thin, rubber layers are represented as capacitors coupling together the larger masses (represented as inductors) of the PZT transducer and the metal head and tail masses. Under low stress bias, the rubber is very compliant and effectively isolates the PZT transducer from the head and tail masses. At very high stress, the rubber stiffens and tightly couples the metal end pieces to the resonating PZT ceramic. For intermediate stresses the rubber acts as an impedance transformer giving parallel resonance of the PZT - rubber - metal - radiation load.

Continuing the biomimetic theme, it is interesting to compare the change in frequency of the tunable transducer with the transceiver systems used in the biological world. The biosonar system of the flying bat is similar in frequency and tunability to our tunable transducer. The bat emits chirps at 30 kHz and listens for the return signal to locate flying insects. To help it differentiate the return signal from the outgoing chirp, and to help in timing the echo, the bat puts an FM signature on the pulse. This causes the resonant frequency to decrease from 30 to 20 kHz near the end of each chirp. Return signals from the insect target are detected in the ears of the bat where neural cavities tuned to this frequency range measure the time delay and flutter needed to locate and identify its prey. Extension of the bat biosonar principle to automotive, industrial, medical and entertainment systems is underway.

PIEZOELECTRIC NONLINEARITY: TUNING THE ELECTROMECHANICAL COUPLING COEFFICIENT

The difference between a smart and a very smart material can be illustrated with piezoelectric and electrostrictive ceramics. PZT (lead zirconate titanate) is a piezoelectric ceramic in which the ferroelectric domains have been aligned in a very large poling field. Strain is linearly proportional to electric field in a fully poled piezoelectric material which means that the piezoelectric coefficient is a constant and cannot be electrically tuned with a bias field. Nevertheless it is a smart material because it can be used both as a sensor and an actuator.

PMN (lead magnesium niobate) is not piezoelectric at room temperature because its Curie temperature lies near 0°C. Because of the proximity of the ferroelectric phase transformation, however, and because of its diffuse nature, PMN ceramics exhibit very large electrostrictive effects.

Electromechanical strains comparable to PZT can be obtained with electrostrictive ceramics like PMN, and without the troubling hysteretic behavior shown by PZT under high fields. The nonlinear relation between strain and electric field (Figure 5) in electrostrictive transducers can be used to tune the piezoelectric coefficient and the dielectric constant.

The piezoelectric d_{33} coefficient is the slope of the strain-electric field curve when strain is measured in the same direction as the applied field. Its value for

Pb(Mg$_{0.3}$Nb$_{0.6}$Ti$_{0.1}$) O$_3$ ceramics is zero at zero field and increases to a maximum value of 1300 pC/N (about three times larger than PZT) under a bias field of 3.7 kV/cm.

This means that the electromechanical coupling coefficient can be tuned over a very wide range, changing the transducer from inactive to extremely active. The dielectric constant also depends on DC bias. The polarization saturates under high fields causing decreases of 100% or more in the capacitance. In this way the electrical impedance can be controlled as well.

Electrostrictive transducers have already been used in a number of applications including adaptive optic systems, scanning tunneling microscopes, and precision micropositioners (Uchino, 1986).

CONDUCTOR FILLED COMPOSITES

Up to this point, our discussion has focused primarily on piezoelectric transducers in which the sensing and actuating functions are electromechanical in nature. But the idea of a smart material is much more general than that. There are many types of sensors and many types of actuators, and many different feedback circuits. Composites consisting of a highly conductive filler powder dispersed in a flexible, insulating polymer matrix are commonly used in electronic applications for thermistors (Hu et al., 1987), pressure sensing elements (Yoshikawa et al., 1990) and chemical sensors (Ruschau et al., 1989).

The electrical properties of these materials are described by conventional percolation theory (Ruschau and Newnham, 1992). When a sufficient amount of conductive filler is loaded into an insulating matrix, the composite transforms from an insulator to a conductor, arising through the continuous linkages of particles, Figure 6. As the volume fraction of filler is increased, the probability of continu-

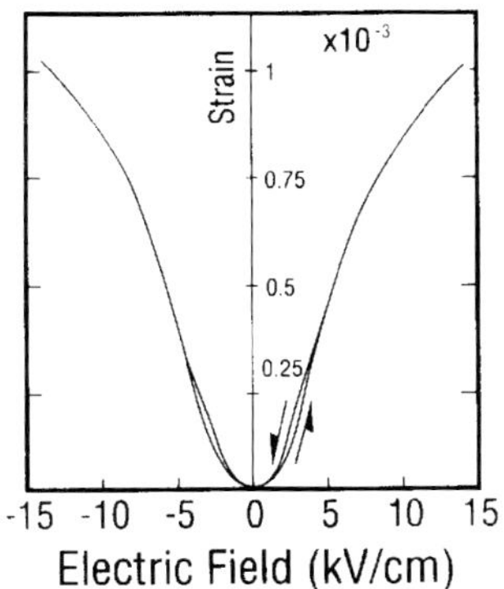

FIGURE 5. Electrostriction in PMN (Pb3MgNb2O$_9$)-based ceramics leads to strains of 0.1%. An attractive feature of these ceramics is the near absence of hysteresis (Cross et al., 1980)

ity increases until the critical volume fraction is reached, beyond which the electrical conductivity is high, closer to the filler material.

Thermistors or temperature dependent resistors, are typically made from doped ceramics ($BaTiO_3$). The desire for improved properties and ease of processing has spurred the development of composite thermistors. This type of composite combines a low resistivity filler powder, such as carbon black, vanadium sesquioxide, or nickel with a high resistivity polymer matrix such as polyethylene or polystyrene. In the dormant room temperature state the thermistor has a resistivity similar to the filler. On being activated by heating, however, the polymer expands pulling the particles apart and causing the resistivity to increase to a value close to the polymer matrix resistivity, an increase as large as eight orders of magnitude (Hu et al., 1987).

Conductive composites exposed to a solvent, in order to produce chemical sensing capabilities, have also been shown to take advantage of percolation phenomena. A conductive composite which shows some sensitivity to chemical vapors undergoes a four-step resistance change during exposure to the solvent. First, adsorption of solvent onto the surface of the composite results in a small, quick increase in resistance. Secondly, as the solvent diffuses through the sensor sample, the number of percolated chains decreases and the remaining chains become less conductive; a gradual increase in resistance occurs during this step. Third, when the last remaining percolated linkage is disrupted, there is a large increase in resistivity, usually several orders of magnitude; this conductor-to-insulator switch is the most important step for application purposes. Finally , as more solvent diffuses into the composite, the resistivity saturates at a level determined primarily by the conductivity of the solvent/polymer solution, although some effects from the filler are evident (Ruschau et al., 1989).

SMART ELECTROCERAMIC PACKAGES

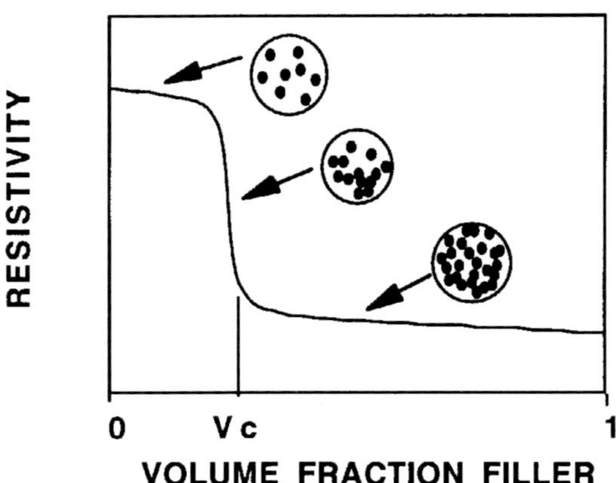

FIGURE 6. Percolation theory as applied to conductive composites, showing the development of conductive pathways with an increase in volume fraction of filler.

Many sensors and actuators can be fabricated in the form of multilayer ceramic packages. Until recently multilayer packages consisted of low permittivity dielectric layers with metal circuitry printed on each layer and interconnected through metallized via holes between layers. Buried capacitors and resistors have now been added to the three-dimensional packages, and other components will follow shortly. Smart sensors, adaptive actuators, and display panels, together with thermistors and varistors to guard against current and voltage overloads, are next in line for development (Newnham, 1988).

Integration and miniaturization of electroceramic sensors and actuators is an ongoing process in the automotive and consumer electronics areas. Multilayer packages containing signal processing layers made up of low-permittivity dielectrics and printed metal interconnections are in widespread production. Further integration with embedded resistors and capacitors are under development, and it seems likely that intelligent systems will make use of the same processing technology. Tape casting and screen printing are used most often. Varistors, chemical sensors, thermistors, and piezoelectric transducers can all be fabricated in this way, opening up the possibility of multicomponent, multifunction ceramics with both sensor and actuator capabilities. Silicon chips can be mounted on these multifunctional packages to provide all or part of the control network. Processing is a major challenge because of the high firing temperatures of most ceramics, typically in the range 800°C to 1500°C. Differences in densification shrinkage and thermal contraction, together with adverse chemical reactions between the electroceramic phases, create formidable problems. Nevertheless, the rewards for such an achievement are substantial. An all-ceramic multifunction package would be small, robust, inexpensive, and sufficiently refractory to withstand elevated temperatures.

The next logical step is to combine the sensor and actuator functions with the control system. This can be done by depositing electroceramic coatings on integrated circuit silicon chips. Electroceramics have a vital role to play in intelligent systems, and many new developments will take place in the coming decade and the next century.

REFERENCES

Bond, C.E. 1979. "Biology of Fishes," Saunders College Publishing Co. Philadelphia.

Burke, S. and J.E. Hubbard. 1987. "Active Vibration Control of a Simply-Supported Beam Using a Spatially-Distributed Actuator," IEEE Control Systems Magazine 7, 25-30.

Cross, L.E., Jang, S.J., Newnham, R.E., Nomura, S., and K. Uchino. 1980. "Large Electrostriction Effects in Relaxor Ferroelectrics", Ferroelectrics 23 187-193.

Edberg, D.L. 1987: "Control of Flexible Structures by Applied Thermal

Gradients," AIAA Journal, 25, 877-883.

Fukami, A., Yano, M., Tokuda, H., Ohki, M. and R. Kizu. 1994. "Development of Piezoelectric Actuators and Sensors for Electronically Controlled Suspension", Int. J. of Vehicle Design, *15* 348-357.

Ghandi, M.V. and B.S. Thompson. 1989. "Smart Materials, Structures, and Mathematical Issues," ed. by C.A. Rogers, Technomic Publishers, Lancaster, PA, 63-68.

Gururaja, T.R., Schulze, W.A., Cross, L.E., Newnham, R.E., Auld, B.A., and J. Wang. 1984. "Resonant Modes in Piezoelectric PZT Rod-Polymer Composite Materials, "Proc. of IEEE Ultrasonic Symposium, 523-527.

Hathaway, K.B. and A.E. Clark. 1993. "Magnetrostrictive Materials", Mat. Res. Soc. Bull. *18* 34-41.

Hu, K.A., Moffat, D., Runt, J., Safari, A., and R.E. Newnham. 1987. "V3O3-Polymer Composite Thermistors." J. Am. ceram. Soc. 70: 583-585.

Klicker, K.A., Biggers, J.V. and R.E. Newnham. 1981. Composites of PZT and Epoxy for Hydrostatic Transducer Applications," *J.Amer.Cer.Soc. 64,* 5-9.

Moyle, P.B. and J.J. Cech, Jr. 1988. "Fishes: An Introduction to Ichthyology," Prentice Hall, Englewood Cliffs, N.J.

Newnham, R.E. 1988. "The Golden Age of Electroceramics," Adv. Ceram. Mat. *3,* 12-16.

Newnham, R.E., Xu, Q.C., Kumar, S., and L.E. Cross. 1989. "Smart Ceramics," *J.Wave-Matl.Int. 4,* 3-10.

Rivin, E.I. 1983. "Properties and Prospective Applications of Ultra Thin Layered Rubber-Metal Laminates for Limited Travel Bearings," *Tribology Int. 16,* 17-25.

Rogers, C.A. 1990. "From the Editor," *Journal of Intelligent Material Systems and Structures l, 3.*

Rogers, C.A., Liang, C., and D.K. Barker. 1989. In "Smart Materials, Structures, and Mathematical Issues," ed. by C.A. Rogers, Technomic Publishers, Lancaster, PA, 39-62.

Rosenweig, R.E. 1985. "Ferrohydrodynamics," Cambridge University Press, New York.

Ruschau, G.R., Newnham, R.E., Runt, J., and B.E. Smith. 1989. "O-3 Ceramic-Polymer Composite Chemical Sensors, Sensors and Actuators, 20: 269-275.

Ruschau, G. and R.E. Newnham. 1992. "Critical Volume Fractions in Conductive Composites", J. Comp. Mat. 26: 2727-2735.

Ruschau, G.R., Yoshikawa, S., and R.E. Newnham. 1992. J. Appl. Phys. 72:953-959.

Schetky, L. 1979. "Shape Memory Alloys," *Scientific American 241,* 74-82.

Sullivan, R.J. and R.E. Newnham. 1992. "Composite Thermistors", pp. 351-367, in: C.N.R.Rao (Ed.) *Chemistry of Advanced Materials,* Blackwell Scientific Publications London, U.K.

Suzuki, Y., Uchino, K., Gouda, H., Sumita, M., Newnham, R.E., and A.R. Ramachandran. 1991. J. Japan Ceram. Soc., *99,* 1135-1140.

Taguchi, M., 1987:"Applications of High-Technology Ceramics in Japanese Automobiles," *Adv. Ceram. Mat. 2,* 754-762.

Takagi, T. 1989. "The Concept of Intelligent Materials and the Guidelines on R&D Promotion", Japan Science and Technology Agency Report.

Tsuka, H., Nakomo, J., and Y. Yokoya, 1990: "A New Electronic Controlled Suspension Using Piezoelectric Ceramics,"IEEE Workshop on Electronic Applications in Transportation.

Uchino, K. 1986. "Piezoelectric/Electrostrictive Actuators," Morikita Publishers, Tokyo.

Uchino, K. 1993. "Ceramic Actuators: Principles and Applications", Mat. Res. Soc. Bull. *18*, 42-48.

Urick, R.J. 1975. "Principles of Underwater Sound," McGraw Hill, New York.

Wadhawan, V.K., Kernion, M.C., Kimura, T. and R.E. Newnham. 1981. "The Shape Memory Effect in PLZT Ceramics," *Ferroelectrics 37*, 575-578.

Xu, Q.C., Belsick, J., Yoshikawa, S., Srinivasan, T.T., and R.E. Newnham, 1991: "Piezoelectric Composites with High Sensitivity and High Capacitance for Use at High Pressures," IEEE Transactions, UFFC *38*, 634-639.

Xu, Q.C., Newnham, R.E., Blaskiewicz, M., Fang, T.T., Srinivasan, T.T., and S. Yoshikawa. 1990. "Nonlinear Multilayer Composite Transducers." Presented at the 7th International Symposium on Applications for Ferroelectrics, June 6-8, Urbana, IL.

Yoshikawa, S., Ota, T., and R.E. Newnham. 1990. "Piezoresistivity in Polymer-Ceramic Composites," J. Am. Ceram. Soc., 73: 263-267.

Chapter Fourteen

FERROELECTRICS

L. ERIC CROSS and
SUSAN TROLIER-MCKINSTRY

Materials Research Laboratory
Pennsylvania State University
University Park, PA 16802

INTRODUCTION

The term ferroelectricity was first used by Valasek(1) to emphasize the analogy between the nonlinear hysteretic high field dielectric response of Rochelle Salt (sodium potassium tartrate tetrahydrate, NaK $C_4H_4O_6 \cdot 4H_2O$) when electric field is applied along the 'b' axis of a single crystal, and the well known magnetic hysteresis behavior of ferromagnetic iron. More recently the concept has been generalized (2,3) and the term ferroic is now used to characterize materials that exhibit crystal phases which show twin (domain) structures that can be modified, i.e. the domains can be reoriented, by magnetic, electric or elastic stress fields, or combinations of such fields.

The independent and dependent variables and the resultant hysteretic responses for the primary ferroic ferromagnetic, ferroelectric and ferroelastic systems are shown in Figure 1. In the ferroelectric, the domain states differ in orientation of a spontaneous electric polarization, they are equilibrium states but can be switched one to another by a realizable electric field. Spontaneous polarization (P_s), remanent polarization P_r and coercive field E_c are all defined by analogy to the similar better known magnetic parameters.

Clearly it is the reorientability between alternative polar equilibrium states which distinguishes the polar ferroelectric from the larger class of pyroelectric crystals in the 10 polar point symmetries.(4)

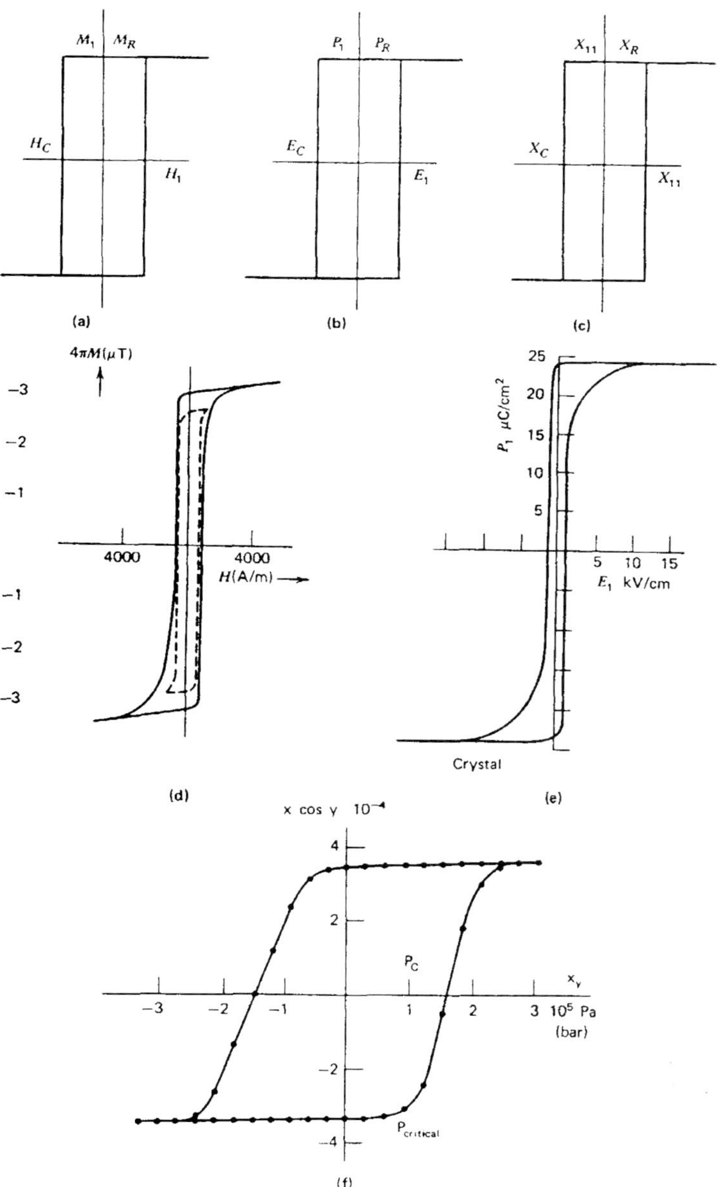

FIGURE 1. Primary ferroics:
(a) Ferromagnet.
(b) Ferroelectric.
(c) Ferroelastic.
(d) Ceramic cobalt ferrite, a ferrimagnet.
(e) Barium titanate single crystal, a ferroelectric.
(f) Lead phosphovanadate single crystal, a ferroelastic.

In many (not all) ferroelectrics, the spontaneous electric polarization in the domain is a decreasing function of temperature, going to zero at a phase transition temperature T_c, called the Curie point. Above T_c many ferroelectrics exhibit very high temperature dependent dielectric susceptibility η such that one diagonal component of the susceptibility tensor follows a Curie Weiss law

$$\eta_{ii} = C/(T-\theta)$$

where C is the Curie Constant and θ the Curie Weiss temperature, again terminology directly derived from magnetism. It must be noted however that in the dielectric case the constant C is some 5 to 7 orders of magnitude larger than the Curie Constant of the conventional ferromagnet.

In practice this startling difference leads to the fact that high useful susceptibilities (permittivities) persist over a wide temperature range into the paraelectric phase, and this intrinsic softness is exploited in many capacitor applications.

SYMMETRY AND SYMMETRY CONSTRAINTS IN FERROELECTRICS

Ferroelectricity is a property of crystals, there are no ferroelectric glasses or liquids (although we do have ferroelectric liquid crystals). The appearance of a polarization vector in a crystal will certainly modify the point symmetry and create a new phase. If the resulting phase is to be ferroelectric, there must be more than one equivalent equilibrium orientation for the polar vector, and it must be switchable between states by a realizable electric field.

All point symmetries other than 1, which has no symmetry to lose, may be prototypic (paraelectric) phases out of which ferroelectric species may occur. A complete cataloguing of all possibilities has been given by Shuvalov(5) and Aizu(6) and compact notations were presented for describing the state shifts. The tabulation for the cubic point groups is given in Table 1 taken from Shuvalov. Taking for example one of the common species from point group m3m, the symbol

m3m (3) D4F 4mm

means that prototype m3m has a ferroelectric state (F) which will have (3) equivalent orientations (6 vector directions) which are definite (D) along the 4 fold axis (4), giving rise to six domain states each in tetragonal point symmetry 4mm.

A consequence of the general symmetry arguments was the realization that there could be ferroelectric species in which the polar vector is reorientable but not reversible. An example is in the cubic group $\bar{4}$ 3m, i.e. $\bar{4}$ 3m(12/2)AmFm. Again $\bar{4}$ 3m is the prototype group. In the ferroelectric species there are 12 vector directions for the ferroelectric polarization; (12/2) indicates the orientation is re-directable but not reversible; (A) indicates arbitrary, but in the plane of the mirror (m), and the ferroelectric phase has monoclinic m symmetry.

In Shuvalov's table the corresponding simple form is the solid surface which would be formed by the normals to the full family of polar directions. In the "nature of the phase change" column, Roman II indicates that the Curie transition may be either 1st or 2nd order thermodynamically , Roman I indicates that the change must be first order (discontinuous.) The small symbols ‖ or ⅄ at the end of the letter descriptors indicate whether the sub lattices in the different domain states are parallel ‖ or canted ⅄.

The full tabulation shows that polar point groups may in fact be prototypic, though now all possible ferroic species must be re-directable and irreversible.

It should be stressed that the symmetry classification is a tabulation of the possible, i.e. all actual ferroelectrics of any symmetry must fall within this grouping. However, whether any structure of a given symmetry is actually prototypic depends on details of its atomic architecture which are not controlled by the crystal symmetry.

TABLE 1

Kind of transition-kind of ferroelectric phase	Corresponding simple form of initial phase	Order of transition
m3m(3)D4F4mm ‖	hexahedron (cube)	II
m3m(4)D3F3m ‖	octahedron	II
m3m(6)D2Fmm2 ‖	rhombic dodecahedron	I
m3m(12)A4Fm ‖	tetrahexahedron	I
m3m(12)A*2Fm ‖	trigonal trioctahedron	I
m3m(12)A**2Fm ‖	tetragonal trioctahedron	I
m3m(24)AlFl ‖	hexoctahedron	I
4 3m(3)D$\overline{4}$Fmm2 ⅄	hexahedron (cube)	I
4 3m(4/2)$\overline{\text{D}}$3F3m	tetrahedron	I
4 3m(12/$\overline{2}$)A*mFm	tetragonal tritetrahedron	I
4 3m(12/$\overline{2}$)A**mFm	trigonal tritetrahedron	I
4 3m(24/$\overline{2}$)A1F1	hexatetrahedron	I
432(3)D4F4 ‖	hexahedron (cube)	II
432(4)D3F3 ‖	octahedron	II
432(6)D2F2 ‖	rhombic dodecahedron	I
432(24/2)A1F1	pentagonal triotahedron	I
m3(3)DmFmm2 ‖	hexahedron (cube)	II
m3(4)D3F3 ‖	octahedron	II
m3(G)AmAm ‖	pentagonal dodecahedron	I
m3(12)A1F1 ‖	didodecahedron	I
23(3)D2F2 ‖	hexahedron (cube)	I
23(4/2)D3F3	tetrahedron	I
23(12/2)A1F1	pentagonal tritetrahedron	I

Ferroelectric species accessible from cubic point symmetries (from Shuvalov 5). The notation gives the symmetry of the prototype (Hermann-Maugain notation), the number of orientation (domain) states, the orientation with respect to a symmetry element of the prototype group, the ferroelectric F operation and the symmetry of the resultant domain state.

PEROVSKITE MATERIALS

The simple cubic perovskite structure, which is the high temperature form for many mixed oxides of the ABO_3 type, was one of the first simple structures to exhibit compounds with ferroelectric properties and it is still the most important prototype.

The simple high symmetry form (point group m3m: See Figure 2) is made up of a three dimensional array of orthogonal corner linked oxygen octahedra. The smaller octahedral B position in the centers of the octahedra are occupied by highly charged small cation such as Ti, Sn, Zr, Nb, Ta, and W. The larger 12 fold coordinated A sites are occupied by lower charged, larger cations such as Na, K, Rb, Ca, Sr, Ba, Pb, etc. The simple compounds $BaTiO_3$, $PbTiO_3$, $NaNbO_3$, $KNbO_3$, $AgNbO_3$, WO_3 all exhibit ferroelectric phases. It is interesting to note that the halide perovskite structures such as $KMgF_3$ which are almost identical in dimension to the oxides have completely uninteresting, low, linear dielectric properties.

The perovskite oxides are particularly interesting in polycrystal ceramic form. The high prototypic m3m point symmetry leads to many equivalent differently oriented domain state polarization directions, thus in the randomly axed polycrystal it is possible to thread the polarization vector through the structure without major discontinuities.

As an aside, note that although the symmetry restrictions would permit two monoclinic and a triclinic domain families with 24, 24 and 48 equivalent states, which would be fantastic in ceramics, such states have never been overtly identified.

$BaTiO_3$

The prototype crystal structure for $BaTiO_3$ is the cubic perovskite structure shown in Figure 2. In this structure, the 12-coordinated Ba ions sit on the corners

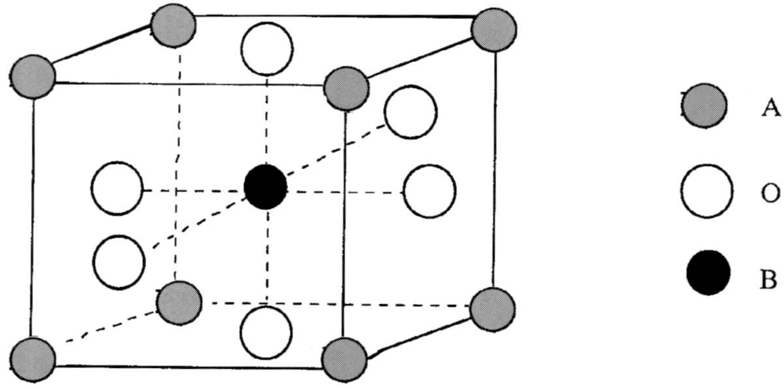

FIGURE 2. Simple cubic perovskite structure for ABO_3 mixed oxides like $SrTiO_3$.

of the unit cell, the 6-coordinated oxygen ions at the face centers, and the Ti ions at the center of the unit cell. As is the case for many of the useful oxide ferro-electrics, the primary structural feature associated with the ferroelectricity is the TiO_6^{8-} octahedron. At temperatures above ~130°C, the octahedron is undistorted, and the material possesses symmetry group m3m. On cooling through the Curie point, however, the Ti^{4+} ion displaces closer to one of the adjacent oxygen along the [001] direction. This results in an elongation along the c axis, contraction along the a and b axes, and most importantly, development of a spontaneous polarization along the ferroelectric axis. This tetragonal structure is stable to ~ 5°C, at which point the Ti^{4+} ion shifts again, along one of the <110> directions of the original cubic unit cell. This leads to an orthorhombic structure with the polarization oriented along the direction of Ti^{4+} displacement. Finally at ~-90°C, the TiO_6^{8-} octahedra distort again, with the Ti^{4+} ion (and hence P_s) displacing along one of the original <111> directions. Each of these symmetry changes corresponds to a first order displacive transition.

Associated with each of the transitions in $BaTiO_3$ is a maximum in the low field dielectric permittivity as shown in Figure 3. Fundamentally, each transition marks a temperature where two phases are similar in energy. On application of an electric field, then, the material can be distorted between the two different polytypes. This, in turn, leads to high polarizability near transition temperatures.

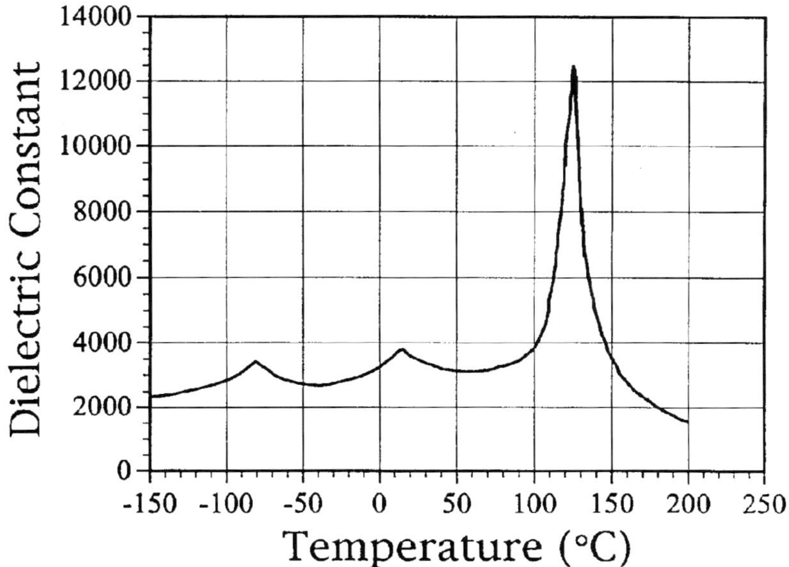

FIGURE 3. Temperature dependence of the dielectric permittivity of polycrystalline $BaTiO_3$ (after Kinoshita and Yamaji (11)).

The high dielectric permittivity even between phase transitions in ferroelectric BaTiO$_3$ is tied to the presence of the spontaneous polarization and to the existence of domain walls. The high permittivity of the material, coupled with the ability to tailor the dielectric response by compositional modifications is responsible for the widespread use of BaTiO$_3$ in high volumetric efficiency capacitors.

$Pb(Zr_{1-x}Ti_x)O_3$

The most important of the piezoelectric ceramics are found in the lead zirconate - lead titanate solid solutions series. Like BaTiO$_3$, these materials are based on the perovskite crystal structure. The phase diagram for Pb(Zr$_{1-x}$Ti$_x$)O$_3$, abbreviated PZT, is shown in Figure 4. Near a Zr:Ti ratio of 52:48 is a roughly temperature independent boundary marking the boundary between tetragonal and rhombohedral distortions. At this morphotropic phase boundary (MPB), strongly piezoelectric compositions can be prepared. The reasons for this are as follows: on the Ti-rich side of the boundary, a tetragonal phase with 6 allowed spontaneous polarization directions is stable; on the Zr-rich side a rhombohedral structure with 8 potential polarization directions is favored. At the MPB, then,

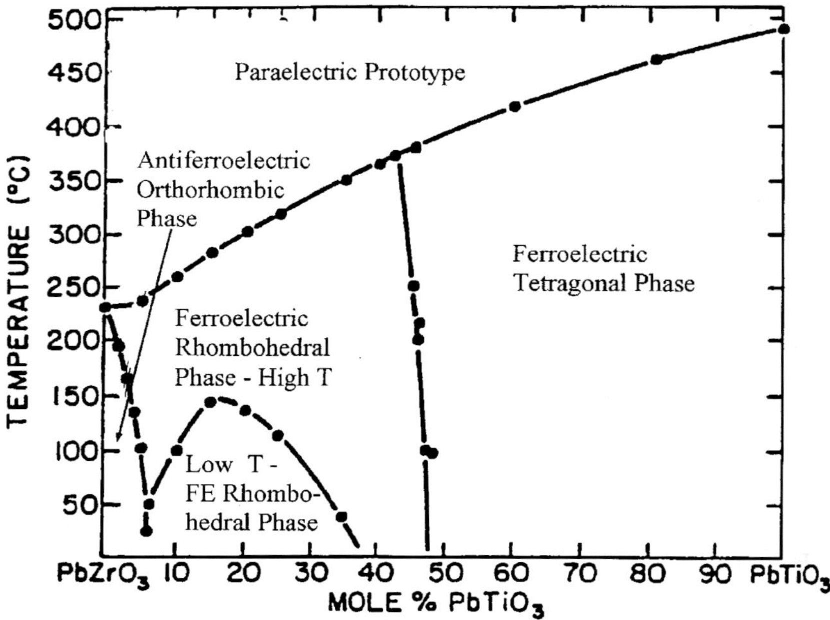

FIGURE 4. Phase diagram of the lead zirconate titanate solid solution (After Jaffe, Cook, and Jaffe (12)). The small region in which a ferroelectric orthorhombic region is stable is not shown.

14 polarization directions are roughly equivalent in energy. Thus, when an electric field of sufficient magnitude is applied such that the domains can be re-oriented, it is very likely that each grain in a randomly-oriented polycrystalline ceramic will have a polarization direction aligned quite well with the field. Consequently, materials near the MPB can be poled more efficiently than most other piezoelectric ceramics. This is reflected in a maximum in the piezoelectric coefficients and the piezoelectric coupling factor at the MPB. As a result, PZT is the most widely used of the piezoelectric ceramic compositions. Only when special properties such as high mechanical quality factor, temperature stability, or low hysteresis are required is PZT supplanted by other materials for electromechanical transducers.

Through proper doping, the electromechanical properties of PZT can be optimized for specific applications. In particular, acceptor doping of PZT leads to the introduction of additional oxygen vacancies into the lattice. These, in turn, lead to the formation of defect dipoles between the aliovalent cation dopant and the oxygen vacancy (i.e. Fe_{Ti}' - $V_O^{\cdot\cdot}$, where Kroger-Vink notation has been employed). This dipole can re-orient as the oxygen vacancy moves. During typical poling operations, the oxygen vacancy is sufficiently mobile that the defect dipole re-orients with the applied field. During use, these aligned defect dipoles provide an internal electric field that stabilizes the direction of the spontaneous polarization, and so minimizes depoling. Soft PZT, on the other hand, is prepared by donor-doping. The excess positive charge due to the dopant is typically accommodated by creating lead vacancies. The resulting defect dipoles in these materials are quite difficult to re-orient given the large diffusion distances involved. Thus, they are generally randomly oriented in the ceramic. As a result, the domain walls in a soft PZT are quite mobile. It has been shown by Cross and co-workers that most of the increase in the piezoelectric properties which occurs when PZT is donor-doped is due to the mobility of the domain walls.(7)

USES OF FERROELECTRIC MATERIALS

Over the last 40 years, ferroelectric materials have found increasing use in applications such as capacitors, piezoelectric or electrostrictive transducers, pyroelectric sensors, and electrooptic devices. The result is a multi-billion dollar world-wide market.

Dielectrics

In a capacitor, the ability to store charge is directly proportional to the dielectric permittivity (ε_r) of the dielectric material used. Given this, ferroelectrics are widely used as dielectrics due to the extremely large relative permittivities which can be achieved (up to ~10,000). By contrast, most ceramics have relative permittivities in the range of 4 - 20. The high dielectric constant of ferroelectric materials is primarily a result of the combination of dipolar polarizability and domain wall contributions to the total permittivity.

Equally important in the application of ferroelectrics as capacitors, though, is the degree to which the properties can be tailored through compositional modification. As an example, consider $BaTiO_3$, since about 90% of all of the high permittivity capacitors manufactured today are based on $BaTiO_3$ compositions. The dielectric response of undoped $BaTiO_3$ was shown previously in Figure 3. $BaTiO_3$ is particularly attractive for capacitive applications because the presence of 3 ferroelectric transitions keeps the permittivity high over a wide temperature range. While the high permittivity is advantageous in capacitors, the temperature dependence of the dielectric constant is problematic. Several approaches are utilized in order to decrease the temperature dependence of the permittivity. First, the position of the transition temperatures can be modified by appropriate isovalent substitutions. Figure 5, for example, shows the effect of several different cations on the Curie temperature of $BaTiO_3$. With the exception of Pb, most substitutions move the cubic-tetragonal transition down in temperature. Thus, for extremely high permittivity values, it is possible to center the permittivity maximum near room temperature. Secondly, some additives aren't incorporated well into the $BaTiO_3$ structure, so that they tend to segregate at grain boundaries. This introduces a low dielectric constant phase into the material which decreases the overall permittivity but helps to decrease the temperature dependence somewhat. Thirdly, the temperature dependence can be further minimized by making the

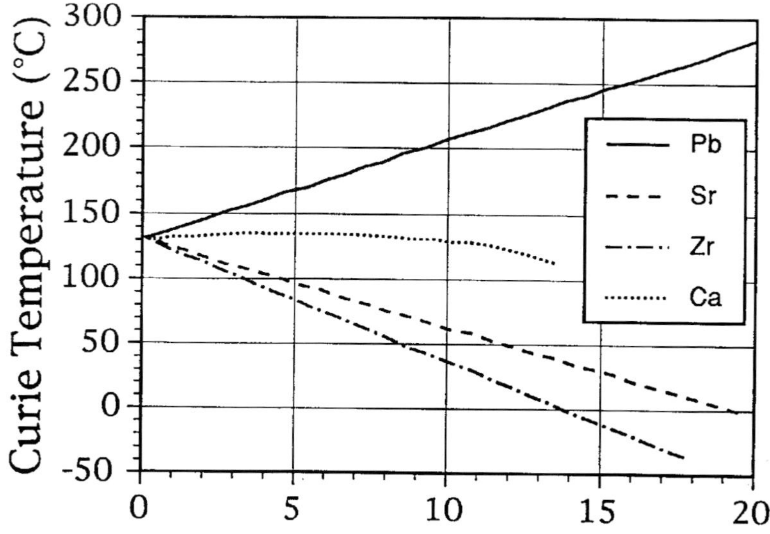

FIGURE 5. Dependence of T_c on the doping in $BaTiO_3$. Data taken from reference 13.

material compositionally heterogeneous through preparation of core-shell geometries, in which, for example, the Ba:Sr ratio is graded through the radius of the grains. When this occurs, different regions of the material undergo T_c at different temperatures, so that the permittivity peak is broadened. Finally, Zr substitution for Ti decreases T_c, but increases the transition temperatures for the other two phase transitions. Thus, the three transitions can be made to nearly converge by proper compositional modification. This leads to a large dielectric constant peak over a broad temperature range.

Currently, many capacitors are prepared as multilayers to maximize the volumetric efficiency of the charge storage. Significant areas of research in this field include:

- decreasing the layer thickness of individual layers. Currently, 3 μm thick multilayers are available commercially, with 1 μm layers being studied in the research phase.
- replacing the standard precious metal electrodes (usually silver-palladium alloys) with base metal compositions such as Ni to reduce the cost of the capacitors.

Another area where significant research effort is directed in capacitors is the integration of high dielectric constant materials with silicon circuitry. Until recently, the standard capacitors used in integrated circuitry were either SiO_2 or silicon oxynitride thin films. However, the dielectric constants of both of these materials are quite low (ε_r ~4 and ~9, respectively). Thus as the linewidths of active devices have decreased, it has been hard to achieve a comparable reduction in capacitor size, even through the use of extremely thin layers (down to 30 Å!) or trench geometries. The situation is particularly severe in dynamic random access memories (DRAM), where the size of the capacitor is now larger than that of the memory element itself. Consequently, it is becoming imperative to integrate higher dielectric constant materials with Si circuitry. As a first attempt at this, several companies have switched to Ta_2O_5 (ε_r ~18-25), but for extremely high integration levels, an even higher dielectric constant will eventually be required. For this purpose, considerable research is now being directed at development of low loss paraelectric $(Ba_{1-x}Sr_x)TiO_3$ thin films. Thin film ferroelectrics can be deposited by a wide variety of techniques, including magnetron sputtering, ion-beam sputtering, chemical vapor deposition, pulsed laser deposition, or through sol-gel routes. In most cases, thin film ferroelectric show somewhat lower dielectric constants and higher coerceive fields than do bulk ceramics of the same composition.

Piezoelectrics and Electrostrictors

Piezoelectricity and electrostriction entail a coupling between electrical and mechanical energy in materials. That is, if a stress is applied to the material it responds with an electrical polarization. Conversely, if an electric field is applied, the material undergoes a strain. Fundamentally, the piezoelectric effect is the linear relationship between the applied electric field and the resultant displacement. Electrostriction, on the other hand, implies proportionality between the square of the electric field and the shape change, and is observed only when piezoelectricity is absent. In practice, both of these relations are applicable over a limited electric field range. For example, when a large alternating electric field is applied to excite piezoelectric displacement, hysteresis in the strain is observed due to domain wall motion. While electrostrictive materials show much lower levels of hysteresis in the strain, it has been found that the strain is properly proportional to the square of the polarization rather than the square of the electric field at high field levels.

Piezoelectric materials find widespread use in both sensor and actuator applications. One example which demonstrates both of these possibilities is the transducer used in biomedical imaging. When an alternating electric field is applied to a piezoelectric transducer, it undergoes an alternating shape change, and can be used to generate sound waves. When these are coupled into a body, the sound wave propagates until it is reflected from a discontinuity. When the reflected sound wave arrives at the transducer, it is reconverted to an electrical signal, and these are, in turn, used to build up a picture of the body. This has significantly increased the ability of medical personnel to image the human body non-invasively. Interestingly enough, since the human body is largely water, the requirements on the piezoelectric element used as the imager are very similar to those required in piezoelectrics used in underwater imaging (i.e. sonar for naval applications). Piezoelectric materials are also used for precision positioning applications, including for the movement of microscope stages and in the motors for autofocusing cameras. We are currently seeing widespread incorporation of piezoelectric materials into commercial products, particularly by Japanese companies. Typical examples include automobile suspension systems which adjust real-time to road roughness, sensors for image stabilization in hand-held camcorders, and many others.

Electrostrictive materials are useful in many of the same types of applications where piezoelectric materials are attractive. One of the significant advantages to the electrostrictor is the smaller level of hysteresis in the strain. This permits precision position control without as much feedback as is required in piezoelectric designs. Of particular note is the use of electrostrictive compositions in adaptive optics systems for positioning of components. In particular, the articulating fold mirrors which play a major role in the correction to the optics of the Hubble Space Telescope are positioned using doped lead magnesium niobate electrostrictive actuators. This has resulted in the restoration of the intended function of the telescope.(8) Similarly, electrostrictive actuators are used in the control of deformable mirrors. Telescopes based on such systems can be corrected real-time for the presence of atmospheric distortion in the images, resulting in the capability for ground-based instruments to have the same resolution as space-based telescopes over the visible frequency range.

Again, much of the current work in the field is directed towards miniaturization of piezoelectric devices, both in bulk ceramic and in thin film form. In the case of the latter, the capability of processing thin films piezoelectrics such as $Pb(Zr_xTi_{1-x})O_3$ on platinum-coated silicon allows the integration of the active sensing or actuator elements directly with on-chip driving or receiving electronics (including all of the necessary amplification). This can significantly increase the sensitivity of devices such as microphones and pyroelectric detectors.(9,10) From the standpoint of actuation, it is clear that as devices are miniaturized, piezoelectric devices permit comparatively higher energy densities, compared to other means of generating displacement. Thus, as active elements are integrated, electromechanical devices become progressively more attractive. Similarly, miniaturization of large strain bulk actuators has permitted piezoelectric and electrostrictive transducers to be utilized in such products as miniature optical scanners and silent wrist watch alarms.

Pyroelectrics

In ferroelectric crystals and films the strong temperature dependence of the spontaneous electric polarization leads to the possibility of very strong pyroelectric effects, as the pyroelectric coefficient p is defined by

$$p = \partial P_s / \partial T.$$

For a thin sheet of crystal or ceramic, poled normal to the sheet and blackened on one surface, absorbed long wavelength infra-red radiation will change the temperature and the pyroelectric current may be used to monitor the radiation level. Such current mode detectors can be very fast and are usable as power meters for pulsed CO_2 lasers working at 10 μm.

If the radiation is chopped at a frequency $\omega \sim 30$ Hz, and the element is thin and very well thermally isolated, the thermal sensitivity is

$$\Delta T \propto \frac{1}{[g^2 + w^2 C^2 A^2 t^2]}$$

where g is the effective thermal conductor to the environment, C the volume specific heat, t the detector thickness, and A the surface area.

In isolated point detectors it is easy to provide excellent thermal isolation by mounting the element on an ultra thin polymer "drum head," and very inexpensive, high sensitivity thermal IR detectors made from a wide range of ferroelectric materials are commonly used in industry.

For imaging, where the individual element is down to a pixel size of ~ 25 μm, the conflicting requirements of electrical interconnection and refined thermal isolation provide challenging problems. Present systems use hybrid technology in which the detector plane is a very thin sheet of poled ferroelectric ceramic, bump bonded to the silicon integrated interrogation circuit. With an f1 lens system, such imaging systems give resolution of 0.07°K in the object plane and are now being marketed in the United States for law enforcement and in the United Kingdom to firefighters because of the enhanced ability of long wave IR to see through smoke.

Currently there is a strong push to move to integrated systems using microelectro-mechanical systems technology to form air bridge thermal isolation on the silicon chip itself. Major problems concern the conflicting processing needs of ferroelectric films and silicon microcircuits, but the prize would be a system which could be fitted in automobiles to provide drivers with useful night vision.

Electro-optic Applications

Many ferroelectric crystals are high band gap insulators and are quite transparent to visible light. The very high electrical polarizability means that the constituent ions move more from their equilibrium position when electric fields are applied than ions in simple solids. Thus whilst the polarization related electro-optic constants are quite normal, the field related constants can be unusually high for low frequency fields.

For polarized light traveling through the crystal, the beam with its electric vector along the field direction will experience a different velocity to that polarized in the orthogonal direction so that on reconstruction the beams will interfere, producing an amplitude modulation of the light. Large aperture optical modulators using this principle have been constructed in ferroelectric potassium dihydrogen phosphate, and its deuterated analog. However, probably the most widely used systems are guided wave structures fabricated on the surface of crystals of lithium niobate or barium strontium niobate ferroelectrics. The light guiding structure is produced by indiffusing a cation which locally raises the refractive index, so that the light is confined along a fine line of the indiffused species. Guided wave modulators, beam couplers, and beam splitters are useful optical components produced using this technology.

The polar symmetry of the ferroelectric means that the domains have non-zero third order optical constants, and for the higher optical fields in laser light they can produce significant second harmonic light, i.e. irradiation with a laser operating at 1 μm in the infra-red will produce a harmonic in the visible at 5,000 Å wavelength.

In most crystals, due to dispersion, the harmonic light travels with a different velocity than the fundamental and the phase requirements for energy transfer are destroyed after a short travel distance. For the ferroelectric, the sign of the nonlinear constant depends on the sign of the ferroelectric polarization, and by breaking the crystal into a suitably periodic domain structures a condition of quasi phase matching may be achieved over much larger lengths, and higher conversion efficiencies can be achieved. Currently there is much work on such quasi phase matching systems in $LiNbO_3$ to up-convert the present highly efficient red light emitting diode lasers to emit in the blue-green, where there are currently few efficient solid state sources.

Electric polarization in the ferroelectric domain leads to unusual charge separation when defect centers in the crystal are ionized by optical irradiation. In suitable crystals the ejected carriers will be re-trapped at new locations, leading to large local space charge fields. This, in turn, leads to a modulation of the optical refractive index through the high electro-optic effect. These photorefractive effects are now being widely researched for a range of unique optical applications: four wave mixing, optical amplification, optical beam fanning for laser protection and holographic information storage.

In the optical arena there is a ferment of research activity, but so far only the more primitive guided wave structures have effected significant market penetration.

REFERENCES

1. J. Valasek. 1921. Phys. Rev. *17*, 475.
2. R.E. Newnham and L.E. Cross. 1974. Mat. Res. Bull. *9*, 927.
3. R. E. Newnham and L.E. Cross. 1974. Mat. Res. Bull. *9*, 1021.
4. J.F. Nye. 1957. *Physical Properties of Crystals,* Claredon Press, Oxford.
5. L.A. Shuvalov. 1970. J. Phys. Soc. Japan *38*, 285.
6. K. Aizu. 1969 J. Phys. Soc. Japan *27*, 387.
7. X.L. Zhang, Chen, Z.X., Cross, L.E. and W.A. Schulze. 1983. J. Mat. Sci. *18* 968.
8. J.L. Fanson. 1995. Materials for Smart Systems, MRS Symp. Proc. *360* 109.
9. E.S. Kim, Ph.D. 1990. Thesis, University of California - Berkeley.
10. E.S. Kim and R.S. Muller. 1987. IEEE Elec. Dev. Lett. *EDL-8* 467-468.
11. K. Kinoshita and A. Yamaji. 1976. J. Appl. Phys. *47* 371.
12. B. Jaffe, W.R. Cook Jr. and H. Jaffe. 1971. *Piezoelectric Ceramics,* Academic Press, India.
13. J.M. Herbert. 1985. *Ceramic Dielectrics and Capactors,* Gordon and Breach Scienctific Publishers, New York.

Chapter Fifteen

INTEGRATED CERAMIC PACKAGING

C.A. RANDALL[1] and J.P. DOUGHERTY[2]

[1]Department of Materials Science and Engineering
Steidle Building
Pennsylvania State University
University Park, PA 16802

[2]Department of Electrical Engineering

INTRODUCTION

The technology drive in microelectronic packaging is towards smaller, faster and cheaper devices with increased performance. The increase of circuit density per integrated circuit (chip) requires an increase in the number of leads connecting the chip with other functional components. The transmission of signals in shorter times requires the signal speed to increase, higher frequencies to be used and distances to be reduced along the transmission line. Increases in power and the associated dissipation of thermal energy also have to be managed as increasing temperature limits the lifetime of the chip. Additionally, thermal stresses are created due to differences in the thermal expansion behavior of the wide variety of materials used and these stresses can lead to reliability problems, especially at interfaces. The materials engineering of substrates and components to manage the signal processing of the integrated chip is a continuing challenge.(1)

This review introduces the technical needs for the advancement of cofired and integrated ceramics in microelectronic packaging. The important properties of the various key material components both organic and inorganic are outlined and contrasted with the future engineering requirements.

THE MARKET AND OPPORTUNITY

The world wide consumption of advanced ceramics reached 10 billion dollars in 1991. Of these ceramics 75% are used for electronic applications and 25% for structural applications. Twenty percent of the electronics market is associated with packaging, and this continues to grow. So, there exists an exciting opportunity for ceramic based electronic packaging in the next 20 years. This increase will be driven by the future developments in electronic consumer products. *Business Week* (2) recently outlined some of the potential revolutions in electronic systems early in the next century:

Year 2002 Personal computers become so powerful they will compete with present day CRAY's. Also, the capability of running any software, regardless of the type of P.C. will exist.

Year 2005 Smart televisions will control the viewing selections based on your previous preferences and suggest new programs according to your tastes.

Year 2008 Telephones will incorporate chip modules that instantaneously translate for callers speaking in different languages.

Year 2011 Computers will manage corporate departments, such as accounting and logistics.

Although the projected dates may prove to be inaccurate, the continued advancement of electronic devices into the foreseeable future is a clear trend. The giga bit chip technology, is anticipated to start around 2005 as summarized in Table 1. The cost of wafer fabrication is of great concern to the semiconductor corporations as the technology complexity increases. Therefore, in the future, the cost per transistor or circuit is predicted to increase as the feature size continues to decrease into the submicron regime.

The integrated multichip module (MCM) is a cost-effective solution to this trend. However, the circuit density on the packaging material must be maintained and the signal speeds must be equivalent to the chip speeds, together with minimum interference to limit corruption of the signal as discussed below.

TABLE I

Semiconductor Advances for the Next Two Decades (after Tummula[3])

	1994	1999	2005	2011
DRAM Capacity, MB/GB	16MB	250MB	4GB	64GB
Microprocessor Speed, Mhz	150	400	600	800
Smallest Feature Size, Microns	0.5	0.25	0.12	0.08

SUBSTRATE MATERIALS AND REQUIREMENTS

Basic Structure:

The material engineering of substrates has been a major focus for electronic packaging (4). Substrates are required for two main types of chip modules. A single chip module (SCM) and a multichip module (MCM), where up to 100 chips can be housed. Figure 1 shows the cross-sectional schematic diagram of a ceramic MCM, a so called MCM-C showing the transmission lines between adjacent dielectric layers and interconnected through via holes.

Generally, a substrate material requires a suitable combination of mechanical, thermal, chemical, and electrical properties. The ideal substrate material properties are listed in Table 2, and the advantages, indicated by "√", and average properties designated "—".

Dielectric Constant and Loss of a Substrate:

The dielectric constant must be as low as possible to limit propagation delays and in high density packages, limit cross-talk (6).

The pulse propagation speed is given by the expression:

$$v = \frac{c}{\sqrt{\varepsilon}}$$

where ε = dielectric permittivity of the insulating layer.

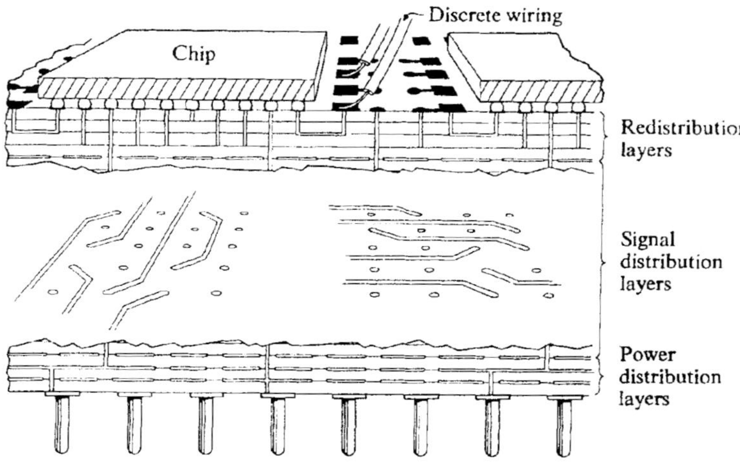

FIGURE 1. Ceramic Multilayer Multichip Module —MCM-C.

c = speed of light
v = pulse propagation velocity

Therefore, to enhance the pulse propagation speed, the substrate requires a low dielectric constant. In addition, since a short pulse is composed of a range of high frequency components of 1GHz and above, therefore, it is necessary for the dielectric constant of the material to be independent of frequency to limit pulse broadening. Figure 2 demonstrates the influence of substrate dielectric constant

TABLE 2 (5)
Ideal Substrate Wish List

Property	Ceramic	Polymer
Low Dielectric Constant (1.0)	—	√
Low Dielectric Loss (DF=0.000001)	√	—
High Insulation Resistance	√	√
High Thermal Conductivity (400 W/m K)	√	
Coefficient of Thermal Expansion (3.0 or 6.0 ppm/C)	√	√
Metallization (Adhesion)	√	—
Flexibility		√
Smooth Surface	√	√
Low Thickness	—	√
Large Size	—	√
Low Moisture Adsorption	√	—
Chemical Resistance	√	
Non-flammable	√	
Recyclable and Environmentally Compatible	—	—
Hole Drilling	—	√
Integration Capabilities	√	—
Low Manufacturing Costs	—	√

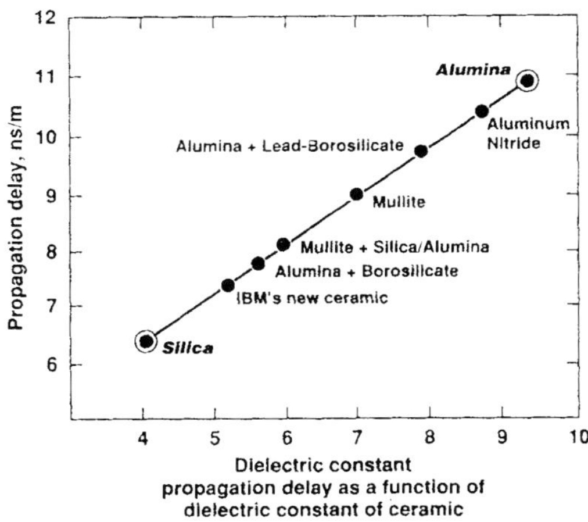

FIGURE 2. Propagation delay as a function of dielectric constant of several oxide ceramics of interest.

on propagation delay times. There is a retardation of the signal propagation speed with increasing dielectric constant.

Cross-talk is an undesirable interference phenomenon that introduces noise into the circuit (7-8). The origin of this interference is the mutual inductance (L) and capacitance coupling (C) between neighboring transmission lines. The equivalent circuit for coupled capacitance and inductance between adjacent transmission lines is represented in Figure 3. In the development of higher density and higher speed circuits, the transmission lines become closer ~µm and the mutual inductive coupling increases with increasing frequencies; therefore crosstalk becomes a greater problem. For example, a substrate permittivity of 9.5 (standard alumina) and a signal rise time of 200 pS, cross-talk ratios as high as -8dB have been estimated for adjacent transmission lines with characteristic impedances of 50 ohms. To limit this problem, ground planes must be moved closer relative to the signal traces. To maintain the line impedance of 50 ohms, a reduction of dielectric constant in the substrate is required; thus allowing higher microstrip line densities within the package, and limiting the cross-talk.

Polymers have intrinsically lower dielectric constants (2-3.5) and losses than ceramics or glass-ceramics. Two important polymer materials used in packaging are polytetrafluroethylene (PTFE) which has a dielectric constant of 2, and poly-imides which have dielectric constants in the range 2.7-3.2. The lowest dielectric constant ceramic material known is fused quartz or silica, SiO_2 with ε ~3.9 and the substrate material with the highest dielectric constant is silicon carbide, SiC, $\varepsilon = 40$. The most commonly used ceramic substrate is based on alumina, $\alpha\text{-}Al_2O_3$, which has a dielectric constant of 9.5. Ceramic materials compete with the polymer substrate materials in terms of dielectric behavior in glass-ceramic composite

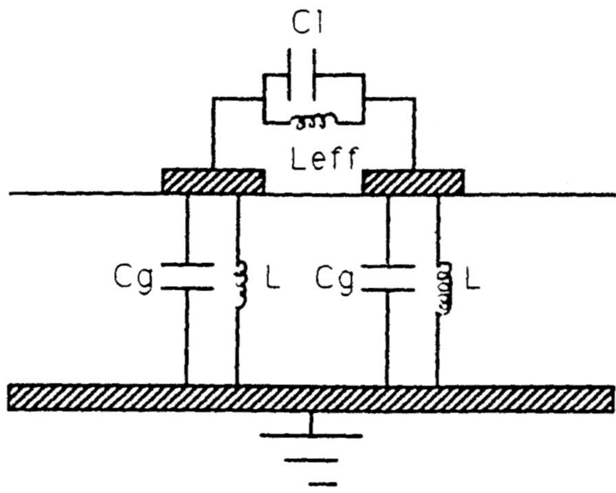

FIGURE 3. The schematic view of the load capacitance and the interconnect capacitance.

systems which have been developed for low-fire multilayer packages and have dielectric constants $\varepsilon \sim 5$. To further reduce the dielectric constant, glass ceramics are being developed that contain up to 10% porosity. N E C (Nippon Electric Company) has reported a SiO_2-borosilicate glass-cordierite porous substrate material with $\varepsilon \sim 3.4$. The porous composite systems require further studies to optimize the balance between dielectric properties and mechanical strength (9).

Alumina based substrates will remain a competitive material in the foreseeable future, especially for the SCM devices, owing to a combination of strength, heat dissipation and dielectric properties, and an extensive technology know-how base. Other ceramic based materials include aluminum nitride, AlN, and beryllium oxide, BeO, these have lower dielectric constants compared to alumina, 8.8 and 6.8 at 1 MHz, respectively, but more importantly they have excellent thermal conductivity.

Thermal Properties:

Thermal conductivity is of ever increasing importance in the thermal management of increased chip densities and faster switching times. The heat dissipation levels out of the package are up to ~ 100 W/cm². This requires the use of ceramics with high thermal conductivities and/or complex cooling processes to transfer heat away from the chips. Polymers typically have lower thermal conductivities than ceramics—approximately 0.2 W/mK compared to 36 W/mK for alumina (6). Attempts have been made to raise the thermal conductivity of polymers by a composite strategy. However, even by adding a second phase to the polymer, such as Al_2O_3 powder in polyimide, the improvement in thermal conductivity is small (0.15 W/mK to 0.38 W/mK). Accompanying this small enhancement in thermal conductivity is an increase in the average dielectric constant. (10).

The thermal conductivity enhances the attractiveness of ceramic materials for high density, high power applications. Some ceramic substrate materials with a high thermal conductivity and relatively low dielectric constants include BeO and C (diamond) have higher thermal conductivities, but other manufacturing issues limit their appeal at the present time. For example, BeO has a thermal conductivity of 250 W/mK but is highly toxic which makes processes such as hole-drilling for vias become expensive. Diamond has the highest thermal conductivity of any substance, ~ 2000 W/mK in single crystals; chemical vapor deposition (CVD) of diamond films still preserves high values (800-1400 W/mK). But high costs and difficulties in metallization limit the advance of diamond substrate manufacturing at this time.

Of the high thermal conductivity ceramics, the clear, cost effective favorite is AlN with a thermal conductivity~ 150-230 W/mK, ten times better than Al_2O_3. However, at the present time, Al_2O_3 substrates are an order of magnitude cheaper than AlN, but as demand increases for higher thermal conductivity materials, the use of AlN will grow in both SCM and MCM applications.

An extra advantage of AlN, in contrast to Al_2O_3 and polymers, is its excellent thermal expansion matching to silicon, as compared to a number of packaging materials. In general, thermal expansion differences within a package result in undesirable stresses during operation which can limit reliability. The thermal-expansion issues could limit some polymer-based substrates applications when higher temperature operating conditions are required, e.g. 'under the hood' automobile circuits, as polymers have high coefficients of thermal expansion compared to silicon. The thermal expansion differences ultimately lead to reliability problems in delamination and wire bonding. In addition, if a material has thermal expansion matching to Si or GaAs, it provides a design advantage to high density connections via "flip-chip" bonding between the chip contacts and the package striplines. With flip-chip connections, there is an increase in the input/output (I/O) terminals via an array of interconnections, compared to the connection density from the perimeter of a chip via wire bonding or TAB (Tape-Automated Bonding). Also in terms of heat management there exists the benefit of heat extraction from the back of the chip instead of the substrate (6).

Multilayer Substrates for MCMs:

Multilayer structures offer short path lengths and high density interconnects to the chips. Polymer and ceramic materials are both used in the fabrication of multilayer MCM packages. Multiple polymer layers can be obtained through either spin coating of polyimide solutions and baking the solution in a solid layer or laminating polyimide sheets together. Frequently sputtering is used to metallize the substrate; usually copper or aluminum are the materials of choice. Through either masking, photolithography, or laser writing the transmission line circuitry is outlined. The via hole interconnections are made through laser drilling before back filling of the vias with electroless nickel or metallic pastes. All the processing steps are performed at low temperatures <200°C. This gives the unique possibility of fabricating the packaging on top of a die and chip array, the so-called "chip's first" approach by General Electric (G E) (11). The thermal expansion and thermal conductivity properties limit the polyimide packages to low power applications. The thermal expansion issues not only impact the reliability of the I/O interconnects but also the via hole connections in many layer packages (<6).

MCM-C multilayer ceramic packages are based on thick film screen printing or multilayer tape casting processing techniques. Figure 4 shows an example of a state of the art hybrid thick film based MCM-C package. The wide variety of MCM-C packages made by IBM are shown in Figure 5. The substrates are based on ceramic-glass composite systems. The degree of glass and composition of glass influences the firing temperature. There are two sintering ranges used in the ceramic packaging industry: high fire and low fire. Substrates with high fire properties are 90% or greater Al_2O_3 particles with a silicate glass added as a sintering aid and as an adhesion promoter for the thick films; the typical sintering temperatures are about 1600°C. The conductors are then refractory metals such as tungsten (W) or tungsten alloys with molyb-

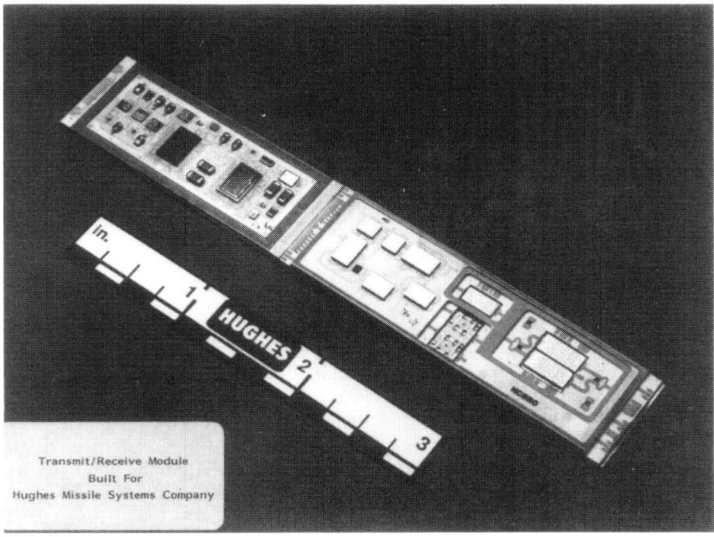

FIGURE 4. Hybrid Thick Film Based MCM-C Package (Courtesy of Hughes Aircraft Company).

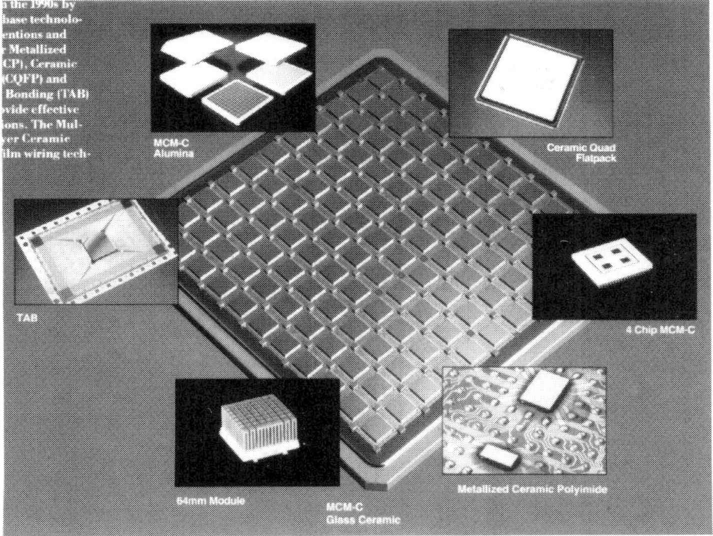

FIGURE 5. A variety of MCM-C packaging alternatives (Courtesy of IBM Technology Products).

denum (Mo), which require firing in a reducing atmosphere.(12) Higher-glass contents lower the sintering temperatures toward 900-950°C, compatible with conventional thick film technology. The so-called L T C C (low temperature cofired ceramic) systems therefore offer the best possibility for a fully integrated package, with buried resistors and capacitor functional components. Figure 6 schematically outlines the intended evolution of the multilayer package. With only stripline and via circuitry connected to the chips, additional signal processing, impedance matching, noise suppression and decoupling capacitor circuitry is surface mounted to the package in close proximity to the chips. Often these surface-mounted components are fabricated through thick or thin film techniques, e.g. multilayer chip capacitors, R-C networks, etc. With future requirements of higher densities, increased speed and higher reliability, the integration of the surface-mounted components into the multilayer package is being investigated.

Substrate development of L T C C substrates have involved two types of composite approaches; (1) glasses and ceramic filler and (2) recrystallizable glass systems. The substrates can be air-fired for noble metal conductors or fired in a reducing atmosphere allowing the use of Cu-conductors. The advantages and disadvantages of L T C C packaging ceramics are summarized in Table 3 (after Gupta) (13).

TABLE 3
L T C C Packaging Advantages - Disadvantages

Advantages	Disadvantages
(i) Dielectric constants lower than AlN and Al_2O_3, typically 5.	(i) The glass-ceramic low-k dielectrics are mechanically weaker than alumina Fracture strength 200 versus 350 MPa.
(ii) Temperature coefficient of expansion can be designed close to Si (3 ppm °C^{-1}), and is also tunable to GaAs (6 ppm °C-1).	(ii) The thermal conductivities are lower than that of alumina (15 versus 25 W/mK).
(iii) With low temperatures (= 900°C) higher conductivity metals such as gold, silver, silver/palladium, and copper can be used.	(iii) The dielectric loss is higher than in polymer and pure phase ceramic materials.
(iv) With the exception of copper, the LTCC can be fired in air, therefore, giving a much simpler co-firing process than in high-fire packages.	
(v) Ability to co-fire capacitive and resistor components.	
(vi) Plating and associated processes are eliminated because chip bonding can be done directly on as-fire conductors.	

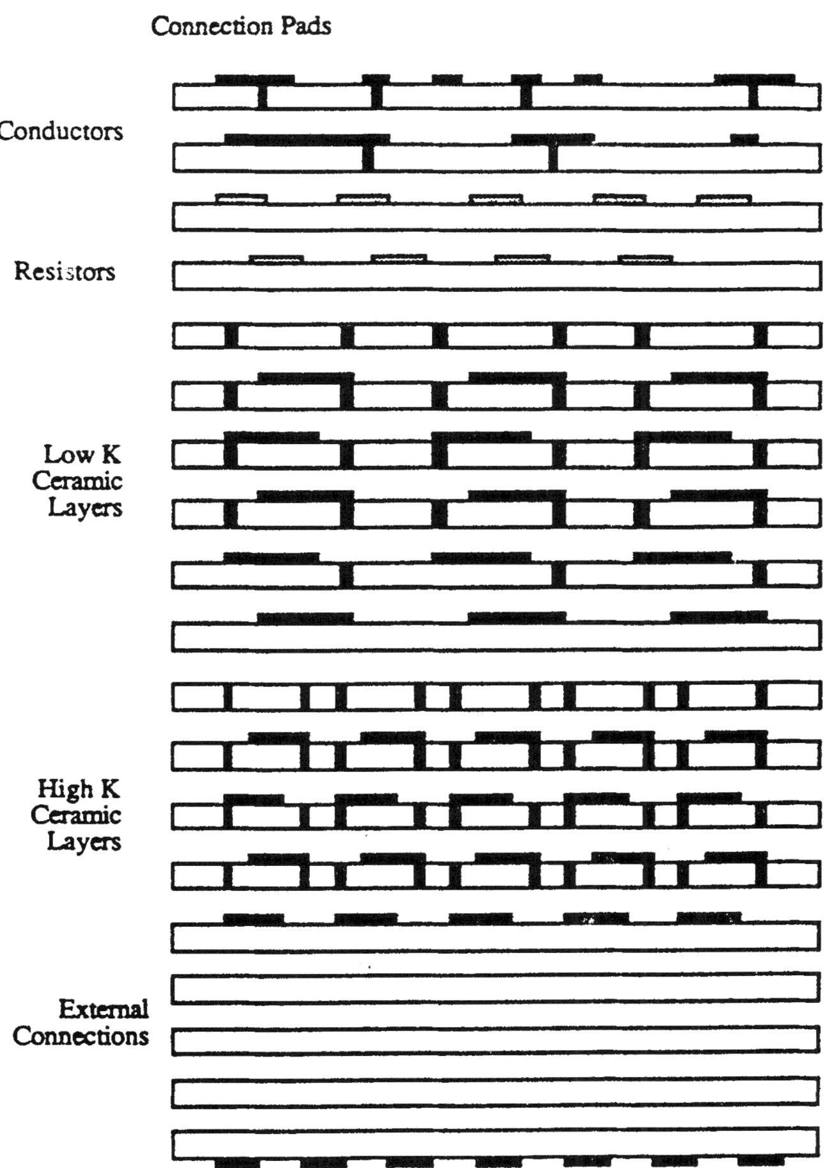

FIGURE 6. Schematic representation of a MCM-C with fully integrated functional components (Utsumi et al.17).

Multilayer Ceramic Processing for L T T C Packaging:

The common ceramic component materials are usually based upon SiO_2 and Al_2O_3 in various volume fractions. The glasses are borosilicates or alumino-silicates with low softening points $T_g<700°C$ enabling sintering at temperatures below 1000°C. The important LTTC materials and their dielectric constant and thermal expansion properties are listed in Table 4. The development of crystallizable glasses presents a different means to engineer the dielectric, thermal and mechanical properties. The sintering process occurs in the glass state which then is converted to a highly crystalline ceramic (~80-90% by volume) with little change in specific volume at a crystallization temperature. Table 5 lists a selection of commercial substrate designs for L T C C packages.

Each powder material is mixed with liquid vehicles (solvent, plasticizers, and binder) to form a slurry and then tapecast into thin green sheets (green tape), typically 10-50 μm thick. The green tape is then cut into sheets typically 100 mm x 70 mm. The via and alignment holes are formed with a punch and die. These sheets are then screen-printed with conductor pastes to form transmission lines and vias. For integrated packages, the internal electrodes for the capacitor and resistor elements can also be printed. If buried resistors are also to be integrated, resistor pastes can be screen-printed onto the metallized green sheets.

In the co-firing of two materials there exists the possibility of sintering damage being caused through differences in the densification rates or chemical interactions between the two materials. Typical defects in multilayer ceramics are delaminations, electrode discontinuities, crack propagation in the ceramic layers, voids, etc. The key to reducing these defects is control of the x-y shrinkage of the substrate during densification. Precise dimensional control is also necessary for automated chip attachment, but the internal stresses created during sintering often cause the substrate to warp. Present approaches to limit this are by the

TABLE 4

Common L T T C Components

Oxide/Glasses	(Dielectric K Constant)	Thermal expansion ppm/°C Coefficient)
Alumina (Al_2O_3)	9.6	6.8
Cordierite ($2MgO \cdot 2Al_2O_3 \cdot 5SiO_2$)	4.5-5.3	2.5
Forsterite ($2MgO \cdot SiO_2$)	6.2	9.8
Mullite ($3Al_2O_3 \cdot 2SiO_2$)	6.6	4.0
Quartz (SiO_2)	3.9	13
Spodumene($Li_2O \cdot Al_2O_3 \cdot 4SiO_2$)	6.0	2.0
Steatite ($MgO-SiO_2$)	5.7	4.2
Borosilicate Glass	4.0	3.3
Pb-Borosilicate Glass	7.0	7.0
Mg-Aluminosilicate	5.0	3.8

application of pressure during densification or constraining the sintering by a metal interlayer (15,16).

A novel method illustrated by the David Sarnoff Research Center was the suppression of the x-y shrinkage in LTTC ceramics via a clamped metal plate. The metal plates are composites of Cu-Mo-Cu or Cu-invar-Cu thermal expansion matched to the glass-ceramic and laminated into the center of the LTCC package. In addition to limiting the sintering shrinkage to the direction normal to the plane of the laminants, the metal plate acts as a heat spreader and also gives added strength to the overall package.

In the Center for Dielectric Studies (CDS) at Penn State, researchers have investigated an alternative method to reduce stresses during densification. A computer controlled infrared furnace/dilatometer system has been developed to manipulate the shrinkage rates during cofiring in a manner that limits sintering damage—a process known as "rate-controlled sintering"(RCS) (16-17).

When two materials are co-fired, e.g. glass-ceramic substrate and Ag-metallization, the different densification behaviors may be reduced by changing the powder reactivity, green densities, agglomeration states, etc. However, it is extremely difficult to control and monitor the complete densification behavior. When two materials undergo co-firing, it is their relative shrinkage rates which contribute to high internal transient stresses creating flaws. By controlling shrinkage rates it was found that the internal stress could be minimized, and with it a

TABLE 5
L T C C Systems and Properties

US L.T.T.C Companies	Glass Matrix	Ceramic Filler	Metallization	Relative Dielectric Constant	Dielectric Loss (1 Mz)	TCE ppm/C
Alcoa	Borosilicate (+dopants)	SiO_2	Au, Ag/Pd	3.9—4.2	<0.003	2.5-3.5
Corning	Crystallizable Glass	Cordierite	Au	5.2	N/A	3.4
DuPont	Alumino Borosilicate	Al_2O_3	Ag, Au	7.8	0.002	7.9
Ferro	Crystallizable Glass	Wollastonite	Ag,Au,Ag/Pd	6.0	0.002	7.0
IBM	Crystallizable Glass	B-Spodumene	Cu	5.0-6.5	N/A	2.0-8.3
IBM	Crystallizable Glass	Cordierite	Cu	5.2-5.7	N/A	2.5-5.5
Tektronix	MgO-CaO-Silicate	Al_2O_3	Ag,Au	5.8	0.0016	4.6
Westinghouse	CaO- B_2O_3. Al_2O_3	SiO_2	Au	4.6	0.001	9.6

reduction in the internal damage of a co-fired package.(15) Figure 7(a-e) shows some of the typical microstructures obtained in a multilayer package as a function of various shrinkage rates in an Ag-green tape co-fired package. The optimized package was fired at a shrinkage rate of 1.5% per minute which produced the most reproducible parts as indicated by the smallest standard deviation in capacitance. Mimicking the idealized heating profiles established via the laboratory RCS furnace could lead to new processing route in cofiring and higher yields.

INTEGRATION OF FUNCTIONAL CERAMICS

As integration levels increase, the problems of co-firing become more prevalent owing to the large number of materials undergoing co-firing within a package. Rate controlled sintering offers one method to overcome these intrinsic problems by finding the best materials engineering compromise to bring the relative transient stress to acceptable levels, and optimizing the densification of each component during co-firing.

The integration of high dielectric materials into a package would permit the loss of surface-mounted chip capacitors. The materials of choice for co-firable high K dielectric permittivities are relaxor ferroelectric materials. NEC research reported a co-fireable dielectric based on the composition 33 wt% $Pb(Fe_{2/3}W_{1/3})O_3$- 67 wt% $Pb(Fe_{1/2}Nb_{1/2})O_3$, this material had a dielectric maximum at about 30°C with a dielectric permittivity ~7000 and dielectric loss ~0.04.(17) Megherhi at Penn State, developed a co-fireable composition based on 0.75 $Pb(Mg_{1/3}Nb_{2/3})O_3$ - 0.25 $Pb(Zn_{1/3}Nb_{2/3})O_3$, and 0.93 $Pb(Mg_{1/3}Nb_{2/3})O_3$ - 0.07 $PbTiO_3$ with small additives of $LiNO_3$ sintering aid. (2 mol %).(18) Tapes of these solid solutions were co-fireable with DuPont 851AT: Al_2O_3-SiO_2-B_2O_3-PbO system, and silver-palladium electrodes at 850 to 950°C. The typical dielectric constants were in excess of 10,000 at room temperature with tanδ ~0.02. The temperature coefficient of capacitance satisfied Z5U specifications (+22% to 50% from 10°C to 85°C). The major extent of the materials development was dedicated to the matching of the densification behavior and the control of chemical interdiffusion to limit the formation of deleterious second phases such as lead silicates and pyrochlores which lowered the dielectric constant of the capacitor material.

Resistor materials based on RuO_2 have also been co-fired at 900°C by NEC.(17) The buried resistor is one of the most difficult functional components to control. Often with thick film techniques, resistors are trimmed to precise values via a laser, this tunability is not possible with a fully buried resistor and therefore requires precise control of the sintering. In general, more work is required to better control the interface between buried component materials and the packaging substrate to preserve the desired capacitance and resistance tolerances.

Another form of integration in MCMs is combining technologies and material types (ceramic and polymers) to get the best of all possible scenarios. An example of this is in an MCM-D being developed by Mitsubishi Materials, where an AlN

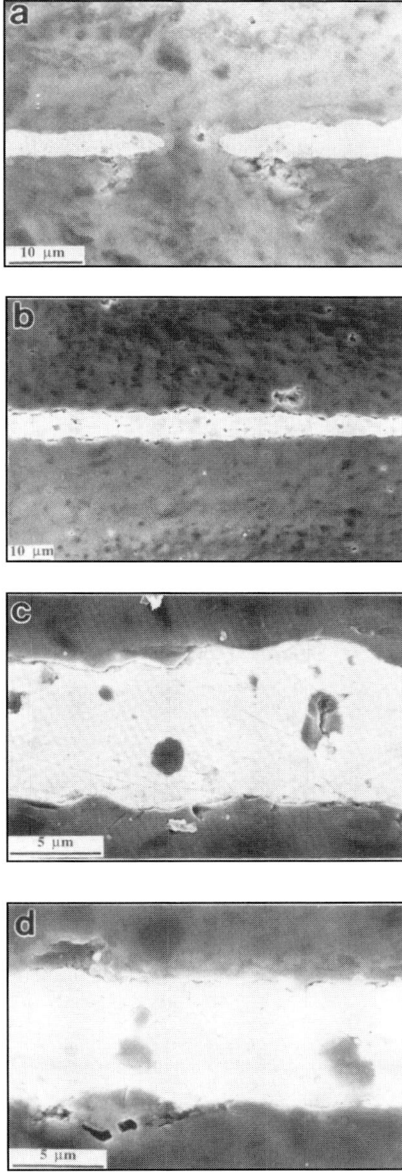

FIGURE 7 (a) and (b) shows examples of microstructural defects obtained at a constant shrinkage rate 0.5% minute[-1]. (a) glass filled conductor film discontinuities, (b) interface porosity located in the silver phase (lighter contrast), (c) microstructure of the buried electrode densified at 1.5% minute[-1] sharing ideal adhesion and sintering between the electrode and dielectric layers, (d) stress induced defects in the dielectric layers during shrinkage at 6% minute[-1].

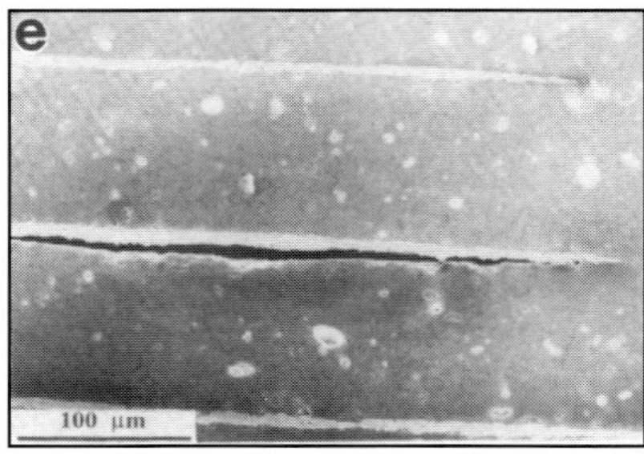

FIGURE 7 (e) large scale delaminations in samples sintered at 25°C/minute close to the buried electrode edges (ref. 15).

substrate is surface treated with a Al_2O_3 and SiO_2 composite layers. This permits good adhesion, planarization, and acts as a chemical diffusion barrier(19). Upon this layer, gold (Au) metallization was applied using a screen printing technique and a metallo-organic Au, to obtain low resistive and planarized bottom electrodes after firing at 650°C. A sol-gel method was used to obtain dielectric layers of $(Ba,Sr)TiO_3$. Layers were planarized using spin coating, after crystallization films with capacitances of 700-900 nF/cm^2 and losses of 0.03 to 0.05 were obtained. A MCM polyimide multilayer structure complete with Cu-transmission lines and vias was then deposited above the ceramic substrates. This structure reduced the number of surface mounted multilayer capacitors and also permitted higher line density and faster transmission close to the integrated circuits. Thermal conductivity was aided through thermal vias in the polyimide coupled to the highly conductive AlN substrate. The resistors were still surface mounted in this package.

Multilayer fired Al_2O_3 substrates are also used in conjugation with polyimide layers in present day packages. A MCM of this type is manufactured by IBM as illustrated in Figure 5. Another form of integration in MCMs is combining technologies and material groups. High speed and high density could be obtained in the polyimide laminants; the ceramic components aid the mechanical strength, thermal expansion and thermal conductivity. Future MCM structures will involve the integration of optoelectronic components into the package with optical fiber connection as well as microwave transmission lines.

SMALL SCALE INTEGRATION

Integrating two or more surface mount electronic components into one is also an active area of research and development in ceramic packaging. We refer to this as small scale integration. At this time small scale integration is successfully satisfying the general trend of miniaturization in electronic components in spite of being a surface mount technology. Three examples are illustrated below:

(i) Multilayer LC-Filter (MLFC)

In 1983 TDK created the first monolithic multilayer chip LC filter. This device combined multilayer inductors and capacitors in a single chip. As in the case of the large scale integration in the MCM-C packages the cofiring of ceramic materials must be matched as closely as possible in terms of densification rates and also thermal expansions. TDK developed a NiCuZn-ferrite cofireable with a TiO_2 capacitor material and silver electrodes at 890°C for 2 hours. The TiO_2 is modified to match the sintering of the ferrite ceramic through the addition of CuO. The ceramic and metal layers are consolidated in the unfired body through systematic patterning using screen printing techniques to optimize the inductance, the electrode design within the ferrite material is helical. The multilayer LC filter has been successfully incorporated into many video cameras and mobile communications where the issues size and weight reduction, higher reliability and lower cost are important.

(ii) Multilayer Varistor Capacitor (MLVC)

Varistors are functional ceramics designed like Zener diodes to protect integrated circuitry from voltage spikes associated with electrostatic discharges. In some cases the rise time of the voltage spikes are faster than the response time of the ceramic varistor in the circuit. Therefore, damage to chips can be incurred by high rise time discharges (~0.5 ns). A methodology to increase the response of the varistor is to raise its capacitance. This can be accomplished through the development of new varistor ceramics such as $SrTiO_3$ based systems or through the integration of capacitor materials with conventional ZnO using multilayer technologies. At Penn State researchers are developing cofirable multilayer capacitor and varistor materials for the MLVC (21). This work is based on relaxor compositions in the $Pb(Mg_{1/3}Nb_{2/3})O_3$-$Pb(Zn_{1/3}Nb_{2/3})O_3$ solid-solution. Figure 8 shows the cofired interface grain structure of the relaxor and the ZnO-Bi_2O_3-CoO varistor materials.

(iii) RC-Networks

Resistor and Capacitor networks are used in the miniaturization of discrete components. These are fabricated either through screen printing pastes of resistive oxides (based on either RuO_2, $CaRuO_3$ or $Bi_2Ru_2O_7$) and dielectric oxides on rigid alumina substrates. In some cases the pastes are cofired and in others a sequence of print-dry-fire steps is adopted. After firing, the resistors are generally laser trimmed to the desired resistance values (1%) and then diced into networks.

Hybrid RC networks for band pass filters and RC diodes have been fabricated by adding thick film resistors, screen printed onto the surface of fired multilayer chip capacitors(22). Figure 9 shows a schematic of an hybrid integrated RC network together with the equivalent circuit. This approach has a number of advantages in controlling the temperature coefficient of capacitance, production yields and reliability.

As the technology and engineering science of ceramic cofiring and post firing thick films continues to develop, the integration of other combinations of functional ceramics will occur such as chemical sensor oxide heterojunctions, electro-optic oxides, pyroelectrics and actuator materials.

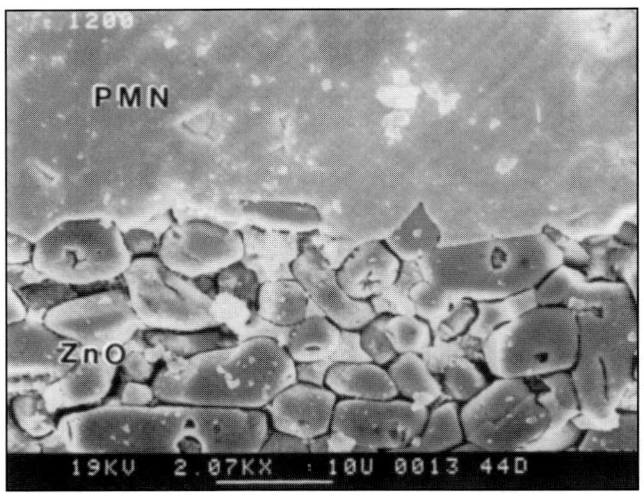

FIGURE 8. Microstructure of the cofired PMN-ZnO interface.

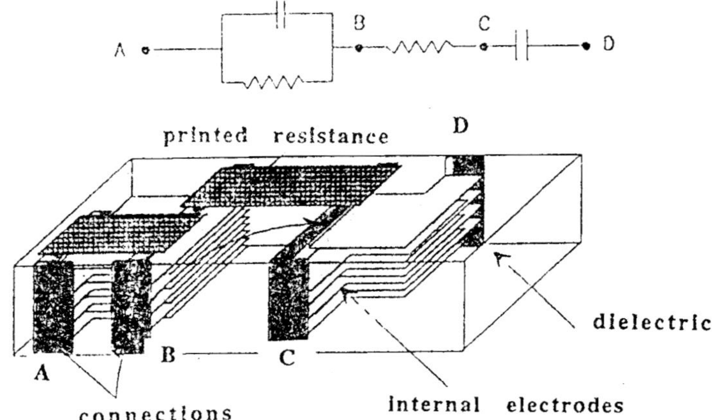

FIGURE 9. Schematic diagram indicating the small scale integration of an RC network with thick film resistors fired onto a previously sintered multilayer capacitor, the equivalent RC network circuit is also shown (ref.22).

SUMMARY AND CONCLUSIONS

In this review some of the complex material science and engineering problems have been outlined in the packaging of integrated ceramics. The difficult issues of material selection with regard to the dielectric, thermal and mechanical properties in the overall optimization of performance in an MCM was outlined. In addition, the processing difficulties were discussed with regard to differences in the densification of cofirable ceramics and the chemical compatibility of the materials. Some of the new developments and strategies to accommodate these problems were discussed. Given the fast rate of development for integrated silicon chips it is necessary for the ceramic integration to keep pace in the size reductions of functional electroceramics. Many exciting challenges exist in the advancement of ceramic science in this area.

ACKNOWLEDGMENTS

Thanks go to IBM and Hughes Aircraft for the use of the photographs showing their respective MCMs. Thanks also to Wes Hackenberger for the micrographs of the RCS packages and Francis Toal for the SEM micrograph of the MLVC interface. Acknowledgements are also extended to Drs. Tom Shrout and Wes Hackenberger for useful suggestions in the development of this chapter, and Joanne Aller for typing this manuscript.

REFERENCES

1. R.R. Tummala. 1991. "Ceramic and Glass Ceramic Packaging in the 1990s," *J. Am. Ceram. Soc.* 74, (5): 895-908.
2. 1994. Business Week, July 4.
3. R.R. Tummala. 1994). "Integrated Substrate—4th International Conference on Electronic Ceramics and Applications," Aachen, Germany: 2, 1097-1103.
4. J.L. Sprague. 1990. "Multilayer Ceramic Packaging Alternatives, IEEE Transactions on Components, Hybrids, and Manufacturing Technology: 13," (2), 390-396.
5. A. Roosen. 1994. "Modern Substrate Concepts for the Microelectronic Industry, 4th International Conference on Electronic Ceramics and Applications," Aachen, Germany: 2, 1089-1096.
6. R.R. Tummala and E.J. Rymaszewski. 1989. "Microelectronics Packaging Handbook," VanNostrand Reinhold Publishers, New York.
7. T. Catt. 1967. "Crosstalk (noise) in digital Systems," *IEEE Trans. Dig. Comp.:* EC-16,(6) 743-763.
8. B.K. Gilbert and W.L. Walters. 1992. "Design Guidelines for Digital

Multichip Modules Operating at High Clock Rates," Int. J of Microcircuits and Electronic Packaging:15(4) 167-173.

9. U. Mohideen. 1987. 'Ultra-low permittivity Porous Silica Thick Films For GaAs IC Packaging, M.S. Thesis, The Pennsylvania State University.

10. D. Reed. 1994. Silica-Polyimide Composites for Electronic Packaging Applications. M.S. Thesis, Pennsylvania State University.

11. C.S. Korman, W. Bicknell, W. Roshen, and R.L. Steigerwald. 1993. "High Density Integrated Power: A New Concept in Power Integration," *Proc. Int. Symp. on Microelectronics,* Dallas, Texas, 622-672.

12. A.J. Blodgett and D.B. Barbour. 1982. "Thermal Conduction Module: A High Performance Multilayer Package," *IBM J.Res.Dev:* 26,30-45.

13. T.Gupta. 1994. "In Search of Low Dielectric Constant Materials for Electronic Packaging," Int. *J. Microcircuits and Electronic Packaging*:17. (10), 80-97.

14. H.T. Sawhill, R.M. Jensen, and K.R. Mikeska. 1990. "Dimensional Control in Low Temperature Substrate for High Performance Computers," *Int. J. of Hybrid Microelectronics: 14* (4) 137-150.

15. W. Hackenberger. 1996. "Reduction of Sintering Damage and Differential Densification in a Low Temperature Cofired Electronic Substrate Using Rate Controlled Sintering," Ph.D. Thesis, Pennsylvania State University.

16. W. Hackenberger, T.R. Shrout, J.P. Dougherty, and R.E. Speyer. 1993. "Sintering Phenomena and Microstructural Development in LTTC Multilayer Substrates," *Proc. Int Symp. on Microelectronics,* Dallas Texas: 215-220.

17. K. Utsumi, V. Shimada, K. Ikeda, and H. Takamizawa. 1986. "Monolithic Multicomponents Ceramic (MMC) Substrate," *Ferroelectrics: 68* 157-179.

18. M. Megherhi. 1992. "Interaction Studies of Lead Magnesium Niobate Based Capacitor Materials For Integrated Ceramic Packaging," Ph.D. Thesis, Pennsylvania State University.

19. S. Toyoda, K. Sugamura, and Y. Kuromitsu. 1995. "Glazed AlN with BSTO Thin Film Capacitors," *Proceedings of Seventh US-Japan Seminar on Dielectric and Piezoelectric Ceramics,* edited by T. Yamamota, Tsukuba Center for Institutes, Tsukuba 121-124.

20. M. Takaya, Fujisawa, A. and Y. Mochizuki. 1990. "The Development of Composite Multilayer Hybrid Components," *Proc. Int. Symp. Micro-electronics,* 747-754.

21. F. Toal, J.P. Dougherty, and C.A. Randall. 1996 "Processing and Electrical Characterization of a Cofirable Multilayer Capacitor-Varistor," *Am. Ceramic. Soc.*, Indianapolis.

22. H. Salze. 1994. "Integrated Passive Functions: Toward a New Generation of Chips," *Proc. 4th Int. Conference on Electronic Ceramics and Applications,* Aachen, Germany, 1055-1060.

The Era of Materials. Edited by S.K. Majumdar, R.E. Tressler, and E.W. Miller © 1998, The Pennsylvania Academy of Science.

Chapter Sixteen

POLYCRYSTALLINE DIAMOND GROWTH BY CHEMICAL VAPOR DEPOSITION

W.A. YARBROUGH and K.E. SPEAR

Department of Materials Science and Engineering
Steidle Building
Pennsylvania State University
University Park, PA 16802

INTRODUCTION AND PRESENT STATUS

One of the more dramatic developments in materials science has been the discovery that diamond can be made by reacting almost any source of carbon, including graphite, with free atoms of hydrogen at low pressures, and at relatively low temperatures. The rapid commercial, technological and scientific developments in the field of diamond synthesis over the past decade have been widely reviewed in the technical literature with Penn State being a center for many of these developments and the source of the first reports of plasma assisted vapor phase diamond synthesis in the U.S. (1-7). This discovery fueled both private investment and major government programs as diamond has a unique set of physical properties making it, in principle, the material of choice for a very wide range of applications. Aside from being the hardest known material, it is optically transparent from microwave frequencies through the infrared and visible, if pure and relatively defect free. It also has the highest thermal conductivity at room temperature of any known solid making it the material of choice for applications for heat dissipation and thermal management. For this reason many companies and government programs have focused on the use of diamond as an electronic packaging material as it also has a high electrical resistivity. It has electronic properties that are in principle superior to either silicon (Si) or gallium arsenide (GaAs). Because of its large band gap (5.45 eV), high breakdown strength ($\sim 10^7$ V/cm), and high electron saturation velocity ($\sim 3 \times 10^7$ cm/sec); as well as its high thermal conductivity, diamond has the largest figure of merit for use in high power or high frequency applications of any known semiconductor. With the discovery of diamond growth by chemical vapor deposition (CVD), and recognizing that a diamond device would perform much better and be more robust, a major effort was made in the mid to late 1980's and early 1990's by governments and private corporations to develop an active device diamond semiconductor technology. As a result of these investments great progress was made in reducing the cost and improving the quality of commercially produced CVD diamond. There are numerous international conferences devoted almost entirely to CVD diamond and its possible impact on solid state electronics is often stressed.

Although progress has been made, and some devices demonstrated, most device fabrication programs remain exploratory and the large markets forecast in the 1980's for diamond (as well as for SiC and some other new electronic materials) never developed. This is largely due to three problems. The first of these is specific to diamond and is the failure to find a suitable n-type doping technology for producing conventional devices. This difficulty can be attributed in part to diamond's metastable and highly refractory nature. Unlike silicon and many other materials, diamond is thermodynamically unstable and many of the doping and annealing techniques used to fabricate devices with other materials are not suitable. The second problem facing the development of CVD diamond based electronics is the absence of a suitable substrate material on which to grow large area single crystal material. Although much effort has been expended, with some progress reported, the goal of a process and materials technology for producing nearly defect free, single crystal material on a scale suitable for large scale manufacturing has remained elusive. The final difficulty is one common to some other advanced electronic materials; e.g., GaAs, GaN, and SiC. This is the cost associated with the advanced processing technologies needed for microfabricating conventional electronic devices. In the 1980's, and into the early 90's, many had hoped that processes widely used for silicon based products would require only simple modification for use with more advanced materials. It is now recognized by many that the development of high temperature, wide bandgap semiconductors like diamond will also likely require new processes and process technologies.

Another reason for engineering new products with CVD diamond comes from its ability to resist physical damage, even in highly corrosive and hostile environments. Its greatest weakness comes from its reactivity with oxygen. However, the rate oxidation in air becomes appreciable only at temperatures of $\sim 700°C$ and higher, and resistance to oxidation can be obtained by the use of protective coatings or passivating layers. This, along with its transparency throughout much of the electromagnetic spectrum, suggests its use as a protective optical window material. Progress towards this goal has made it possible to grow millimeter thick layers of polycrystalline diamond with optical properties that are fully competitive with the best natural gem quality material. Diamond has a single crystal strength of ~ 500 MPa and since polycrystalline materials generally have fracture strengths 2 to 4 times greater than single crystals it was hoped by some that polycrystalline diamond would have strengths of ~2 GPa. Although some samples have approached strengths this high, most thick (>0.5 mm) free standing diamond layers are much weaker, with values from as low as ~200 MPa. Assuming that fracture is controlled by flaws in the grown layer, the estimated flaw or void sizes for strengths this low are of the order of the grain size, i.e. ~100 μm and greater. For these reasons much work remains to be done before CVD diamond can become the major new engineering material that many envisioned in the mid to late 1980's.

Diamond is the hardest material known and a major thrust of much private investment has been the development of diamond coated tools for machining many metals and abrasive construction materials. The pioneers in this effort include most of the major Japanese and U.S. manufacturers of cutting tools and consumable industrial products. These include Norton, Kennemetal, Sandvik, Sumitomo Electric Industries, Mitsubishi Materials Corp., Toshiba Tungaloy and many others. Mechanical hardness alone is not sufficient to guarantee success in a machining application and diamond, being carbon and reactive with many transition metals, suffers from a high mechano-chemical wear rate in the machining of ferrous alloys. This is largely attributed to the formation of numerous stable and metastable carbides at the tool-workpiece interface followed by oxidation of the carbon transferred from the diamond surface to the workpiece metal. Thus, although a diamond coated tool will outlast a conventionally coated tool in many applications, the wear rate of diamond is often large enough to make it rarely cost effective in machining transition metal alloys. Nevertheless diamond coated tools have found several markets where they have proved valuable. A major such market is in the machining of the high silicon-aluminum alloys widely used in the automobile industry. These alloys are difficult to machine as the silicon crystals in a high silicon alloy are highly abrasive and most conventional tools wear rapidly leading to a lower productivity. A diamond coated tool will generally last longer than a conventional tool by a factor of several times. This is particularly important when one considers that many process lines are now automated and the unplanned interruption of one machining operation can result in a productivity loss for an entire manufacturing process line. For this reason a major limitation on the wider use of diamond coating technology has been the problem of obtaining, and maintaining, sufficient adhesion of the diamond coating to the tool base material. Although this problem has been solved for many tool designs and applications it remains a matter of considerable concern. Because of its toughness and fracture resistance the dominant tool material is cobalt cemented tungsten carbide. In the initial stages of the deposition process for diamond, cobalt (as well as some other transition metals) catalyze the formation of graphite instead of diamond. This will normally lead to the formation of a weak graphitic interlayer between the tool base and the diamond coating, causing the premature loss of the diamond coating and leading to unpredictable and greatly shortened tool life.

Thus much work continues on the problem of how to ensure adequate and reliable adhesion during the coating process. For this reason some manufacturers, notably Idemitsu Petrochemical Corp. and others in Japan, have opted to change tool materials, using hot pressed ceramics, commonly silicon nitride, in place of cemented carbides in order to ensure adequate adhesion between the diamond and the tool base material.

THE FAMILY OF CARBON BASED MATERIALS

Solid carbon comes in several phases with the ones of greatest interest being diamond and graphite. These are quite different in that they have both very different crystal structures and very different bonding schemes. The diamond structure is illustrated in Figure 1. Diamond has a cubic crystal structure analogous to sphalerite (cubic ZnS) or the cubic form of silicon carbide. In this structure each carbon atom is bonded to four other carbon atoms in a tetrahedral arrangement. The bonding of each atom to another is through the formation of sp^3 hybridized orbitals and this arrangement leads both to a large bond strength and a highly symmetric arrangement in three dimensions. In contrast graphite has a different electronic bonding arrangement with the formation and overlap of highly directional sp^2 hybridized atomic orbitals, leading to one of the most anisotropic

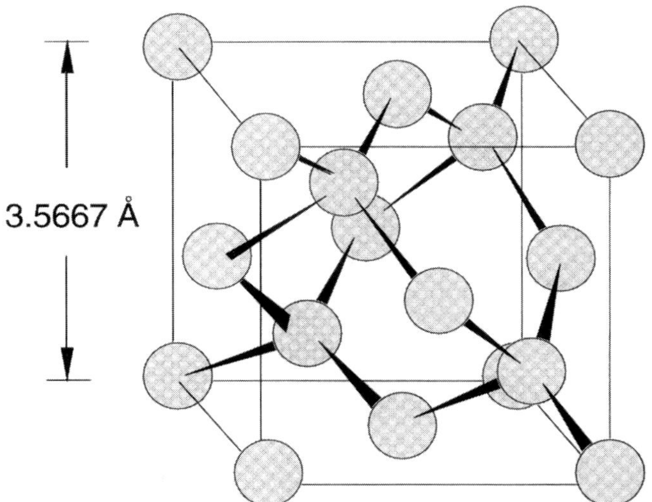

FIGURE 1. The crystal structure of diamond. Shown is the conventional or crystallographic unit cell. Diamond is face-centered cubic with the same structure as silicon at 2/3 of the lattice parameter or silicon.

and heterodesmic structures known in materials science. Although the bonding between individual carbon atoms is stronger in graphite than in diamond, with a single carbon-carbon bond dissociation energy almost twice that in diamond, there are only three such bonds per carbon atom in graphite as opposed to four in diamond. In addition the strong carbon-carbon bonds in graphite all lie in a single plane, the plane of the layers shown in Figure 2. The bonding between the layers is London exchange force (or van der Waals) bonding and thus is relatively very weak. It is the weakness of the interlayer bonds in graphite that give it many of the properties for which it is best known. These include its low in-plane shear

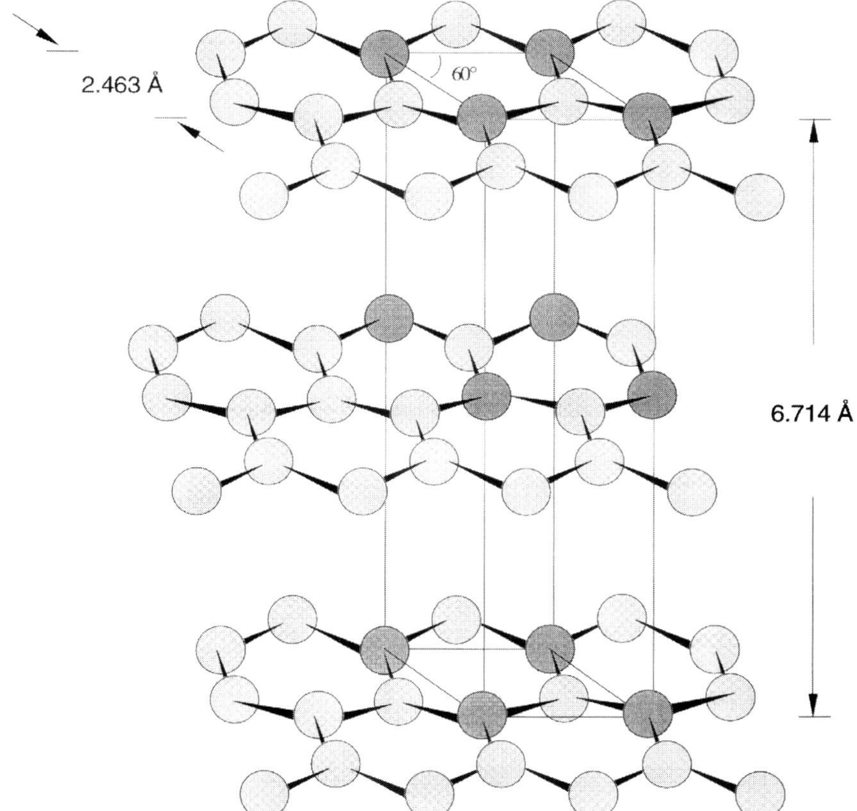

FIGURE 2. Graphite consists of sp² "sheets" of carbon atoms staggered above and below one another. Two different unit cells, one rhombohedral and the other hexagonal (shown in the figure), are known. The difference between them is determined by the stacking sequence of the layers. The unit cell is outlined above with the atoms at the edges and corners darkly shaded.

strength, giving it the ability to be used as a writing material, from which it gets its name, and as a lubricant. While it is impossible to cleave graphite parallel to the c axis, it is relatively easy to separate the layers perpendicular to this axis, and this rather large anisotropy in mechanical properties is also mirrored in its chemical and physical properties. This relative weakness in interplanar bonding also leads to a large family of structural modifications that vary in the stacking sequence of the "sheets". In poorly crystallized forms these layers are also laterally dislocated with respect to each other. The combination of translational and rotational dislocation of layers with respect to each other can result in total disorder along the 'c' axis direction. These materials are sometimes called the "turbostratic" graphites, i.e. graphitic carbons in which the central structural feature remains the stacking of carbon "sheets", sp^2 bonded together in planes, but without any order along the 'c' axis. Because of the marked differences in bonding and structure that lead to large differences in properties between diamond and the graphitic carbons, many tend to think of these almost as unrelated compounds, even though both are simply different phases of carbon.

Although the word graphite is properly reserved for use with either the hexagonal or rhombohedral forms of sp^2 hybridized carbon, there exists a large family of more or less disordered sp^2 hybridized carbon solids which, in part because they can be described as derivatives of graphite, are sometimes referred to as "graphite" or "graphitic" carbon, including some forms of "soot" or amorphous carbon. Other forms of carbon include lonsdaleite, (8,9) a relative of diamond which although it also consists of sp^3 hybridized carbon is hexagonal instead of cubic, cliftonite, (10) a meteoritic form of carbon, and chaoite, (11) a shock fused derivative of graphite. In addition to the crystalline forms there are many different non-crystalline or "glassy" carbons. A simple model of poorly crystallized graphitic material exhibiting disorder along both directions can be seen in Figure 3 where, although some planarity is retained, considerable dislocation and disorientation exists, in some cases with well organized regions on a scale of 1 to 100 nanometers, hence the use of the term "nanocrystalline" graphite by some workers in the field. A reasonable model involves not only the disruption of the planar network along an 'a' axis direction but also a bending of the normally planar sp^2 bonded sheets. This gives rise to an "entangled ribbon" model which has been proposed for some "glassy" carbons (12). In this model, "ribbons" consisting of stacks of sp^2 bonded graphitic sheets rap around and through each other, with branching in both the 'c' and 'a' axis directions. That a significant deviation from planarity for fully sp^2 hybridized carbon is possible is evidenced by the stability of "buckminsterfullerene" or C_{60}. Its structure is similar to that of a soccer ball, and consists of a completely enclosed structure of sp^2 hybridized carbon atoms (13).

Some of the non-crystalline carbons can have properties, notably hardness, that rival those of diamond. These can be prepared as hard, abrasion resistant thin films and are called the "diamond-like" carbons or DLC. Unlike the crystalline phases these materials have no well defined structure and can include large amounts of hydrogen. Being non-crystalline their structure, properties and range of utility varies widely with their composition and method of preparation. The DLC materials can be prepared as optically transparent, or nearly transparent thin coatings and debate continues in the scientific community as to whether or not any of these materials may contain true crystalline diamond on a scale of tens of nanometers. The major limitations on the use of these materials is that they are commonly available only in thin film form (typically less than 1 μm in thickness) and normally have large internal stresses. In addition all DLC materials are thermally unstable, decomposing to graphite or a more graphitic structure with the loss of desirable properties at temperatures of 200 to 400°C.

THE TECHNOLOGY OF DIAMOND GROWTH

The vapor deposition of high quality diamond crystals came as a major surprise to most scientists because diamond was well known as a high pressure form of carbon with its synthesis having been first achieved using high pressure high temperature methods (14). In spite of this there were always those who believed that diamond might be made from hydrocarbons at low pressures and this effort has a long history with a substantial patent literature (15-19). In most of the original work, carbon monoxide or various organic compounds, e.g. methane, acetone, methyl mercaptan, propane, etc., were used and although some variation in the methods exist, e.g., composition, pressure and the presence or absence of specific catalysts, all essentially involved the thermal decomposition of the precursor gas

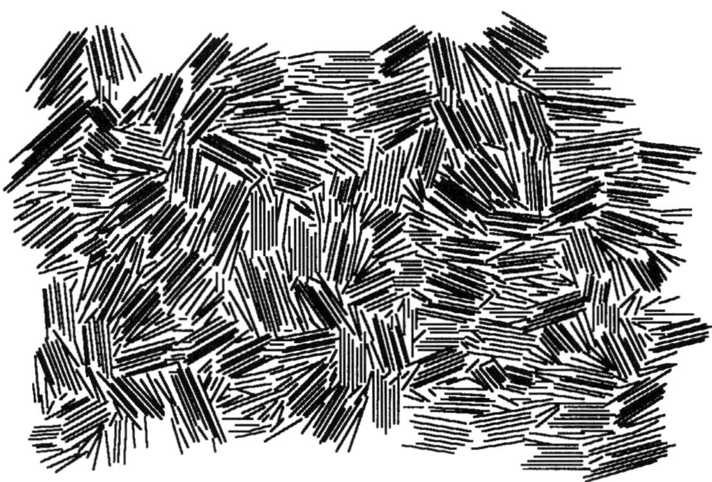

FIGURE 3. A schematic illustrating "nanocrystalline" graphite. Only very short range order along the "c" axis of graphite is observed.

to deposit carbon on diamond or other related powders (19,20). Some diamond growth was reported with all of these methods but the eventual nucleation and growth of graphite could not be avoided. Because of diamond's thermodynamic metastability many thought the effort to make diamond at low pressure was foolish and early reports were often met with skepticism. Indeed some of the pioneers in vapor phase diamond synthesis, e.g. J. C. Angus at Case-Western Reserve University and B.E. Knox at Penn State University found their efforts, and their early reports of diamond synthesis ridiculed. In spite of this many pioneers persisted in their efforts and were tantalizingly close to the steady state synthesis of diamond throughout the 1960's and 1970's. For example, the formation of various hydrocarbons from the reactions of water or hydrogen with carbon were carried out at Penn State throughout the 1960s by Penn State fuel scientists, P.L. Walker, F J. Vastola and J.P. Wightman. These researchers used microwave plasmas, now commonly used to grow diamond, to enhance the kinetics of gas phase reactions. In addition to forming numerous hydrocarbons they found that a hard and inert deposit was formed on the walls of their apparatus. In the 1970s Bruce Knox, Associate Professor of Materials Science, remembered these results and repeated the experiments as part of a program studying carbonaceous coatings. He studied the formation of deposits from various hydrocarbon-hydrogen mixtures and found that the nature of the microwave plasma deposit was independent of the hydrocarbon precursor and depended only on the H:C ratio in the gas. Several deposits produced in this work showed a diamond diffraction pattern using electron diffraction. However, a few weak forbidden reflections raised doubt as to whether graphite might be present. The coatings project supporting this work was discontinued, and because of lingering doubts as to whether diamond had indeed been produced as well as the skepticism of others, the necessary work to demonstrate diamond growth by low pressure plasma synthesis was not carried out (6).

During this same period, pioneering work was also being pursued in the Soviet Union. This culminated in 1977 with a summary and seminal publication of their work (21). This summarized the work of a decade and suggested that in the presence of free atoms of hydrogen the formation of graphite might be prevented. Although details were not reported, the authors specifically mentioned the "catalytic" production of atomic hydrogen using a tungsten filament and reported that they had combined the growth of diamond with the local production of atomic hydrogen to obtain "sufficiently thick" epitaxial layers of diamond. The preparation of diamond on non-diamond substrates, without detectable graphite formation, was subsequently reported in another publication (22) and in it the critical role of atomic hydrogen and the use of electric discharges (plasmas) and thermal methods for its production were discussed. These results were rapidly confirmed by Japanese scientists (23-25) in a series of papers with greater experimental detail, initiating a new era in the synthesis of diamond.

These reports were generally ignored in the Western world until in 1984 Penn State Prof. Rustum Roy visited Japan and was shown vapor deposited diamond crystals similar to those seen in Figure 4. The obvious presence of a cubic growth habit, with well defined {100} facets, was convincing evidence that well crystallized diamond was indeed being deposited. On his return to Penn State Prof. Roy organized a team of researchers that included K.E. Spear, R. Messier, A.R. Badzian, T. Badzian and B. Simonton to reproduce the results he had seen in Japan. With funding from the Office of Naval Research the team at Penn State demonstrated the growth of diamond using plasma assisted chemical vapor deposition and reported this success in February of 1986 (7). Recognizing the potential importance of this work the Office of Naval Research then initiated a major program to pursue the development of CVD diamond science and technology. One of the ancillary benefits of the demonstration of well crystallized low pressure diamond synthesis under CVD conditions has been to help usher in a new age of metastable materials synthesis by vapor deposition. Partly because of the history of diamond research, many current and future researchers may be spared the skepticism faced by the early pioneers in diamond vapor deposition.

Throughout this same period, work also progressed in the deposition of hydrogenated and hydrogen-free amorphous or poorly crystallized carbon films, many of which have properties much more like those of diamond than those of graphite, hence the use of the term "diamond-like" carbon. These films were also prepared from hydrocarbon gases using low pressure plasmas and some from carbonaceous ions using ion beams (26-28). Much of this work has been amply

FIGURE 4. Diamond crystals grown by hot filament assisted chemical vapor deposition. A cubic growth morphology with clearly defined {100} facets was convincing early evidence that the material formed was indeed well crystallized diamond.

reviewed (29-32). There have been some spectacular claims made for the synthesis of diamond using these methods (33), but many of these claims have later been abandoned (34), often unfortunately without the fanfare that accompanied the original report. Hence there arises a need to differentiate carefully between those conditions and methods known to produce well crystallized diamond and those for which claims have been made, but for which confirming experimental evidence is unavailable. A complete list of the methods used for the reproducible and stable growth of well crystallized diamond is extensive, but these can be generally classified under four major headings; plasma assisted chemical vapor deposition (35-37), thermally assisted (or "hot filament") chemical vapor deposition (38,39), reactive vapor deposition (e.g., combustion methods), (40-42) and various combinations of these (43). All of these methods have in common a number of experimental observations. Among these are (44):

1. *Gas activation.* Some means of gas activation is required for achieving appreciable diamond growth rates. Activating the gas prior to deposition increases diamond rates from ~nm/hour to μm/hr. Chemical activation in halogen containing systems may be possible at relatively low temperatures.

2. *Independence of activation method and nature of carbon containing precursor.* Good quality CVD diamond has been produced utilizing a wide variety of activation methods and carbon sources, including graphite and non-crystalline carbons (45).

3. *Hydrogen is commonly required for efficient growth.* Deryagin and coworkers and Angus first proposed that atomic hydrogen had to be present. Hydrogen in excess of that introduced as part of a hydrocarbon precursor gas may not be needed. Some recent reports however, have claimed that efficient diamond growth can be achieved without the introduction of hydrogen, either independently or bound with the carbon precursor. Interestingly, however, oxygen is a major constituent of the claimed process.

4. *Diamond precursor species.* The most abundant carbon containing precursors in typical diamond growth systems are methyl radicals and acetylene molecules.

5. *Oxygen enhances the quality of the diamond deposited.* Oxygen added in small amounts has been shown to improve the quality of the diamond deposited, however, there are conflicting reports on its effect on growth rate.

6. *Co-deposition of diamond and non-diamond carbon.* Under conditions commonly used, vapor deposited diamond will contain defects with some amounts of graphitic or non-diamond carbon.

7. *Growth rate increases with increasing temperature.* The growth rate of diamond increases with increasing substrate temperature until, at a temperature that varies with the method used, the deposition of graphite and non-diamond carbon dominates. The optimum temperature for most deposition processes is ~1000°C.

8. *Substrate surface treatments.* A substrate pretreatment of scratching with submicrometer diamond powder is commonly used to enhance diamond nucleation.

9. *Crystallite morphology.* Octahedral and cubo-octahedral crystals are commonly formed with {100} and {111} facets dominating. Twinning on the {111} planes frequently occurs and five fold twins about a defect axis are observed.

These common features suggest a common chemistry for the growth process, with the choice of method dependent upon considerations of efficiency, convenience, cost, and applicability to the problem at hand. The major scientific, as well as technological, advance came with the development of various methods of supplying additional energy or external activation to the source gases. With these methods it became possible to grow diamond continuously, and without the eventual nucleation and growth of graphitic carbon. Each of these major classifications can be further subdivided by various parameters such as pressure, power consumption, capital cost, and maximum attainable growth rate. Unfortunately, much of the literature reports only linear growth rates, without reference to the area over which diamond can be grown. Hence, direct comparisons often cannot be made and the methods for which the highest growth rates are reported are not necessarily the most efficient. The highest growth rates have been obtained using high power density, atmospheric pressure plasma processes, where gas phase temperatures of ~5,000 to 8,000°C are readily obtained (46). The highest reproducible growth rates over reasonable areas are in the range of 100 to ~900 um/hr using two plasma torches with hydrocarbon introduction downstream of the discharge (47). Other difficulties exist with making direct comparisons between the different methods. Such parameters as nucleation density, stability of operation, area and uniformity of deposition, and ultimate thickness of a continuous coating are often not reported. In addition, many of these methods are not fully optimized with attention being focused on how to increase the area of uniform deposition, how to reduce deposition temperatures without significant sacrifices in growth rate, and how to deposit on non-planar or convoluted surfaces. Success on many of the technological questions will depend on progress with related scientific issues.

THE CHARACTERIZATION OF DEPOSITED CARBON

A large number of relatively novel or unusual methods have also been claimed to yield diamond and it becomes necessary to determine which of the numerous forms of carbon may in fact contain crystalline diamond. Unfortunately, the determination of when a carbon coating obtained by either chemical vapor deposition (CVD) or physical vapor deposition (PVD) is diamond can represent a non-trivial challenge. If one considers only those reports of essentially pure carbon coatings and films, containing essentially no hydrogen (~1 at. % or less), there are many prepared using methods in addition to those listed above which show evidence for the presence of diamond (48-52). Many of these also show evidence for the presence of other carbon phases, including non-crystalline or highly disordered carbons. It becomes necessary to differentiate between those deposits which are clearly well crystallized diamond, and those which may contain diamond, but which consist primarily of various non-diamond carbons. This differentiation between diamond and what is better termed diamond-like carbon often remains an issue in the materials community, with significant uncertainty in many cases. Indeed, much of the earlier skepticism concerning vapor phase diamond growth, was overcome only when crystals of sufficient size and perfection to clearly show the presence of facets with cubic symmetry had been achieved.

Two means of characterization have been most commonly used for differentiating the different carbon materials. One technique is diffraction analysis, either by x-ray (XRD) or selected area electron diffraction (SAD) in the transmission electron microscope. The relatively weak x-ray atomic scattering factor for carbon often makes it difficult to detect diffraction maxima using XRD. The diffraction lines are often weak, greatly broadened or show asymmetry resulting from strain, defects, or small crystallite size. In addition a strong crystallographic orientation is often present in these films, (53-57) and the relative intensities from an as-deposited film will not often correlate with a randomly oriented reference pattern. Although SAD has been used to argue the presence of diamond, its sensitivity is sufficiently great to risk that the diamond seen is only a small fraction of the carbon present. Significant errors and uncertainties arise as a result of the extremely small volume of material being sampled and the strong probability of texture or preferred orientation in vapor phase deposited thin films (58-60). A second technique which has proved quite useful is Raman spectroscopy. Diamond has a characteristic Raman peak at 1332 cm^{-1} (61,62). Those carbon phases, coatings or layers which do not exhibit x-ray diffraction consistent with the presence of diamond, or its characteristic Raman peak, but which importantly do have many of the physical properties of diamond including hardness, chemical inertness, low intrinsic optical absorption, etc. are those which have become known as diamond-like carbon (27-32).

GROWTH KINETICS AND MECHANISMS

Because of the very high activation energy for carbon diffusion in either graphite or diamond, the reaction chemistry that leads to diamond growth is almost certainly a gas-surface chemistry that does not include the bulk diffusion of either reactants or products. The surfaces of diamond and graphite are hydrogenated with structures and compositions very different from those of the bulk phases and most believe that understanding the growth of diamond is essentially a problem in understanding carbon surface chemistry in the presence of atomic hydrogen. Consistent with this is also a free radical mechanism for the gas growth of diamond which can be written as a simple chain of surface hydrogen abstraction and addition reactions. These occur between various gaseous species and the solid surface, the subscript "d" denoting a carbon atom at the diamond surface. Many different growth mechanisms have been proposed, however, regardless of the specific mechanism, two general features are commonly agreed upon (44). First, the principal element of diamond growth is the formation of an active site, i.e. a surface sp^3 radical, followed by the addition of a carbonaceous specie to this vacant surface site. Critical to this is clearly the production of atomic hydrogen by some means of gas activation (eqn. 1).

$$H_2 \rightarrow H\cdot + H\cdot \tag{1}$$

$$C_dH + H\cdot \rightarrow C_d\cdot. + H_2 \tag{2}$$

Second, that the number density of carbon surface radicals is determined primarily by a balance between the abstraction of hydrogen from surface C-H bonds by gaseous hydrogen atoms (eqn. 2) and the addition of hydrogen atoms to these surface radical sites (eqn. 3):

$$C_d\cdot. + H\cdot. \rightarrow C_dH \tag{3}$$

The surface radical concentration is also determined in part by other reactions, including the addition of carbonaceous species. However if their concentrations are small, and the temperature of the substrate in the range of ~800 to 1100°C, it is generally accepted that kinetic competition between reactions (2) and (3) determine the concentration of vacant surface sites. Since (2) has a significant activation energy (~5 to 10 kcal/mole) while (3) does not, this competition is temperature sensitive. This explains in part why the growth rate of diamond should increase with increasing temperature. As the atomic hydrogen concentration increases, the rate of abstraction (2) increases and the concentration of vacant surface sites asymptotically approaches a constant determined by the ratio of the rate constants for (2) and (3).

As the relevant reaction chemistry for almost any CVD process, and almost certainly that relevant to diamond, is a gas-surface chemistry, the structure and composition of the solid surface becomes all important in controlling the phase and microstructural development of the solid. The three major low index surfaces of diamond are the (111), (110) and (100). The lattice terminated structures of these are shown in Figures 5, 6 and 7. Any detailed reaction mechanism that hopes to explain all of the phenomena that have been observed in diamond growth must ultimately be written with respect to one or more of these surfaces. All of the mechanisms and reaction chemistries that have been proposed for diamond suffer from major uncertainties with respect to the structures and chemistries of these

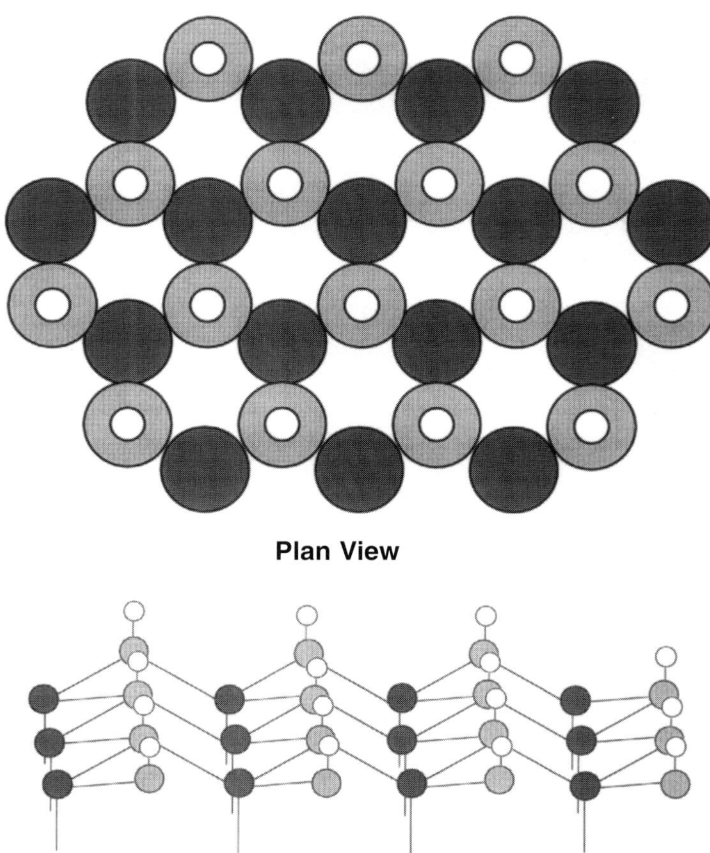

FIGURE 5. Unreconstructed and hydrogen terminated (111) surface of diamond. The lattice is terminated by hydrogen bonded to the surface carbon atoms, avoiding any reconstruction of the surface. The top image (plan view) is the view seen by an observer above the surface. The bottom perspective view illustrates the "puckered" nature of this surface and more clearly shows the hydrogen atoms lying in a plane above the carbon surface.

surfaces as uncertain steric and possibly electronic effects exist. Indeed one of the great frontiers in materials science is in the understanding and control of surface and interfacial phenomena for the synthesis and formation of new materials.

The surface of diamond during growth is obviously labile and its steady state structure or composition is generally different from that of the truncated lattice. An understanding of the steric, chemical and electronic properties of the surface is critical for any complete model of diamond growth. For example, work on modeling the chemisorption or chemical addition of the methyl radical to the diamond (111) surface has led to the conclusion that a significant steric interference is likely between neighboring methyl groups (63). In addition, severe steric

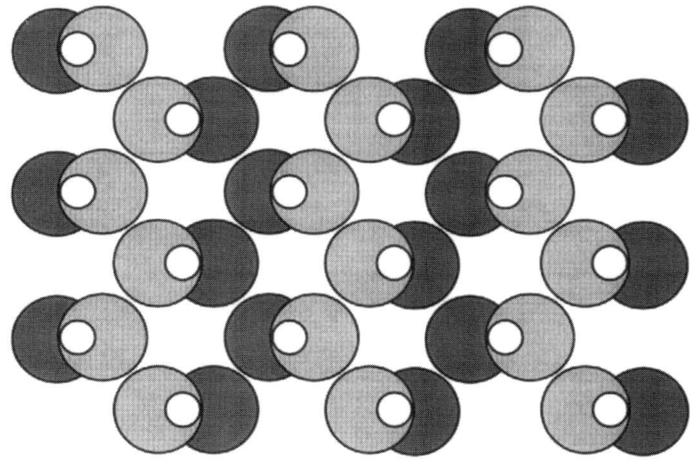

Plan View

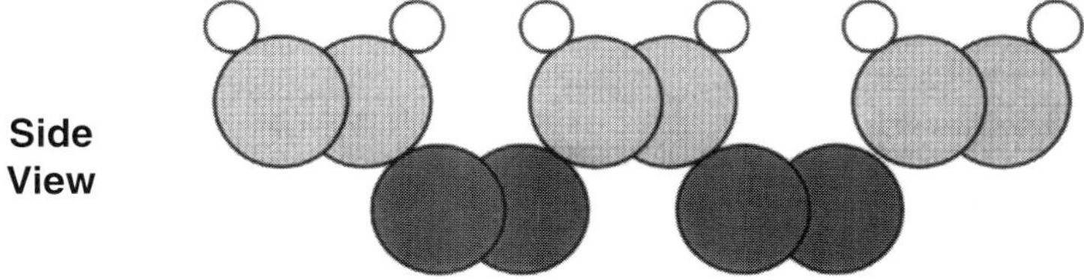

**Side
View**

FIGURE 6. Unreconstructed and hydrogen terminated (110) surface of diamond. As with the (111) surface shown in Figure 5, the carbon atoms in this surface are in the same positions as in the diamond lattice. Each carbon atom has bound to it a hydrogen atom, terminating the lattice without reconstruction.

interferences exist between neighboring hydrogen atoms on a dihydrogen terminated diamond (100) surface. Indeed most doubt it can exist for this reason. Rather the surface observed using scanning tunneling microscopy is most like that of Figure 8. In this structure, referred to as a 2x1 reconstruction, the surface carbon atoms are displaced from their bulk lattice positions and bonded together in pairs to form chains. The problem with the dihydrogen terminated, unreconstructed surface is illustrated in Figure 7 which shows views of a dihydrogen terminated and unreconstructed diamond (100). Using bond lengths of 1.54 Å for

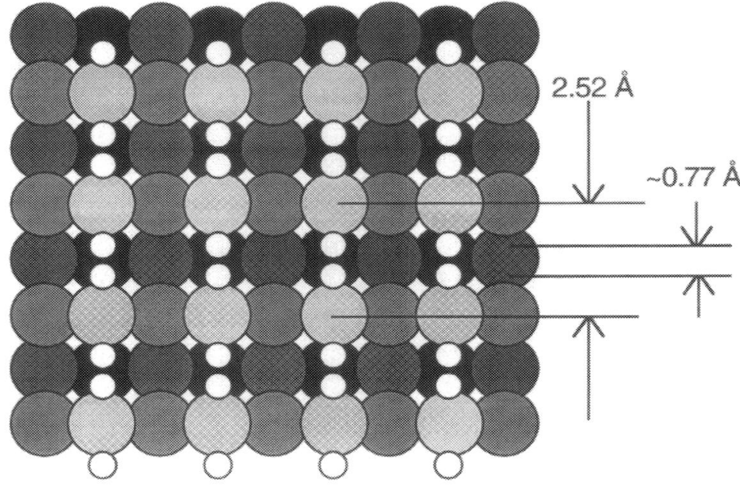

Plan View

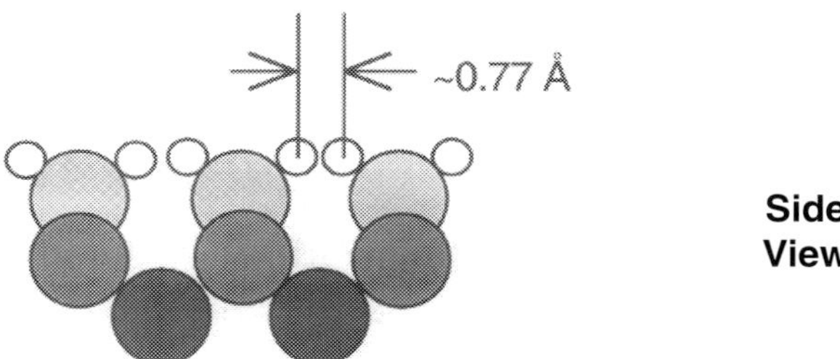

Side View

FIGURE 7. Unreconstructed and hydrogen terminated (100) surface of diamond. The structure is not believed stable because of the substantial steric repulsion expected between the neighboring hydrogen atoms on the unreconstructed surface. Rather it is believed by most researchers that the (100), even if hydrogen terminated, reconstructs to a 2x1 structure as shown in Figure 8.

the carbon-carbon bonds and ~1.10 Å for the carbon-hydrogen bonds, with perfect tetrahedral bond angles of 109.5°, one calculates an internuclear hydrogen-hydrogen spacing on this surface of ~0.77 Å, or nearly the same as that in the H_2 molecule, 0.74 Å. As this would be a non-bonding interaction, very significant steric repulsion between neighboring hydrogen atoms would be expected. This argues that the surface free energy of the 2x1 reconstructed surface is much lower and it becomes the preferred structure. This problem does not arise in the case of the diamond (110) surface, pictured in Figure 6. The hydrogen terminated (110) consists of parallel "chains" of alternating carbon atoms, each passivated by a single hydrogen. The simple hydrogen terminated (110) surface does not have the problems noted above and either methyl radical or acetylene based mechanisms

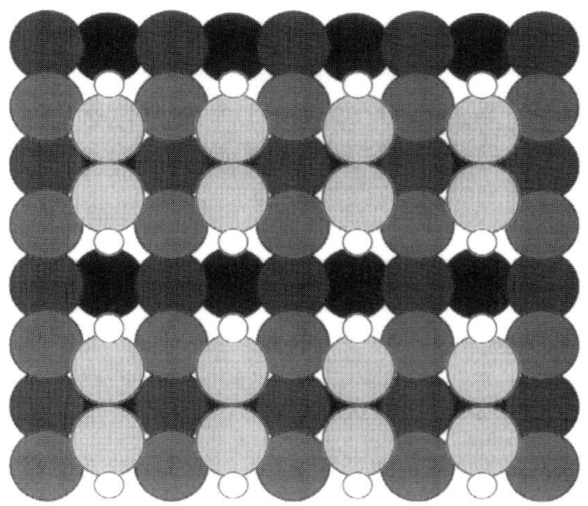

Plan View

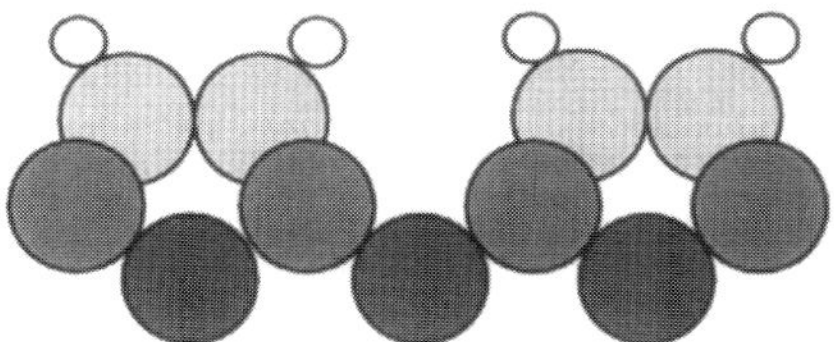

Side View

FIGURE 8. Illustrated is the atomic arrangement of a 2x1 reconstructed hydrogen terminated (100) diamond surface. Direct evidence has been reported for this structure for the 2x1 (100):H surface.

can be postulated for this surface. Nucleation of a new "chain" on the (110) surface can occur by the condensation of adjacent methyl radicals, by the addition of an acetylene molecule across adjacent vacant surface sites, or by the formation of a three carbon atom bridge structure between second nearest neighbors in an underlying [011] "chain". Mechanisms based on growth in this direction are attractive as is known from both careful x-ray study of polycrystalline layers, grown by both plasma and hot filament techniques (54,64,65), and from measured growth rates on single crystal diamond substrates. Vapor phase diamond growth in many cases occurs at the greatest rate along the direction normal to this (011) surface.

NUCLEATION AND MICROSTRUCTURAL DEVELOPMENT

The nucleation of diamond on non-diamond substrates is important to many existing and proposed applications and consequently is of great theoretical and experimental interest. Of particular interest to many is the heteroepitaxial growth of single crystal diamond. More generally the ease of nucleation, and its density, can effect the properties of polycrystalline films through its influence on growth morphology and its evolution in time. That many CVD and other uniaxially grown crystalline materials exhibit a columnar-like growth morphology, shown schematically in Figure 9, has been well known for many years (66). This is generally understood to result from the presence of randomly oriented, discrete nucleation sites on the substrate surface. As these individual crystals grow, grain boundaries are formed between them, and their lateral expansion is prevented or inhibited by the presence of neighboring crystals. If surface and bulk diffusion

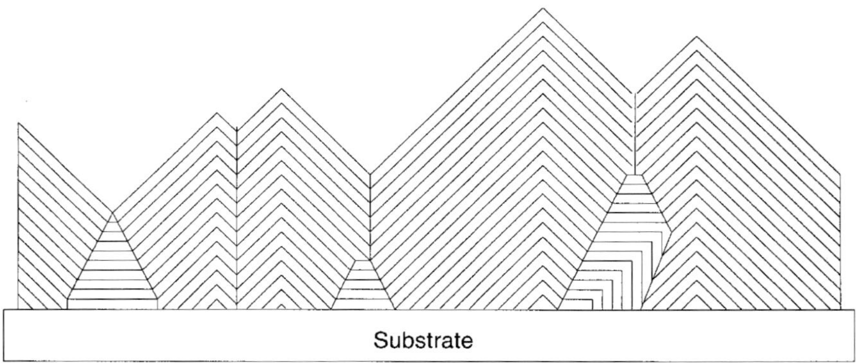

FIGURE 9. Shown is a model for the microstructural development of uniaxially grown polycrystalline layers. Randomly oriented nuclei of the solid are first formed at the surface of the substrate. As these crystals grow, those that are favorably oriented (have their dominant growth axis normal to the substrate) will expand vertically and horizontally at the greatest rate and will determine the texture of the layer.

are sufficiently inhibited, i.e. Ostwald ripening, recrystallization or sintering are slow relative to growth by vapor phase transport, crystals grow dominantly along the axis normal to the plane of the layer. If the initial nuclei are randomly oriented, some will be favorably oriented and many others more or less unfavorably oriented. As growth continues, those with favorable orientations will grow more rapidly, extending above and expanding laterally at the expense of those less favorably oriented. As the morphology of the coating or layer thus produced matures, it will of course begin to exhibit the crystallographic texture of its dominant growth axis, typically a <110> texture or an "off axis" or vicinal <100> texture in vapor phase diamond (54,65). Thus in general no special property need exist to rationalize the development of crystallographic texture other than the idea that crystals grow more rapidly along some directions than along others.

Several factors can complicate this picture seriously. The first is that the dominant growth axis need not necessarily be a low order integral multiple of the unit cell axes, i.e. in a large number of cases it will be a vicinal direction and the faces produced as a result of this will have vicinal indices. It was pointed out by Buckley in his classic text on crystal growth that all too often the existence of vicinal growth habits is ignored when growth theories are formulated (66). In such cases crystals may grow, not by a simple ledge or "kink-step" growth mechanism on a low index face as is often pictured, but rather through a competition between layer nucleation and extension. Another complication is that in the simple description above, renucleation, that is the formation of a randomly oriented nucleus on a previously existing crystal surface, was ignored. Renucleation may also occur at grain boundaries and at other defects in the crystal surface. If the renucleation rate is high relative to the linear growth rate for an individual crystal, and these nuclei have no preferred orientation, then very fine grained more or less randomly oriented coatings and layers may result. If these renucleation sites have a preferred orientation associated with them, then the layer or coating may develop a crystallographic texture, but it may not represent the dominant growth axis. Finally there is the practical problem of deciding when a given morphology is "mature", that is when the growth process with respect to microstructural evolution has reached a steady state.

The model pictured in Figure 9 is supported experimentally for diamond by atomic hydrogen etching experiments at Penn State (67,68). This model supposes that those grains or crystals that have their dominant growth axis aligned perpendicular to the plane of the layer should have the highest growth rate. This implies that growth is fastest on the (110) surface and leads to a simple explanation for the dominantly octahedral and cubo-octahedral crystal morphologies noted earlier. The work at Penn State used highly <110> textured thick (~1 mm) free standing layers that had been polished optically smooth. The simple model implies two results. The first is that the smaller grains (< 80 µm in diameter) exposed at the polished surface should be those more or less misoriented relative to the rest of the layer. The second is that the rate of crystal growth should be at its greatest on the defect free (110) surface. Both these predictions were confirmed in these studies. Figure 10(a) is an optical photomicrograph of the surface of the polished layer after etching with atomic hydrogen generated using a microwave plasma. The intercrystalline boundaries and twin defect planes were found to etch or gasify preferentially as these regions likely contain highly defective, and possibly non-crystalline, carbon. The large roughened grains outlined by dark boundary lines all have approximately the same orientation leading to very similar surface features. The smallest grains, such as the one visible at the center of Figure 10(a), are tilted away from this dominant axis. These grains remained relatively smooth and highly reflective as they developed low angle ledges instead of becoming rough and microfaceted. Shown in Figure 10(b) is a similar polished sample on which diamond was regrown. Randomly oriented grain boundaries and twin defect planes are clearly visible with extensive microfacetting of the surface. The straight parallel crevices are due to (111) twin planes emerging at the surface. Note the meeting together of 5 sets of these at angles of approximately 70° at the center of Figure 10(b). This is known as "five-fold" twinning, a common occurrence in diamond as discussed later in this chapter.

Although a wide range of substrate materials have been used for diamond deposition, the experimental database necessary for developing a detailed theory for nucleation remains limited. The substrates on which diamond has been reproducibly grown can be divided into three general categories. These are: 1) lattice matched, chemically compatible substrates, e.g. diamond and cubic boron nitride; 2) carbides and materials capable of forming interfacial carbides, e.g. Si, Mo, SiC, Si_3N_4, Al_2O_3 or WC; and, 3) non-carbide forming substrates such as Pt, Ni, Cu and Au. Most of the available experimental data is for silicon as this has been the most commonly used substrate. It has generally been observed that the initial process with carbide forming substrates, mainly silicon, is the reduction of any surface oxide and the formation of a carbide or carbidic carbon layer (69,70). Indeed Badzian suggested that diamond nucleation on silicon occurs heteroepitaxially on β-SiC (71). However a silicon carbide interlayer is not always observed (69,72) and the nucleation of diamond on β-SiC is not any more efficient than on many other substrates (73). In addition, when surface carbide formation was observed it was formed very rapidly, while diamond nucleation in the absence of a surface pretreatment required an exceedingly long incubation period (74), suggesting at least some additional requirement.

It was discovered early that rapid nucleation on silicon and many other substrates was obtained by polishing or abrading the surface with a fine diamond powder or paste prior to deposition. Although the typical procedure used calls for removing residual diamond by various cleaning methods, its complete removal can only rarely be demonstrated or assumed. Figure 11 shows diamond growing along scratches (produced by diamond) in the surface of a silicon wafer. The correlation in many experiments between the scratches left by polishing with diamond powder, and the subsequent nucleation and growth of diamond crystals is unmistakable. Hence, two major schools of thought have been advanced to explain the efficacy of polishing with diamond.

(a)
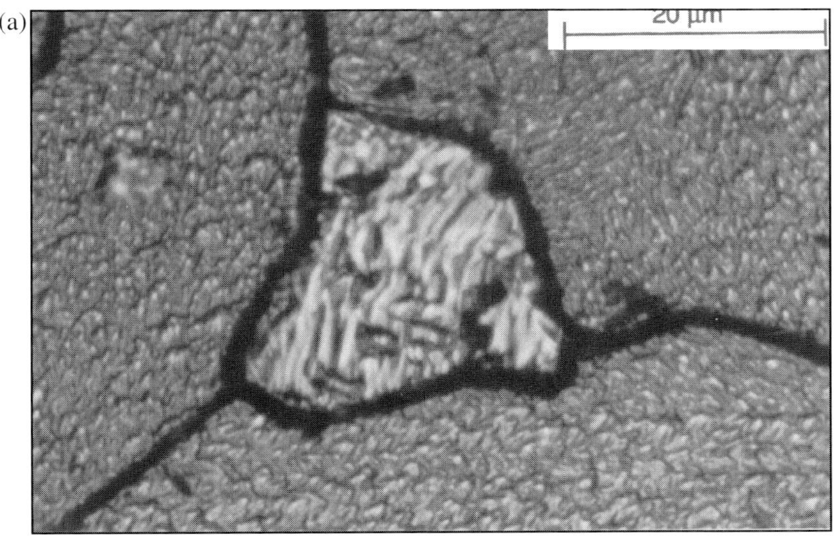

(b)
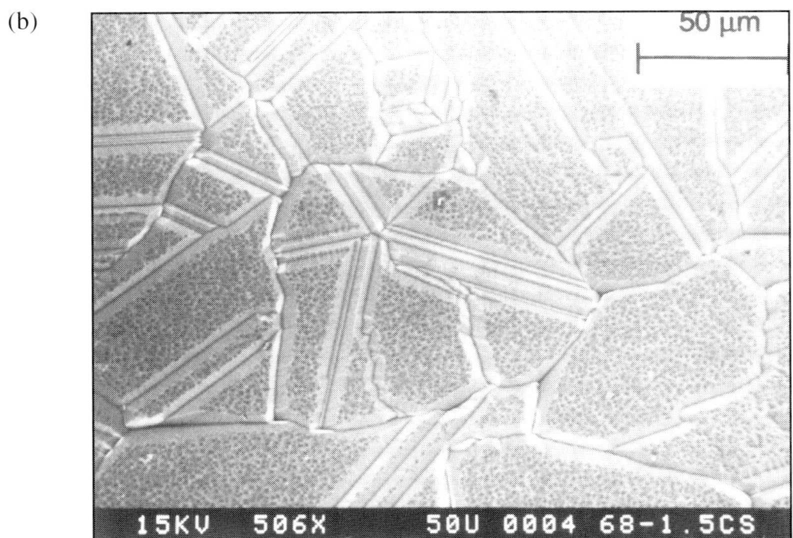

FIGURE 10. (a) Pictured is an optical photomicrograph of a polished polycrystalline diamond surface etched using atomic hydrogen. The boundaries between crystals are preferentially etched and appear as interconnected dark irregular lines. The small grain at the center is bright because it developed low angle ledges rather than becoming rough and microfaceted. (b) This is a scanning electron photomicrograph of a similar surface on which additional diamond was grown. Both irregular grain boundaries and twin planes are clearly revealed as these become faceted. The highest growth rate occurs along the <110> direction.

The first argues, particularly in the case of single crystal silicon, that the mechanical and crystallographic damage done by the polishing enhances nucleation by creating high surface free energy damage sites on the surface. A variation on this latter idea is the suggestion that polishing, just prior to deposition, removes surface contaminants and/or strongly adherent films which if not removed may inhibit nucleation. The presence of scratches or other non-uniformities in an otherwise highly planar and very uniform surface has been long known to greatly enhance the nucleation of many phases. The second of these maintains simply that polishing with diamond leaves diamond particles, and/or an ill defined carbonaceous residue in and on the surface. The subsequent "nucleation" observed is simply the result of diamond growing on diamond or nucleating on a disordered carbonaceous layer. The nucleation of diamond on graphite has been studied (75) and it was found, in keeping with much graphite chemistry, that nucleation occurs dominantly on the prism plane surfaces, with little or no nucleation on the basal plane surface. In addition diamond readily nucleates on nanocrystalline or surface damaged graphite, and the greater the level of surface damage, the higher the density of nucleation (75). The nucleation of diamond at the prism plane edges of graphite can readily be modeled through a simple exten-

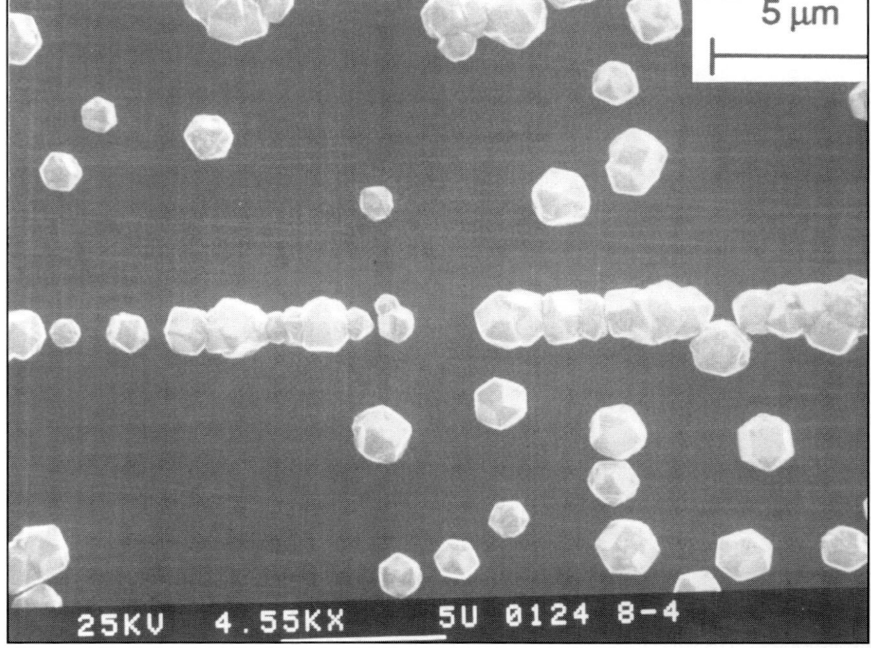

FIGURE 11. Photomicrograph showing diamond nucleated preferentially along scratches in the surface of diamond polished silicon.

sion of its known hydrogenation chemistry (76). Figure 12 shows a small fragment of the graphitic carbon plane with edge termination by hydrogen. Hydrogenation is believed to occur through the addition of hydrogen at the edges with the formation of sp³ hydrogenated carbon atoms, followed by the loss of a methyl radical leading to the formation of methane (77). If instead of rupturing a carbon-carbon bond, methyl radicals are added to the graphitic plane, for example at a {11$\bar{2}$1} plane, a 5,6-dimethyl-1,3-cyclohexadiene unit would be formed as shown in the figure. The equilibrium conformation of 1,3-cyclohexadiene is known to deviate from planarity, but with only small energy differences for relatively large angular deviations. (78). This picture for the nucleation of diamond onto a {11$\bar{2}$1} plane surface of graphite also easily leads to the formation of multiply twinned diamond crystals, a very commonly observed phenomenon in the nucleation of diamond (79,80). Hence a disordered, amorphous or nanocrystalline carbon at

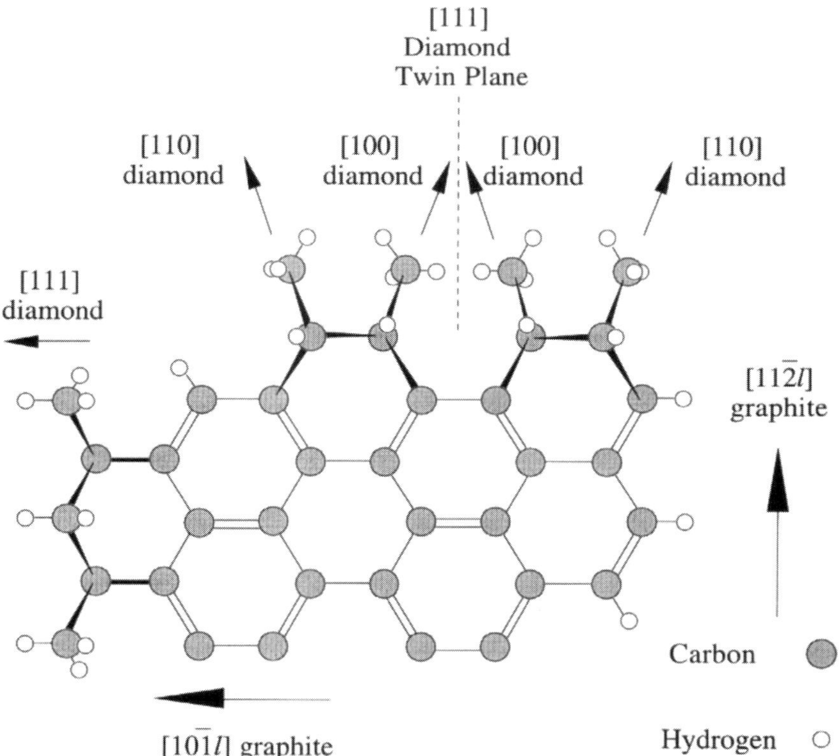

FIGURE 12. Model for diamond nucleation on graphite. Shown is small fragment of the graphitic carbon plane with sp3 carbon deposited along its edges, illustrating how the prism plane edges of graphite would provide a template for the nucleation and growth of diamond.

the substrate surface is likely to enhance the nucleation of diamond. It has been reported that the deposition of a diamond-like carbon (81), or a hydrocarbon oil coating (82), enhances the nucleation of diamond. Whether these results can be interpreted as support for such a theory, or whether their efficacy is due solely to an increased local surface concentration of carbon, is still open to question.

Polishing with other materials besides diamond does not have nearly the same efficacy as diamond. For example, in the hot wire deposition of diamond a series of silicon wafers were polished using various abrasives and then cleaned using an ultrasonic bath and a highly alkaline detergent. The abrasives tested included submicron cubic boron nitride (Boraxon™, General Electric Co., Worthington, Ohio), α- and β- silicon carbides, tantalum carbide, tungsten carbide, aluminum oxide, silicon nitride, and others. The treated samples were placed in a tantalum hot wire deposition system and attempts made to grow diamond on each. In the case of the diamond polished sample nucleation was observed within 5 minutes using previously conditioned tantalum wire. With all of the other abrasive materials, no evidence of nucleation could be observed even after 30 minutes. In an independent set of experiments at Penn State using microwave plasma assisted deposition the effects of seeding silicon surfaces with various powders was studied (74). Without seeding, or diamond polishing the surface, diamond nucleation could be observed only after 8 to 10 hours in the reactor. With the various powders tested only diamond produced evidence of growth immediately, with all other powders requiring greater periods of time in the reactor before a weight increase and evidence of growth could be obtained. The most provocative experiment, however, was to polish a 3 to 5 mm wide path across a 2.54 cm diameter silicon wafer with diamond with a small cotton swab, followed by 15 minutes in an ultrasonic bath with a strongly alkaline detergent. The wafer was then rinsed, using an additional 15 minutes in an ultrasonic bath with deionized water. Defining the path polished across the wafer as the east-west direction, a second path was polished across the wafer in the north-south direction using submicron Boraxon™ (cubic boron nitride) powder as was done with the diamond. This wafer was then placed in the same tantalum filament reactor used for the earlier polishing experiments and diamond grown on the wafer. High density nucleation was obtained only on the regions of the wafer which had been polished with diamond, and which had not been over polished with boron nitride. Moreover when the region of intersection between the two polishing steps was examined by scanning electron microscopy it could be seen that although some diamond nucleation took place in this area, it was correlated with the scratches running east and west, the diamond polishing direction, and not with the scratches running north and south. This sample is shown in Figure 11. These experiments lend strongly support the thesis that it is a residual carbonaceous byproduct of polishing with diamond, that enhances nucleation. Whichever explanation one prefers for the effect, it nevertheless remains true that many surfaces which one might wish to coat cannot be polished or damaged in any way and still retain their value, e.g. in semiconductor electronics. Consequently a current research topic in many laboratories addresses the means by which nucleation can be controlled and enhanced without this step.

An important goal of much of the research in vapor phase diamond synthesis is the heteroepitaxial growth of diamond, e.e. the growth of single crystal diamond on other materials. Obviously large area epitaxial material is important to many potential uses and although success has been claimed many of these reports have later been corrected, indicating the considerable difficulty that exists, not only in achieving heteroepitaxy, but also in thin film characterization. Much effort has centered on the attempt to grow diamond heteroepitaxially on one or more of the FCC transition metals such as Ni, Co, or Fe which have lattice parameters close to that of diamond (82). The deposition of diamond on other transition metals, notably platinum, has also been studied, and in most all experiments it is reported that the initial carbon deposits are graphite or graphitic in nature and that only after the surface of the metal is covered with carbon can diamond be nucleated and grown. It is for example the presence of substantial amounts of cobalt in the surfaces of cobalt cemented tungsten carbide, and the formation of a graphitic interlayer, which is blamed for the oft cited difficulty in growing adherent diamond layers on cemented carbide cutting tools. Hence the failure to obtain diamond heteroepitaxy on these metals might be attributed, at least in part, to an inability to stabilize sp^3 hybridized carbon, relative to sp^2 hybridized carbon, at transition metal surfaces. This implies that not only is a reasonable lattice match necessary, but also that fully saturated sp^3 hybridized carbon structures, as obvious precursor structures to diamond, should also be stable at the interface under the conditions necessary for growth.

Diamond heteroepitaxy has, very likely, been achieved on high pressure synthesized cubic boron nitride (84-86). In this case not only is the substrate of the same structure as diamond (ignoring differentiation between the positions of nitrogen and boron in the lattice) with nearly the same lattice constant (3.62 Å for c-BN vs. 3.57 Å for diamond), but also possesses a compatible chemistry. Illustrative examples include the formation of the carborane compounds such as 1,5-dicarbaclovopentaborane where hydrogenated sp^3 hybridized carbon is bonded to boron, or alklyl substituted ammonia like species as in trimethylamine. Indeed nitroborane chemistry is often considered analogous to the chemistry of carbon (87). Hence the answer to the challenge of diamond heteroepitaxy may well lie in the design of a suitable structure which simultaneously provides a lattice match and permits the stabilization of sp^3 hybridized carbon by atomic hydrogen. Such an interlayer might obviously be cubic boron nitride, however the vapor phase growth of well crystallized cubic boron nitride is itself problematic with only the preparation of poorly crystallized, nanocrystalline material having been achieved to date.

TWINNING AND DEFECT FORMATION

A major difficulty, noted earlier, that limits the much wider application of diamond technology is that, although very high quality crystals have been grown, none are known that are entirely free of twin defects. Diamond, like many other cubic crystals, readily twins along its (111) planes. The formation of twins during growth interrupts the translational symmetry of the crystal lattice along all directions except the normal (the direction perpendicular) to the twin plane itself. Figure 13 illustrates the geometry in two dimensions of a (111) twin in diamond. This type of defect is easily introduced as the energy or strain associated with its formation is very small and similar to the difference in conformational energies for the "boat" and "chair" forms of cyclohexane. The angle of deviation for the dominant growth axis is 70.5°. A five fold repetition of this leads to a total rotation of the <110> direction of 352.5°, or nearly 360°. An illustration of this "five-fold" twin structure is visible in Figure 10(b). It is the formation of (111) twins

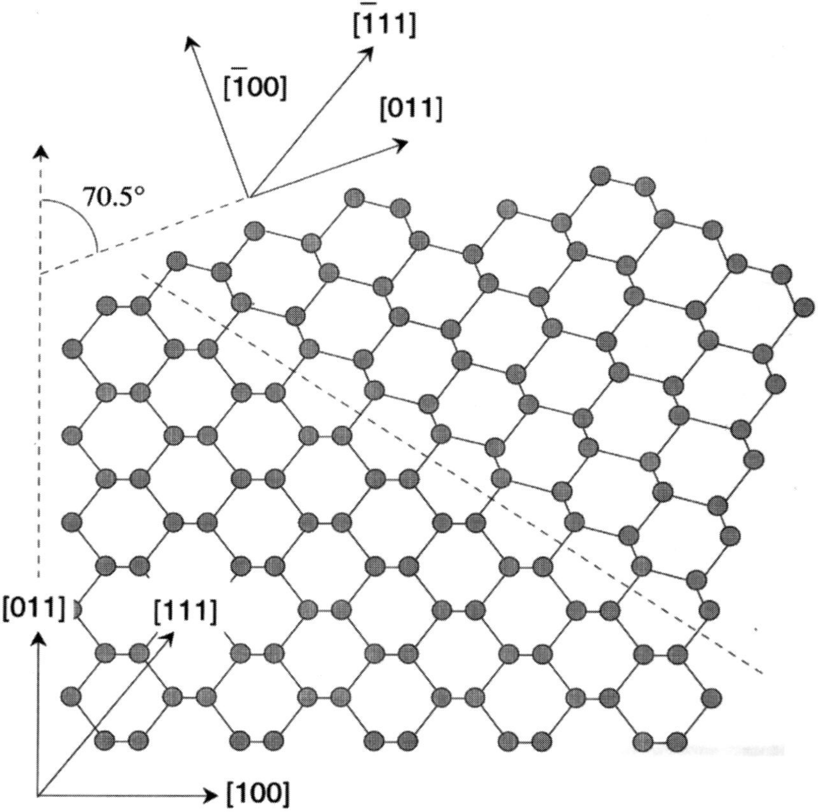

FIGURE 13. Pictured is the diamond crystal structure viewed along the [110] direction and perpendicular to a (111) twin plane (dashed line). While the lattice remains coherent across the twin plane, the other directions are rotated through 70.5 degrees.

that can lead to the formation of so-called "penetration" twins such as that shown in Figure 14.

The periodic repetition of these twins along <111> direction, alternating the rotation of the lattice at each twin would lead to the formation of one or more of many possible hexagonal polytypes just as is commonly seen with SiC. Researchers at Penn State have calculated the expected Raman active vibrational modes for some of these arrangements and have shown that some "anomalous" Raman scattering peaks might be attributed to the formation of diamond polytypes by a periodic twinning mechanism (88). This phenomenon is potentially serious for many applications, not only because of the obvious disruption of the translational symmetry of the lattice, but also because it can lead to large incommensurate regions and voids in the crystal. The formation of such a void can be viewed in two dimensions by projection of the lattice onto a (110) plane as shown in Figure 15. Although the lattice is coherent across the plane of the twin, it will be incoherent at other locations, leaving voids which in turn may encapsulate impurities or act as strength limiting flaws. In addition it should be noted that the "five-fold" twin does not permit the lattice to close upon itself. Rather there is of necessity a ~7.5° "gap" generated. This is commonly relieved either through the formation multiple twins with some strain, as can be seen in Figure 10(b) or by encapusulation of impurities and graphitic carbon.

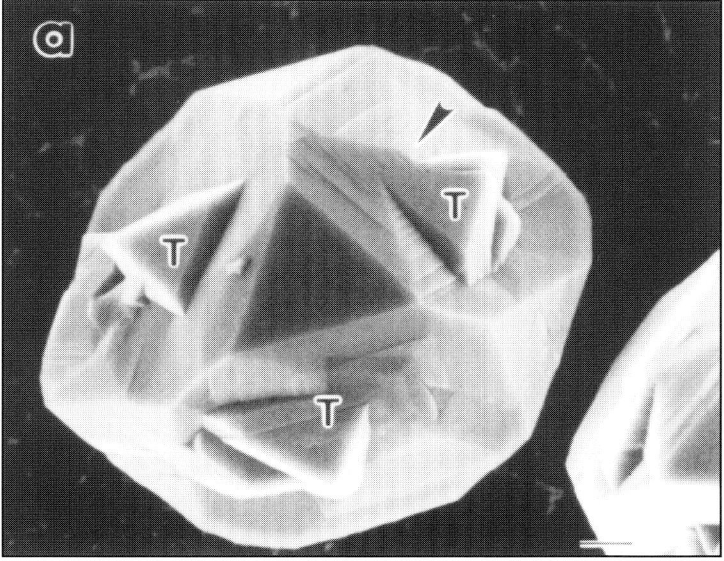

FIGURE 14. Shown is a vapor phase grown diamond crystal exhibiting "penetration twins" marked with the letter T. These result from twinning during growth along a <111> direction and are most easily seen on the (100) faces (arrow) of cubo-octahedral crystals. Scale bar in the photomicrograph is 5 micrometers.

FUTURE DIRECTIONS

Clearly much work remains to more fully understand and exploit the synthesis of diamond by chemical vapor deposition. Among the issues likely to continue to be of active interest into the next century are:

1. Is there carbon diffusion on the diamond surface during the growth process? In recent years some have suggested that some experimental results are best explained if carbon can diffuse from site to site on the diamond surface at rates that may be appreciable. If this is true then many ideas concerning the dominant growth mechanisms(s) will require modification.

2. Are liquid phase growth methods for well crystallized diamond also possible? Many suspect this may be possible and research is being pursued at Penn State to explore whether or not it may be possible to produce diamond using hydrothermal, electrochemical or other means of transporting carbon with the net growth of diamond.

3. Is atomic hydrogen the only species that leads to the preferential growth of diamond instead of graphite? Some reports suggest that the halogens, oxygen, and other elements are capable of performing roles similar to those attributed to atomic hydrogen.

4. How does surface chemistry of diamond growth in the presence of potential electronic dopants (e.g., nitrogen, boron, phosphorous, etc.)

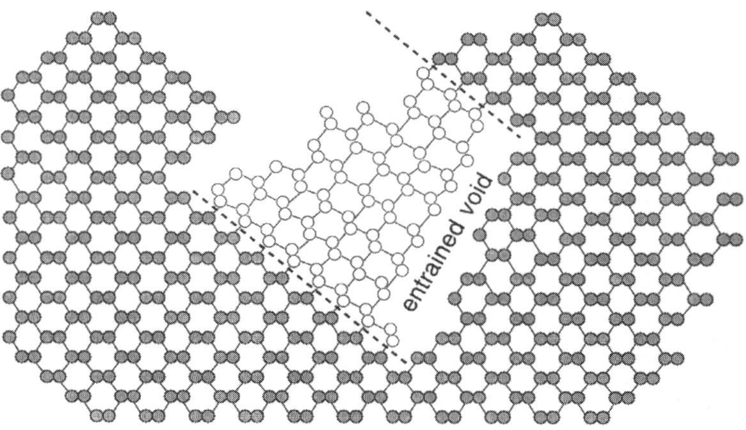

FIGURE 15. The formation of (111) twins during growth can lead to intracrystalline voids. This produces sites in which impurities may be trapped and which can act as strength limiting flaws and lead to high local stresses. The unshaded atoms belong to one twin while the shaded atoms belong to the parent crystal.

proceed? Very important technologically is the issue of how to best incorporate dopant elements in order to achieve the highest possible carrier concentration without defect compensation.

5. Can the lessons learned in the growth of diamond be carried over to the growth of other important metastable crystalline phases? A major goal of many programs worldwide is the growth of cubic boron nitride, the sphalerite structure analog to diamond, that is also a candidate high temperature semiconductor material. With further knowledge and understanding of the relevant surface chemistries, superlattice structures may be possible employing this material along with diamond.

6. Can the generation of voids and defects be avoided in the growth of thick free standing layers? Diamond offers very significant promise as an advanced engineering material provided fully dense layers can be produced economically.

7. Can enzymes, proteins, conducting or passivating polymer and oligomeric materials be bonded to the diamond surface to produce novel electro-chemical and biophysical or biomedical sensor materials? Diamond is one of the few truly bio-neutral materials with many possible applications in medicine. Progress in exploiting the unusual properties of this material in medicine will also likely depend upon a clear understanding of the relevant surface chemistry.

The answers to these and many similar issues are expected to produce many new products and processes as well as advance the science of materials. It is hoped that the many advances expected in the coming new century with novel processes and materials like vapor phase deposited diamond will help enrich and improve our lives.

REFERENCES

1. *Synthetic Diamond: Emerging CVD Science and Technology.* 1994. (Karl E. Spear and John P. Dismukes, eds.) John Wiley and Sons, Inc., New York, March, 654 pages.
2. K.E. Spear. 1989. "Diamond, Ceramic Coating of the Future," *J. Am. Cer. Soc., 72*(2), 171-191.
3. W.A. Yarbrough. 1992. "Vapor Deposited Diamond - Problems and Potential," *J.Am.Cer.Soc., 75* (12), 3179-3199.
4. W.A. Yarbrough. 1991. "Current Research Problems and Opportunities in the Vapor Phase Synthesis of Diamond and Cubic Boron Nitride," *J.Vac.Sci. and Technol., A9* (3), 1145-1152.
5. W.A. Yarbrough and R. Messier. 1990. "Critical Issues and Problems in the Chemical Vapor Deposition of Diamond," *Science, 247,* 688-696.
6. K.E. Spear. 1987. *Earth and Miner. Sci.,* The Pennsylvania State University, 46 (4), 53-59.

7. A.R. Badzian, B. Simonton, T. Badzian, K.E. Spear, R. Messier and R. Roy. 1986. "Vapor Deposition Synthesis of Diamond Films," *SPIE Proceedings Vol. 683: Infrared and Optical Transmitting Materials* (R.W. Schwartz, ed.) SPIE - The International Society of Optical Engineering, Bellingham, WA, pp. 127-138.

8. C. Frondel and U.B. Marvin. 1967. *Nature, 214,* 587.

9. F.P. Bundy and J.S. Kasper. 1967. *J. Chem. Phys., 46*(9), 3437.

10. G. Vdovykin. 1973. *Space Science Reviews, 14,* 758.

11. A. El. Goresy and G. Donnay. 1968. *Science, 161,* 363.

12. G.M. Jenkins, K. Kawanmura and L. Ban. 1972. *Proc. Roy. Soc. A, 327,* 501.

13. H.W. Kroto, J.R. Heath, S.C. O'Brien, R.F. Curl and R.E. Smalley. 1985. *Nature 318,* 162.

14. F.P. Bundy, H.T. Hall, H.M. Strong and R.H. Wentorf. 1955. *Nature, 176,* 51.

15. W.G. Eversole. 1962. "Synthesis of Diamond," U.S. Patents 3, 030, 187 & 3, 030, 188 (17 Apr.).

16. H.J. Hibshman. 1968. "Diamond Growth Process," U.S. Patent 3, 371, 996 (5 Mar.).

17. John C. Angus. 1971. "Manufacture of Synthetic Diamonds," U.S. Patent 3, 630, 677 (28 Dec.).

18. John C. Angus and N.C. Gardner. 1972. "Process for the Catalytic Growth of Metastable Crystals from the Vapor Phase," U.S. Patent 3, 661, 526 (9 May).

19. E.C. Vickery. 1973. "Process for Epitaxial Growth of Diamonds," U.S. Patent 3, 714, 334 (Jan. 30).

20. S.P. Chauhan, Angus, J.C. and N.C. Gardner. 1976. *J. Appl. Phys. 47* (11), 4746.

21. B.V. Derjaguin and D.V. Fedoseev. 1989. *The Growth of Diamond and Graphite from the Gas Phase,* Nauka, Moscow, 1977. English language translation: *Surface and Coatings Technology,* B.D. Sartwell and A. Matthews, eds., *38* (1,2), 131.

22. B.V. Spitsyn, Boulilov, L.L. and B.V. Derjaguin. 1981. *J. Cryst. Growth, 52,* 219.

23. S. Matsumoto, Y. Sato, M. Kamo and N. Setaka. 1982. *Jpn, J. Appl. Phys., 21,* L183.

24. M. Kamo, Sato, Y., Matsumoto, S. and N. Setaka. 1983. *J. Crys. Growth, 62,* 642-644.

25. S. Matsumoto. 1985. *J. Mat. Sci. Letts., 4,* 600-602.

26. S. Aisenberg and R. Chabot. 1971. *J.Appl. Phys., 42,* 2953.

27. S. Aisenberd. 1976. "Film Deposition," U.S. Patent 3, 961, 103 (1 June).

28. L. Holland and S.M. Ojha. 1976. *Thin Solid Films, 38,* L17.

29. *Properties and Characterization of Amorphous Carbon Films.* 1990. John J. Pouch and Samuel A. Alterovitz, eds., Materials Science Forum, Vols. 52 & 53, Trans Tech Publications, Brookfield, VT.

30. J.C. Angus, P. Koidl and S. Domitz. 1986. *Plasma Deposited Thin Films,* J. Mort and F. Jansen, eds. (CRC Press, Cleveland, OH), pp. 89-127.

31. J. Robertson. 1986. *Advances in Physics. 35* (4), 317.

32. H.-C. Tsai and D.B. Bogy. 1987. *J. Vac. Sci. Tech., A5* (6), 3287.

33. J.L. Robertson, S.C. Moss, Y. Lifshitz, S.R. Kasi, J.W. Rabalais, G.D. Lempertm and E. Rapoport. 1989. *Science, 243,* 1047-1050.

34. S.C. Moss letter. 1989. *Science, 244,* 1029.

35. S. Matsumoto. 1985. *Proceedings Seventh International Symposium on Plasma Chemistry,* Vol.1, C.J. Timmermans, ed., Einhoven, July, 79-84.

36. N. Fujimori, T. Imai and A. Doi. 1986. *Vacuum, 36* (1-3).

37. B. Singh, Y. Arie, A.W. Levine and O.R. Mesker. 1988. *Appl. Phys. Lett., 52* (6), 451-452.

38. T. Kawato and K. Kondo. 1987. *Jpn. J. Appl. Physics, 26* (9), 1429.

39. A. Feldman, E.N. Farabaugh and Y.N. Sun. 1987. *Laser Induced Damage in Optical Materials,* H.E. Bennett et. al., eds. (NIST Spec. Publ. 75 6, U.S. Government Printing Office, Washington, DC, 1988), pp. 246-255.

40. Y. Hirose and N. Kondo. 1988. Paper No. 29a-T-1, Japan Applied Physics Society Meeting, March 29, Tokyo, Japan.

41. L.M. Hanssen, W.A. Carrington, J.E. Butler and K.A. Snail. 1988. *Mat. Lett. 7,* 289.

42. W.A. Yarbrough, M.A. Stewart and J.A. Cooper, Jr. 1989. *Surface and Coatings Technology, 39/40,* 241.

43. N. Fujimori, A. Ikegaya, T. Imai, K. Fukushima and N. Ohta. 1989. In *Proceedings of the First International Symposium on Diamond and Diamond-Like Films,* J.P. Dismukes et. al., eds. (The Electrochemical Society, Pennington, NJ.), pp. 465-474.

44. K.E. Spear and Michael Frenklach. 1994. "Mechanisms for CVD Diamond Growth," Chapter 8 in *Synthetic Diamond: Emerging CVD Science and Technology,* K.E. Spear and J.P. Dismukes, eds., John Wiley and Sons.

45. B. Cline, W. Howard, H. Wang, K.E. Spear and M. Frenklach. 1992. "Cyclic Deposition of Diamond: Experimental Testing of Model Predictions," *J. Appl. Physics, 72* (12), 5926-5940.

46. A. VonEngel. 1983. *Electric Plasmas: Their Nature and Uses,* Taylor and Francis Ltd., New York and London, 162.

47. N. Ohtake, H. Tokura, Y, Kuriyama, Y. Mashimo and M. Yoshikawa. 1989. In *Proceedings of the First International Symposium on Diamond and Diamond-Like Films,* J.P. Dismukes et. al., eds. (The Electrochemical Society, Pennington, NJ.), pp. 93-105.

48. M. Sokolowski, A, Sokolowska, B. Gokieli, A. Michalski, A. Rusek and Z. Romanowski. 1979. *J. Cryst. Growth, 47,* 421-426.

49. H. Vora and T.J. Moravec. 1981. *J. Appl. Phys., 52* (10), 6151.

50. Y. Namba and T. Mori. 1985. *J. Vac. Sci, Technol., A3* (2), 319.

51. T. Mori and Y. Namba. 1983. *J. Vac. Sci. Technol., A1* (1), 23.

52. M. Kitabatake and K. Wasa. 1985. *J. Appl. Phys., 58* (4), 1693.

53. Walter A. Yarbrough, and Rustum Roy. 1988. *Diamond and Diamond-Like Materials Synthesis,* G.H. Johnson, A.R. Badzian and M.W. Geis, eds.

(Materials Research Society, Pittsburgh, PA, pp. 33-38.

54. Ch. Wild, N. Herres, J. Wagner, P. Koidl and T.R. Anthony. 1989. In *Proceedings of the First International Symposium on Diamond and Diamond-Like Films,* J.P. Dismukes et al., eds. (The Electrochemical Society, Pennington, NJ.), pp. 283-295.

55. M. Kamada, S. Arai, A. Sawabe, T. Murakami and T. Inuzuka. 1988. Paper 1-10, *First International Conference on the New Diamond Science and Technology, Program and Abstracts,* JNDF, Tokyo, October 24-26, 1988, pp. 96-97.

56. K. Kobashi, K. Nishimura, Y. Kawate and T. Horuchi. 1988. *Phys. Rev. B. 38* (6), 4067.

57. Y. Sato. 1988. *New Diamond 1988,* (Japan New Diamond Forum, Tokyo, Japan) pp. 30-35.

58. S.B. Rice and M.M.J. Treacy. 1988. *Mat. Res. Soc. Symp. Proc., 115,* 15.

59. B.E. Jacobson. 1982. Ch. 7 in *Deposition Technologies for Films and Coatings,* R.F. Bunshah, et al., eds., Noyes Publications, Park Ridge, N.J., pp. 289-333.

60. R.W. Gretz. 1966. *Vapor Deposition,* C.F. Powell, J.H. Oxley, and J.M. Blocher, eds. (John Wiley, New York), pp. 149-189.

61. S.A. Solin and A.K. Ramdas. 1970. *Phys. Rev. B.,* 1(4), 1687-1698.

62. D.S. Knight and W. White. 1990. *J. Mater. Res.,* 4(2), 385.

63. S.J. Harris.1991 *New Diamond Science and Technology,* Proc. of the 2nd Intl. Conf. on the New Diamond Science and Technology. R. Messier and J. Glass, eds. (Materials Research Society, Pittsburg, PA).

64. K. Kobashi, K. Nishimura, K. Miyata, K. Kumagai and Y. Kawate. 1990. In *Diamond and Diamond-Like Films,* edited by J.P. Dismukes et al. (Electrochemical Society Proc., 89-12, Pennington, N.J., 1989) pp. 296-305. Also J. Mater. Res., 5 (11), 2469.

65. E.D. Sprecht, R.E. Clausing and L. Heatherly. 1990. *J. Mater. Res., 5* (11), 2351.

66. H.E. Buckley. 1951 *Crystal Growth,* John Wiley and Sons, New York, pp. 262-269.

67. G.D. Barber, W.A. Yarbrough and Kevin Gray. 1995. "Microstructural Evolution and Flaws in CVD Diamond," in *Applications of Diamond films and Related Materials: Third International Conference,* NIST special pub. 885, A. Feldman, Tzeng, Y., Yarbrough, W.A., Yoshilawa, M. and M. Murakawa, eds., (U.S. Dept. of Commerce, Washington, DC, 1995), pp. 551-556.

68. C. Murphy. 1996. B.S. Thesis in Ceramic Science and Engineering, Penn State University.

69. B.E. Williams, J.T. Glass, R.F. Davis, K. Kobashi and Y. Kawate, 1988. In *Diamond and Diamond-Like Materials Synthesis:* Extended Abstracts, (Materials Research Society, Pittsburgh, PA, pp. 99-102.

70. R.W. Collins, Y. Cong, Y.-T. Kim, K. Vedam, Y. Liou, A. Inspektor and

R. Messier. 1989. *Thin Solid Films, 181*, 565.

71. A.R. Badzian and T. Badzian. 1988. *Surface Coatings Tech.,* 36, 283.

72. H. Kawarada, et al., 1988. In *Diamond and Diamond-Like Materials Synthesis: Extended Abstracts,* (Materials Research Society, Pittsburg, PA), pp. 89-92.

73. T.M. Hartnett, M.S. Thesis. 1988. The Pennsylvania State University.

74. P.K. Bachmann, W. Drawl, D, Knight, R. Weimer and R.F. Messier. 1988. *Diamond and Diamond-Like Materials Synthesis: Extended Abstracts* (Materials Research Society, Pittsburgh, PA) pp. 99-102.

75. J.J. Dubray, C.G. Pantano and W.A. Yarbrough. 1990. In *Diamond and Diamond-Like Films and Coatings,* NATO Advanced Study Institute, II Ciocco, Castelvecchio Pascoli, Italy, July 22 - August 3.

76. D.W. McKee. 1981. "The Catalyzed Gasification Reactions of Carbon," in *Chemistry and Physics of Carbon,* edited by P.A. Walker and P.A. Thrower, (Marcel Dekker, New York, N.Y.), pp. 1-118.

77. A. Tomita and Y. Tamai. 1974. *J. Phys. Chem., 78,* 2254.

78. A.I. Kitaigorodsky. 1973. *Molecular Crystals and Molecules,* Academic Press, New York, pp. 402-404.

79. G-H. M. Ma, Y.H. Lee and J.T. Glass. 1990. *J. Mater. Res., 5* (11), 2367.

80. J. Narayan. 1990. *J. Mater. Res., 5* (11), 2414.

81. K.V. Ravi, M. Peters, L. Plano, S. Yokata and M. Pinneo. 1988 *First International Conference on the New Diamond Science and Technology, Program and Abstracts,* 24 to 26 October 1988. (JNDF, Tokyo), pp. 32-33.

82. A.A. Morrish. 1992. Personal communication, Naval Research Laboratory, Washington, D.C.

83. R.F. Davis, J.T. Glass, G. Lucovsky and K.J. Bachmann. 1987. *Growth, Characterization and Device Development in Monocrystalline Diamond Films,* Annual Report, ONR Contract No. N00014-86-K-0666, North Carolina State University, Raleigh, N.C.

84. T. Murakami, S. Koizumi, K. Suzuki and T. Inuzuka. 1989. "Epitaxial Growth of Diamond on High Pressure Synthesized c-BN Particles, "Paper No. 27p-E-8. Fall Meeting of the Japan Society of Applied Physics.

85. W.A. Yarbrough, A. Kumar and Rustum Roy. 1987. "Effect of Substrate Preparation in the Vapor Phase Synthesis of Diamond," Paper No. N3.8, Fall Meeting, Materials Research Society, 30 Nov-5 Dec., Boston, MA.

86. R. Haubner. 1990. *Refractory Metals and Hard Materials,* 9(2), 70.

87. F.A. Cotton and G. Wilkinson. 1966. *Advanced Inorganic* Chemistry, 2nd ed. (Interscience, New York, N.Y.) pp. 287-289.

88. K.E. Spear, A.W. Phelps and W.B. White. 1990. "Diamond Polytypes and their Vibrational Spectra," *J. Mater. Res.,* 5(11), 2277-2285.

Chapter Seventeen

THE SURFACE AND STRUCTURE OF GLASS

CARLO G. PANTANO*

Department of Materials Science and Engineering
Steidle Building
Pennsylvania State University
University Park, PA 16802

INTRODUCTION

The surface of glass is fundamental to the fabrication of glass products, to their appearance, and to their performance. The surface tension of molten glass is responsible for its formability and smooth surface. The surface of glass is the point of entry for water vapor, acids and other corrosive environments. And perhaps of most practical importance, the surface is the primary site for defects that reduce the mechanical strength of glass. At the same time, the surface provides the means to modify and improve the appearance and properties of the glass through staining, ion exchange and dealkalization. The surface of glass is also an excellent substrate for the deposition of optical thin films and decorative coatings.

The properties of glass surfaces are intimately dependent upon the glass composition and structure. The primary objective of this presentation is to explain and demonstrate these relationships. The ability to measure and/or model the composition and molecular structure of the glass surface provides great insight about these effects. This is especially true in the case of the alkali oxides (Li_2O, Na_2O and K_2O) whose ionic radius and distribution in the glass structure can be used to explain many composition effects. The chemical structure and properties of soda-lime-silicate glasses will be addressed, specifically.

---

*Professor Carlo G. Pantano was named the 1994 Dominick Labino Lecturer by the Glass Art Society. This chapter is based upon the presentation he made at their annual meeting: Pacific Crosscurrents, Oakland, CA, March 1994.

In many instances, the ability to tailor surface properties by changing the bulk glass composition is limited by melting and forming requirements, or cost. Thus, the modification of surface properties through surface treatment is of great practical and commercial significance. Some of the surface treatments used for the control of chemical, mechanical and optical properties of glass, as well as for color effects, will also be described. Likewise, some of the surface-sensitive analytical techniques developed over the last twenty years will be employed to directly demonstrate these effects.

CHEMICAL STRUCTURE AND PROPERTIES OF GLASS

The technical basis for understanding the properties and behavior of oxide glasses - especially at the surface - is structure. Here, the term structure refers to the atomic arrangement. But in contrast to most solids where the atoms are arranged in an orderly *crystalline* array, the atoms in glass are more *randomly* distributed. The nature of this structure depends, primarily, upon the composition of the glass.

The development of computer modeling has greatly extended our ability to simulate the processes of melting, homogenization and quenching-in of the glass structure. The basis for these computer simulations is rather simple. An assemblage of atoms - with the desired composition - is created in the memory of the computer, and then the melting process is simulated by setting the atoms into motion. There is a direct relationship between temperature and the average velocity of the atoms. In most cases the temperature is chosen to be sufficiently high that complete mixing and homogenization of the various atoms occurs. The desired output from this simulation is the relative position of the various atoms in the system - that is, the atomic arrangement - at any desired time or temperature in the process. This can be followed throughout the simulation by the solution of Newton's laws of motion using mathematical functions that describe the forces of attraction and repulsion amongst the various atoms. Upon cooling the system (by step-wise lowering of the average velocity of the atoms), the final equilibrium configuration of the glass is quenched-in. This configuration can be output in a variety of ways.

Figures 1 and 2 are examples of these computer simulations. Figure 1 shows the bulk structure obtained after melting a very pure silica (SiO_2) glass. This is not so easy in practice, but relatively straightforward by computer simulation. Figure 2 shows the surface structure for a pure silica glass and a sodium-silicate ($Na_2O \cdot 3SiO_2$) glass; this output shows a projection of atoms at the surface only. These two structure models reveal the most important chemical features of oxide glasses: the *network* and the *non-bridging oxygen site.*

The *network* of the glass is due, primarily, to the SiO_2 content. Thus, it is not surprising that Figure 1 is a pure network glass. Every atom is bonded into the network through primary chemical bonds. The silicon atoms are bonded to 4 oxygen atoms (to form silica tetrahedra); the oxygen atoms are all bonded to two silicon atoms.

The presence of alkali (e.g., Li_2O, Na_2O, K_2O), alkaline - earth (e.g., CaO, MgO, BaO), and other metal oxides (e.g., PbO, ZnO) in a silicate melt drastically alters the network structure represented in Figure 1. Primarily, these metal oxides create breaks in the 3D network structure. These breaks are termed *non-bridging*

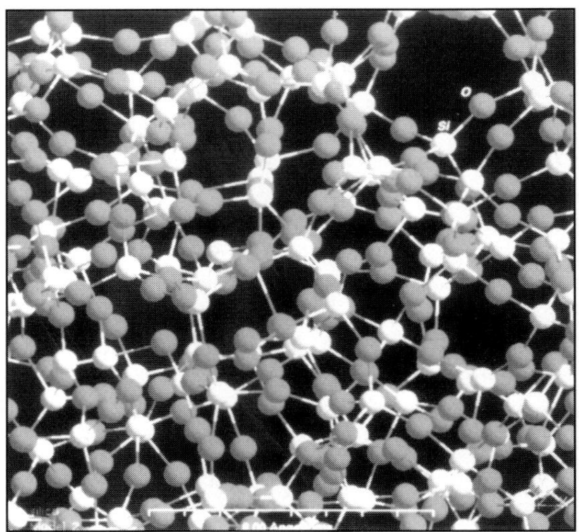

FIGURE 1. Molecular dynamics computer-simulation of the *bulk network structure* of pure silica (SiO_2) glass; provided by Y. CaO and A.N. Cormack, Alfred University.

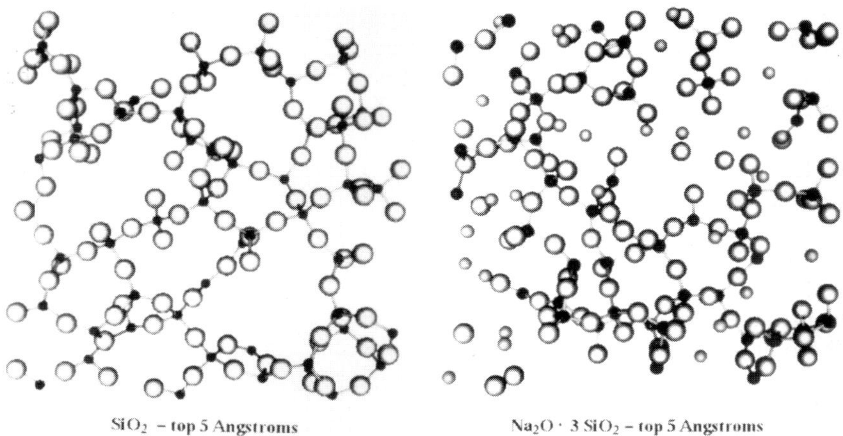

FIGURE 2. Molecular dynamics computer-simulation of the surface atoms (2-dimensional plane view) for: (a) a pure silica (SiO_2) glass, and (b) a sodium-silicate ($Na_2O \cdot 3SiO_2$) glass; provided by S.H. Garofalini, Rutgers University.

oxygens. Whereas all the oxygen atoms in the pure SiO_2 glass bridge between two silicon atoms (see Figure 2a), there are some oxygens in the modified glass which are bonded to only one silicon atom (see Figure 2b). The modifier atoms (e.g., Li, Na, K, Ca, Mg, Ba, Pb, etc.,) are always located in the vicinity of a non-bridging oxygen. These atoms do not form primary bonds in the network, and in general, they are only weakly bonded into the structure of the glass. Nevertheless the network must conform to the size of the modifier atom (e.g., K^+ is twice the diameter of Li^+, Pb^{++} is 20% larger in diameter than Ca^{++}).

The computer simulations of the surface suggest that in the case of pure silica, the outermost atomic layer is not a random mixture of Si and O atoms, but rather, is dominated by oxygen atoms. This is probably a reflection of the chemical stability of the silica tetrahedra (wherein oxygen atoms shield the silicon). In the case of the sodium-silicate, the simulations suggest that the outermost atomic layer is dominated by the sodium atoms. These simulations were experimentally verified using a method termed low-energy ion scattering spectroscopy (ISS). In this experiment, very low energy helium ions were scattered off the clean glass surface in a vacuum. The amount of energy loss that occurs during the scattering event reveals the identity of the scattering atoms. The lower spectra in Figure 3 shows the predominance of Na (scattering sites) on the clean glass surface. Similar experiments for $K_2O \cdot 3SiO_3$ revealed that K atoms also shield the glass surface. In the case of pure silica, the glass surface appeared to be an oxygen monolayer. The high concentration of alkali in the outermost monolayer accounts for the hygroscopic/reactive nature of alkali-containing glasses.

The molecular structure of the glass directly influences all properties of the glass. Consider, first, the viscosity and thermal expansion behavior. The connectivity of the network is a primary factor in determining the glass viscosity. The pure silica glass, in Figure 1 and Figure 2a has an exceedingly high viscosity because every atom must break primary chemical bonds during viscous flow. On the other hand, the modified glass in Figure 2b has a less connected network structure due to the presence of Na_2O and its associated non-bridging oxygens. Thus, viscous flow - which requires the relative motion of the atoms - can occur without breaking as many primary (bridging oxygen) network bonds. Obviously, with higher concentrations of modifier oxides, more non-bridging oxygens will be created, and thereby the viscosity will be further reduced. One non-bridging oxygen is produced per one metal oxide molecule added; e.g., one molecule of Na_2O produces one break in the network, one molecule of Li_2O produces one break in the network, and one molecule of CaO produces one break. In this sense, all the modifier oxides are equivalent in their ability to lower viscosity; that is, one non-bridging oxygen (break) per molecule.*

However, there is a secondary effect of the size and ionic charge of the metal cation, itself. The potassium ion (K^+) is twice the size of the lithium ion (Li^+). Thus, the K ion requires more space in the structure, and during viscous flow, the relative motion of the network is more difficult near this large ion relative to a small ion such as lithium; thus, one molecule of Li_2O lowers the viscosity more than one molecule of K_2O. In this sense, it is easy to understand that water (H_2O) - which is chemically analogous to Li_2O, Na_2O and K_2O - will greatly lower the viscosity of molten glass. One molecule of H_2O produces one break in the structure, but the H ion is infinitely small in size! Thus, it does not hinder relative motion of the network at all. Likewise, glassforming additives such as CaO,MgO, PbO also produce one break in the structure. But these metal cations (Ca^{+2}, Mg^{+2} or Pb^{+2}) have twice the positive charge of Li^+, Na^+, K^+ and H^+, so relative motion of the negatively charged ends of the broken network are more restricted. Therefore, MgO, CaO and PbO do not lower viscosity (per molecule) to the same extent (per molecule) as Li_2O, NaO, K_2O and H_2O. Consider, finally, fluorine (F_2). It is an especially potent additive for lowering the melt viscosity. This is because one molecule of F_2 creates one break in structure, but does not require any charge-compensating cation (charged metal atom); thus, the fluorine atom *caps* the broken-ends of the network.

The effect of glassforming additives upon the coefficient of thermal expansion (CTE) can also be explained using these simple structure models. The thermal

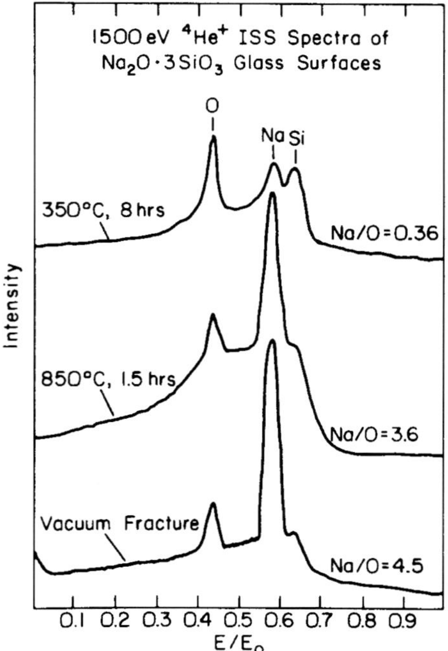

FIGURE 3. Ion-scattering spectra (ISS) that show changes in the monolayer surface composition of a glass due to heating in vacuum.

*Note: in weight percentages, the more common parameter, the *numbers* are different; for example Li_2O produces 70% more non-bridging oxygens than Na_2O per unit weight! This is the reason 1 weight percent Li_2O has a bigger effect on properties than 1 weight percent Na_2O.

expansion of glass is due to an increase in the vibrational amplitude of the atoms with increasing temperature; i.e., the more exaggerated the atomic vibration, the more space or volume it requires. The CTE of pure silica glass is one of the lowest for any material because all of the atoms are bonded into a network, and therefore, large vibrational displacement of the atoms is restricted. The addition of almost any glass forming oxides to silica (except TiO_2) increases the CTE because the associated breaks in the network (that is, non-bridging oxygens) permit greater vibrational amplitudes. If these breaks are associated with large metal atoms (e.g., K), the increase in CTE per molecule is greater than with small metal atoms (e.g., Li). This is simply due to the difference in size and ionic field-strength of these metal atoms. Likewise, the metal atoms with +2 charges (i.e.,Ca^{+2}, Mg^{+2} and Pb^{+2}) do not increase the CTE as much as those with +1 charges (i.e., Li^{+1}, Na^{+1} and K^{+1}) because of their higher in field-strength.

SURFACE TENSION

The surface tension of glass influences melting, fining, homogenization, forming and heat treatment. In melting, the surface tension determines the extent to which the molten glass wets batch particles, and thereby, influences their rate of dissolution in the melt. In fining, it determines the stability of bubbles. The sulphates, nitrates and chlorides used in most batch materials serve to lower surface tension and enhance the wetting and fining behavior of the glass during the initial stages of melting. The *relative* surface tension of the oxides affects their propensity to mix and homogenize the melt. Those oxides with a high surface tension, such as aluminum oxide, do not mix easily and often promote the formation of striae and cord. Similarly, two glasses with very different surface tensions are difficult to fusion weld for this reason. More obvious is the role of surface tension on forming, heat-treatment and flame polishing. It is this force that acts to spheroidize the glass gather or blown object. It is also the force that rounds the sharp edges of glass during heat-treatment and smooths the surface of any glass during flame polishing.

A schematic, model of a glass droplet is shown in Figure 4. The shaded objects represent the silica tetrahedra which constitute the molecular network structure of the glass. (Figure 1 shows, in more detail, that each silica tetrahedra is made of a central silicon atom bonded to 4 oxygen atoms at the corners of the tetrahedra). The arrows are used to denote the forces that bond the silica tetrahedra to one another. Note that in the *bulk* of the glass droplet, these forces average (or balance) out due to bonding in all dimensions. But at the surface, the tetrahedra have *dangling-bonds,* and so, the forces are unbalanced. In essence, atoms and molecules at the surface are attracted towards the bulk where all their bonding requirements can be satisfied. This *surface tension* drives any liquid to spheroidize*; for example, a glass gather, a water drip on a faucet, or a droplet on a surface. If the surface tension of a liquid is high, wetting is limited (it beads-up).

If inhomogeneities with a high surface tension are present in a liquid, they will try to spheroidize, and hence, will resist mixing and homogenization (e.g., Al_2O_3 cord and striae). Surface tension acts against the expansion of bubbles or the stretching of fibers and films.

Another effect of surface tension is that it tends to preferentially concentrate certain species in the surface. The net forces at the surface (in Figure 4) are essentially atoms - in this case oxygen atoms at the corners of the silica tetrahedra - looking for other atoms to bond with. Due to the 3-dimensional nature of the silica tetrahedra, it cannot orient itself in any way at the surface to satisfy the bonding demands of the oxygen atoms at the corners: one or more is always present at the free surface. It is in this regard that boron-oxide plays a unique role.

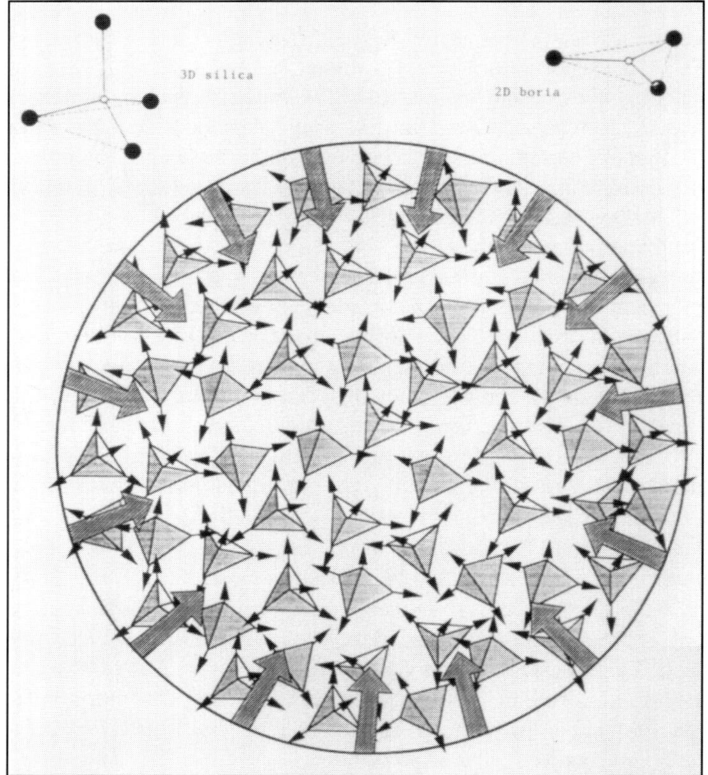

FIGURE 4. Schematic representation of a spherical glass gob (projected in 2-dimensions) which demonstrates that the origin of *surface tension* is the unbalanced atomic and molecular forces at the free surface (large arrows); the small arrows represent the atomic/molecular forces acting between the silicate tetrahedra which average to zero within the bulk of the glass.

This molecule forms 2-dimensional triangular units in the glass structure; that is, a triangle with a central boron atom and three oxygen atoms at the corners. Obviously, this 2-dimensional molecule can orient itself so that the chemical bonds lie in the 2-dimensional surface. Thus, it is not surprising that boron-oxide lowers the surface tension of silicate glass, and for this reason a small concentration of boron-oxide in the bulk can exert a big effect on surface tension. It tends to concentrate in the surface during forming and annealing because the 3-dimensional silicon-oxide tetrahedra are readily displaced by the more 2-dimensional boron-oxide triangles at the surface.

Fluorine also has a potent effect in lowering surface tension. It, too, concentrates in the surface but for a different reason than boron-oxide. Fluorine atoms substitute for oxygen atoms at the corners of the silica tetrahedra, but in contrast to oxygen atoms which demand primary bonds with two other atoms, fluorine seeks only one bond. It is an ideal termination of the network molecular structure at a free surface. Thus, it lowers the surface tension (by limiting the total number of unbalanced forces at the surface), and concentrates in the surface by preferentially displacing oxygen atoms at the surface.

Finally, the effect of metal oxides that contain very large metal atoms - specifically, K_2O, As_2O_3, PbO and BaO - should be mentioned. These large atoms (K, As, Pb and Ba) tend to lower surface tension because they can polarize - that is, they can distort their large, but deformable, electron clouds to adapt to the less than ideal bonding environment at the free surface. They, too, tend to preferentially concentrate in the surface.

This segregation effect is not limited to free surface (glass/vapor interface), but is especially evident at glass/solid interfaces. Figure 5 compares the x-ray photoelectron spectra (XPS) of two glass surfaces: one created by fracture and the other created by casting against a carbon mold. Note that the cast surface (wherein thermal segregation could occur) has a composition that is enriched in B and As relative to the bulk composition (revealed by the fracture surface). This represents one example where the processing of a glass can create a surface whose composition (and corresponding properties) differ relative to the nominal bulk values.

EVAPORATION/VOLATILIZATION

The terms *evaporation* and *volatilization* are used synonymously in glass technology, and refer to the preferential loss of species in the glass or glass surface through vaporization. The practical significance of evaporation/volatilization appears in melting and in forming. During melting, the composition of the batch

---

*In fact, surface tension also drives solids to spheroidize, but due to their infinitely high viscosity/rigidity, only atoms and molecules at or near the surface of a solid may respond to force; the net effect - if any - is *smoothing* of the solid surface.

can change continuously if one or more species are especially susceptible to this effect. The composition changes are small in magnitude - typically fractions of a percent - but the specific elements which are influenced are those whose concentration is critical to some properties. During forming, evaporation/volatilization further alters the composition of glass, but the alteration is confined to the surface region (since at the lower temperatures of forming, the evaporated surface species are not as readily replenished from the bulk of the glass as they are during the high temperature melting process.) In forming, the effects of even fractional changes in composition can influence the quality of glass product. Consider, for example, the gathering of a glass object (either in hand gathering or in a mechanized gob-feeder). The thin layer of modified glass on the surface will acquire a different refractive index. The continued gathering and forming of the article will (re)distribute this modified glass in different regions of the object where it is often visible as cord or striae. If there is color in the glass that is especially sensitive to composition and redox, these optical effects are further exaggerated. Similarly, the final surface of the glass object - if subject to evaporation/volatilization - can exhibit a unique set of properties due to compositional modification at the surface. In the case of chemical durability (which depends directly upon the alkali content of the glass surface), the effect can be beneficial because alkali oxides such as Na_2O are especially susceptible to evaporation/volatilization. Thus, the chemical

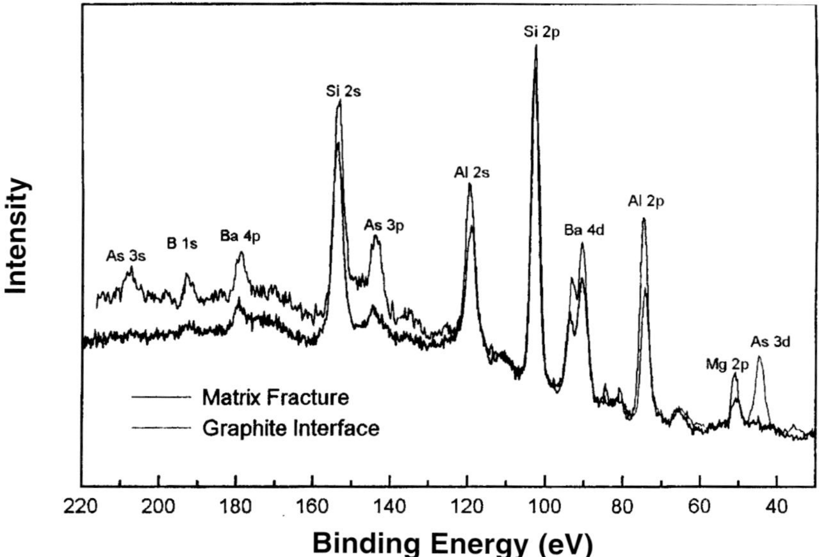

FIGURE 5. X-ray photoelectron spectra (XPS) which compare the bulk composition (lower) with the composition of a surface created by casting against a graphite mold (upper); the glass is barium aluminosilicate with minor additions of boron-oxide and arsenic-oxide.

durability of the surface is improved by the decrease in concentration of these species (as discussed in the next section). Also of practical significance is the effect of evaporation/condensation of corrosive species (alkali, boron and fluorine) upon corrosion and degradation of the furnace and surrounding refractories.

The fundamental origin of evaporation/condensation is evident upon inspection of the structure models in Figures 4 and 6. With increasing temperature, the corresponding increase in vibrational energy causes atoms and molecules at the surface to be desorbed or to vaporize through reaction with the gas phase; this is the essence of evaporation/volatilization. And clearly, the phenomena becomes more probable with increasing temperature. Those species which are weakly bonded into the molecular structure are most susceptible. Thus, it is not surprising that the evaporation/volatilization of network species such as SiO_2 and Al_2O_3 is not significant in the range of melting temperatures. But the alkali species, which are bonded only through the non-bridging oxygens, are more readily desorbed. The volatility of alkali species increases in the order Li_2O, Na_2O and K_2O; Na_2O is about twice as volatile as Li_2O, and two or three times less volatile than K_2O. Similarly, fluorine - which is bonded through only one corner of the silica tetrahedra - is quite volatile. It is also important to point out that the relative volatility of the glass constituents also depends upon their concentration in the surface. Recall that some species - B_2O_3,F and large atoms such as K and Pb - tend to segregate into the glass surface because they lower the surface

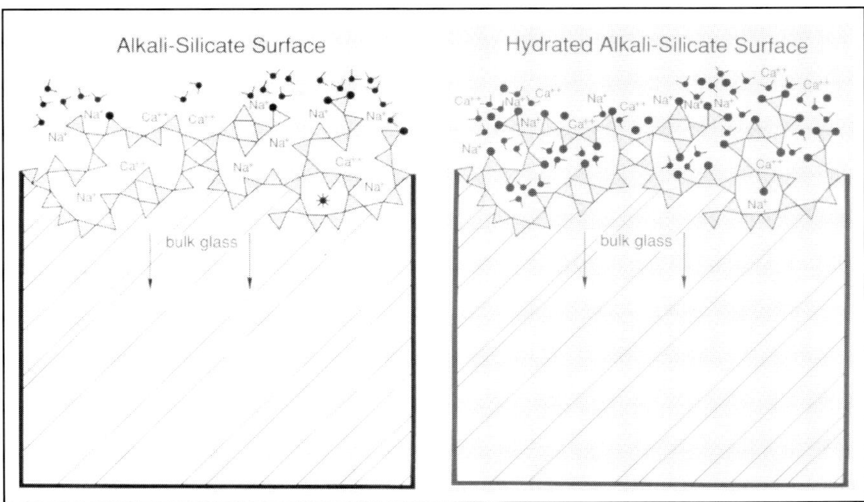

FIGURE 6. Schematic representation, *in cross-secion*, of a soda-lime-silicate (Na_2O-CaO-SiO_2) glass surface: (a) before ion-exchange and weathering (with water molecules adsorbed to a fresh surface, and (b) after *weathering* where an in-depth *ion-exchange* reaction has occurred; note the penetration of water (and protons) *into* the glass surface, and transfer of Na and Ca to the outermost surface layer of the glass.

tension. Their higher concentration in the surface magnifies the quantity evaporated. Moreover, the preference of the surface for these species enhances the continuous replenishment of their vacated surface sites (by out-diffusion from the bulk of the glass). The net effect is that the evaporation/volatilization of surface active species is especially prevalent during both melting and forming. This effect is verified by the data presented in Figure 3 (upper two spectra). Annealing of the glass surface at low temperature (350°C) for a long-time (8 hours) depletes the Na in the surface due to volatization. At higher temperature (850°C), the volatization rate is even higher, but the depletion of Na in the surface is balanced by the out-diffusion that can occur in this temperature range. (The higher volatization rate was independently verified by collecting the Na vapors on a cold plate).

It is to be noted, finally, that the volatility of most species in glass is influenced by the presence of H_2O and SO_2. In most cases, the evaporation rate is enhanced by the presence of these molecules either in the glass or the atmosphere above the glass. The evaporation/volatilization of B_2O_3 is especially dramatic in the presence of water vapor.

CHEMICAL DURABILITY

Chemical durability refers to the resistance of a glass or glass surface to chemical attack by reactive gases, water, acids or bases. The *weathering* of glass refers specifically to the degradation of a glass surface due to attack by the atmosphere; water vapor (humidity) is the most common reactant in this case. Initially, it causes the surface to haze or bloom, but in time, a more permanent dimming, pitting, flaking or crazing of the glass surface may occur. The latter effects cannot be cleaned-off; they are irreversible (unless the surface is completely dissolved away in hydrofluoric acid whereby the original finish is also lost). The *corrosion* of glass usually refers to dissolution or mass-loss; it may or may not be accompanied by a visible degradation of the surface. It is most prevalent in hydrofluoric acid, alkaline (basic) solutions, and very hot water. There is no single test or parameter to characterize the chemical durability of a glass. It depends upon the temperature, the nature of the environment in contact with the glass (vapor vs solution, acid vs base), the ratio of glass surface-area to environment (e.g., a small volume of water in contact with glass is more corrosive than a large volume), the condition of the glass surface (smooth versus ground), and of course, the glass composition.

After many years of study, it is now an accepted fact that the weathering and corrosion of silicate glasses depends, fundamentally, upon a process termed *ion-exchange*. There is still some debate about the mechanism of ion-exchange, but the net effect is clear. Modifier-ions (Li^+, Na^+, K^+, Ca^{++}, Ba^{++}, Pb^{++}) in the glass are exchanged for hydrogen (H^+) or hydronium (H_3O^+) ions present at the glass surface due water adsorption or contact with an aqueous solution. The ion-exchange process between glass and water is described, schematically, in Figure 6. The result is a layer on the glass surface wherein the modifier ions have been replaced by hydrogen ions and water molecules. This process is also called alkali leaching, preferential leaching or surface film formation.

During the weathering reaction that occurs in humidity, the alkali builds-up on the surface (as shown in Figure 6b); this is the visible bloom often observed on fresh glass after some storage in a humid atmosphere. It can be cleaned-off, but the leached layer will still be present within the glass surface. Often, this leached layer creates a dimming or iridescence of the surface. The ion-exchange process may continue after cleaning, but it will be slowed-down because ion-exchange must proceed across the leached layer. If the (initial) bloom is not removed, it can further react with humidity to create a more and more alkaline film on the glass surface. (Recall that glass tends to dissolve in alkaline solution). Thus, there is a time when the initial leaching reaction (and bloom) trigger a secondary reaction when pitting, dissolution and crazing of the surface occurs due to the build-up of alkalinity on the surface. Obviously, it is greatly accelerated in very humid environments and in situations where cleaning, rinsing or condensation-run off are prevented. This is also the origin of crizzle, glass disease, and other long-term degradation effects observed in *old* glass objects. Although cleaning the glass (including the continuous removal of surface alkali by, for example, rain on window glazing) does not prevent leaching and film formation, it ***does prevent*** the onset of the more degrading secondary corrosion reactions.

Again, surface-sensitive analytical techniques have helped to develop these fundamental concepts. Figure 7 presents secondary-ion mass spectroscopy (SIMS) depth profiles that verify the schematic models presented in Figure 6. Figure 7a is the composition-depth profile for a *hazed* piece of alkali-lead-silicate glass (commonly termed "crystal"), while Figure 7b presents the profile obtained after washing-off the haze. It is evident that the haze is a .05-.10μm thick surface film which is enriched in sodium and potassium (probably hydroxides). It is obvious that the sodium and potassium were leached from depths of the order of 200 to 300 nm; that is, regions depleted of sodium and potassium are apparent in the subsurface. The corresponding penetration of hydrogen to depths of the order of 200 to 300 nm indicates that water in the ambient atmosphere is probably responsible for this corrosion reaction. This glass surface, even after the cleaning (Figure 7b), is permanently hydrated and depleted of sodium and potassium.

The ion-exchange/leaching reaction also plays a role in the corrosion of glass in liquids (versus the humid environments described above); for example, the degradation of glassware in the kitchen, or glasses during processing of lenses and optical components. Consider the contact of household glassware with water (e.g., during washing); initially, this causes the preferential leaching of alkali-ions (as described above). It is to be emphasized, though, that in the case of these liquid environments (versus the humidity case described earlier), alkalinity ***does not*** build-up on the surface. It is continuously dissolved away by the liquid environment in contact with the glass surface. But the leached surface layer continuous to grow until it is visible to the eye (e.g., dimming or iridescence), or until it becomes

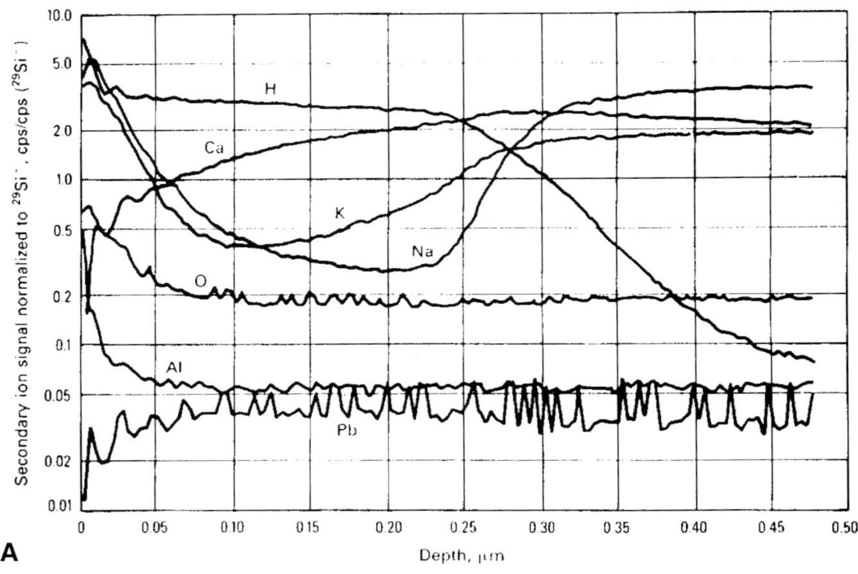

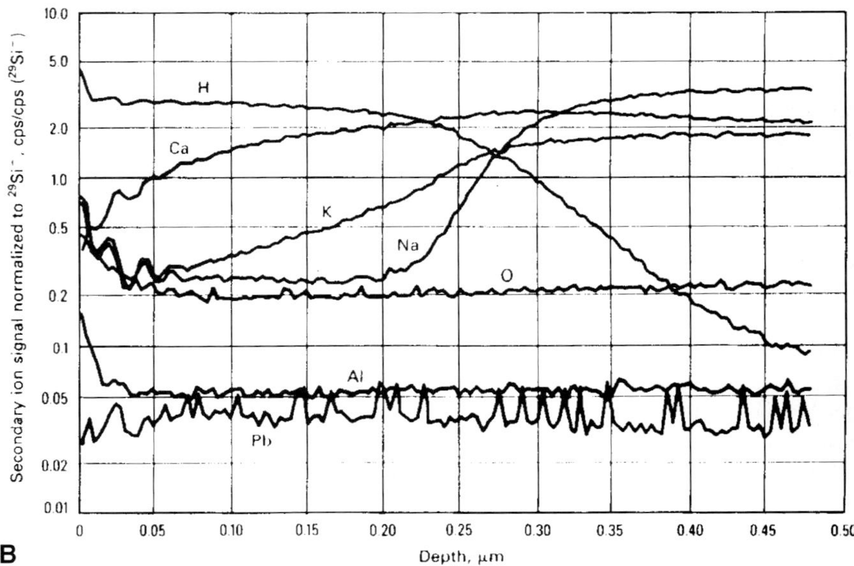

FIGURE 7. Composition-depth profiles obtained using secondary-ion mass spectroscopy (SIMS); the profiles in (a) are for the weathered surface of an alkali-lead-silicate; the profiles in (b) were obtained after cleaning.

so thick that it crazes, crizzles or flakes, or until it becomes a sponge for dirt and soil and renders the glass surface difficult to clean. These effects are more obvious in kitchens where an automatic dishwasher is used - primarily because of the very hot water. The rate of ion-exchange reaction is very sensitive to temperature; for every 10 to 15°C increase in the temperature, the reaction rate doubles! Thus there is no question that the hot-water used in commercial or industrial cleaning of glassware, as well as in hot-water automatic dishwashers, greatly accelerates the visible appearance of the leached surface layer. (Incidentally , the use of alkaline solutions - especially for automatic dishwashers - is designed to dissolve away any visible leached surface layer; recall, again, that silicate dissolves in alkaline solution).

One additional example should suffice to exemplify the fundamental significance of ion exchange and leaching reactions at glass surfaces. Here, reference is made to the action of acids (primarily, the strong mineral acids such as hydrochloric, nitric and sulfuric.*) The common feature of acids is the high concentration of hydrogen and hydronium ions (H^+ and H_3O^+). This greatly enhances the rate of the ion-exchange/leaching reaction, and thereby, the rate of surface layer formation. These acids are commonly encountered in manufacturing processes, and when in contact with glass, can rapidly cause dimming and iridescence due to the rapid growth of leached surface layers. Figure 8 shows the time dependence of this acid leaching reaction for the case of a commercial barium-lead-silicate glass in 1N HCl at 20°C. Obviously, this is a problem that must be avoided in the production of optical glasses and components. In some cases, it is necessary to dissolve-off the surface layer in HF or alkaline solution as a final cleaning step.

But in some cases, the controlled acid leaching of glass surfaces can be of benefit. The depletion of alkali-species in the leached surface layer lowers the refractive index of the surface; this is the origin of dimming and loss of luster. If the thickness of the layer can be controlled, and the durability of the layer itself is adequate, an anti-reflection film can be formed. This is much less expensive than other methods for rendering glass surfaces anti reflective (specifically, the deposition of vacuum coatings on the glass), and is used in a number of commercial *anti-reflective* glass products including liquid crystal displays, solar-cell cover glasses, and picture-frame glazing.

These concepts and examples, together with the schematic structure model in Figure 6, can be used to explain the composition dependence of ion exchange/leaching. It is immediately evident that the alkali concentration in the glass is the primary factor. It is only the alkali (Li^+, Na^+ and K^+), and to a much lesser extent the alkaline earths (Mg^{++}, Ca^{++} and Ba^{++}) and lead (Pb^{++}), which react with humidity and water. Certainly, the more of these species that are present in the glass, the greater will be

---

*The only exception is hydroflouric acid (HF) which causes glass to *dissolve* because it directly attacks the network Si atoms. HF acid is a special case when discussing silicate glasses because the fluorine-ion creates a unique situation when in contact with any silicon-containing materials; that is, the formation of very stable silicon tetrafluoride vapors (SiF_4) and soluble silicon hexafluoride anions ($SiF_6^=$).

the extent of weathering and corrosion. Moreso, the presence of these species creates breaks in the network structure, and so, the migration of protons (H^+) and hydronium ions (H_3O^+), and alkaline ions - during ion exchange - is facilitated. In fact, this explains why aluminum-oxide (Al_2O_3) and boron-oxide (B_2O_3) improve the chemical durability of glass; the addition of even small quantities of these species "ties-up" the breaks created by the alkali species. Thus, the migration of the ion-exchanging species is slowed down. Similarly, the effect of the ion size can be considered. A glass containing (only) K_2O versus (only) Na_2O has poorer chemical durability because the leaching of the large K^+ ions leave bigger holes in the structure than the smaller Na^+. Thus, the continued exchange of ions, as well as the swelling due to water penetration, is greatly facilitated in the K_2O glass. Similarly, glasses with (only) Li_2O have better durability than Na_2O (only) glasses because the Li^+ ion leaves a very small void. In this regard, it is interesting to consider the effect of small substitutions (or mixtures) of these alkali and alkaline earth species. Note (in Figure 6) that alkali and alkaline earth species in glass create channels in the structure where the non-bridging oxygen breaks are created. The substitution of a small quantity of K or Ba (large ions), for Na or Ca (smaller ions), randomly introduces "blockages" in the channels that impede ion exchange and water penetration. Thus, when a small quantity of large ions are mixed with the smaller ions the durability of the glass is often improved (relative to a drastic reduction in durability when K_2O or BaO are used to completely replace Na_2O and CaO).

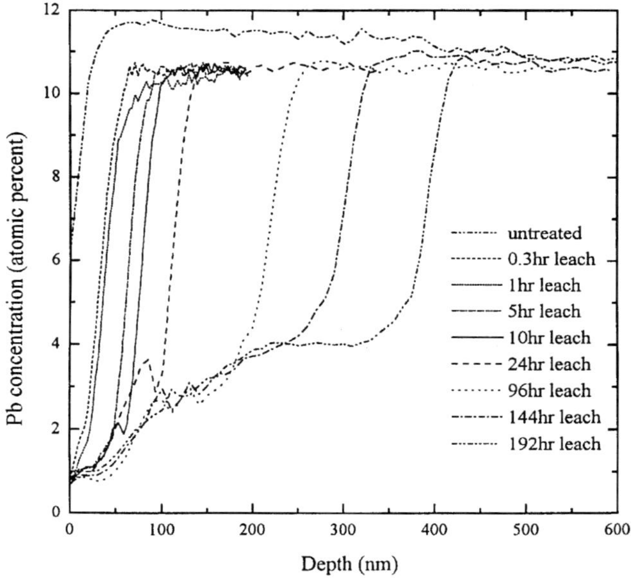

FIGURE 8. Lead depth profiles (obtained with SIMS) as a function of leaching time in 1N Hcl

SURFACE TREATMENTS

The phenomena of ion-exchange was discussed in the previous section with specific reference to chemical durability; that is, the exchange of modifier ions in the glass for protons (H^+) or hydronium ions (H_3O^+) in an aqueous solution or adsorbed film of water vapor. However, ion-exchange is a generic process that can, in fact, be used to engineer the appearance and properties of a glass surface by exchanging modifier ions with other atomic species. Here, the application of ion-exchange for the coloring of glass, for the strengthening of glass, and for the dealkalization of glass, will be briefly described.

The *staining* of glass is an art that has been practiced for many years. Usually, it refers to the coloration of a glass surface by ion-exchange of silver (Ag) or copper (Cu) ions from vapors, melts, or pastes with alkali ions *in* the glass structure. It is different from coloring methods that deposit a layer of colored material *on* the glass surface. Rather, ion-exchange staining modifieds the composition, structure and optical properties of the glass surface. The exchange of Cu^+ (in a CuCl vapor) with Na^+ (in the glass surface) leaves Cu^+ ions in the glass. These ions create a yellow color. If the glass is subsequently heated in hydrogen, the Cu^+ ions in the glass surface change to Cu^o (neutral atoms) and the glass turns red. Similarly, the exchange of Ag^+ ions in a melt (of $AgNO_3$) or a paste (of $AgSO_4$ or AgCl) with Na+ ions in the glass leaves Ag^+ ions in the glass. The Ag^+ ions produce no color. But if the Ag ion-exchanged glass is heated in hydrogen (or if the glass composition includes chemical reducing agents), the Ag^+ ions change to Ag^o atoms and the glass turns amber to yellow. In contrast to ion-exchange with protons (H^+) and hydronium ions (H_3O^+) which are small and mobile even near room temperature, these staining reactions require elevated temperatures to facilitate transport in and out of the surface.

The *chemical strengthening* of glass depends upon ion-exchange of large ions (e.g., K^+ in a KNO_3 melt) with smaller ions in the glass (e.g., Li^+). This causes the large ion to be "stuffed" into the glass surface. The net effect is the creation of a compressive stress in the glass surface which can counteract the tensile stresses that cause cracks to propagate (and the glass to shatter). This process is the basis of commercial products including eyeglass lenses; in this case, the process is typically carried-out by the optician after grinding and finishing the lens. Again, it requires elevated temperatures.

The *dealkalization* of finished glassware depends upon the exchange of protons (H^+) in an acidic vapor (such as HCl) with alkali ions in the glass. In contrast to the H ion-exchange leaching that occurs during corrosion in water vapor or acid solution, this dealkalization treatment with acidic vapors is carried-out at high temperature. This causes the alkali-leached surface layer to "compact", and thereby, to limit subsequent reaction with water, water vapor or acid at lower temperature. This process is used to improve the weathering resistance of glass containers in storage or shipment under humid conditions, as well as to enhance their corrosion resistance when used for pharmaceutical containment.

SUMMARY

The primary objective of this presentation has been to describe relationships between glass composition and molecular structure, and in turn, the way the molecular structure influences or controls properties. The discussion has been qualitative and conceptual even though detailed quantitative descriptions of these characteristics and properties are available (see Bibliography below). Nevertheless, these simple concepts, that is, the *silicate network, non-bridging oxygens, ion size and charge effects, and ion-exchange* - are sufficiently intuitive for the artist, engineer or scientist to predict or describe, at least qualitatively, the practical behavior of any glass composition.

ACKNOWLEDGMENT

The author gratefully acknowledges the efforts of all those students who contributed to his studies of glass surfaces. He further acknowledges the financial support of the Division of Materials Research of the National Science Foundation, the Air Force Office of Scientific Research and the Office of Naval Research.

BIBLIOGRAPHY

1. H. Rawson. 1980. *Properties & Applications of Glass,* Elsevier.
2. A.K. Varshneya. 1994. *Fundamentals of Inorganiz Glasses,* Academic Press.
3. D.E. Clark, Pantano, C.G. and L.L. Hench. 1979. *Corrosion of Glass,* Books for Industry.
4. M.B. Volf. 1984. *Chemical Approach to Glass,* Elsevier.
5. M.B. Volf. 1988. *Mathematical Approach to Glass,* Elsevier.
6. J. Eales. 1977. *Manual and Databook of Glass Technology Calculations,* Ordentliah.
7. N.P. Bansal and R.H. Doremus. 1986. *Handbook of Glass Properties,* Academic Press.
8. D.C. Boyd and J.F. MacDowell. 1986. *Commercial Glasses,* Am. Ceram. Soc.
9. M. Cable and J.M. Parker. 1992. *High-Performance Glasses,* Chapman and Hall.
10. C.G. Pantano. 1989. *Glass Surfaces,* Reviews of Solid State Science.
11. C.G. Pantano. 1985. *Surface Chemistry in Relationto the Strength and Fracture of Silicate Glass,* Strength of Inorganic Glass.
12. C.G. Pantano, *"X-Ray Photoelectron Spectroscopy of Glass"* in Experimental Techniques of Glass Science (ACS, Ohio, 1993) pp129-160.

Chapter Eighteen

BIOMATERIALS

P.W. BROWN

Department of Materials Science and Engineering
Steidle Building
Pennsylvania State University
University Park, PA 16802

INTRODUCTION

A very wide suite of materials is used to replace hard human tissues and there is an extensive body of literature describing biomaterials (e.g. 1-4). Biomaterials include metals, ceramics and polymers. Each class of material is selected because of the properties exhibited. A ceramic, such as alumina, is selected for the ball and cup of a hip joint prostheses to take advantage of its high hardness and low coefficient of friction. Alternatively porcelain is selected for dental overlays because it can be processed to bond to noble metals, because it exhibits hardness values compatible with those of teeth, and because it exhibits translucency reminiscent of teeth. Excepting amalgam, metals are almost universally used in applications where fracture toughness is an important consideration. Typical applications for metals include their use as stems and balls in hip joint prostheses, and as pins, screws and plates to fix fractures. Metals commonly used include cobalt chrome alloys, Ti and its alloys, and selected stainless steel. Although polymeric materials may thought to be most appropriately used as soft tissue prostheses, ultrahigh density polyurethane is used as cups in which metallic hip joint prostheses articulate, arcylics are used as "cements" to bond joint prostheses to bone, and a variety of ionic polymers are used in dental applications. While the use of composites is relatively common in dental restoratives (5), there are few examples where composites are designed for the replacement of hard tissues based on their ability to approximate the mechanical properties of hard tissue they replace.

Advances in materials technology will eventually lead to the development of advanced composite materials to replace hard tissues. Depending on application, these composites will need to attain mechanical properties at least comparable to those of the bone or tooth structures being replaced. Composites offer advantages over individual materials, such as metals or ceramics, in that they can be tailored to distribute stresses in ways which are much closer to those of native hard tissues. Depending on the requirements specific to the individuals involved, it may be desirable that such composites remodel when used as bone substitutes. Remodelling is the term used to characterize the cellular-mediated process by which bone is resorbed and replaced. However, regardless of such a need, the composites should be biocompatible and support bone intergrowth in some instances. These requirements are most readily met if the inorganic constituent used is compositionally close to bone and tooth. Table 1 lists the compositions of bone and tooth structures. Excepting enamel, bone and tooth are composites comprised primarily of hydroxyapatite and collagen, although these are present in differing proportions depending on the specific hard tissue. Enamel can be approximated to an hydroxyapatite ceramic in that its non-mineral content is low.

Bones and teeth are microstructurally complex. The morphology of bone is illustrated in Figure 1 (7) and it may be generally characterized as cancellous or cortical. Tooth structures are illustrated in Figure 2 (8) and include enamel, dentin, and pulp. Dental pulp is frequently referred to as the nerve of a tooth. For present purposes it will be excluded from further consideration as a hard tissue. Common to these hard tissues is hydroxyapatite. Therefore, the selection of hydroxyapatite as the mineral component of a synthetic composite is desirable. The biocompatibility of synthetic apatite is well documented, making it an attractive candidate for the mineral constituents of such composites. To realize optimal composite properties a polymeric non-mineral constituent is also likely to be required. Because

TABLE 1

Compositions of enamel, dentin and bone (after [6])

Composition (wt %)	Enamel	Dentin	Bone
calcium	36.5	35.1	34.8
phosphorus (as P)	17.7	16.9	15.2
Ca/P	1.63	1.61	1.71
Na	0.5	0.6	0.9
K	0.08	0.05	0.03
Mg	0.44	1.23	0.72
CO_3	3.5	5.6	7.4
F	0.01	0.06	0.03
Cl	0.3	0.01	0.07
mineral	97	70	65
organic*	1.5	20	25
absorbed H_2O	1.5	10	10

* primarily collagen

of the presence of polymer, there are limits on the temperatures which such composites can be fabricated. If the polymeric constituent is a polypeptide (e.g. collagen), the fabrication temperature needs to be quite low. It would be most desirable if composites could form by self-assembling cement-like reaction *in vivo*. In this instance composite structure developes at 37.4°C. If this were the case, a composite could be constituted in the clinical setting. Property development *in vivo* is common with dental restoratives but less familiar for orthopaedics. It is highly desirable in both application areas to eliminate the need for a preform, thereby allowing an implant to be accommodated to the shape of a defect. If produced in monolithic form, HAp may have mechanical integrity and, therefore, can emulate the functions of hard tissues and serve as substrates on which cellular functions can proceed. In addition, HAp formed *in vivo* does not rely on preforms. Therefore, it can serve a variety of needs in orthopaedics, plastic and reconstructive surgery, and dentistry. Because of these clinical advantages, this paper will limit its consideration to hydroxyapatite capable of being formed *in vivo*.

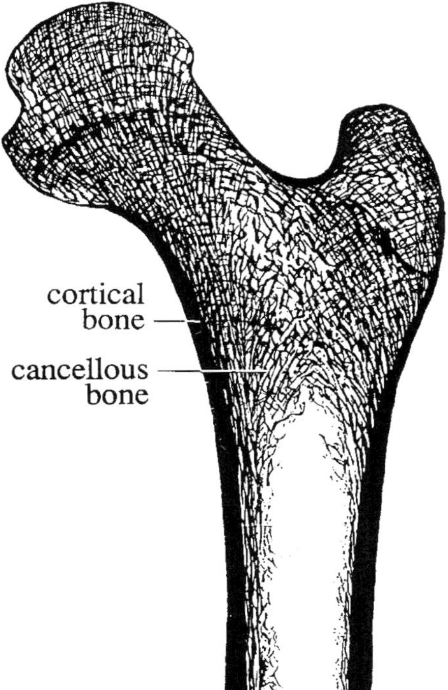

FIGURE 1. A cross-section through a femur showing an interior region of low density cancellous or trabecular bone. The bone which defines the bulk shape of the bone is more dense compact or cortical bone (after Crouch (7)).

HYDROXYAPATITE

Stoichiometric HAp is $Ca_{10}(PO_4)_6(OH)_2$. However, HAp is not a compound of fixed composition. Rather, it exists over a compositional range which extends from a Ca/P ratio of 1.67 to a Ca/P ratio of 1.5. The composition of calcium-deficient HAp having the latter Ca/P ratio is $Ca_9HPO_4(PO_4)_5OH$. Compared to stoichiometric HAp, the structure of calcium-deficient HAp (Ca/P of 1.5) is attained by creating a vacancy on a Ca site and an hydroxyl site and by protonating a phosphate group. Thus calcium-deficient HAp is substantially more soluble than stoichiometric HAp. The Ksp values are 10^{-85} and 10^{-117}, respectively (9). As will be discussed, this is an important consideration if it is desirable that HAp be capable of resorption.

HAp is only one of a series of calcium phosphates. Table 2 lists the various calcium phosphates in the order of increasing basicity. Of these calcium phosphates, only HAp, $CaHPO_4·2H_2O$, and $Ca_8H_2(PO_4)_6·5H_2O$) are found *in vivo*. It is widely believed that $Ca_8H_2(PO_4)_6·5H_2O$ precedes the formation of HAp ion bone mineralization (10). It is found in dental calculi and is an occasional constituent

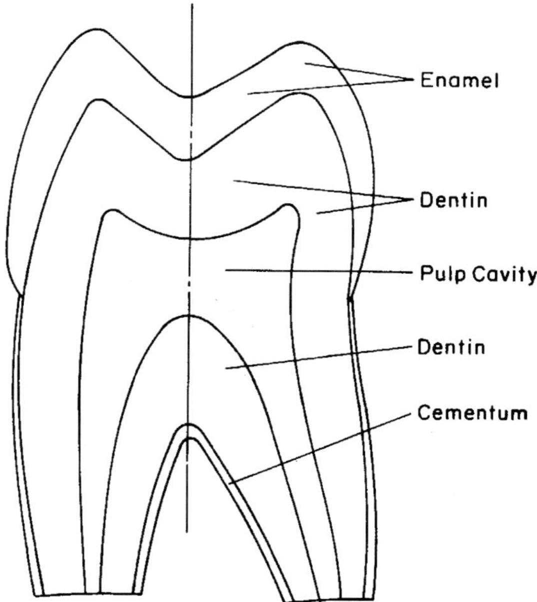

FIGURE 2. A cross-section through an incisor. The exterior surface of the tooth is sheathed by a layer of enamel about 100 µm in thickness. Underlying the enamel layer is the dentin. Dentin is more porous than enamel and contains a much higher proportion of organic material. The dentin surrounds the pulp (after Provenza (8)).

of kidney stones. $CaHPO_4 \cdot 2H_2O$ occurs in kidney stones and is a product of tooth decay. Mg-substituted β-tricalcium phosphate also occurs in dental caries and in kidney stones. Finally, β-tricalcium phosphate has been observed in calcifying heart valves (11) while amorphous calcium phosphate has been reported to be present in soft tissue calcification (6). Recently it has been demonstrated that suitable combinations of these calcium phosphates, such as $CaHPO_4 \cdot 2H_2O$ and $Ca_4(PO)_2O$, can react by an acid-base reaction at low temperature to form HAp. The formation of HAp in this manner is the principal topic of this paper.

Surprisingly little work has been done to establish the compositional relationships among the calcium phosphates. Brown and Chow (12) illustrated the solubility curves for a variety of calcium phosphates, Figure 3. Each curve in this solubility diagram represents the variation in the solubility of a particular calcium (defined as the negative log of the Ca ion concentration) with the pH. Acidic calcium phosphates are highly soluble and are not shown. α-tricalcium phosphate does not have a true solubility and is not shown. While tetracalcium phosphate does not have a true solubility, its rate of dissolution shows a strong pH dependence and this "pseudosolubility" curve is shown. The diagram can be interpreted in terms of the phase rule in that at constant temperature and pressure invariance in the ternary system (the compositions of all calcium orthophosphates can be described as combinations of CaO, P_2O_5, and H_2O) involves equilibrium between 3 phases. Accordingly, intersections of the solubility curves are invariant points. As this solubility diagram shows, HAp is the stable calcium phosphate over the nor-

TABLE 2
Calcium Orthophosphates

FORMULA	NAME	C/P	STABILITY
$Ca(H_2PO_4)_2 \cdot H_2O$	Monocalcium Phosphate Monohydrate	0.5	stable
$Ca(H_2PO_4)_2$	Monocalcium Phosphate	0.5	no true solubility
$CaHPO_4 \cdot 2H_2O$	Dicalcium Phosphate Dihydrate	1.0	metastable
$CaHPO_4$	Dicalcium Phosphate	1.0	stable
$Ca_8H_2(PO_4)_6 \cdot 5H_2O$	Octacalcium Phosphate	1.33	metastable
α-$Ca_3(PO_4)_2$	Tricalcium Phosphate	1.5	no true solubility
β-$Ca_3(PO_4)_2$	Tricalcium Phosphate	1.5	metastable?
$\sim Ca_9(PO_4)_6 \cdot 3H_2O$	Amorphous Calcium Phosphate	1.5	metastable
$Ca_9(HPO_4)(PO_4)_5OH$, $Ca_{10}(PO_4)_6(OH)_2$	Hydroxyapatite	1.5 to 1.67	stable
$Ca_4(PO_4)_2O$	Tetracalcium Phosphate	2.0	no true solubility

mal range of biological pH values. A significant exception is the invariant point involving HAp and $CaHPO_4 \cdot 2H_2O$. This point is achieved in the formation of dental caries. Depression of the local pH by bacterial activity can be tolerated by tooth enamel until the pH reaches about 4.6. HAp is the stable calcium phosphate over the normal range of biological pH values. A significant exception is the invariant point involving HAp and $CaHPO_4 \cdot 2H_2O$. This point is achieved in the formation of dental caries. Depression of the local pH by bacterial activity can be tolerated by tooth enamel until the pH reaches about 4.6. HAp is in metastable equilibrium with $CaHPO_4 \cdot 2H_2O$ at this pH. The stable equilibrium involves HAp and $CaHPO_4$; however, the kinetics of $CaHPO_4 \cdot 2H_2O$ formation result in the formation of this solid. With continued generation of acid by bacteria, the HAp decomposes in favor of $CaHPO_4 \cdot 2H_2O$ and the formation of a caries lesion initiates.

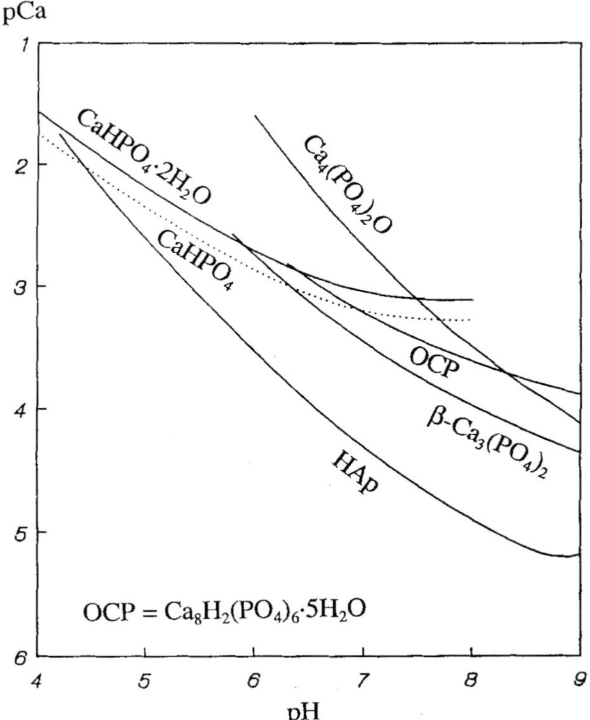

FIGURE 3. Solubility curves of selected calcium phosphates (after Brown and Chow, 1987). The composition of the solution from which hydroxyapatite precipitates at metastable equilibrium determined by the intersection of the relevant solubility isotherms when mixtures of solid $Ca_4(PO_4)_2O$ and $CaHPO_4 \cdot 2H_2O$ or $CaHPO_4$ are used as reactants. (Reproduced courtesy of the J. Am. Ceram. Soc., See Reference 23).

SUBSTITUTED APATITES

The solubility diagram is also useful in comparing the stabilities of substituted apatites. The inorganic constitutent in bones and teeth is a substituted HAp having the generalized composition: $Ca_{(10-x-y)}Na_{x[\ ]y}(HPO_4)_a(CO_3)_b(PO_4)_{(6-a-b)}(OH,F,_{[\ ]},CO_3)_n$ where n $\leq$ 2 and [ ] denotes a vacant lattice site, is the inorganic constituent in bones and teeth. Bone apatite has a Ca/P of about 1.7, but in bone apatite about 1 of the 6 phosphate groups is replaced by carbonate. LeGeros (6) reported the carbonate contents of biological apatites vary depending on the function of the tissue, Table 1. Figure 4 (after 13) shows the relative solubilities of hydroxyapatite, carbonate-substituted apatite and fluorapatite. These solubility differences are biologically significant and the low temperature formation of these compositions is desirable. Replacement of hydroxyl by fluoride reduces apatite solubility. This reduction depresses the pH of the invariant point with $CaHPO_4\cdot2H_2O$. As a consequence, topical treatment with fluoride-containing toothpastes confers to teeth an increased resistance to the formation of dental caries. Alternatively, the higher solubility of carbonate apatite facilitates the remodelling of bone. Bone-resorbing cells, called osteoclasts, function by attaching to a bone surface and

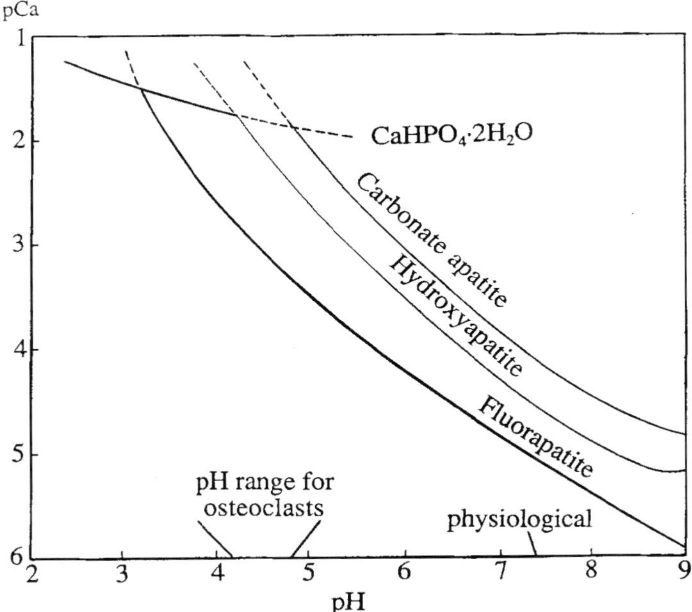

FIGURE 4. Solubiity curves of hydroxyapatite, fluorapatite, and carbonated hydroxapatite (schematic) illustrating the relative solubilities of these compounds and the relative pH values of their invariant points with $CaHPO_4\cdot2H_2O$.

producing HCl. This depresses the pH and significantly increases the solubility of bone apatite. Carbonate apatite (bone apatite) shows sufficient solubility to dissolve at appreciable rates even at pH values higher than that of the invariant point with $CaHPO_4 \cdot 2H_2O$. Thus, bone resorption can occur in the absence of the formation of a second phase. Because it is highly advantageous if the resorbability of synthetic apatite is tailorable, it is desirable to achieve compositional control. Although stoichiometric hydroxapatite and fluorapatite monoliths can be produced by conventional solid state processes, carbonate apatite and calcium-deficient hydroxyapatite decompose at temperatures below which they can be sintered to the densities needed to achieve the requisite mechanical properties. Thus apatite compositions approximating that of bone cannot be conveniently produced by compaction and sintering.

FORMATION OF HYDROXYAPATITE AT PHYSIOLOGICAL TEMPERATURE

Brown and Chow (12) discovered that HAp could be formed in aqueous solution at low temperature by the following reaction:

$$\overset{H_2O}{2CaHPO_4 \cdot 2H_2O = Ca_4(PO_4)_2O} \Rightarrow Ca_{10}(PO_4)_6(OH)_2. \qquad \text{EQ. [1]}$$

However, the rate of this reaction tends to be quite slow. Clinical expectations in dentistry are that reactions reach a substantial extent of completion in less than 30 minutes, whereas those in orthopaedics are that reaction be complete in a few hours. Thus, it is critical to establish the factors which influence the rates of HAp formation. Two examples of reactions we have investigated by which calcium-deficient and stoichiometric HAp can be formed are:

$$\overset{H_2O}{6CaHPO_4 + 3Ca_4(PO_4)_2O} \Rightarrow 2Ca_9(HPO_4)(PO_4)_5OH + H_2O, \qquad \text{EQ. [2]}$$

$$\overset{H_2O}{6CaHPO_4 + 6Ca_4(PO_4)_2O} \Rightarrow 3Ca_{10}(PO_4)_6(OH)_2 \qquad \text{EQ. [3]}$$

If HAp is to be formed at clinically relevant rates at low temperature, reactions having the characteristics of peritectics should be avoided. Although the solubility diagrams shown in Figures 3 and 4 are useful, they are limited in that the nature of the invariant points shown cannot be determined. To establish the nature of the relevant invariant points and to establish the mechanistic paths controlling HAp formation, the compositional relationships among the stable calcium phosphates at 25°C is shown in Figure 5 (14). The construct used was developed for representing equilibria between phases of greatly differing solubilities (15-17).

Concentrations are plotted as $[\text{conc.}]^{1/N}$. The value of N selected depends only on the scaling requirements. In Figure 5 the mole fractions of CaO and P_2O_5 in the solid calcium phosphates and in solution are plotted as their 5th roots. The solubility curves for the stable calcium phosphates and for calcium hydroxide are shown. The invariant point between stoichiometric HAp and $Ca(OH)_2$ is the equivalent of a eutectic $(L \Leftrightarrow S_1 + S_2)$. The remaining invariant points are the equivalent of peritectics $(L + S_1 \Leftrightarrow S_2)$. These relationships strongly influence the formation of HAp at physiological temperature.

The figure also shows that the Ca/P ratio in HAp depends on solution composition and that calcium-deficient, not stoichiometric, HAp is in equilibrium with solid $CaHPO_4$. Using the concept of local equilibrium, the local effects in solution due to initial dissolution of $CaHPO_4$ can be treated as independent of those during dissolution of $Ca_4(PO_4)_2O$ in the analysis of HAp formation by EQ's [2] and [3]. $CaHPO_4$ dissolution introduces calcium and phosphate to the solution in a molar ratio of 1:1. This is illustrated as occurring along the path indicated in the figure. Eventually, the solution composition will reach a point on the solubility isotherm for HAp. Although the solution is saturated with respect to

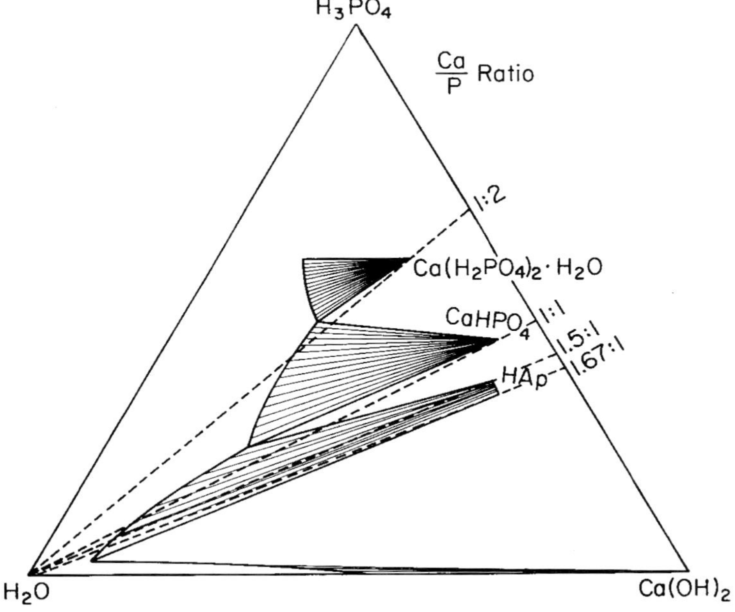

FIGURE 5. Ternary system: $CaO\text{-}H_3PO_4\text{-}H_2O$. Solubility data are expressed schematically consistent with mole fractions being plotted as their 5th roots (14). The diagram shows that HAp is not a compound of fixed composition and that the invariant points between HAp and $Ca(OH)_2$ is a eutectic while the other shown are peritectics.

HAp at this point, it remains undersaturated in $CaHPO_4$. Continued dissolution will supersaturate the solution in HAp and, eventually, a critical degree of supersaturation will be reached. HAp will then precipitate. However, until the solution composition reaches the invariant point involving $CaHPO_4$ and $Ca_4(PO_4)2O$, it is still undersaturated in $CaHPO_4$. Therefore, continuing $CaHPO_4$ dissolution will result in HAp formation. The HAp which forms nucleates and grows on the solid surfaces available to it, namely on the surfaces of the dissolving $CaHPO_4$. The highest degree of supersaturation will also occur near the surfaces of the dissolving $CaHPO_4$ crystals and they appear to provide sites for heteroepitaxial nucleation of HAp. Growth of these nuclei creates a situation where HAp overgrowth isolates one of the reactants from the solution. The consequence is the rate of reaction eventually becomes controlled by the rate of $CaHPO_4$ dissolution which is diffusionally controlled. Therefore, several weeks are required for complete reaction. Constituting the reactants in ways which minimize the effects of these peritectic reactions permits phase-pure HAp to be formed in periods as short as one hour.

Since the original work by Brown and Chow, the formation of HAp from a variety of calcium phosphate-based reactants has been demonstrated. Included are acid-base reactions in which the acidic constituent is monocalcium phosphate monohydrate ($Ca(H_2PO_4)_2 \cdot H_2O$) (18) as shown in EQ [4] and hydrolysis reactions in which tricalcium phosphate ($Ca_3(PO_4)_2$) or octacalcium phosphate ($Ca_8H_2(PO_4)_6 \cdot 5H_2O$) convert to HAp as shown in EQ [5] and EQ [6a-b], respectively. The formation of HAp using formulations containing these compounds have been investigated (19,20)

$$Ca(H_2PO_4)_2 \cdot 2H_2O + 2Ca_4(PO_4)_2O \xrightarrow{H_2O} Ca_9HPO_4(PO_4)_5OH + 2H_2O \quad \text{EQ. [4]}$$

$$3\alpha\text{-}Ca_3(PO_4)_2 \xrightarrow{H_2O} Ca_9HPO_4(PO_4)_5OH \quad \text{EQ. [5]}$$

$$Ca_8H_2(PO_4)_6 \cdot 5H_2O + Ca(OH_2) \xrightarrow{H_2O} Ca_9HPO_4(PO_4)_5OH + 6H_2O \quad \text{EQ. [6a]}$$

$$Ca_8H_2(PO_4)_6 \cdot 5H_2O + 3Ca_4(PO_4)_2O \xrightarrow{H_2O} Ca_{10}(PO_4)_6(OH)_2 + 5H_2O \quad \text{EQ. [6b]}$$

Although bone mineral does not form by the reactions listed in EQ's [1]-[6], there are striking similarities with natural mineralization. The reactions shown occur isothermally, under steady state conditions, can occur at a pH near physiological and at low ionic strength, and are directed epitaxially by the presence of proteins. The kinetics of such reactions are also modulated by the presence of proteins (21, 22). In addition, the HAp which forms in this manner is monolithic; it is not a powder.

SOLUTION CHEMISTRY AND HAP FORMATION

If HAp-based implants are to be formed *in vivo* by the types of reactions illustrated in EQ's [1]-[6], attainment of steady state is important in avoiding variations in solution chemistry which could lead to conditions which are cytotoxic. Figure 3 shows that both $CaHPO_4$ and $CaHPO_4.2H_2O$ appear to form an invariant point with $Ca_4(PO_4)_2O$. While these cannot be true invariant points because $Ca_4(PO_4)_2O$ does not have a true solubility, for the present purpose it is reasonable to approximate them as such. In this event, these pseudo-invariant points are metastable with respect to the solubility curve for HAp. Thus, the mechanism of HAp formation can be considered in two steps. Dissolution of reactions should result in solutions having compositions which can be approximated to these metastable invariant points. However, the supersaturation of solutions with respect to HAp, provides the driving force for its formation. If the dissolution and precipitation events are coupled, then the pH of the solution in which HAp is forming should be about 7.8. This is significant in that a pH of 7.8 is close to the physiological pH of 7.4. Therefore HAp formation *in vivo* should not create conditions in solution which would be aggressive to adjacent tissues. Analysis of the steady state reaction also provides a means for exploring the relationship between the steady condition produced, the rate of HAp formation and its microstructure. Figure 6 (23) shows the variations in pH when $CaHPO_4$ and $Ca_4(PO_4)_2O$ react at mole ratios ranging from 1:1 (to produce stoichiometric HAp) to 2:1 (to produce calcium-deficient HAp). As these data show, the steady

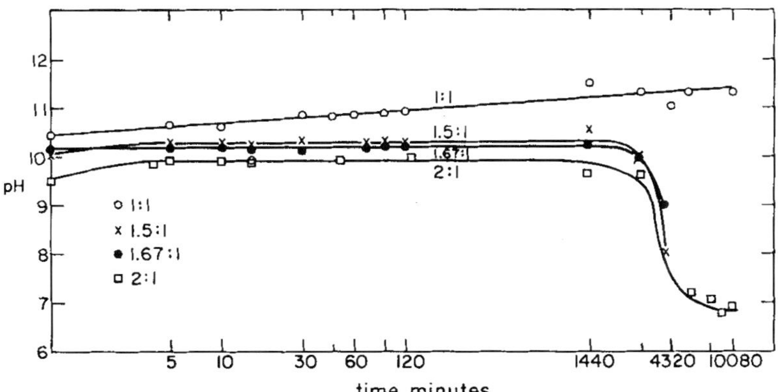

FIGURE 6. The variations in pH during the formation of HAp by reactions of $CaHPO_4$ and $Ca_4(PO_4)_2O$ depending on the molar ratio of these reactants (23). HAp formation occurs at steady state and the pH values only decrease after about 24 hours when the $Ca_4(PO_4)_2O$ has been consumed. The pH values always exceed that of the pseudo-invariant point shown in Figure 3. (Reproduced courtesy of the publisher).

state pH is significantly different from 7.8 indicating that the steady state condition is not associated with the pseudo-invariant point. Rather, the pH is elevated indicating the dissolution of the acidic constitutent to be rate limiting. Recognizing the peritectic character of the invariant point between $CaHPO_4$ and HAp, the steady state pH can be maintained at values which would not be considered as aggressive to adjacent tissues (24).

A steady state condition in solution can be controlled by controlling the rates at which the precursors dissolve through particle size effects or by solution chemical means such as common ion effects (25, 26). Our work has shown that substituted apatites, involving substitutions which strongly influence solubility, can also be produced as monolithic forms. HAp compositions, which include the range of those reported for hard tissues, can be produced by carrying out the reactions illustrated EQ.'s [2] and [3] in the presence of sodium, fluoride and carbonate ions. For example, it has been demonstrated that fluorapatite and fluoride-substituted HAp can be formed (27). It has also been shown that monolithic apatites occur, in which paired substitution of sodium for calcium and carbonate for phosphate can be formed (21).

KINETICS OF HAP FORMATION

From the standpoint of clinical use each of the reactions shown in EQ's [1]-[3] have advantages and disadvantages. Reactions between physical mixtures of $Ca_4(PO_4)_2O$ and $CaHPO_4$ or $CaHPO_4 \cdot 2H_2O$ tend to occur slowly, often requiring several days for complete reaction (18,28). From a clinical standpoint this is unacceptably slow. Alternatively, the use of reactants compatible with HAp remains attractive because the formation of persistent intermediates does not occur. $CaHPO_4$ and $CaHPO_4 \cdot 2H_2O$ are compatible with HAp. Our recent work has demonstrated that HAp formation can occur rapidly by reactions involving the reactants. For example we have observed the formation of phase-pure HAp in less than one hour at 37.4°C.

Because $Ca_4(PO_4)_2O$ and $CaHPO_4$ or $CaHPO_4 \cdot 2H_2O$ can react slowly, the use of other acidic reactants has been explored. $Ca(H_2PO_4)_2 \cdot H_2O$ and $Ca_4(PO_4)_2O$ react rapidly. The reaction to form calcium-deficient HAp was illustrated in EQ [3]. However, the reaction shown in EQ [3] is oversimplified in that HAp does not form directly. Rather the reaction again occurs in two steps. In the first of these $Ca(H_2PO_4)_2 \cdot H_2O$ is consumed in the formation of DCPD as an intermediate:

$$2Ca(H_2PO_4)_2 \cdot H_2O + 4Ca_4(PO_4)_2O \xrightarrow{\text{H}_2\text{O}} 6CaHPO_4 \cdot 2H_2O + 3Ca_4(PO_4)_2O \quad \text{EQ. [8]}$$

In the second step, calcium-deficient HAp is formed as shown in EQ [1b]. Because this second step can be slow, the advantage of using $Ca(H_2PO_4)_2 \cdot H_2O$ as a reactant can be less than expected. However, the use of $Ca(H_2PO_4)_2 \cdot H_2O$ as a reactant does result in rapid set because of the formation of $CaHPO_4 \cdot 2H_2O$ as an intermediate.

PHYSICAL PROPERTIES OF HAP PRODUCED AT PHYSIOLOGICAL TEMPERATURE

Enamel in teeth contain only a relatively small proportion of collagen. Because of their high mineral content and lower porosity compared to bone, teeth are harder, stronger and more brittle. Low-temperature formation has typically resulted in hydroxyapatite monoliths which exhibit relatively poor mechanical properties. For example, compressive strengths of hydroxyapatite prepared in this way range from 34-51 MPa (29, 30). The principal factor which limits the compressive strengths of HAp formed using the reactants described is that densification by the conventional method of sintering is precluded. Because of this it is critical to minimize the proportion of water present. This can be accomplished in 2 ways. The first is to limit the amount of water (whether as waters of hydration or as protons) in the precursor phases. For this reason, the use of $CaHPO_4 \cdot 2H_2O$ as a reactant limits the strength of the HAp produced because 2 moles of water are liberated for each mole of stoichiometric HAp formed. A second way to maximize the compressive strength of the HAp which forms is to minimize the proportion of water used when constituting the precursor mixtures. This can be illustrated using Figure 7 which shows the relationship between the weight ratio at which the water and the solid reactants were mixed (liquid-to-solids ratio) and the porosity which remains when HAp is formed from $CaHPO_4$ and $Ca_4(PO_4)_2O$ according to EQ [2]. This curve does not pass through the origin; even if calcium-deficient HAp was formed from these reactants at "zero" liquid-to-solids ratio, one mole of water would be liberated per every 2 moles of HAp formed as shown in EQ [2].

We have produced HAp monoliths which develop compressive strengths of 175 MPa within 24 hours (31). Figure 8 shows the relationship between compressive strength and porosity for sintered HAp. It also shows the HAp we produce at 38°C has nominally the same strengths as sintered HAp provided the porosities are comparable. Although a compressive strength of 175 MPa is somewhat below that reported for enamel, it is in the range of values reported for dentin. This indicates that high-temperature processing is not needed to produce HAp having mechanical properties approaching those of native HAp.

HYDROXYAPATITE-BASED COMPOSITES

Mechanical integrity of skeletal structures is optimized with respect to weight minimization and the ability to rapidly deliver calcium to support metabolic processes. The latter two factors require porous structures. The mechanical disadvantages of the presence of pores in bone, which are regarded as flaws from the standpoint of fracture mechanics, are overcome by the incorporation of a large proportion of collagen reinforcement. The conflicting requirements of mechanical integrity and porosity make it unlikely that a single material would be capable of closely approximating the mechanical functional requirements of bone. Indeed, none have been found. The alternative is to consider a composite which would exhibit adequate mechanical properties. Our previous work has shown that HAp-based composite materials likely to exhibit biocompatibility can be formed at physiological temperature and in a pH range compatible with conditions *in vivo*. Two such composite systems have been investigated: HAp-gelatin (32) and HAp-collagen (22). Physical property thus far achieved in hydroxyapatite-collagen composite are flexural strengths of 7.5 to 12.5 MPa, Young's moduli from 200 MPa to 1.7 GPa, work of fracture 0.51 KJ/m², and stable crack growth. This work of fracture may be compared with pure hydroxyapatite prepared under similar conditions (i.e. elevated porosity) which exhibits a value of 0.07 KJ/m² and fails catastrophically (33).

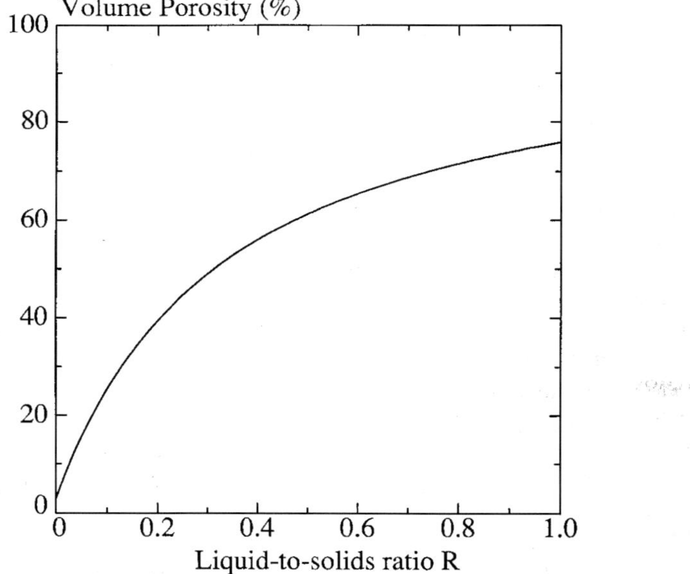

FIGURE 7. A plot showing the variation in the porosity of HAp formed by reaction of CaHPO₄ with Ca₄(PO₄)₂O depending on the proportions of the liquid and solid reactants (31). (Reproduced courtesy of Chapman and Hall, See Reference 31).

SUMMARY

This paper reviews various aspects of the low-temperature formation of hydroxyapatite. This overview was written with a number of objectives in mind. The first was to summarize chemical routes to the formation of HAp monoliths and composites. A second objective was to illustrate the relationships between the phase equilibria, which govern the dissolution and precipitation of HAp and related biologically relevant calcium phosphates, the mechanisms which influence the kinetics of HAp formation, and the variations in the compositions of the aqueous solutions in which the reactions occur.

HAp formation in the manner described in the paper is advantageous in that the precursors can be mixed when required and the reactions forming HAp can occur *in vivo*. This eliminates the need for preforms and permits geometrically

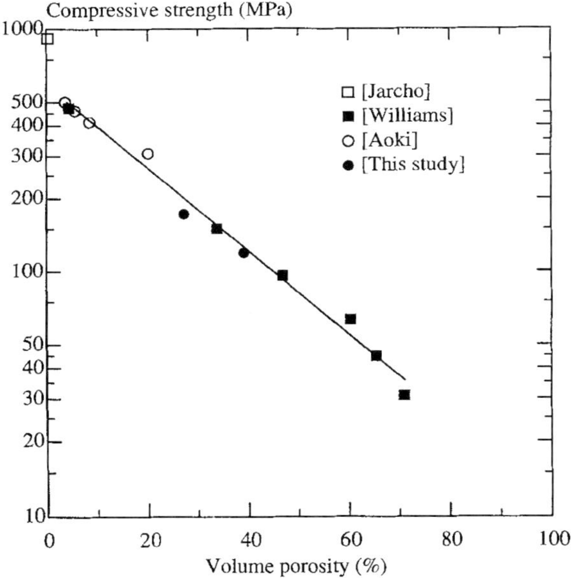

FIGURE 8. The compressive strength of HAp depending on its porosity (31). The data in this figure were obtained in 2 ways. Literature values were obtained for various preparations of sintered HAp. Experimental data were obtained for HAp formed by a reaction which can be approximated to EQ. [2] by varying the liquid-to-solids ratio. The highest compressive strength obtained for HAp prepared according to EQ. [2] was approximately 175 MPa. This is significant because this value is in the range of compressive strength values reported for dentin and while actually exceeding that for bone. Comparison of the strength data for sintered HAp and that produced at low-temperature shows that low temperature HAp and sintered HAp exhibit equivalent compressive strengths at equivalent porosities. (Reproduced courtesy of the Chapman and Hall, See Reference 31).

complex defects in bone and teeth to be filled. Beyond this, HAp compositions which cannot be produced by solid state reactions at high temperature can be prepared. In general, HAp implants fabricated by high-temperature processing, while biocompatible, are unlikely to resorb and, therefore, will remain in place throughout the lifetime of the individual. This is highly desirable if ways can be found to use HAp in the filling of dental caries. Depending on the age and health of the individual, absence of resorption may not necessarily be undesirable for HAp-based synthetic bone preparations. However, it is frequently desirable that an implant be resorbable; the advantage is that it will be replaced over time by native bone. Thus, the formation of monolithic HAp *in vivo* having a composition tailored to a specific need, under conditions which are not harmful to adjacent tissue, and at clinically relevant rates provides the clinician with a class of biomaterials not presently available. The formation of apatites which may exhibit these attributes has been the subject of this paper.

ACKNOWLEDGEMENT

The financial support of NIDR Grant DE09421 is gratefully acknowledged.

REFERENCES

1. J. Black. 1992. *Biological Performance of Materials,* Marcel Dekker, New York.
2. J.B. Park. 1992. *Biomaterials Science and Engineering*, Plenum Press, New York.
3. G. Heimke. 1989. "Advanced Ceramics for Biomedical Applications," Angew. Chem. *101*(1), 111-16.
4. R.G. Craig, O'Brien, W.J. and J.M. Powers. 1987. *Dental Materials, Properties and Manipulation,* Mosby, St. Louis.
5. A.D. Wilson and J.W. Nicholson. 1993. *Acid-base Cements*, Cambridge University Press, Cambridge.
6. R.Z. LeGeros. 1991. *Calcium Phosphates in Oral Biology and Medicine,* Karger, Basel.
7. J.E. Crouch. 1978. *Functional Human Anatomy,* Lea & Febiger, Philadelphia, p. 86.
8. D.V. Provenza. 1986. *Oral Histology,* Lea & Febiger, Philadelphia, p. 269.
9. F.C.M. Driessens. 1983. "Formation and Stability of Calcium Phosphates in Relation to the Phase Compositions of the Mineral in Calcified Tissues," In: *Bioceramics of Calcium Phosphate,* K. deGroot (ed). CRC Press, Boca Ration, FL.
10. G.H. Nancollas. 1982. "Phase Transformations During Precipitation of Calcium Salts," 79-100, *Biological Mineralization and Demineralization,* G.H. Nancollas, Ed. Springer-Verlag, Berlin.

11. B. Tomazic. 1994. Characterization of Mineral Phases in Cardiovascular Calcification," *in Hydroxyapatite and Related Materials,* P.W. Brown and B. Constantz, Eds. CRC Press.

12. W.E. Brown and L.C. Chow. 1988. "A New Calcium Phosphate, Water-Setting Cement," *Cements Research Progress-1987,* P.W. Brown, Ed., 351-79, American Ceramic Society, Westerville, OH.

13 W. Stumm and J.J. Morgan. 1981. *Aquatic Chemistry,* Wiley.

14. P.W. Brown. 1992. "Phase Relationships in the Ternary System CaO-P_2O_5-H_2O at 25°C," J. Am. Cer. Soc. *75,* 17-22.

15. P.W. Brown. 1989. "Phase Equilibria and Cement Hydration," in *The Materials Science of Concrete,* J.P. Skainy, Ed., Am. Ceram. Soc.

16. P.W. Brown. 1990. "The System Na_2O-CaO-SiO_2-H_2O," J. Am. Ceram. Soc. *73*(11) 3457-63.

17. P.W. Brown, Gulick, J. and J. Drumm. 1993. "The Ternary Diagram MgO-H_3PO_4-H_2O at 25°C" J. Am. Ceram. Soc. *76*(6), 1558-62.

18. M.T. Fulmer, Martin, R.I. and P.W. Brown. 1992. "Formation of calcium-deficient hydroxyapatite at near-physiological temperature," J. Mater. Sci: Matls in Medicine *3,* 299-305.

19. O. Bermudez, Boltong, M.G., Driessens, F.C.M. and J.A. Plannell. 1994. "Development of an octacalcium phosphate cement," J. Mater. Sci.: Matls in Medicine *5,* 144-47.

20. O. Bermudez, Boltong, M.G., Driessens, F.C.M. and J.A. Plannell. 1994. "Development of some calcium phosphate cements from combinations of α-TCP, MCPM and CaO," J. Mater. Sci.: Matls in Medicine 5, 160-63.

21. R.I. Martin and P.W. Brown. 1994. "Formation of Hydroxyapatite in Serum," J. Mater. Sci.: Matls in Medicine *5,* 96-102.

22. K. TenHuisen, Martin, R.I., Klimkiewicz, M. and P.W. Brown. 1995. "Hydroxyapatite-Collagen Composites," J. Biomed. Mater. Res. *29,* 803-10..

23. P.W. Brown, Hocker, N. and S. Hoyle. 1991. "The Solution Chemistry of Hydroxyapatite Formation at Low Temperature," J. Am. Ceram. Soc. *74,* 1848-55.

24. R.T. Martin and P.W. Brown. "Aqueous Formation of Hydroxyapatite," J. Biomed. Mater. Res., submitted.

25. R.I. Martin and P.W. Brown, "Hydration of Tetracalcium Phosphate," Adv.Cem.Res. 5 (19), 119-25 (1993).

26. M. Fulmer and P.W. Brown. 1993. "Effects of Na_2HPO_4 and NaH_2PO_4 on Hydroxyapatite Formation," J. Biomed. Mater. Res. *27,* 1095-1102.

27. M. Fulmer and P.W. Brown. 1992. "Low-Temperature Formation of Fluoroapatite in Aqueous Solution," J. Am. Ceram. Soc. *75,* 3401-07.

28. P.W. Brown and M. Fulmer. "Effects of electrolytes on hydroxyapatite formation at 25° and 38°C," J. Biomed. Mater. Res., accepted.

29. L.C. Chow, Takagi, S., Constantino P.D. and C.D. Friedman. 1991. "Self-Setting Calcium Phosphate Cements," *Specialty Cements with Advanced Properties,* B.E. Scheetz et al., Eds, Matl. Res. Soc. Proc. 179, 3-24.

30. Y. Fukuse, Eanes, E.D., Takagi, S., Chow, L.C. and W.E. Brown. 1990. "Setting Reactions and Compressive Strengths of Calcium Phosphate Cements," J. Dent. Res. *69*, 1852-56.
31. R.I. Martin and P.W. Brown. 1995. "Mechanical Properties of Hydroxyapatite Formed at 38°C," J. Mater. Sci.: Matls in Medicine *6*, 138-43.
32. K. TenHuisen and P.W. Brown. 1994. "The Formation of Hydroxyapatite-Gelatin Composites at 38°C," J. Biomed. Mater. Res. *28*, 27-33..
33. G. Lewis. 1990 "The Fracture Toughness of Biomaterials. V. Bone," J. Mater. Educ. *12*, 197-222.

Chapter Nineteen

TOUGHENING MECHANISMS IN CERAMICS

DAVID J. GREEN

Department of Materials Science and Engineering
Steidle Building
Pennsylvania State University
University Park, PA 16802

INTRODUCTION

There has been substantial improvements over the last 25 years in the strength and fracture toughness of ceramics and they are now being considered for many structural applications, especially at high temperatures. The keys to this progress have been an improved understanding of the mechanisms involved in the fracture of these brittle materials and the development of new toughening mechanisms. The aim of this chapter is to review the important concepts involved in these fracture mechanisms and the way in which they relate to the strength of ceramics.

Most people have had first hand experience of the brittle failure process associated with ceramics, such as dropping a plate or breaking a glass. This type of brittle failure process can persist to high temperatures, is usually catastrophic and can occur with little warning. Moreover, it often involves only small amounts of energy and appears to proceed with very little permanent deformation of the material. Similar problems are encountered when ceramics are subjected to rapid temperature changes and the thermal stresses can induce damage or failure. The lack of toughness in ceramics is invariably a key question in technological applications that involve structural and thermal stresses. It is critical, therefore, to understand how the brittle fracture process relates to the structure of the material, especially the mechanisms that are involved in the crack propagation process. The main thrust of this approach has been to understand the way in which the structure of the material influences fracture at the microscopic level. The design of *microstructures* through the appropriate processing of a material is used to control the fracture resistance of the material.

Although brittle failure has been a key factor in the use of materials since primitive man fashioned his first stone tools, an understanding of this process has only been developed in this century and for ceramics within the last 25 years. During this latter period there has been dramatic progress in the development of high strength ceramics and the more ductile metals. It is interesting to note that ceramics are available with toughness and strength values that exceed cast iron, a metal that played such a key role in the Industrial Revolution.

FRACTURE MECHANICS AND STRENGTHENING

The basis for the understanding of the mechanics and thermodynamics of brittle failure was pioneered by Griffith (1) and this was later developed into the field of linear elastic fracture mechanics by Irwin (2). For many brittle materials, this approach assumes there are microscopic defects or flaws in the material. These defects are assumed to exist in the form of cracks and the analysis of the mechanics leads to a simple relationship between strength and toughness, i.e.,

$$\sigma_f = K_{IC}/[Y\sqrt{c}] \qquad (1)$$

where σ_f is the fracture stress, c is the critical crack size, Y is a geometric constant that depends on the loading and the crack shape and K_{IC} is the critical stress intensity factor. In the current ceramic literature, this latter parameter is generally referred to as the fracture toughness and is sometimes given the symbol T.* The elegance of linear elastic fracture mechanics is that it allows fracture to be described in terms of the overall energy changes in the system or by describing the nature of the intense stress field near a crack tip and that these approaches are explicitly linked.

One can use Equation 1 as a guide to the general features that one would utilize to increase the strength of a ceramic material. Clearly one approach would be to increase the fracture toughness value without changing the critical flaw size. It has been found that toughening mechanisms are available in ceramics and usually involve manipulating and tailoring the microstructure. The mechanisms that can lead to toughening can be categorized into three types, and these are shown schematically in Figure 1. *Crack tip interactions* can occur when a crack

---

*In some texts an alternative parameter, the crack (or fracture) resistance, R, is used to describe the failure criterion.

tip interacts directly with an obstacle in the microstructure. In *crack tip shielding,* events are triggered by the high stresses in the crack tip region in a way that reduces the overall amplitude of these stresses. In the third category, ligaments of some type are left behind the advancing crack tip. This *crack bridging* makes it more difficult for the crack faces to open and hence hinders crack propagation.

A second approach to increase strength would be to determine the source of the defects that lead to the *formation of critical cracks* and then to find ways to eliminate or reduce the size of these flaws. It is important that we recognize that ceramics contain various populations of defects that compete as failure origins and within these populations, the defects lead to critical cracks of various sizes and severity. Thus, the strength of ceramics is best represented as a statistical distribution that reflects these variations in defect type and size. In order to reduce the critical flaw size, there are two basic approaches, i.e., one can improve techniques by which the material was made (improved processing) or one can

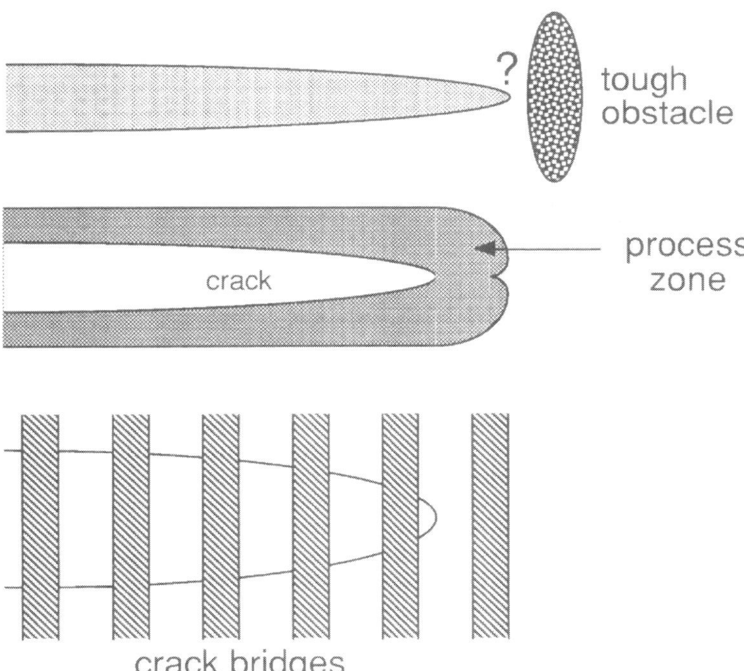

FIGURE 1. Schematic of various toughening mechanisms. In crack tip interactions (top), a crack confronted with a tough obstacle must find a by-pass mechanism. In crack tip shielding (middle) various processes can be activated by the high stresses near the crack tip and this process zone shields the crack from the applied stresses. The lower illustration shows crack bridging, in which unbroken ligaments are left behind the crack tip.

find techniques to eliminate materials that will have the largest critical cracks. This latter approach generally involves the use of *non-destructive evaluation*, to 'find' the most important defects or *proof testing* (stressing all components prior to use) and eliminating those that fail or those that give a signal (e.g., acoustic emission) that the failure process is underway. Clearly these approaches do not account for strength degradation in service and one must devise other schemes to prevent and detect such damage.

The final approach to increasing strength can also be deduced from Equation 1 but is more subtle. This approach attempts to utilize macroscopic residual stresses in a material to reduce the applied stress in the vicinity of the fracture origin. In ceramics, surface flaws are often a major flaw population and thus, techniques that place the surface in compression reduce the effective tensile stress acting on these flaws. In glasses, for example, thermal and chemical tempering processes that lead to surface strengthening have been utilized for many years.

The fracture mechanics approach developed for ceramics and as stated in Equation 1, has been found in some cases to be rather simplistic, especially in the description of the more recently developed high toughness ceramics. For example, in some cases, the fracture resistance of a material may change as a crack grows. Thus, the toughness of a material may no longer be stated as a single number but rather as an 'attribute' of a material. For materials in which fracture toughness increases with crack length, this behavior is often referred to as a *rising R (or T)* curve. An important consequence of this behavior is that stregnth can become less sensitive to flaw size than stated in Equation 1. This effect also leads to differences in the behavior of large and small cracks in a material (3). Indeed, small artificial cracks for studying fracture, are now often introduced into a ceramic using a diamond indenter. This type of approach, often known as *indentation fracture mechanics,* has allowed greater insight into the fracture process and most of the progress using this approach has occurred since 1980. Another effect that will not be discussed here is that cracks, under some circumstances, can propagate at stresses less than that predicted by Equation 1. This *sub-critical crack growth* is often associated with *stress corrosion* or high temperature *inelastic deformation.* Clearly, this can be a key issue in determining the lifetime of a ceramic under stress but it is also beyond the scope of this chapter. For a detailed scientific description of these more advanced topics, interested readers are referred to the excellent book by Lawn (4).

As indicated earlier, the two key issues in increasing the strength of ceramics are the control of toughening mechanisms and the processes by which defects form into critical cracks. It is the purpose of this chapter to discuss the mechanisms involved in the former of these two topics. This type of understanding is a useful demonstration of the techniques that are used by material scientists to control the properties of materials and the means by which ceramics are being made with increased mechanical reliability.

TOUGHENING MECHANISMS

As pointed out in the last section, there are three types of toughening mechanisms that are of current interest in ceramics, viz., crack tip interactions, crack tip shielding and crack bridging. These mechanisms usually involve the interaction of a crack with microstructural features that were incorporated into the material by the material engineers and scientists involved in the development of the material.

CRACK TIP INTERACTIONS

The primary aim of this type of mechanism is to place obstacles in the crack path to impede crack motion. These obstacles could be second phase particles, whiskers, fibers or possibly, regions that are simply difficult to cleave. One would expect that stress concentrations or residual stresses associated with such obstacles would play a role in this process. There are two means by which the crack can by-pass such obstacles. In one case, although the crack is pinned by the obstacle, it can by-pass the obstacle by *bowing* around either side of it, remaining on virtually the same plane (Figure 2). In the other case, the crack could attempt to completely avoid the obstacle by *deflecting* out of the crack plane. The deflection of the crack front can be accomplished by tilting of the crack path or twisting of the crack front (Figure 3). In a real situation, a combination of bowing and deflection may occur but here they will be discussed separately.

Crack Bowing

This mechanism has been analyzed theoretically and it was shown that crack bowing should lead to an increase in fracture toughness (5, 6). There is considerable

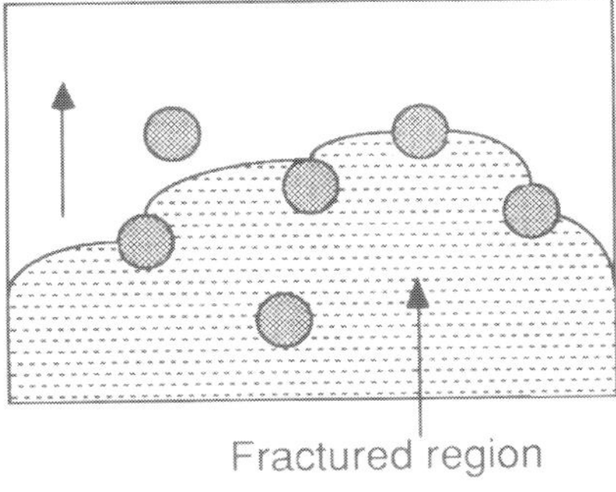

FIGURE 2. Schematic of position of a crack front as it traverses through the cross-section of a material. The crack front can bow as it interacts with tough obstacles in its path.

direct fractographic evidence for such bowing and this has been summarized previously (7). An example of a crack bowing round an inclusion in glass is shown in Figure 4. In this figure the crack front fringes were produced by ultrasonic fractography. The fringe spacing represent crack front positions at one microsecond time intervals (7). The theoretical analyses of the crack bowing mechanism assume the obstacles are impenetrable but one would expect that the strength and toughness of such obstacles to be a key issue. For example, one might expect either the obstacle could fail before the bowing process is complete or the obstacles

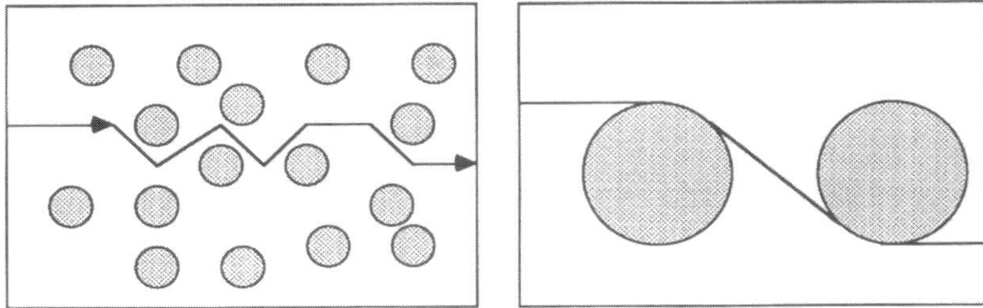

FIGURE 3. Schematic of crack deflection; a) crack path tilts to avoid obstacles and b) crack front twists to by-pass obstacles.

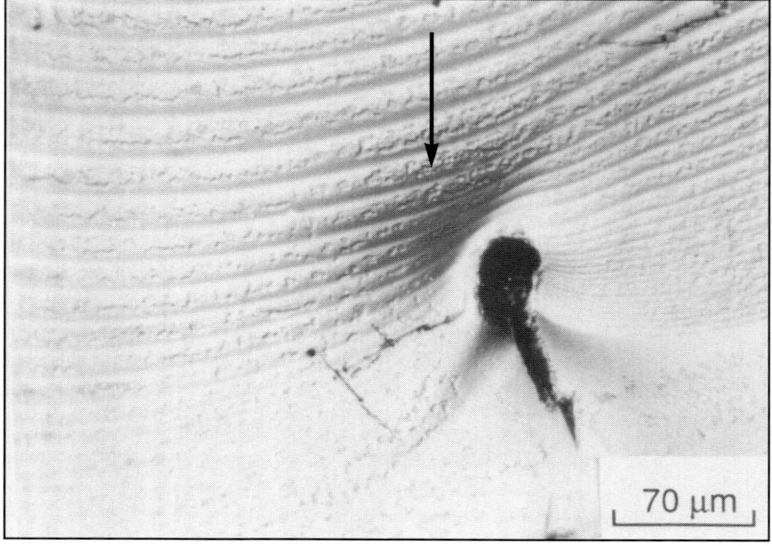

FIGURE 4. Interaction of a crack front with an inclusion in glass showing crack bowing. The fracture surface markings were produced by ultrasonic fractography (frequency 1 MHz). Crack direction is indicated by the arrow.

would left behind as unbroken ligaments behind the crack tip. In this latter case, crack bowing can be seen as a precursor to crack bridging. Two other important aspects that were not considered in the original analysis are the influence of the local stress fields around the obstacle and the interaction between the bowing crack front and nearby obstacles. In some cases, the former effect can allow the crack front to by-pass obstacles by allowing the crack front to deflect out of plane.

Crack Deflection

If a crack is deflected out of the plane that is normal to the applied tensile stress, the crack will no longer be subjected to the maximum tensile stress and the crack faces will involve shear displacements. This behavior, known as *mixed mode fracture*, occurs on the microscopic scale and is expected to increase the difficulty of crack propagation. Two types of deflection have been analyzed: tilting of the crack about an axis parallel to the crack front and twisting about an axis normal to the crack front. The overall deflection process is manifested as roughness of the final fracture surface. The reorientation of the crack plane leads to a reduction of the crack extension force on the deflected portion and hence an increase in the fracture resistance (8). The crack deflection mechanism was analyzed by evaluating the stress field components associated with the deflected portion. These components were then combined with a mixed mode failure criterion and compared to the crack resistance of the matrix that surrounds the obstacle. The fracture mechanics analysis demonstrated that it is the twist component that contributes most to the increase in toughness (8). For a random array of obstacles, it was also shown that the toughening increment depends on the volume fraction and shape of the particles (8). For a given volume fraction of obstacles, shaped as rods, discs or spheres, rods were found to be the most effective in increasing toughness, as shown in Figure 5. As seen in this figure, it was also found that most of the toughening develops for volume fractions of obstacles <0.2. The toughening predicted for rod-shaped particles is shown schematically in Figure 6 and rods with large aspect ratios were found to impart maximum toughness. This effect is a result of the increase in twist angle that occurs with increasing aspect ratio (8). The fracture mechanics analysis of crack deflection does not account for local stress fields or local changes in the crack resistance as the crack by-passes an obstacle. Measurement of fracture toughness values and crack deflection angles in various ceramics have been shown to be consistent with the predictions of the crack deflection mechanism. Figure 7 shows a micrograph of the crack deflection process in a silicon carbide platelet-alumina composite.

CRACK TIP SHIELDING

The high stresses near a crack tip can trigger various events and reactions in a *process zone*. In some cases, these events decrease the stress amplitude in this region below that expected from the applied loads. In this sense, the process zone 'shields' the crack tip region from the applied crack extension forces. There are two primary mechanisms that have been identified as leading to crack tip shielding, viz., transformation toughening and stress-induced microcracking.

Transformation Toughening

An elegant approach to the toughening of ceramics is the use of a stress-induced phase transformation and there has been several reviews on this subject (9-12). The most successful utilization of transformation toughening has been in the use of zirconia-based materials. Zirconia has a high temperature tetragonal phase that usually undergoes a martensitic phase transformation to monoclinic

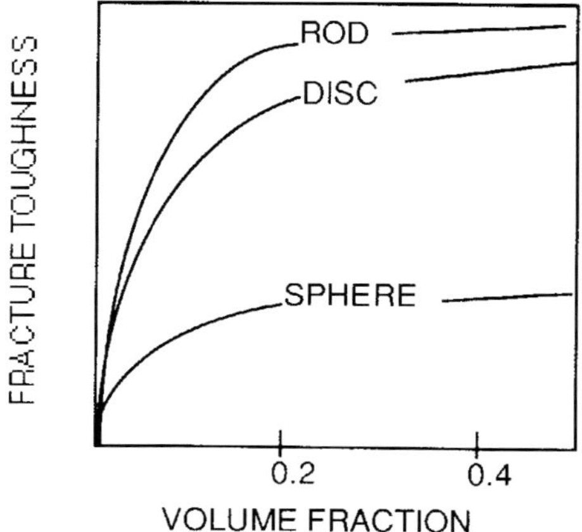

FIGURE 5. Toughening by crack deflection is estimated to increase as the obstacle shapes change from spheres to discs and rods.

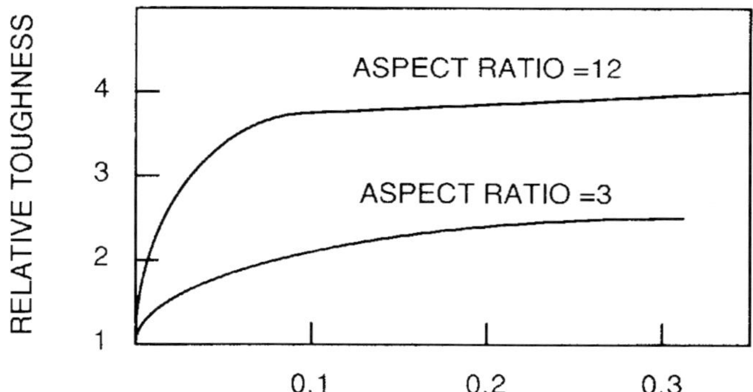

FIGURE 6. Toughening by crack deflection is estimated to increase as the aspect ratio of rods increases.

during cooling (~1000°C). The transformation to the monoclinic phase involves a 3 to 5% volume increase and a simple shear of ~16%. These large transformation strains can invoke large transformation stresses in a material. Thus, when one considers the thermodynamics of the transformation, the strain energy due to the transformation needs to be taken into account. This has the effect that the tetragonal phase, if constrained, can be retained to temperatures below the normal transformation temperature. In addition, undercooling will also be possible if the monoclinic phase is difficult to nucleate. Indeed, it is possible to retain the tetragonal phase to below room temperature provided the zirconia particle size is below a critical value. It is these materials in which the tetragonal phase is retained that are able to utilize the mechanism of transformation toughening.

There are a wide variety of microstructures that can incorporate the transformation toughening effect but most of the work has been performed on materials in which the zirconia has been 1) precipitated in a cubic zirconia matrix to form a two phase microstructure, 2) formed into a particulate composite (e.g., Al_2O_3/ZrO_2) or 3) alloyed into a solid solution (e.g., with Y or Ce as the solute). An example of a composite zirconia microstructure is shown in Figure 8. Fracture toughness measurements on zirconia-toughened materials have shown

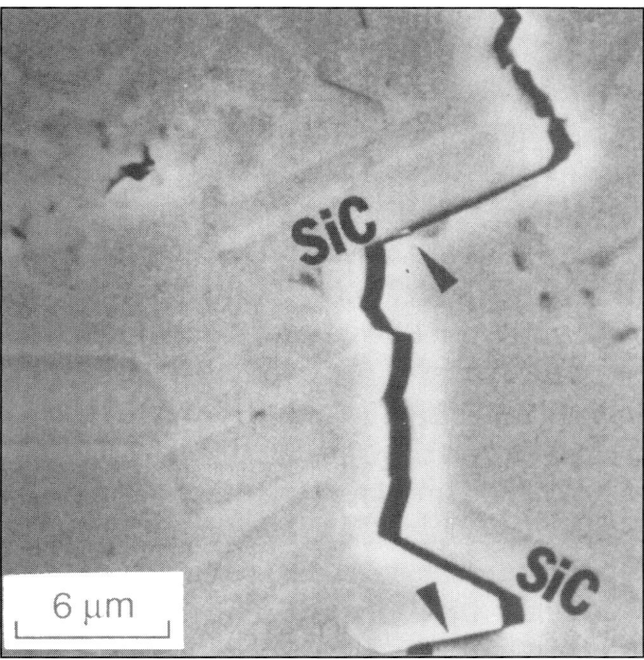

FIGURE 7. Crack deflection caused by SiC platelets in an alumina matrix.

that values in excess of 20 MPa.√m are possible.

Clearly it is important to identify the role of the tetragonal ZrO_2 in this toughening phenomenon and the parameters that control fracture toughness. Figure 9 shows a schematic illustration of a crack surrounded by a transformation zone. The increase in volume caused by the transformation of the particles will induce the zone to dilate but the non-transformed material around the zone will oppose this increase in zone size. The transformation zone will therefore be in a state of residual compression and this will act to close the crack and hence, will make crack propagation more difficult.

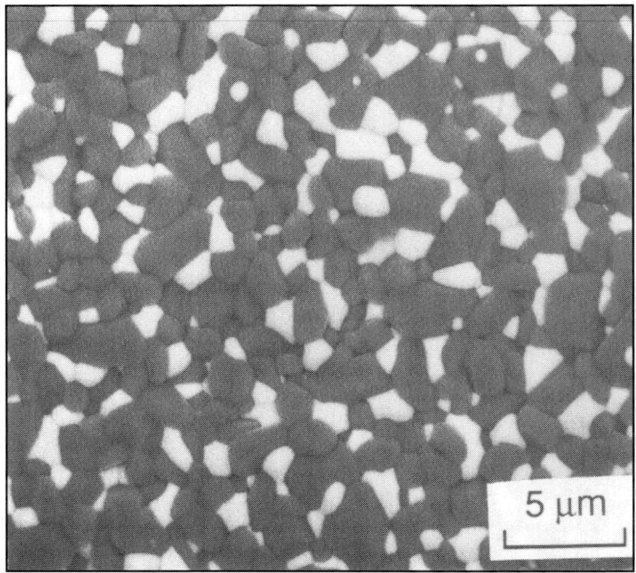

FIGURE 8. Microstructure of an alumina-zirconia composite. The zirconia is in the tetragonal form but transforms to monoclinic under the action of stress.

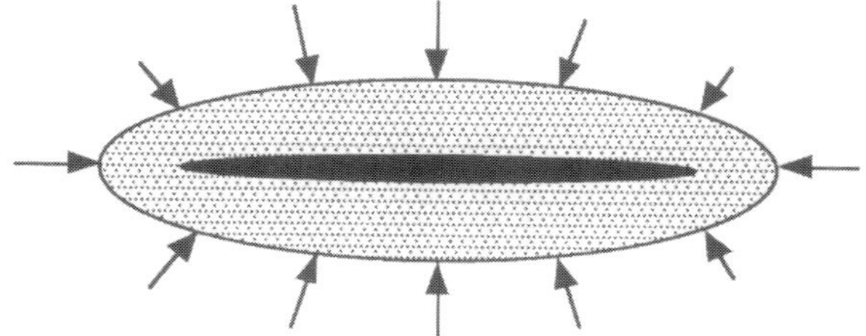

FIGURE 9. Schematic illustration of a crack surrounded by a process zone in which the zirconia has been transformed by stress. The elastic constraint of the surrounding non-transformed material places the zone in compression, acting to close the crack.

There have been various approaches to modeling transformation toughening. One of these considers the influence of the stress field on the stress intensity factor (13), while the other two utilize a thermodynamics approach (14,15). All three approaches came to identical conclusions and thus, give a compelling argument about the way the toughness develops and the parameters that control the process. The analyses first consider a crack tip in which there is only a frontal transformation zone. The toughness increase associated with a frontal zone is often found to be small. It is only when the zone extends behind the crack tip that the toughening occurs, i.e., as the zone extends into the crack wake (Figure 10). Thus, one expects the toughening to increase as the crack propagates (rising R curve behavior). For long cracks, it is found that the toughening reaches an asymptotic limit and the associated toughening (ΔT) is given by

$$\Delta T = CEVe^{T}\sqrt{h}/[1 - \upsilon]$$

(2)

where E and υ are the Young's modulus and Poisson's ratio of the material, e^{T} is the volumetric transformation strain, V is the volume fraction of material transformed by stress and h is the height of the transformation zone (13). The constant C depends on the assumptions involved in the calculations but is usually in the range 0.2 to 0.4 (9).

To utilize transformation toughening, one often wishes to increase the transformation zone by making the critical transformation stress as small as possible, e.g., by optimizing the zirconia particle size. There is, however, a limit to this approach if the transformation zone size approaches the component size. For these cases, the process zone will no longer be constrained and the crack closure effect will be lost. The strength of the ceramic will then become equivalent to the transformation stress. In many ways, transformation toughening is analogous to

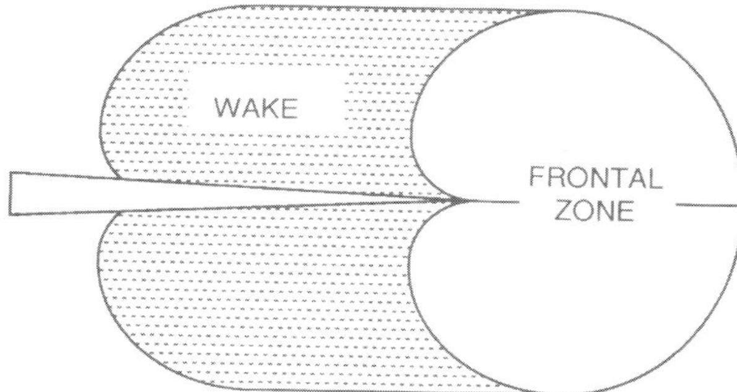

FIGURE 10. The zirconia transformation in a frontal zone does not usually lead to a significant increase in fracture toughness. The major toughening usually occurs as the transformation zone passes into the crack wake.

plastic deformation that is used so effectively in metals. Decreasing the yield stress in metals can increase toughness but in the limit, it will also decrease strength as the fracture stress becomes equivalent to the yield stress. In contract to ductility in metals, however, transformation toughening can become more difficult with increasing temperature as the (tetragonal) zirconia particles become more stable. The low fracture toughness at high temperature for transformation-toughened materials implies that their main applications are probably restricted to moderate temperatures.

The above discussion has emphasized the martensitic phase transformation in zirconia. Other possibilities do exist and these have been explored but without the same success. For example, martensitic phase transformation are known to occur in other ceramic materials. It has also been suggested that ferroelastic transformations and other stress-induced twinning processes can be used to increase fracture toughness (16).

Microcrack Toughening

Ceramics that contain localized residual stresses are known to be capable of microcracking (17). These residual stresses arise in ceramics as a result of phase transformations, thermal expansion anisotropy in single phase materials and thermal expansion or elastic mismatch in multiphase materials. It is also expected that regions of low toughness, such as grain boundaries, would be attractive sites for such cracks. It has been known for some time, that these microcracks can form spontaneously during the fabrication process, provided the grain or particle size is above a critical value. This critical size effect can be analyzed using fracture mechanics and the results for various configurations have been reviewed (17). Ceramics containing microcracks after fabrication have been associated with good thermal shock resistance (18) but it is expected that such materials would have low strengths as the microcracks are likely failure origins. A more attractive proposition, in terms of fracture toughness, is to fabricate materials in which the particle size is below that for spontaneous microcracking but that microcracks could be induced by stress. Calculations that consider microcracks forming under an applied tensile stress indicate that particles down to about one-half of the critical size might be candidates for the formation of stress-induced microcracks (17). In terms of fracture, one would expect that the microcracks would form in a zone around large cracks, much in the same way as a stress-induced phase transformation zone forms (section 3.2.1). An alternative view of the microcrack toughening phenomena would be to consider the increased amount of fracture surface that is associated with the formation of such a zone.

As mentioned above, the shielding process associated with a microcrack zone is, in some ways, similar to that involved in transformation toughening. In the latter case, it was indicated that the dilatation of the process zone was a primary factor in the toughening process. The formation of microcracks is generally associated with residual stress fields and when a crack forms in such a field, it will give rise to a volume increase as the crack opens. Thus, if this volume increase can be determined as a function of the microcrack density, the toughening increment could be predicted simply from the transformation toughening analyses. There are, however, two other effects that are important in the microcrack shielding process. The first of these effects gives rise to additional shielding and is a result of the decrease in the elastic modulus that occurs when a material microcracks. This elastic stiffness (modulus) effect will itself reduce the stresses in the zone and thus, aids in the shielding process. The other effect is based on the realization that the microcracks must degrade the fracture toughness within the zone. That is, the microcracks have degraded the material in the process zone and fracture surface is already available when the main crack propagates. It is usually shown that in the degradation of the material by the microcracks is offset by the shielding modulus effect and thus, again it is the dilation that is the main source of toughening (19). Thus, the asymptotic toughening increment is expected to be similar to that given by Equation 2 except that Ve^T is replaced by the volumetric strain associated with the microcracking, e^M. The analysis also predicts stress-induced microcracking will give rise to a rising R curve behavior.

The overall trends in the fracture toughness increment due to microcracking are very similar to those of transformation toughening. For example, as shown in Figure 11, the toughness is expected to increase with particle size up to the critical particle size for spontaneous microcracking. For larger particle sizes, the material is already microcracked prior to the application of stress and further toughening will only arise if one can increase the microcrack density. The temperature sensitivity could be somewhat different than discussed for transformation toughening depending on the source of the residual stress. If the microcracks are a result of some type of thermal expansion mismatch, then one would expect that increasing temperature would result in a decrease in toughness, as the magnitude of the residual stress decreases. On the other hand, if the microcracks were formed by a phase transformation, the toughening effect would be relatively temperature insensitive unless the transformation reverses. Increasing the volume fraction of the microcracks is clearly attractive but again, interaction effects may increase the likelihood of spontaneous transformation and microcrack linking could be a prime source of pre-existing flaws in a material. The factors that control the microcrack zone size are expected to be similar to those that control spontaneous microcracking and include the magnitude of the residual stress, particle size and the size (and shape) of defects within the residually-stressed regions (17). In a similar way to transformation toughening, one would expect that there will be a limit to the toughening if the microcrack zone size approaches the specimen size and a 'yield-limit' to the strength if the critical microcracking stress becomes too small. Non-linear elastic behavior associated with microcracking has been observed in overaged partially-stabilized zirconia (20, 21) and microcrack toughening has been considered to be an important toughening mechanism in some zirconia systems (9) and in SiC/TiB_2 particulate composites (22).

CRACK BRIDGING

In the discussion of crack tip interactions it was indicated that a crack may by-pass an obstacle, leaving it intact. In such cases, the obstacle is left as a ligament behind the crack tip. In other cases, ligaments may be formed by the mechanical interlocking of grains. These ligaments will make it more difficult to open the crack at a given applied stress and will increase fracture toughness. This mechanism has been shown to be important in frictionally-bonded fiber composites (23, 24), in large-grained Al_2O_3 (25, 26), in polycrystalline ceramics with lenticular grain structures (27) and in whisker- and metal-reinforced ceramics (28, 29).

The addition of continuous carbon or SiC fibers to glasses or ceramics has been studied since about 1970 and it was found, in some cases, that unidirectional fiber composites do not undergo catastrophic failure in tensile loading (30). This 'ductile' type of behavior for a material composed of two brittle components is particularly attractive for structural applications and is illustrated in Figure 12. In the optimum materials, the tensile loading behavior is initially elastic until at a particular stress, a crack passes through the matrix. This crack, however, by-passes the fibers and leaves them available for load carrying and the fibers completely bridge the crack. Further loading causes the formation of regularly-spaced, bridged, matrix cracks until at or near the peak load, the fibers fail. The ensuing failure, however, is not catastrophic as the fibers continue to pull out of the matrix. This pull-out process also consumes energy and contributes to the toughness. In these materials, the final failure is not the result of

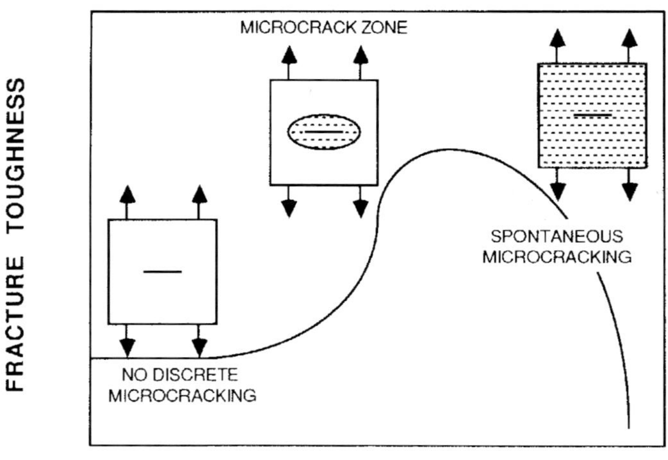

FIGURE 11. Fracture toughness increases with particle size in a microcrack-toughened ceramic unless the size exceeds the critical size for spontaneous microcracking.

the propagation of a single crack and thus, a fracture toughness value cannot be defined. It is, however, possible to use fracture mechanics to describe the conditions at which the first crack passes through the matrix (24). In this analysis, it was shown under some circumstances that the matrix cracking stress (σ_0) is independent of the pre-existing flaw size and is a material property. The analysis showed that σ_0 depends on the matrix fracture toughness, the interfacial shear strength, the volume fraction and radii of the fibers and the elastic constants of the two components. Techniques to increase σ_0 can be ascertained from the analytical expression, but if this stress approaches the fiber strength, fiber failure will accompany matrix cracking and there will be a transition to a more brittle type of failure (12). For these cases, a bridging zone of limited extent is formed behind the crack and moves with the crack. Such a zone can still, however, contribute significantly to the fracture toughness but the 'ductile' failure process shown in Figure 12 will be lost. A key feature of fiber-reinforced composites is the need to control the interface between the fiber and the surrounding matrix. This interface needs to possess a low toughness so that an approaching crack will face a debonded fiber or be deflected by the fiber, i.e., it must not pass through the fiber. This is a major challenge to the material, because fiber failure may intervene prematurely and the bridging will be lost.

The emphasis in the above discussion was on fully-bridged cracks but it is now realized that partial bridging can be a useful toughening mechanism. In

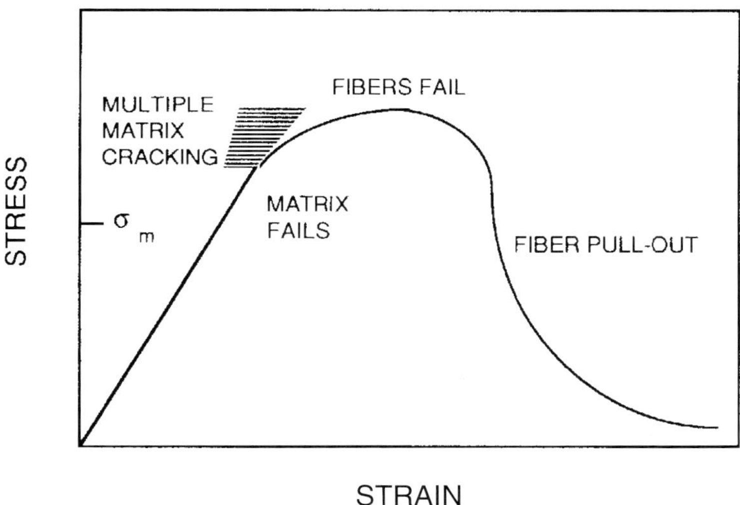

FIGURE 12. For a fully bridged crack, the matrix can undergo multiple cracking and the final failure can involve fiber pull-out.

order to determine the effect of crack bridging on fracture toughness, it is necessary to know the force-displacement relationships for the ligaments (4,12). The bridges in these materials may be ductile or brittle, as shown in Figure 13.

Ductile bridges are usually metal particles or ligaments and toughening in the cemented carbides (ceramic-metal composites) and in ceramics produced by the Dimox™ process is thought to occur by this mechanism (12). The increment in crack resistance from this type of bridging is given by (12).

$$\Delta R = \chi V \sigma_Y r \tag{3}$$

where r is the radius if the bridge, V is the area fraction of bridges across the crack plane, σ_Y is the yield stress and χ is a parameter that depends on the failure strain of the bridge. If the bridges remain bonded χ is approximately unity but for partially-debonded bridges χ may approach 8 (12).

For brittle bridges, weak interfaces again become important and residual stresses due to thermal expansion anisotropy or mismatch appear to play a role in bridge formation. Brittle bridges must undergo some debonding so that significant displacements can occur at the bridge sites. It has been suggested that the increment in crack resistance is a result of three terms that involve the energy in bridge failure, debonding and frictional pull-out (12), i.e.,

$$\Delta R = \frac{V d S^2}{E} + \frac{4 \Gamma d V}{r(1-V)} + \frac{2 \tau V L^2}{r} \tag{4}$$

where d is the debond length, S is the bridge strength, Γ is the interfacial fracture

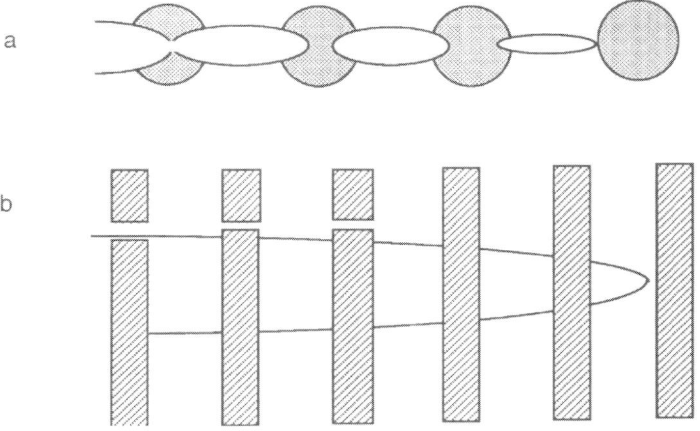

FIGURE 13. Partial crack bridging produced by a) ductile or b) brittle obstacles.

energy, E is Young's modulus, L is the pull-out length and τ is the shear resistance of the interface after debonding.

CONCLUSIONS

In the last 25 years, ceramics have been developed with fracture toughness and strength values that are one order of magnitude greater than found in the more traditional ceramic materials. The basic mechanisms have been identified by a careful combination of microstructural design, new processing techniques, observations of the fracture process, new property measurement techniques and micromechanical modeling. In a sense, there has been the advent of a new field of scientific endeavor. There is, however, much remaining to be explored. Scientists have really just touched on the ways that materials can be combined into composite structures that are designed at all structural scales; from the atomic to the macroscopic. No doubt new fracture mechanisms will be identified, ways will be found to combine mechanisms and it will be only limited by the imagination of the materials scientist.

ACKNOWLEDGEMENTS

The author would like to thank John Hellmann for critically reading the manuscript and to Pat Nicholson, Dave Embury, Dick Hoagland and Fred Lange for their advice and encouragement in this scientific area.

REFERENCES

1. Griffith, A.A. 1920. The phenomena of rupture and flow in solids. *Phil. Trans. Roy. Soc.*, (London), A221:163-98.
2. Irwin, G.R. 1958. Fracture, pp. 551-589 in *Handbuch der Physik,* vol. 6. Springer-Verlag, Berlin.
3. Cook, R.F., Lawn, B.R. and C.J. Fairbanks. 1985. Microstructure-strength properties in ceramics, *J. Am. Ceram. Soc.,* 68: 604-15.
4. Lawn, B.R. 1993. *Fracture of brittle solids - second edition,* Cambridge University Press.
5. Lange, F.F. 1970. Interaction of a crack front with a second phase dispersion, *Phil. Mag.,* 22: 983-92.
6. Evans, A.G. 1972. The strength of brittle materials containing second-phase dispersions, *Phil. Mag.,* 26: 1327-44.
7. Green, D.J. 1976. *Crack-particle interactions in brittle composites,* Ph. D. Thesis, McMaster University, 1976.
8. Faber, K.T. and A.G. Evans 1983. Crack deflection processes - I. Theory,

Acta Metall., 31: 565-76.

9. Evans, A.G., and R.M. Cannon. 1986. Toughening of brittle solids by martensitic transformations, *Acta Metall.,* 34: 761-800.

10. Heuer, A.H. 1987. Transformation toughening in ZrO_2-containing ceramics, *J.Am.Ceram. Soc.,* 70:689-98.

11. Green, D.J., Hannink, R.H.J. and M.V. Swain. 1988. *Transformation Toughening of Ceramics,* CRC Press, Boca Raton, Florida.

12. Evans, A.G., 1990. Perspective on the development of high-toughness ceramics, *J.Am. Ceram. Soc.,* 73: 187-206.

13. McMeeking, R.M. and A.G. Evans. 1982. Mechanics of transformation toughening in brittle materials, *J. Am. Ceram. Soc.,* 65 : 242-46.

14. Marshall, D.B., Drory, M.D. and A.G. Evans. 1983. Transformation toughening in ceramics, pp. 289-307 in *Fracture Mechanics of Ceramics,* Vol. 6, R.C. Bradt et al., (eds.), Plenum Press, New York.

15. Budiansky, B., Hutchison, J. and Lambroupolos, J.C. 1983) Continuum theory of dilatant transformation toughening in ceramics, *Int. J. Solids Struct.,* 19: 337-55.

16. Virkar, A. V. and R.L.K. Matsumoto. 1986. Ferroelastic domain switching as a toughening mechanism in tetragonal ZrO_2, *J. Am. Ceram. Soc.,* 70: 279-89.

17. Green, D.J. 1983. Microcracking mechanisms in ceramics, pp. 457-78 in *Fracture Mechanics of Ceramics,* Vol. 5, R.C. Bradt et al., (eds.), Plenum Press, New York.

18. Hasselman, D.P.H. 1969. Unified theory of thermal shock resistance of ceramic materials, *J. Am. Ceram. Soc.,* 52: 600-604.

19. Evans, A.G. and K.T. Faber 1984. On the crack growth resistance of micro-cracking brittle materials, *J. Am. Ceram. Soc.,* 67: 255-60.

20. Green, D.J., Embury, J.D. and P.S. Nicholson. 1973. Fracture toughness of a partially stabilized ZrO_2 in the system CaO-ZrO_2, *J. Am. Ceram. Soc.,* 56: 619-23.

21. Green, D.J. and P.S. Nicholson. 1973. Microstructure development and fracture toughness of a calcia partially-stabilized zirconia, pp. 541-54 in *Fracture Mechanics of Ceramics,* Vol. 2, R.C. Bradt et al., (eds.), Plenum Press, New York.

22. Gu, W-H., Faber, K.T. and R.W. Steinbrech, Microcracking and R curve behavior in SiC-TiB_2 composites, *Acta Metall.,* 40: 3121-28.

23. Aveston, J., Cooper, G.A. and A.Kelly. 1971. Single and multiple fracture, pp. 15-26 in *Properties of Fiber Composites.* IPC Science and Technology Press Ltd., Surrey, England.

24. Marshall, D.B., Cox, B.N. and A.G. Evans.1985. The mechanics of matrix cracking in brittle-matrix fiber composites, *Acta Metall.,* 33: 2013-21.

25. Steinbrech, R., Khehans, R. and W. Schaarwachter. 1983. Increase of crack resistance during slow crack growth in Al_2O_3 bend specimens, *J. Mater. Sci.,* 18: 265-70.

26. Swanson, P.L., Fairbanks, C.J., Lawn, B.R., Mai, Y-W and B.J. Hockey.1987.

Crack-interface grain bridging as a fracture resistance mechanism in ceramics, *J. Am. Ceram. Soc.*, 70: 279-89.

27. Lange, F.F. 1979. Fracture toughness of Si_3N_4 as a function of initial $\sigma-$ phase content, *J. Am. Ceram. Soc.*, 62: 428-30.

28. Becher, P.F. 1991. Microstructural design of toughened ceramics, *J. Am. Ceram. Soc.*, 74: 255-69.

29. Evans, A.G. and R.M. McMeeking. 1986. On toughening of ceramics by strong reinforcements, *Acta Metall.*, 34: 2435-41.

30. Marshall, D.B. and A.G. Evans. 1985. Failure mechanisms in ceramic-fiber/ceramic matrix composites, *J. Am. Ceram. Soc.*, 68: 225-31.

Chapter Twenty

NONOXIDE STRUCTURAL CERAMICS AND ADVANCED CERAMIC FIBERS

R.E. TRESSLER

Department of Materials Science and Engineering
Steidle Building
Pennsylvania State University
University Park, PA 16802

INTRODUCTION

Advanced structural ceramics have been the subject of intense research and development efforts worldwide for some 20 years with the emphasis shifting somewhat away from monolithic ceramics to toughened ceramics (see Chapter 19 for a more detailed discussion of toughened ceramics and ceramic matrix composites (CMCs) in the last five years.) These efforts have been stimulated by the recognition that the substitution of lower density, higher heat resistant, lower cost (potentially) ceramics for metals in many high temperature applications would result in major energy savings. In addition, the ability to operate at higher temperatures with better performance (higher strength to weight ratios, for example) than is possible with metals would enable the design of much improved machines. The use of CMCs in the hot sections of the large gas turbine planned for the High Speed Civil Transport is such an application which would result in a thrust to weight ratio improvement over current engines of almost 2:1. Of course, a whole new design is then possible in which many additional improvements are incorporated leading to this large payoff.

Although the future looks bright for this class of highly engineered advanced materials, *it has for many years now.* In fact, predictions for sales of these materials (see for example Ref. 1) has shown very marked increases always about five years later than the date of the prediction. The major obstacles to reaching the goals continue to be (1) cost and reproducibility of fabrication, and (2) mechanical reliability. The community of materials developers and fabricators has simply underestimated the difficulty in reducing costs and in achieving adequate mechanical reliability. In addition, the design and engineering communities have not been adequately educated about the special methodologies that are required for design with these materials (probabilistic approach to strength rather than deterministic, for example).

Nevertheless, the worldwide markets for these materials are growing. The majority of the sales volume is still oxide structural ceramics for wear resistant and erosion resistant components in the mineral processing industry, the transportation industry (including automotive), metal forming industry, etc. where alumina ceramics dominate particularly the newer toughened aluminas. The toughened zirconias have made inroads and have penetrated markets which were once the stronghold of cemented carbide materials (wire drawing dies, metal forming dies, cutting edges, etc.). Silicon nitride and whisker toughened aluminas with high strength and high toughness have become very important cutting tool materials in specialized niches (such as cutting and machining of hard alloys such as cast irons and nickel alloys, and interrupted cutting of hard alloys). Specialized markets exist for silicon carbide ceramics, silicon nitride ceramics, boron carbide (armor, nuclear reactor parts and nozzles), aluminum nitride (electronic packaging and armor), and for other nonoxide ceramics.

The silicon nitride and silicon carbide families of materials dominate the developmental materials markets particularly for high temperature applications because of their high thermal conductivities, high hardnesses, low densities, low thermal expansivities, and potential for high strength and high toughness with appropriate microstructural tailoring. These desirable properties are derived from their covalent bonding and their crystal structures. In contrast to most oxides which tend to be more ionic in bonding and often have closepacked oxygen crystal structures, these materials have stronger bonding, and thus, are more refractory and have very low intrinsic atomic diffusivities. These features are highly desirable for stiff, creep resistant (resistant to permanent deformation), structural materials at high temperature. So they are excellent structural materials in monolithic form and also as potential reinforcement particles, whiskers, or continuous fibers for high temperature composites.

But their use at high temperatures in oxidizing environments, where they are thermodynamically unstable since SiO_2 is much more stable, is permitted because they form a SiO_2 scale or skin which forms an impediment to further oxidation since oxidant must diffuse through this very impermeable glassy scale. It is this silica-forming capability which sets the Si_3N_4 and SiC families of ceramics apart from other carbides and nitrides such as B_4C, TiC, TiN, ZrC, ZrB_2, etc. which form crystalline, friable oxidation products which are not protective. However, there are serious limitations to the protectiveness which has been the subject of considerable research especially at Penn State (see for example Ref. 2).

This chapter will concentrate then on the silicon carbide and silicon nitride families of advanced structural ceramics emphasizing the developments of the last ten years. Since in monolithic form these materials fail catastrophically, their properties for some key applications can be improved by making CMCs which fail "gracefully." Thus, the recent developments in CMCs emphasizing the role of silicon carbide and silicon nitride as matrix materials and fibers will be discussed. Oxide fibers are also being researched for reinforcement fibers, so they will also be briefly reviewed.

DEVELOPMENTS IN SILICON CARBIDE-BASED CERAMICS

The starting point for most ceramics is a suitable powder which can then be mixed with sintering aids and densified in a high temperature firing step, and some of the most important SiC ceramics are indeed fabricated in this way. But because of the covalency of SiC and the resultant very slow atomic diffusion (along with unfavorable surface energies), alternative routes to making silicon carbide ceramics have been historically more important than classical powder forming and sintering route although major advances in the last 15 years have led to advanced silicon carbide ceramics made by the sintering route. Powdered SiC is often an important ingredient in processing by alternative routes although reactive melt forming of SiC and deposition of SiC from the vapor phase to make dense parts do not rely on SiC powders. In this discussion SiC powder formation and ceramic fabrication from powders will be discussed first followed by the alternative routes to forming structural components of SiC.

Silicon Carbide Powder

Silicon carbide first found widespread use as an abrasive grain and as a component in grinding wheels because of its high hardness, refractoriness, and low cost of production. The invention of the Acheson Furnace process was the key event in making SiC grain widely available at a reasonable cost for these applications. Indeed the SiC products became known as Carborundum after the company that pioneered many of the early applications (See Ref. 3 for the first 100 years history of this interesting company). The carbothermic reduction of SiO_2 ($SiO_2 + 2C = SiC + SiO_2$) is performed in large furnaces by using graphite electrodes to pass high electric current through the compact of C and SiO_2, thereby heating the assemblage to temperatures in excess of 2700°C resulting in high yields of α-SiC crystals and agglomerates which are separated from unre-acted material and crushed and milled to the desired size of particles. The alpha form of SiC refers to the hexagonal or rhombohedral polymorphs of SiC in which tetrahedra of SiC are stacked in an ACAC sequence with the polymorphs resulting from systematic faults in this stacking of the layers of tetrahedra. Silicon carbide also has a beta form which is cubic and results from an ABCABC stacking of the layers of SiC tetrahedra (Figure 1). In both the alpha polymorphs and the beta poly-morph the structure is polar so that the $\{000\bar{1}\}$ and $\{\bar{1}\bar{1}\bar{1}\}$ planes are carbon termi-nated and the $\{0001\}$ and $\{111\}$ planes are silicon terminated. These planes show dramatically different oxidation rates particularly at lower temperatures, a phenomenon that is not understood in detail today (4). The impact of this polarity of structure on SiC crystals as semiconductor devices is discussed in Chapter 22.

The beta SiC structure is found in powders formed at low temperatures (below ~1400°C) or in CVD deposited SiC deposited in the same temperature range while the alpha form invariably results from sintering or heating to temperatures in excess of ~2000°C. Thus, both forms are important in the field of structural ceramics. Although the lower cost Acheson furnace prepared α-SiC is the common SiC powder, vapor phase formed β-SiC powder (e.g., by the reaction of gaseous silicon hydrogen chlorine compounds with hydrocarbons or halogenated hydrocarbons) is used for additive powders in composites and as a starting powder for some sintered parts because a very strong patent position by the Carborundum Co. in sintered α-SiC ceramics has excluded other companies from using that approach.

Reaction Formed Silicon Carbide Ceramics

The oldest form of fired ceramics containing silicon carbide are vitreous bonded or clay bonded ceramics in which the SiC grains are mixed with clay or glass frit (powdered glass), formed to shape, and fired in air so that the oxide binder forms a glass which bonds the particles usually in a porous ceramic body. These materials are used for grinding wheels and refractory applications.

Another variant uses powdered silicon in the mixture which is fired in an ammonia atmosphere so that silicon nitride and/or silicon oxynitride is formed and bonds the SiC grains together again in a porous material. This type of material resists attack by molten metals and is more refractory than some of the oxide bonded products. A variant of this process is reaction sintering, but it is covered below since it involves molten metal infiltration and has been adapted to fabricate dense high performance structural ceramics.

Sintered Silicon Carbide

In the 1970's the Carborundum Company and the General Electric Company both discovered how to sinter very fine grain SiC powder to a dense, strong ceramic with the use of boron and carbon as sintering aids. Because the Carborundum patent used α-SiC powder while the G.E. patent used β-SiC powder, the Carborundum patent was more practical. In addition, if the G.E. process was carried out without very careful controls (zero alpha grains present, for example), large alpha grains formed during the sintering process and inhib-ited densification and the large grains became strength limiting defects resulting in lower strength material. With their strong process and product protection

Carborundum proceeded to develop a commercial market for high, strength, corrosion resistant, creep resistant SiC parts. Their sintering aid, boron, was a particularly fortuitous choice since the oxidation product, B_2O_3 merely dissolved in the SiO_2, the SiC oxidation product, forming a protective borosilicate glass which is almost as attractive as SiO_2 (5).

Other companies attempted to market sintered alpha SiC ceramics with Al_2O_3, Al, Al_4C_3, AlN, as sintering aids, but the oxidation products tended to be crystalline and subject to cracking which made the oxide scale nonprotective, thus eliminating one of the major advantages of SiC over other nonoxides.

This invention spurred a large effort in the development of prototype components for internal combustion engines (gas turbine hot section components, turbocharger rotors for reciprocating engines, valves and valve train components, for example) and tubular components for advanced heat exchangers and radiant tube burners, for example. Although there is some commercial activity in tubular components, the higher performance applications (gas turbine engine parts) have not developed as originally envisaged, primarily because truly high strength material could not be consistently made in component form at the same time when the silicon nitride technology was able to produce higher performance components. In addition silicon nitride ceramics had the intrinsic advantage of having a thermal coefficient of expansion of about one half of that of silicon carbide. This difference translates into much lower thermoelastic stresses in a given transient or steady thermal gradient in spite of the higher thermal conductivity of silicon carbide ceramics. Thus, from a thermomechanical performance standpoint, state-of-the-art sintered silicon nitride ceramics are generally superior to state-of-the-art silicon carbide ceramics, and are thus, the material of choice for highly stressed parts or frequently thermally cycled parts. However, in certain applications the higher thermal conductivity of SiC ceramics is the critical design criterion (such as heat exchangers and radiant tubes) where heat flux is an important performance factor and silicon carbide ceramics are chosen over silicon nitride ceramics.

Since the toughness of these sintered silicon carbides is lower than the state-of-the-art silicon nitrides, various toughening schemes have been studied for these materials. The addition of TiB_2 particles was shown to lead to significant toughening, for example, but in general these measures have not been a commercial success.

Reaction Sintered Silicon Carbide Ceramics

When molten silicon makes contact with carbon at high temperatures (the melting point of Si = 1410°C), silicon carbide forms by an exothermic reaction. Since molten silicon wets carbon (or graphite), it can infiltrate into pores in a porous compact. Many inventions and commercial fabrication processes have taken advantage of these properties to produce reaction sintered silicon carbide. In the most common practice, finely divided carbon is mixed with silicon carbide particles and a binder which pyrolyzes to form carbon (phenolic resin, for example), and the desired object is formed by any of the green forming techniques (slip casting, extrusion, pressing, etc.) The object is placed in a dense graphite holder (boat) surrounded by particulate silicon. When the assemblage is heated in an inert environment or vacuum, first to pyrolyze the binder, then to temperatures to greater than 1410°C, the silicon melts and infiltrates the object reacting with the carbon to form fine silicon carbide particles which bind the larger particles from the original mixture and form a strong, rigid ceramic.

In general, the ceramic contains from 8-30% free silicon and is impermeable and nearly fully dense. For some products, called recrystallized silicon carbide, the object is fired again to a very high temperature (1700-1800°C) which vaporizes the free silicon and causes the silicon carbide grains to grow. This product is very creep resistant to temperatures near the firing temperature, but is porous because the silicon carbide does not densify during this process; the free silicon is merely vaporized.

The major advantages of this reaction sintering technology are that large pieces can be fabricated (tubes 10 ft long and 10 in in diameter are routinely produced and the final parts are nearly the same size as the starting green body unlike sintered parts which shrink 30-40% by volume during firing). Also, high purity silicon can be used with carefully purified silicon carbide and carbon to produce quite pure final products. The Norton company has built a significant business providing semiconductor device fabrication furnace tubes and wafer handling furniture which can be in direct contact with the silicon wafers (6).

Recent innovations, such as the use of very fine grain SiC particles, has produced high strength material which competes directly with the sintered SiC for applications at low and intermediate temperatures (7). Also, innovative one-step forming and processing of shapes with a regular cross-section such as tubes, has resulted in low cost products which can compete directly with nickel alloys in some applications (8). The ability alloy to Si so that at the end of the reaction with free carbon a more refractory silicide remains (e.g. $MoSi_2$) has been developed by a group at MIT (9).

Chemical Vapor Deposited Silicon Carbide

Since SiC is a hard, chemically resistant material and it can readily be deposited by reaction of gases of hydrocarbons and volatile silicon compounds or by decomposition of methyltrichlorosilane (MTS) at elevated temperature, it has been an attractive candidate for coatings for many years. Unfortunately, from the standpoint of coating metals, the TCE of SiC is much too low compared to most metals, so that the coatings have such high residual compressive stresses when cooled that they spall off the metal substrate. But SiC coatings are very important for coating graphite and carbon products, particularly for use in the electronic materials processing industry, since the CVD SiC can be of very high purity. Large pieces of graphite can be coated and complex shapes can be coated. Several companies specializing in these products have been successful by developing very close ties with the manufacturers of the semi-conductor processing equipment and with the end users of this equipment who need replacement parts.

CVD silicon carbide has also been used to protect carbon and carbon-carbon composite parts which are used in the aerospace industry as rocket engine parts and reentry cone hot components. In conjunction with other coating materials (such as B_4C), SiC can provide protection against destructive oxidation of the carbon structure and provide adequate life for short-term missions.

A natural extension of this protection scheme for carbon-carbon composites was to replace the carbon matrix with the oxidation resistant SiC matrix. The French group headed by Professor R. Naslain pioneered the CVI of silicon carbide into carbon fiber preforms and later into SiC-based fiber preforms (10). The firm Societié Europenne Propulsion (SEP) commercialized these materials for aerospace applications. The first mass produced part will be C/SiC panels for the exhaust section of the engine for the French Mirage fighter aircraft where prototype components have been thoroughly tested (11). The major advantage in this application is a large reduction of weight since the ceramic composite components replace superalloy parts.

The graceful failure of these materials and excellent toughness results from the very deliberate tailoring of the interface so that cracks which start in the ceramic matrix are deflected at the fiber matrix interface. Pyrolytic (CVD deposited) carbon has been the interface of choice until now. This French group has recently learned how to treat the surface of the fibers and deposit very thin alternating layers (0.05-0.1 µm) of C and SiC as an interface material, so that stronger interfaces have been developed but retaining the ability to deflect matrix cracks. The new "interphase" material appears more oxidation resistant as well.

The CVI technology has also received much attention in the U.S. with DuPont taking a license to the SEP technology and establishing commercial fabrication facilities. The 3M Co. has pursued the fabrication of large, dense, thin wall tubes (impermeable) and deliberately porous thin wall tubes to be used as hot gas filters (12). These developments are at the early stage of development.

The Thermoelectron Corporation has developed a process to simultaneously CVD deposit SiC while adding particulate material to the deposition surface. By this approach they claim much higher rates of material fabrication and the ability to tailor SiC matrix composites by using particles of other materials including whiskers. The development is at an early stage of exploration (13).

SILICON NITRIDE CERAMICS

Modern silicon nitride ceramics as a family should be considered to be Si-Al-O-N's with other cations than Al substituted, as well, since every commercially available product has oxide additives to aid in the sintering by forming silicate and oxynitride liquids during the densification process. The sintering of silicon nitride powder is at least as difficult as sintering silicon carbide except that it tends to dissociate at high temperature. So at sintering temperatures of 1700-1800°C a nitrogen overpressure must be used to avoid decomposing the Si_3N_4 which would reverse the densification process.

The modern history of silicon nitride ceramic developments is a very interesting study of highs and lows in the technological interest in material (see Ref. 14a and b for historical accounts). The first major program was the British one, which started in the 1950's, in which two engineering ceramics were produced; a reaction-bonded silicon nitride (RBSN) and a hot-pressed silicon nitride (HPSN)

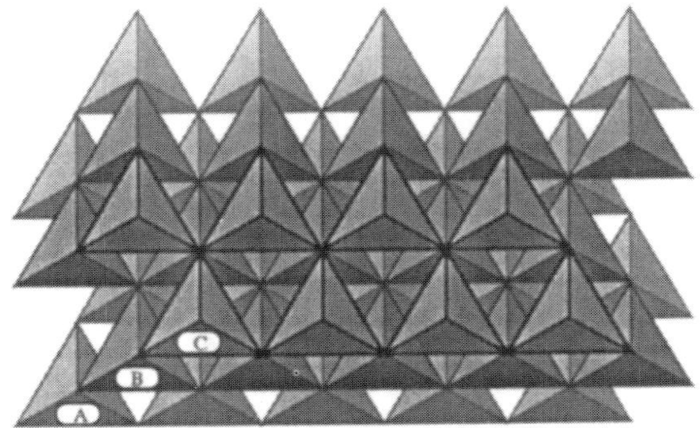

Zincblende, 43m

Wurtzite, 6mm

FIGURE 1. Tetrahedral representation of zinc blende (top) and wurtzite (bottom) crystal structures, SiC_4 tetrahedra can stack with ABC (top) or AB (bottom) periodicity.

with MgO as a densifying additive. The RBSN is quite porous and, thus weak, since it is fabricated by the gas phase nitridation of a powder preformed part. The hot pressed material must be machined to shape after pressing, and, because it is so dense and hard, this diamond machining is prohibitively expensive. Silicon nitride technology was given a boost in 1971 when the U.S. started a major ceramic gas turbine engine program just when the British effort was tapering off. The U.S. effort resulted in pressureless sintered silicon nitrides (SSN's) or gas pressure sintered silicon nitrides in which an overpressure of nitrogen is used to suppress dissociation of the nitride at the sintering temperatures.

FIGURE 2. Photo of components fabricated from silicon carbide (courtesy of the Carborundum Co.).

The Japanese were "fast followers" in this field with a succession of materials programs, and now Japanese firms are major factors in powder production and ceramic component production.

The British introduced the concept of silicon nitride "alloys" and led the development of sialon ceramics in which Al and O were substituted for Si and N in the β-Si_3N_4 structure. Other cations have also been substituted, notably Y^{3+}, to produce a variety of silicons based on the β-Si_3N_4 structure, so called β'-sialons. The α'-Si_3N_4 structure can also accept substitutions leading to a family of α'-sialons. Composites of α' and β' sialons are technologically important. Compositions based on substitutions in Si_2N_2O are called O'-sialons and show better oxidation resistance than the α'- and β'-sialons, but they require more development to be technologically useful.

The state-of-the-art materials are now highly engineered microstructures because of the knowledge gained of sintering additive effects (oxide compounds and level of additive) and temperature and atmosphere effects on the densification process and grain growth and morphology of the grains. The technology resides in several industrial firms (notably Allied-Signal, Dow Chemical, and Norton Co. in the U.S. where Penn State alums have played key roles in the technology development, and in Japan, Kyocera, NGK Sparkplug, and Toshiba) and silicon nitride components are commercially available. The powders are

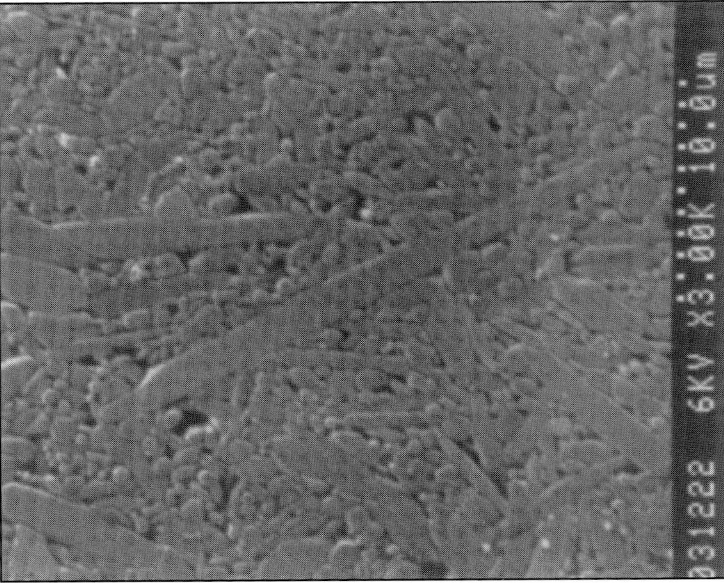

FIGURE 3. Micrograph of "in-situ" toughened silicon nitride (courtesy of the Allied Signal Co.).

available from a small number of vendors (Ube in Japan, Dow Chemical in the U.S., H.C. Starcke in Germany, for example).

The best powders (fewest impurities, lowest oxygen content) are obtained by the di-imide process from liquid NH_3 and $SiCl_4$ which can be very highly purified. Powders are also prepared by nitridation of Si and carbothermal reduction and nitridation of SiO_2. Both α and β powders are available, the desired final microstructure often dictates which powder is used (see discussion below). However, the powders routinely used in production consist of greater than 90% α-Si_3N_4 particles by volume. The cost of the powder continues to be a major obstacle in obtaining finished part costs which will permit the penetration of automotive engine part markets.

It's ironic that the most widely used version of silicon nitride ceramics now are not the ones with the best high temperature properties, which is the criteria which has driven much of the recent development efforts, particularly in the development of materials for gas turbines. The mass produced products rely more on the erosion and wear resistance of the material than the high temperature properties. (See applications discussed below). The ability to reproducibility fabricate the part at a cost effective price has allowed the more mature silicon nitrides and the pressureless sintered ones to penetrate markets. The fabrication processes used in commercial production are proprietary when one focuses on specific products. But, in general, a submicron size Si_3N_4 powder is mixed with oxide additives which react with the native oxide on the surfaces of the silicon nitride powder particles during the densification firing to form a liquid phase which leads to particle rearrangement and the other processes involved in liquid phase sintering. The powder handling is often done in a nitrogen ambient to minimize oxygen uptake.

The forming processes used span the range of typical ceramic powder forming processes - slip casting (with the new wrinkle being gel casting (15)), isostatic pressing, dry processing, etc. The sintering is generally carried out in a nitrogen overpressure or in a hot isostatic press where the component is either sintered to the close pore stage beforehand or "canned" to eliminate intrusion of the pressure medium. High temperatures in the 1700-1900°C range are used to achieve full density.

The developmental silicon nitride materials which hold the promise for revolutionizing high temperature, high performance applications and enabling new designs and new efficiencies in heat engines, for example, are ones where toughening has been incorporated along with creep resistance so that the materials will be dimensionally stable and resistant to catastrophic failure at temperatures approaching 1400°C. The strategy has been to use additives which combine with SiO_2 which are invariably present as a native oxide on the surfaces of fine powders of Si_3N_4 to form very refractory silicates or to use additives (for example, Al_2O_3, and Y_2O_3) in combination so that much of the grain boundary phase, which is liquid at the sintering temperature, crystallizes to form a refractory compound (yttrium aluminum garnet in the above cited case) which provides creep resistance.

The key feature of these new silicon nitrides is highly elongated grains of B-Si$_3$N$_4$ which form by growing during the sintering process at the expense of the fine grains of x-Si$_3$N$_4$ or B-Si$_3$N$_4$. The elongated grains provide so called "self-reinforcement" or "in situ toughening" (Figure 3). The toughness is greatly improved over an equiaxed microstructure and changed in character, that is the crack resistance or toughness increases with the length of the flaw or crack so that as the crack becomes larger the resistance increases up to some "plateau" or steady-state toughness (Figure 4). Also, the creep resistance is dramatically improved by these elongated grains as illustrated in Figure 4 where the creep behavior of hot pressed (fine grain equiaxed microstructure) Si$_3$N$_4$ is compared to a material of the same composition which has been heat treated to grow the elongated grains. The creep rate is decreased by a factor of twenty which is an enormous change as far as high temperature materials are concerned.

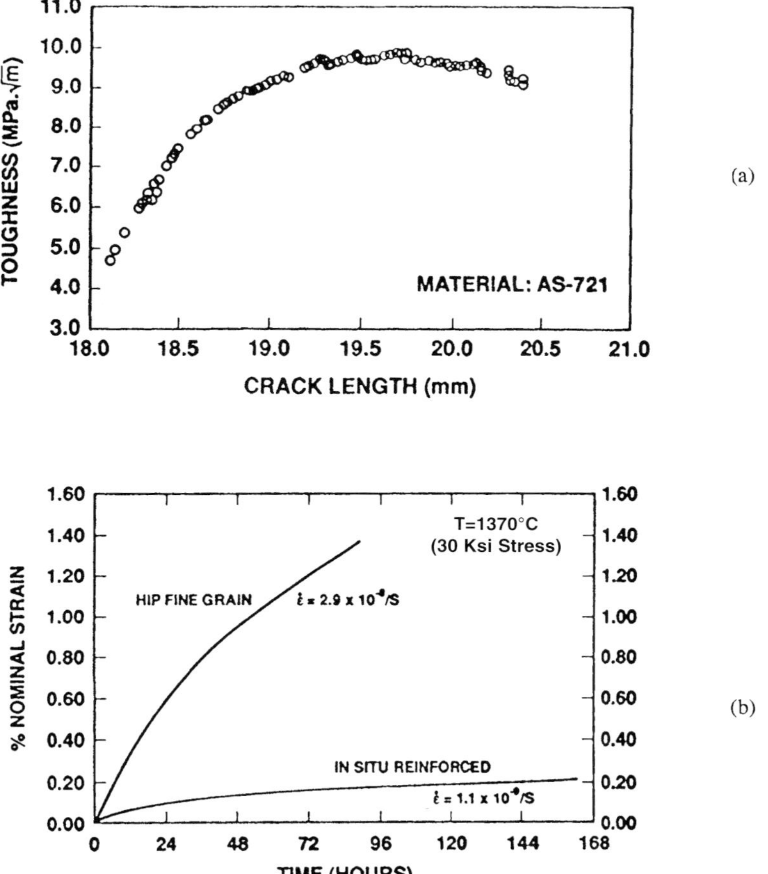

FIGURE 4. (a) Toughness and (b) creep resistance of the silicon nitride shown in Figure (3) (courtesy of the Allied Signal Co.).

These new materials are candidates for the most demanding applications such as land-based, power-generating gas turbine engine blades and stator vanes where the lack of creep (dimensional stability) is critical. Since crack growth and creep are dramatically suppressed in these materials, the long term durability issue (for campaigns of ~20,000 hrs) becomes one of resistance to oxidative corrosion and, particularly, selective corrosion which leads to pitting and the generation of new flaws which can weaken the material and lead to delayed failure (2). This is an area which requires extensive research which is just now beginning.

APPLICATIONS FOR SILICON CARBIDE AND SILICON NITRIDE CERAMICS

Since the myriad of applications for these materials change with time, no attempt is made here to catalogue all or even most of them. Instead, various categories of applications are discussed with the key requirements in the material or component described. Some attempt is made to indicate how these materials compete for the markets and why the particular material is chosen for a given application.

Low Technology Substitutions

In these applications, the ceramic replaces steel or tungsten carbide parts, for example, in an existing application with limited design changes. The market exists

FIGURE 5. Photo of silicon nitride turbine engine parts and turbocharger rotors (courtesy of the Kyocera Co.)

and the ceramic may be chosen to replace the more conventional material because of a performance advantage. The low density of the ceramic is often an advantage if the part rotates or travels in the machinery of which it is a component. Most of these applications for ceramics are taking advantage of the wear and abrasion resistance and resistance to corrosion by solutions such as acids and bases.

Some examples of these low tech substitutions are mechanical seal faces where the superior wear resistance of silicon carbide has made it the water pump seal of choice for many automobiles now. Nozzles such as water cutting nozzles or sand blast nozzles have much longer life when silicon carbide is used. Pump parts such as plungers and sleeves when the fluid is abrasive are now routinely made of silicon carbide. Paper making machine components which are exposed to corrosive liquids and abrasion and wear are fabricated of silicon carbide.

Silicon nitride bearings are being used in high performance applications now because of its excellent wear resistance. Silicon nitride cutting tools are filling important niches (as indicated above) because of excellent wear resistance, refractoriness, and low TCE. It is also being used as a dimensionally stable material in gauges and measuring instruments. In diesel engines silicon nitride has replaced steel as an injector link in the fuel injector. A myriad of precision gears, shafts, sleeve bearings, etc., are now being made of silicon nitride.

High Technology Substitutions

In these applications the ceramics are introduced because they have an impact on overall system performance and they required a significant design and testing iteration, often involving substantial alteration of the overall system. Energy recovery and conversion applications are an example of a category of systems. Here, the erosion and abrasion resistance is also important and, probably, at intermediate to elevated temperatures. The strength retention at high temperature is often important in these applications as is the low TCE. The low density is again important for the moving parts, and corrosion resistance in contaminated gaseous environments such as combustion ambients is often important.

Silicon carbide is being used as burner components and heat exchanger tubes, usually substituting for superalloys and often in tubular form. Radiant tubes for indirect burners for heat treating steel, for example, is an example of an emerging high tech substitution market. Silicon carbide has been demonstrated as valve train components (valves, guides, and cam followers) but silicon nitride is more likely to capture this market if it moves from race engines to automobile and truck engines (cost is still a major barrier). Silicon carbide in the high purity siliconized form or CVD form (as indicated above) is routinely used now in semiconductor processing furnace parts and epitaxial deposition reactors.

Silicon nitride turbocharger rotors are being produced at the rate of some 30,000/month in Japan for use in top grade cars. They are also being tested in off-road and over-the-road diesel engines in this country and Europe. These ceramics are used in glow plugs for diesels, precombustion chambers in car engines and valve train components. Silicon nitride is also being used for high temperature fans (in heat treating furnaces), metal casting components, high temperature springs, etc. Katz (16) has recently estimated the total U.S. Market for silicon nitride ceramics at $26-32 million in 1992 with a prediction of $81-159 million in 2000. He left automotive reciprocating engine parts out of his prediction for 2000, because cost breakthroughs must be made for that market to materialize even though technical success has been demonstrated.

New Technology Applications

This class of applications is the "blue sky" future which involves new or redesigned systems with motivation coming from energy savings, large system performance gains, or cost breakthroughs. In the area of silicon carbide, composites (discussed below) which show graceful failure and durability, will have major applications in energy conversion equipment such as gas turbines for airborne use. High temperature heat exchangers in slagging coal combustion represent potential markets for silicon carbide ceramics and composites. Gas turbine components of silicon nitride, particularly for land based applications, are being studied. These are examples of high performance systems where the ceramic components represent enabling technology to make a step function increase in performance or fuel efficiency.

Silicon carbide in electronics applications for high thermal conductivity chip carriers and as semiconductor devices (single crystals) are currently being researched extensively and could represent major markets.

ADVANCED CERAMIC FIBERS AND CERAMIC MATRIX COMPOSITES

Ceramic fibers for structural reinforcement have been a natural extension of the use of glass fibers to reinforce polymers. Ceramic fibers generally provide greater stiffnesses, resistance to chemical attack during fabrication into metal or ceramic matrix materials, and better strength at intermediate to high temperatures than glass fibers. When used in polymer matrix composites and metal matrix composites, the reinforcement fibers are used primarily to stiffen and strengthen the composite while providing adequate toughness to be useful. In these cases strong bonding between the fiber and matrix are necessary. In the case of ceramic matrix composites the primary reason for using fibers is to *toughen* the material and change the failure mode from "catastrophic" to "graceful" failure so that some early warning of impending failure is possible when the CMC's are used in engineering structures. In this latter case, the cracks which initiate in the matrix must be deflected at the fiber-matrix interface creating debonding and fiber pullout and toughness as the material fails gracefully (Figure 7). Thus, the interface must be carefully tailored to be of lower fracture energy than the fibers, and this property must persist at high temperatures in aggressive environments. The conventional thinking has been that this criterion requires a weak interface which is not optimal for creep resistance at high temperature where the interface should be stronger to transfer stress from the less creep resistant matrix to the creep resistant fibers. Recent work in France (18) has shown that the interface can be much stronger than is typically found in glass-ceramic reinforced with polymer-derived SiC based fibers, for example, and achieve excellent toughness and strength. The effect on creep resistance, particularly after the onset of matrix cracking, is not yet known. But this development is very encouraging and may lead to a much more versatile CMC technology.

In the case of ceramic fibers for polymer matrix composites (PMCs) and metal matrix composites (MMCs) ceramic fibers are being used in the highest performance applications (carbon and boron fiber reinforced polymers and boron and carbon reinforced aluminum). For MMCs, although much research effort has been expended on SiC fiber reinforced titanium alloys and Al_2O_3 fiber reinforced Ni-, Cr-, and Fe alloys commercially used materials have not emerged, in many cases because of the TCE mismatch between the fiber and the matrix which leads to damage and sites for premature failure.

For CMCs the fabrication processes always involve high temperature cycles (T>1000°C), and the largest payoff for applications of these materials are also in high temperature applications. Thus, the fibers must be resistant to change during the fabrication process and the properties of the fiber must be retained at high temperatures for long service campaigns. These property goals for ceramic fibers have been summarized and progress toward achieving them has been chronicled by Tressler and DiCarlo in the Proceedings of the High Temperature Ceramic Matrix Composites Meetings 1 and 2 (19,20). Also, these proceedings provide an excellent in-depth report of the research in this frontier in structural ceramics.

Alumina-based and mullite-based ($3Al_2O_3 \cdot 2SiO_2$) fibers in small diameter form have been available for some 20 years; the former developed by DuPont as candidate reinforcements for metals and the latter developed by the 3M Company for use in high temperature fabrics. Small diameter is important so that the fibers can be woven into 2D and 3D reinforcement architectures to resist specific stress states in much the same way that carbon fibers are woven for high performance polymer and carbon matrix composites. In the 1980's, Nippon Carbon Company commercialized the pioneering technology developed by Yajima and co-workers (21) for fabricating Si-C based small diameter fibers from polycarbosilane polymers, Nicalon™ fibers. Other developments in Japan and the U.S. provided a suite of these polymer derived fibers as candidates. Early work at Penn State showed that the strength and elastic moduli of these polymer derived fibers were substantially better at high temperature than were the properties of the oxide fibers (Figure 6). Subsequent work there and at NASA's Lewis Research Center showed that the creep resistance and rupture lives of these SiC based fibers were superior to the polycrystalline oxide fibers; the latter meeting the performance criteria at temperatures above 1100°C while the former were limited to temperatures below 1000°C.

It is now clear that the much higher ionic diffusivites in the oxide materials than in the SiC-based materials causes these polycrystalline oxides to be intrinsically less creep resistant than the SiC materials at comparable grain sizes and grain morphologies. Thus, recent efforts have concentrated on optimizing the polymer derived SiC-based fibers in the sense of eliminating an amorphous Si-O-C phase which results from oxygen curing, a crosslinking step in the polymer processing before py rolysis. By substituting a gamma radiation or electron beam radiation crosslinking process, the oxygen level has been reduced to below 0.5%. The Nippon Carbon Company is now offering Hi-Nicalon™ fibers which are nanocrystalline assemblages of β-SiC crystals and turbostratic carbon (~30 vol%) which are quite creep resistant to temperatures greater than 1200°C. The latest work at NASA Lewis and Laboratoire des Composites Thermostructuraux (LCTS) in Bordeaux, France, has shown that further heat treatments (1 hr, 1600°C in Ar, for example) does not significantly alter the room temperature strength and dramatically improves the creep resistance so that use temperatures approaching 1400°C may be possible (20). Their research is now concentrating on reducing the amount of carbon in the fiber to improve the oxidation resistance.

In a parallel development the Carborundum Company is providing in a pilot plant scale a fine grain polycrystalline SiC fiber prepared by a powder route and sintered with B and C additives, mimicking their monolithic alpha SiC material. In this case, the room temperature strengths are not as high as that of the polymer derived materials but the creep resistance is better, offering application temperatures

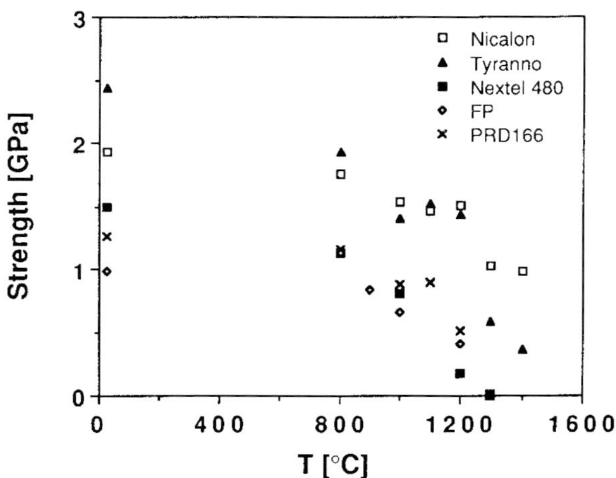

FIGURE 6. Strength versus test temperature for small diameter advanced oxide and nonoxide ceramic fibers (Ref. 19).

of 1400°C and perhaps higher. Other competitor fibers which are not as far along in the development cycle are being developed by Dow Corning Company, and the University of Florida.

The research on oxide fibers has turned to single crystals and highly elongated grain microstructures and polycrystals of the materials with the lowest diffusivities. Single crystal C-axis oriented sapphire fibers of large diameter were developed in the early 80's and were shown then to be the most creep resistant material available for oxidizing environments (22). However, crack growth which results in delayed failure and short rupture lives limits the attractiveness of this fiber even if development efforts are successful in producing a cost effective small diameter fiber. Research on directionally solidified eutectic fibers has shown some promise but the work is at an early stage. The 3M Company has tailored the microstructure of an alumina-mullite composite fiber (Nextel 720) so that elongated mullite grows at the grain boundary with larger alumina grains than in their Nextel 610 fiber which yields much better creep resistance than the aluminum fibers and a candidate for use perhaps at 1100°C. Early work in polycrystalline yttrium alumina garnet fibers has shown promise for this low diffusivity material.

Fibers based on silicon nitride have not been as successful so far as the SiC-based fibers because of the deleterious effects of oxygen and the difficulty in crystallizing the fiber to the extremely small grain sizes needed for high strength. Recent work in Germany has been more encouraging in this regard but the

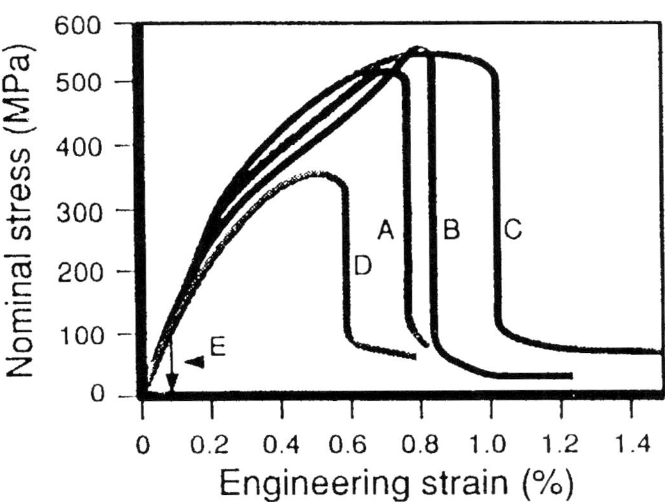

FIGURE 7. Graceful failure of a continuous fiber reinforced ceramic matrix composite (Ref. 17).

required high temperature properties have not been demonstrated as of yet.

The CMC technology is at an early stage. As mentioned above, the only fully commercialized continuous fiber CMC is the CVI SiC/C or SiC/SiC materials which are being produced for aerospace applications in Europe. Although C/C composites are often thought not to be CMCs they are the precursor to CVI CMCs, and they are fully commercialized for aerospace applications and for high performance brakes, particularly aircraft brakes (11). The CVI process is intrinsically slow and therefore, expensive since the capital equipment used is very expensive. Thus, the use of this technology in less demanding, industrial applications has been very slow because of the cost barrier. The SiC/SiC composites are being aggressively pursued in the U.S. as porous thin wall hot gas filter tubes where the total weight of the material required is very small, and therefore, possibly cost effective. Similarly, dense impermeable CVI SiC/SiC tubes are being tested in radiant burner applications, again where the total weight of material is very small compared to the much thicker walled monolithic ceramics.

SUMMARY

This brief survey of nonoxide structural ceramics, fiber, and ceramic matrix composites provides a snapshot of the most important ceramics based on silicon carbide and silicon nitride. The processing technology with the important properties achieved by these processes are reviewed in the silicon carbide and silicon nitride families of materials. Silicon carbide ceramics are much more important as commercial products today primarily because a viable technology has been available for nearly a century, and reaction bonding and reaction sintering of these powders are well developed technologies which provide complex and near net shape parts with large size capability. The silicon nitride technology is now well developed and the materials are being used widely in low technology, low temperature applications which capitalize on their wear, erosion, and corrosion resistance. Turbochargers are being commercially produced now in Japan but breakthroughs in cost of production must be achieved before there is widespread use of these materials in automotive markets.

Ceramic fibers have been the subject of intense research effort in the last five years, because there was not a suitable, small diameter high strength fiber available for reinforcing ceramics for use above ~1100°C. As a result of these efforts, better oxide fibers are now available, but most importantly, silicon carbide based fibers are now available with adequate creep resistance and strength to be used in high performance ceramic matrix composites up to 1400°C. Ceramic matrix composite systems are being used (CVI SiC matrix) in very high performance applications such as exhaust flaps in fighter aircraft engines. Widespread use in industrial applications is predicted if costs can be reduced by a factor of five to ten and long term durability in combustion ambients can be demonstrated.

REFERENCES

1. White, G. and D.W. Olson. 1990. *The New Materials Society: Challenges and Opportunities,* Vol. 1: New Materials Markets and Issues, pp. 2.1-2.49. U.S. Bureau of Mines, Washington, D.C.
2. Tressler, R.E. 1994. Theory and Experiment in Corrosion of Advanced Ceramics, pp. 3-22. In: K.G. Nickel (Ed) *Corrosion of Advanced Ceramics.* Kluwer Academic Publishers, The Netherlands.
3. *The Carborundum Company: The First 100 Years* (A Commemorative History). 1991. The Carborundum Company, Niagara Falls, NY.
4. Ramburg, C.E., Spear, K.E. and R.E. Tressler. 1995. Oxidation Behavior of Single Crystal and CVD Silicon Carbide. *J. Electro. Chem. Soc.* 142 (11): L214-216.
5. Costello, J.A. and R.E. Tressler. 1981. Oxidation Kinetics of Hot-Pressed and Sintered Alpha Silicon Carbide. *J. Am. Ceram. Soc.* 64(6): 327-331.
6. Tressler, R.E. and B. Foster. 1984. Silicon Processing with Silicon Carbide Furnace Components. *Solid State Tech.* 27(10): 143-146.
7. Carroll, D.E., Tressler, R.E., Tsai, Y. and C. Near. 1986. High Temperature Properties of Siliconized Silicon Carbide Composites, pp. 75-788. In: R.E. Tressler, G.L. Messing, C.G. Pantano, R.E. Newnham (Eds.) *Tailoring of Multiphase and Composite Ceramics.* Plenum Publishing, New York, NY.
8. Darroudi, T., Tressler, R.E. and M.R. Kasprzyk. 1993. Low-Cost Melt-Formed Siliconized Silicon Carbide Radiant Tube Materials, *J.Am. Ceram. Soc.* 76 (1): 173-179.
9. Chiang, Y.M., Messuer, R.P., Terwilliger, C.D. and D.R. Behrendt. 1991. Reaction Formed Silicon Carbide. *Mater. Sci. and Eng.* VA144: 73-74.
10. Naslain, R. and F. Langlais. 1986. CVD Processing of Ceramic-Ceramic Composite Materials, pp. 145-164. In: R.E. Tressler, G.L. Messing, C.G. Pantano, R.E. Newnham (Eds.) *Tailoring of Multiphase and Composite Ceramics.* Plenum Publishing, New York, NY.
11. Lamicq, P.J. and J.F. Jamet. 1995. Thermostructural CMC's: An Overview of the French Experience, pp. 1-12. In: R. Naslain and A.G. Evans (Eds.) *High Temperature Ceramic Matrix Composites I,* Ceramic Transaactions, Vol. 57, The American Ceramic Society, Columbus, OH.
12. Pysher, D.J. Mechanical Behavior of Ceramic Composite Hot Gas Filters after Exposure to Severe Environment. Presented at 95th Annual Meeting of The Amer. Ceram. Soc., Cincinnati, OH, May 2, 1995 (Paper SII-17).
13. Reagan, P., Scoville, A.N. and R. Leaf. 1986. Method of Forming Composite Articles from CVD Gas Streams and Solid Particles or Fibers, U.S. Patent 5,154,862.
14. a. Jack, K.H. 1993. Sialon Ceramics: Retrospect and Prospect, pp. 15-28. In: I.W. Chen, P.F. Becker, M. Mitomo, G. Petzov, T-S. Yan (Eds) *Silicon Nitride Ceramics* (MRS Synposium Proceedings, vol 287).
    b. Komeya, K. Progress in Silicon Nitride Ceramics in Japan, pp. 29-38 in 14(a).

15. Omatete, O.O., Janney, M.A. and R.A. Strehlow. 1991. Gelcasting - A New Ceramic Forming Process. *Am. Ceram. Soc. Bull.* 70 (10): 1641-1644.

16. Katz, R.N. Applications of Silicon Nitride Based Ceramics in the U.S., in Ref. 14, pp. 197-208.

17. Evans, A.G. 1990. Perspective on the Development of High Toughness Ceramics. *J. Am. Ceram. Soc.* 73 (2): 187-206.

18. Rebillat, F., Lamon, J., Lara-Curzio, E., Ferber, M.K. and T.M. Besmann. 1995. Interfacial Characterization of SiC.SiC Composites with Multilayer Interphases, pp. 311-316. In: R. Naslain and A.G. Evans (Eds.) *High Temperature Ceramic Matrix Composites I,* Ceramic Transactions, Vol. 57, The American Ceramic Society, Columbus, OH.

19. Tressler, R.E. and J.A. DiCarlo. 1993. High Temperature Mechanical Properties of Advanced Ceramic Fibers, pp. 33-50. In: R. Naslain, J. Lamon, D. Doumeingts (Eds.) Proceedings of Int. Conf. on Ceramic Matrix Composites (HT-CMC), ECCM-6, Bordeaux, France. Woodhead Publishing Ltd., Cambridge U.K.

20. Tressler, R.E. and J.A. DiCarlo. 1995. Creep Rupture of Advanced Ceramic Fiber Reinforcements, pp. 141-155. In: R. Naslain and A.G. Evans (Eds.) *High Temperature Ceramic Matrix Composites I,* Ceramic Transactions, Vol. 57, The American Ceramic Society, Columbus, OH.

21. Yopina, S., Okamura, K., Hoyoshi, J. and M. Omeri. 1976. Synthesis of Continuous SiC Fibers with High Tensile Strength. *J. Am. Ceram. Soc.* 58(7-8): 324.

22. Tressler, R.E. and D.J. Barber. 1974. Yielding and Flow of C-axis Sapphire Filaments. *J. Amer. Ceram. Soc.* 57 (1): 13-19.

Chapter Twenty-One

HIGH TEMPERATURE STRUCTURAL ALLOYS AND INTERMETALLICS

DONALD A. KOSS

Department of Materials Science and Engineering
Steidle Building
Pennsylvania State University
University Park, PA 16802

INTRODUCTION

The evolution of high temperature structural materials in the last forty years has been stimulated primarily by the development of gas turbines for the aerospace industry. In such applications, the design requirements for materials are very demanding. For example, materials currently must withstand creep deformation in aggressive gaseous environments at temperatures exceeding 1200°C. Future improvements in engine operating performance will require the use of materials at even higher temperatures. In response, the materials community has developed a family of high performance alloys, the so-called "superalloys," to meet these demands. While the superalloys can be based on either iron, cobalt, or nickel, the Ni-base superalloys are especially prominent because of their superior elevated temperature properties. Thus, they remain the high temperature material of choice for the critical rotating parts in gas and steam turbines as well as many other elevated temperature applications. Despite the extensive research and development of Ni-base superalloys (see ref.'s. 1-4 for reviews), further improvements in their use temperatures is limited by melting temperatures in the range of 1400°C.

In view of the temperature limitations inherent in Ni-base superalloys, there has been a concerted effort to explore intermetallic compounds as an alternate structural material for high temperature structural applications. The presence of phases based on intermetallic compounds, such as the Ni_3Al-based phase in nickel base superalloys, has long been acknowledged as a source of strengthening alloys. However, researchers since 1800's have recognized that, by themselves, intermetallic compounds possess properties much different than the metals on which they were based. For example, in 1890, Martens measured the hardness of Cu_3Sn to be twelve times that of Cu or Sn (5). Despite their high hardness, intermetallic compounds tended to remain a scientific curiosity chiefly because of their strong tendency to be brittle at low temperatures. Nevertheless, many intermetallic compounds possess high melting temperatures and are likely to be creep resistant; in addition, at least those intermetallics based on Al (i.e., the aluminides) may also exhibit oxidation resistance. In view of their relatively low densities, there is considerable interest in aluminides as a new structural material for elevated temperature, weight-critical applications. Of the many aluminides, alloys based on the following compounds show considerable promise for high temperature structural applications: (1) Ni_3Al, (2) NiAl, and (3) TiAl.

This chapter will briefly review the development of current Ni-base superalloys, as an example of the "state of the art" against which other potential elevated temperature materials, such as Ni_3Al and NiAl alloys, must be measured. As illustrations of new materials based on intermetallics, we also describe developments in Ni_3Al, NiAl, and TiAl alloys. It might be noted that the potential success of intermetallics to displace conventional, existing alloys in the near future probably rests with the TiAl alloys, which currently seek to extend the range of operation of conventional Ti-based alloys from ~550°C to temperatures in the 650° - 700°C range for TiAl alloys. If successful, this would be a significant achievement and probably stimulate development of other intermetallic alloys.

THE NI-BASE SUPERALLOYS

The development of aircraft powered by jet engines in the 1940's dramatically increased the demand for high temperature structural materials with a combination of elevated temperature strength and surface stability in an aggressive, gaseous environment. That demand led to the development of the "superalloys," which are based in Ni, Co, or Fe and which are intended for elevated temperature service characterized by relatively severe mechanical stressing and requiring high surface stability (1). The successful development of the superalloys has provided an enabling technology for the development of the current generation of high performance gas turbines for both aerospace and land-based applications. After 50 years of alloy and process developments, these alloys are an excellent example of the success of alloy and process developments for superior properties. The following provides a brief overview of Ni-base superalloys, which possess the higher temperature capabilities than either Fe-base or Co-base superalloys. For detailed reviews of all superalloys, see references 1-4.

Ni-base superalloys are commonly used as structural materials at temperatures in excess of 1100°C, which is in excess of 80% of their absolute melting temperature (~1400°C). Note that, due to cooling schemes, the gas temperatures in a turbine may be well in excess of 1200°C. This remarkable feat is a result of a blessing from Mother Nature and a rich history of both alloy and process development. The evolution of superalloys have their roots in austenitic stainless steels and the discovery the addition of Al would improve creep strength due, as would be later identified, to the formation of coherent precipitates of the intermetallic phase gamma prime (Ni_3Al) in the fcc, disordered gamma-phase matrix. Subsequent alloy development over many years has led to a large number of Ni-base superalloys which, while complex in their chemistries, still remain alloys which are based on a gamma solid-solution matrix which is strengthened primarily by coherent, relatively stable, ordered gamma-prime particles.

As will be discussed later, the gamma-prime phase (Ni_3Al) has a Ll2-type structure, which is typical of Cu_3Au; Ni_3Al remains ordered to its melting temperature of about 1355°C. In most Ni-base superalloys, gamma-prime particles reside as the equilibrium phase in the form of discontinuous, usually cubical particles which are coherent with the fcc gamma-phase matrix; the coherence and equilibrium nature of the particles contribute significantly to their degree of thermal stability, which is far superior to most precipitation-hardened alloys. Also unlike most particle-hardened alloys, the volume fraction of gamma-prime particles is usually quite high, ranging from 0.2 to 0.7. Given that these coherent particles must be sheared by mobile dislocations for plastic deformation, the yield stress of Ni-base superalloys is understandably very sensitive to deformation behavior of the gamma-prime phase. As shown in Figure 1, the temperature dependence of the gamma-prime phase is most uncommon in that it increases with increasing temperature (see "Ni_3Al and Its Alloys" for discussion). The result is that Ni-base superalloys, due to the high volume fraction of shearable gamma-prime particles, possess high strengths at elevated temperatures, as suggested by Figure 1. This figure also suggests it should be possible to increase the strength of a Ni-base superalloy by increasing the volume fraction of gamma-prime through Al additions. This is typically done for cast alloys, since the high volume-fraction gamma-prime alloys are difficult to work even at elevated temperatures. Further increases of alloy strength can be imparted by solid solution strengthening additions aimed to strengthen the gamma-prime particles (especially Ti and Nb, which partition to it. It is also possible to solid solution strengthen the gamma-phase matrix using Co, Cr, Fe, Mo, and/or W.

Since Ni-base superalloys are typically used in creep-critical applications, factors such as grain boundary sliding or premature cavitation at grain boundaries become a concern. The role of grain boundaries in creep strength and fracture has been addressed in detail in Ni-base superalloys. Perhaps the most obvious route in inhibiting grain boundary sliding is to use stable, discontinuous particles residing at grain boundaries. The typical polycrystalline alloy usually contains $M_{23}C_6$ carbides located as discontinuous particles on grain boundaries;

these carbides usually form due to the presence of Cr, Mo, W, or Hf. As a result, creep rupture strengths are improved presumably by inhibiting grain boundary sliding. In addition, it has been found that B and Zr (in trace levels) change grain boundary chemistry, improving creep strengths. Finally, grain boundary "properties" can also be improved dramatically by small amounts of Hf which promotes resistance to premature creep fracture.

While alloy development and the use of vacuum metallurgy techniques to achieve composition control has been successful in developing high strength polycrystalline Ni-base superalloys, the creep strength remains limited by the presence of grain boundaries. Furthermore, their relatively high modulus resulted in significant plastic-strain damage under low cycle fatigue conditions, such as thermo-mechanical fatigue (TMF). The grain-boundary and elastic modulus problems were further addressed by directional solidification (DS) processing to obtain cast alloys with elongated, aligned grain structures and with lower elastic modulus. The "DS superalloys" have a lower longitudinal elastic modulus which, when combined with the elongated grain shape, result in improved creep and TMF strength. Figure 2 shows an example of the improved TMF behavior of DS superalloys as compared to conventionally cast counterparts. The final step in this

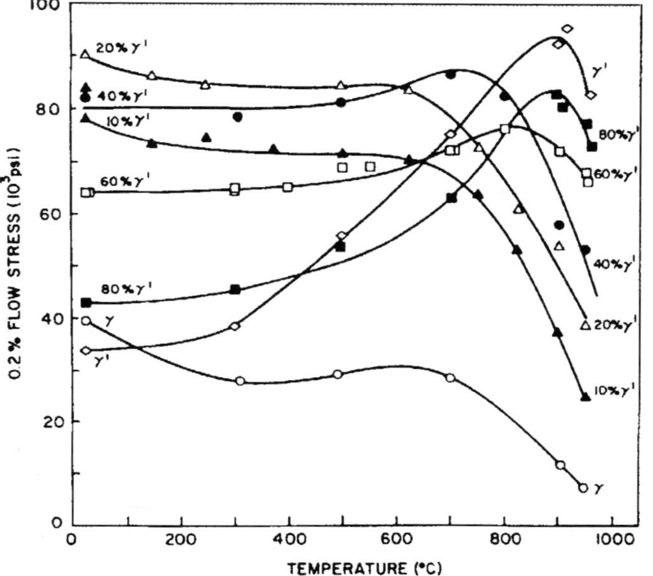

FIGURE 1. The temperature dependence of the yield stress of Ni-Cr-Al alloys containing various volume fractions of the gamma-prime phase. After ref. (6).

evolution was to eliminate grain boundaries entirely and to use solidification processing to obtained oriented single crystals. This approach has the further advantage that alloy chemistries may be tailored without any concern for grain boundaries issues (for example, B, Zr, and Hf are not needed). The resulting single crystal alloys have a somewhat higher melting temperature and, therefore, use a higher temperature to couple with their high strengths. Thus, the current benchmark for performance as a structural material, albeit under primarily uniaxial stress states, where elevated temperature strength is the principal design criterion.

While we have emphasized the issue of strengthening Ni-base superalloys, resistance to environmental degradation and surface stability is also critical. This topic is discussed in depth in another chapter of this book. In the present context,

TABLE 1

Chemical compositions of two Ni-base Superalloy Chemical Composition (wt %)

Alloy	Use form	Cr	Co	No	W	Ta	Nb	Al	T	C	B	Zr
IN738LC	Poly xtal	16	8.5	1.7	2.6	1.7	0.9	3.4	3.4	0.1	0.01	0.05
PWA1480	Single xtal	10	5		4 12	5.0	1.5					

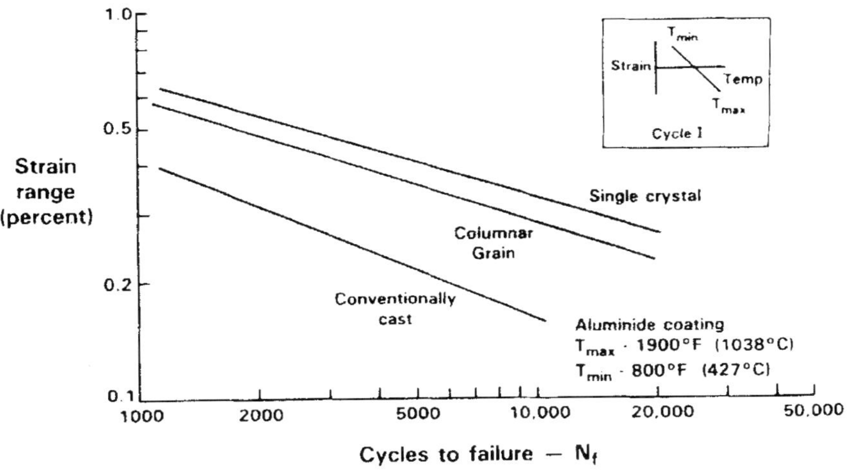

FIGURE 2. The thermo-mechanical fatigue response of conventionally cast, polycrystalline Ni base superalloy as compared to directionally solidified columnar grain and single crystal superalloys. After ref. (7).

it is important to recognize that the combined presence of both Al and Cr act synergistically to promote a protective external oxide scale, usually a mixture of Al_2O_3 and Cr_2O_3. Furthermore, Cr is also very important in imparting resistance to hot corrosion attack due to sulfidation. Thus, in addition to Al additions, which are necessary for gamma-prime formation, typical Ni-base superalloys would be expected to contain Cr.

We may summarize our discussion of Ni-base superalloys by considering typical chemical compositions, and the relative complexities, of the common, cast polycrystalline Ni-base superalloy, IN 738, and a single crystal counterpart, PWA-1480. Even though our overview has not examined all of the considerations that make Ni-base superalloys a success, most of the "chemical mix" of these two alloys may be readily understood in terms of the strengthening and oxidation/corrosion demands placed on these materials. It should be noted that, because of the absence of grain boundaries, the single crystal alloys, such as PWA-1480, tend to be compositionally more simple.

NI₃Al AND ITS ALLOYS

Of all of the intermetallic compounds, the nickel aluminide, NI₃Al, has been the most widely studied because of its technological as well as scientific significance (see ref.'s 8-16 for example). As described in the previous section, the gamma-prime phase, which is basedon Ni_3Al, is the principal source for the elevated-temperature strengthening of Ni-base superalloys. Since many of these superalloys contain in excess of 50 vol pct of the discontinuous gamma-prime particles, it is logical to consider the development of intermetallic alloys based on gamma-prime as the matrix.

As previously mentioned, Ni_3Al exhibits an unusual temperature dependence of its yield strength, which, as shown in Figures 1 and 3, increases with increasing temperature to ~600°C. This unusual yield strength behavior has been related fact that low temperature deformation on the $\{111\}$ glide planes consists of motion of superdislocations which may cross slip onto $\{100\}$ planes; in this case, the screw dislocations dissociate and form nonplanar cores, rendering them sessile and hindering $\{111\}$ slip. Since the cross-slip process is thermally activated, increasing temperature increases the frequency of cross-slip onto $\{100\}$ planes, making $\{111\}$ slip more difficult and increasing the yield stress with increasing temperature as seen in Figures 1 and 3. At very high temperatures, there is sufficient thermal energy to activate cube $\{100\}$ slip and the yield stress decreases rapidly due to a thermally activated change of slip system. It should be noted that other intermetallic compounds which share the $L1_2$ structure of Ni_3Al also exhibit the above positive temperature dependence of the yield strength behavior.

In order to develop a structural alloy based on Ni_3Al, its yield strength must be increased. Two obvious routes exist: alloy stoichiometry and solid solution hardening. Like most intermetallics, Ni$_3$Al has a yield stress which is sensitive to alloy stoichiometry. Aluminum-rich deviations from stoichiometry results in significant hardening, especially at elevated temperatures. Solid solution hardening is especially strong with the addition of elements such as Zr and Hf, which are transition-metal elements with large atom sizes; this effect is illustrated in Figure 3.

While Ni$_3$Al possesses sufficient slip systems for polycrystalline deformation, it tends to be brittle at room temperature in polycrystalline form. Although the fracture path is intergranular, no appreciable impurity segregation is found at the grain boundaries. The susceptibility to intergranular fracture is due in part to be environmentally assisted crack growth caused by moisture in the laboratory air. The H$_2$O is believed to react with Al in the alloy, forming Al$_2$O$_3$ and releasing atomic H,

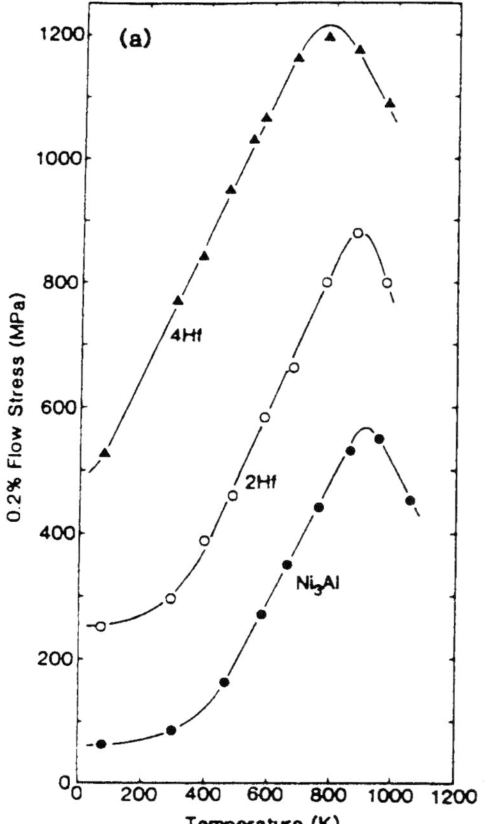

FIGURE 3. The temperature dependence of the yield strength of Ni$_3$Al and Ni$_3$ (Al,Hf) alloys. After ref. (17).

which causes crack growth along the grain boundaries. In addition to the susceptibility to hydrogen embrittlement, Ni₃Al also has a poor grain-boundary cohesive
strength, rendering it susceptible to intergranular fracture even in the absence of
hydrogen. Studies in Japan as well as the United States have shown that the susceptibility to intergranular fracture is very sensitive to boron. Small additions of B
($\cong 0.1$ at pct.) have been found to be very effective in suppressing intergranular fracture and improving ductility of Ni₃Al containing less than 25% Al. This beneficial
effect is related to the segregation of B to grain boundaries which enhances the
grain boundary cohesive strength of Ni₃Al. Several other alloying additions also
impart ductility improvement, although not as strongly as boron; these include Zr,
Fe, and Mn.

Exposure at elevated temperatures to air atmospheres can also embrittle Ni₃Al
due to oxygen-induced intergranular fracture. This embrittlement process
involves crack growth due to stress-assisted oxygen penetration along grain
boundaries. Again, methods to inhibit such embrittlement have been developed;
these include preoxidation, development of columnar grain structures, and the
addition of Cr at levels of ~8 at pct.

Based on a combination of strengthening and ductilizing considerations,
Ni₃Al alloys have been introduced with compositions (in at. pct.) as follows: Ni
- 14 to 18 Al-6 to 9 Cr - 1 to 4 Mo-0.1 to 1.5Zr/Hf - 0.01 to 0.20B. As discussed,
the Cr imparts resistance to elevated temperature oxygen-induced imbrittlement,
Zr and Hf provide rapid solid solution hardening, B is present to avoid low temperature intergranular fracture, and Mo imparts additional solid solution hardening. Thus, like Ni-base superalloys, Ni₃Al alloys become complex in their
chemistry with additions designed to alleviate certain ductility problems and to
enhance their high temperature strength. With variations in chemistry, these
alloys are available for use in either the cast or wrought conditions. It should be
noted that when compared to many nickel-base superalloys, the Ni₃Al alloys have
good oxidation resistance at elevated temperatures.

Structural applications of Ni₃Al alloys include the following: (1) diesel-engine
turbocharger rotors, (2) high temperature dies for forging and molds for glass
processing, (3) furnace fixtures for heat treating metals in carbonizing and oxidizing atmospheres, and (4) rollers for heat treating furnaces which heat treat
steel slabs. Other applications will no doubt follow. In summary, while Ni₃Al
alloys will not replace Ni-base superalloys, they nevertheless are finding specialized applications.

NiAl AND ITS ALLOYS

When compared to Ni-base superalloys as well as the Ni₃Al alloys, the intermetallic compound NiAl offers several advantages: (a) about 30% lower density,
(b) roughly 6x higher thermal conductivity, (c) the potential for better oxidation
resistance, (d) higher elastic modulus, and (e) increased (~300°C) melting tem-

perature. Thus, there has been considerable interest in NiAl-base alloys for gas turbine applications where weight is a critical factor. Despite their promise, and considerable research, development of NiAl base alloys remains severely limited by their brittle behavior at low temperatures and the difficulty of imparting creep strength without increasing ductile to brittle transition temperature; see ref.'s. 8 - 16, 18, and 19.

The room temperature ductility of near-equiatomic polycrystalline NiAl is usually in the range of 0-2%. This poor level of ductility is a consequence of the fact that, at low temperatures, NiAl prefers to deform by cube {100} <100> slip, which does not provide the necessary five independent deformation systems necessary for grain-to-grain compatibility needed to deform a polycrystalline material. Alloy additions to modify slip behavior, particularly in promoting <111> slip, have failed. Grain refinement has also been unsuccessful in improving room temperature ductility. Instead, significant increases in tensile ductility occur only with the thermal activation of <111> slip; as a result, NiAl typically exhibit a ductile to brittle transition, which occurs at ~400° to 500°C in binary polycrystalline NiAl alloys. As a result, both tensile ductility and fracture toughness of polycrystalline Ni-Al-base alloys remain unacceptably low at room temperature. It might be noted that there has been limited success using alloying additions such as Fe to promote strain-induced martensite as a means of satisfying the grain compatibility requirement; in such cases ~2 to 3% tensile ductility and improved fracture toughnesses at room temperature is possible. There has also been some success in imparting low temperature ductility by macroalloying additions which stabilize a ductile second phase. However, thermal stability of such microstructures at high temperatures limit this approach.

In view of the difficulties with polycrystalline ductility, an alternate approach is to utilize NiAl alloys in a single crystal form. In a research and development at GE Aircraft Engines, Darolia and co-workers have explored the strengthening and fracture behavior of a large number of NiAl-based alloys in the form of oriented single crystals. For crystals tested along a <110> direction, the ductile to brittle transition is lowered from ~450°C to ~200°C as shown in Figure 4. In addition, small additions of Fe (~0.25%) increase the tensile ductility from 1 to 6%. The mechanism for microalloying effect is not known, although it is known that similar additions of Fe to polycrystalline NiAl does not improve ductility.

While progress in improving the low temperature ductility of NiAl alloys remains slow, there have been very significant successes in creep strengthening these alloys. Solid solution hardening (relying on Co, Fe, and Ti) is possible, but the most dramatic strength improvements are achieved using a particle hardening approach. In the case of single crystal NiAl-base alloys, Hf additions can cause the formation of fine-scale, uniformly distributed Heusler phase. As shown in Figure 3b, the Heusler phase can provide a large increment of creep strengthening when compared to equiatomic NiAl single crystal. As a result, the single crystal NiAl + Hf compares favorably with Ni-base superalloys on a basis of creep strength, especially on a density-corrected basis.

In summary, research and development of single crystal NiAl-base alloys continues. Elevated strength properties, particularly as a result of alloying with Hf additions appear to be very promising. Unfortunately, Figure 4a shows that the same alloying additions which benefit creep strength are also likely to increase the ductile

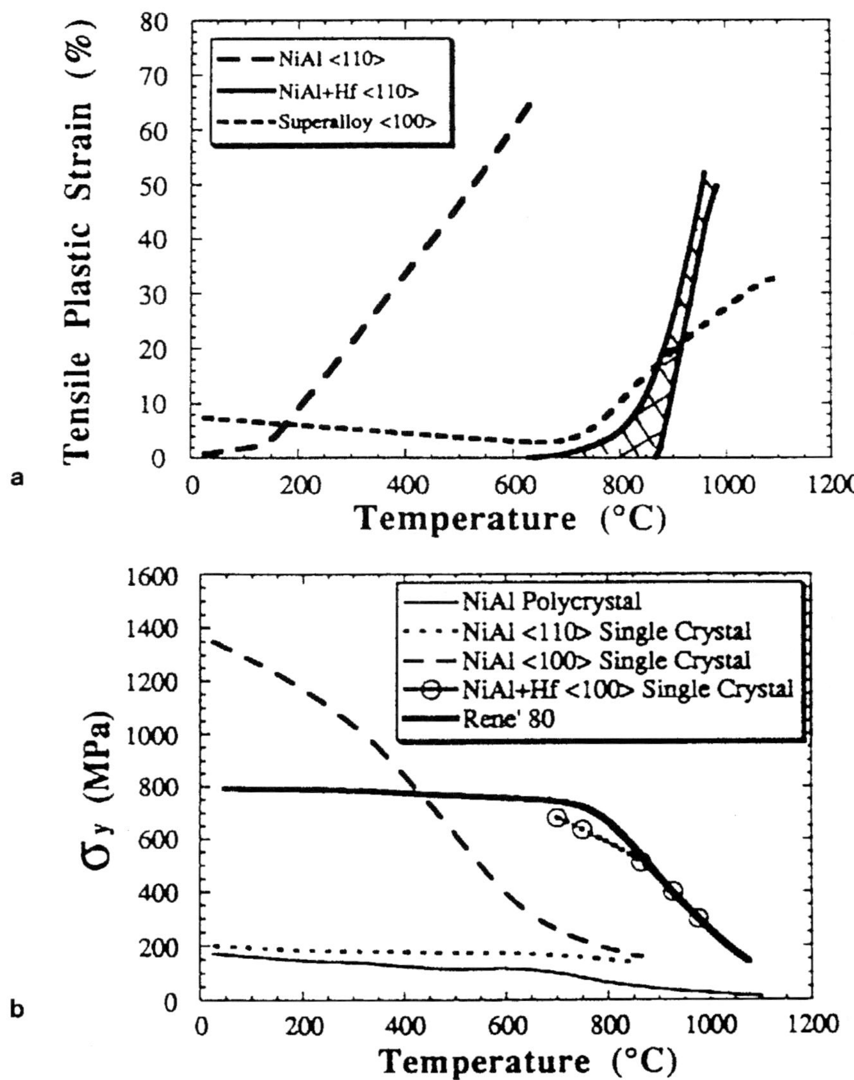

FIGURE 4. The temperature dependence of (a) the strain to failure and (b) the yield stress for NiAl single crystal alloys. Ni-base superalloys and polycrystalline NiAl are included for comparison. After ref. (20).

to brittle transition to well in excess of 600°C; this effect implies a high level of flaw sensitivity of such NiAl + Hf alloys at low temperatures. For example, if binary NiAl single crystals have low toughness values in the range of 4 to 8 MPa$\sqrt{m}$, then it may be anticipated that creep-strengthened NiAl alloys would have even less toughness. Thus, the dilemma persists as to whether to seek either creep strengthening or low temperature ductility and toughness in these alloys.

Test programs are currently evaluating single crystal NiAl alloys for aircraft engine applications. The success, or failure, of this effort will have a very large influence in determining future efforts based on NiAl alloys.

TiAl AND ITS ALLOYS

Alloys based on the intermetallic compound TiAl, or gamma as it is commonly known, have been of interest for elevated temperature structural applications since the 1970's (see ref.'s 1-14, 21). Gamma possesses a low density (~4gm/cc, compared to ~8 to 9gm/cc for Ni-base alloys), a high specific modulus which is ~50% greater than Ti and Ni base alloys when normalized to density, and a potential for high temperature strength given its relatively high melting temperature. Furthermore, TiAl alloys exhibit improved oxidation resistance when compared to Ti alloys and, importantly, gamma alloys are burn resistant. This latter issue is important in those applications where titanium fires may occur due to metal rubbing at elevated temperatures. Like other intermetallics, TiAl also suffers from poor ductility and fracture toughness at low temperatures. Processing, especially wrought processing where hot working is involved, has also posed major problems. Nevertheless, very significant progress has been made as a result of large relatively recent (since ~ 1988) research efforts especially in Japan, U.S. and Germany. Gamma alloys, especially in the cast form, are currently being actively evaluated for both aerospace and automotive applications. The success of these efforts will likely have a large impact on the future of not only gamma alloys but also other intermetallics. This Chapter reviews progress in alloy design and processing considerations of TiAl alloys.

The key to imparting significant levels of low temperature ductility (~4%) to TiAl-based alloys is the recognition that two-phase alloys lean in Al content (<50%) are much more ductile than their single phase counterpart. These alloys contain small amounts of the intermetallic phase, alpha-two, which is based on Ti$_3$Al. The two-phase alloys can be processed to a range of microstructures with differing properties. To begin to appreciate the microstructures which can develop, it is instructive to consider a portion of the Ti-Al shown in Figure 5. While the TiAl phase possesses a wide composition range, two-phase alloys lean in Al are of particular interest. Alloys in the 46 to 49% Al range can be heat treated into either a single-phase alpha temperature region or a two-phase gamma plus alpha region allowing a wide range of microstructures to form on cooling. Two distinctly different microstructures are most common. A duplex structure, consisting of equal amounts of equiaxed grains of the gamma-phase and two-phase grains in which the alpha two and gamma phase exist as alternating lamellae of plates is common; this structure is usually formed by hot working and heat treatment in the gamma plus alpha phase field. Upon subsequent cooling, the primary alpha grains partially transform to the lamellar gamma with the remaining alpha lamellae ordering to the alpha-two phase. In contrast, heat treatment at high temperatures in the alpha-phase field followed by slow cooling allows the entire microstructure to become fully lamellar as the coarse alpha grains transform to first gamma lamellae with alpha-two forming subsequently from the retained alpha phase by an ordering reaction. Obviously, several other structures are possible by varying the processing and heat treatment temperatures.

Typical tensile data show single phase TiAl alloys exhibit less than 1% tensile ductility. However, two-phase Ti-Al alloys with Al contents in the 46 to 49% Al range and with duplex microstructures possess ~2% ductility. The improved

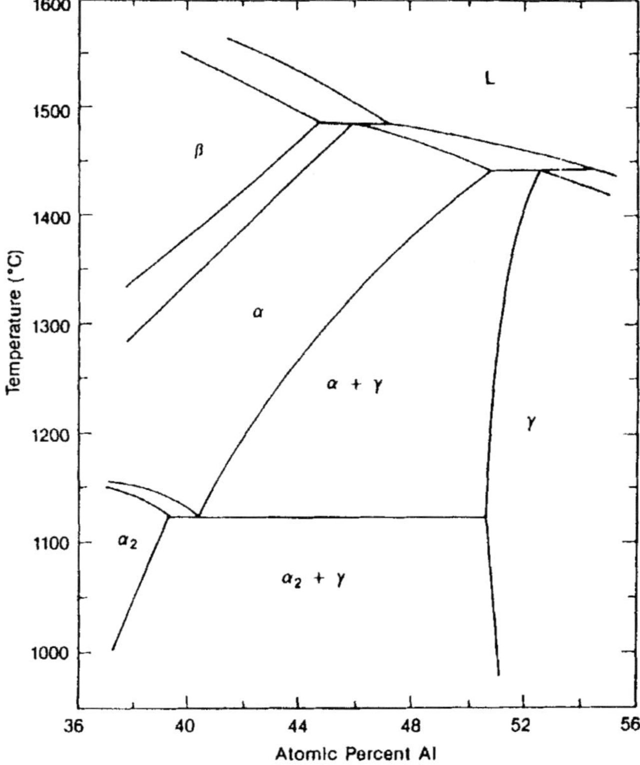

FIGURE 5. A partial Ti-Al phase diagram near the stoichiometric TiAl composition. After ref. (22).

ductility has been primarily to the small grain size in the two-phase alloys and a change in deformation behavior of the gamma phase attributed to the scavenging of oxygen by the alpha-two phase. Further improvements to the ductility (to levels of 3 to 4%) can be obtained by minor alloying additions of V, Mn, or Cr while maintaining the duplex microstructure of these elements. It should be noted that, even with alloy additions, the tensile ductility of TiAl alloys with the fully lamellar microstructure remains small ($\leq$ 1%).

While the tensile ductility of the duplex microstructure is superior, the fully lamellar structure exhibits a better fracture toughness. In this case, the K_{ic} value increases with lamellar grain size with values in the 20 to 30 MPa$\sqrt{}$m range, as compared to 10-15 MPa$\sqrt{}$m for duplex microstructures. This effect is related to the tortuous crack path in the lamellar structures and increased tearing resistance to crack growth caused by the linkage of microcracks ahead of the crack front by large-strain, shear fracture of the remaining ligaments.

The lamellar structure also exhibits superior creep resistance when compared to the duplex (or single phase) structure. As shown in Figure 6, the fully lamellar structure exhibits a time to 0.2% creep which is nearly 100 times greater than the duplex structure. This effect appears to be related to the larger grain size (~10x larger) of the lamellar structure and to the serrated or interlocking nature of the boundaries.

While the oxidation behavior of TiAl alloys in the 45 to 50% Al range is much better than Ti alloys, it can be further improved by additions of elements such as

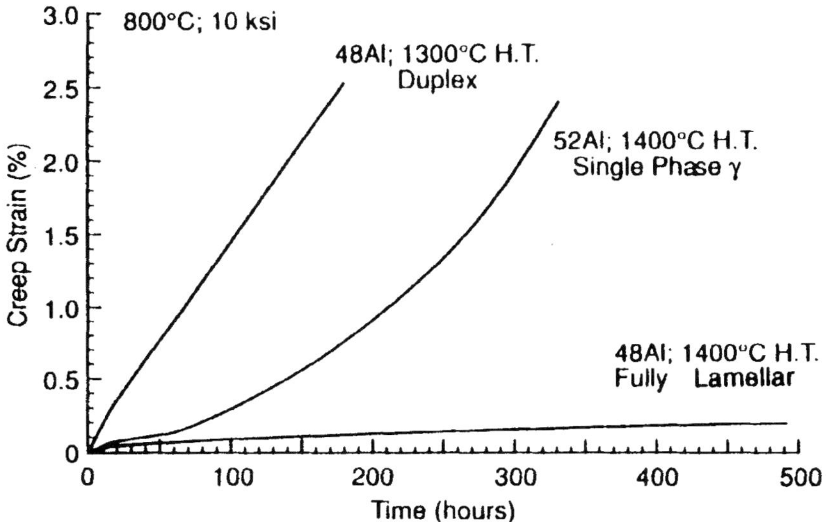

FIGURE 6. A comparison of the creep response of gamma alloys with three typical microstructures. After ref. (22).

Nb. The additions of relatively small amounts of Nb, ~2%, promote the formation of a continuous external scale which is protective. Significantly, this effect permits TiAl alloys containing Nb to be used at temperatures to 800°C without coating.

The combination of the ductility and oxidation considerations described above suggests two-phase alloys with which are lean in Al and contain Nb and either Cr, Mn, or V. Currently, the composition of Ti - 48Al - 2Cr - 2Nb is being considered as a potential gamma alloy for high temperature structural applications. This alloy of interest in both cast and wrought conditions. As will be discussed below, a variation of the two-phase alloy which contains TiB_2 particles for grain-size refinement during casting is Ti-45to47Al-2Mn-2nb-0.8vol%TiB_2. As Figure 7 shows, gamma alloys are an especially attractive alternate to titanium alloys above ~550°C.

While the compositions of gamma alloys may be comparatively well defined, their microstructures are quite sensitive to processing. Two routes are being

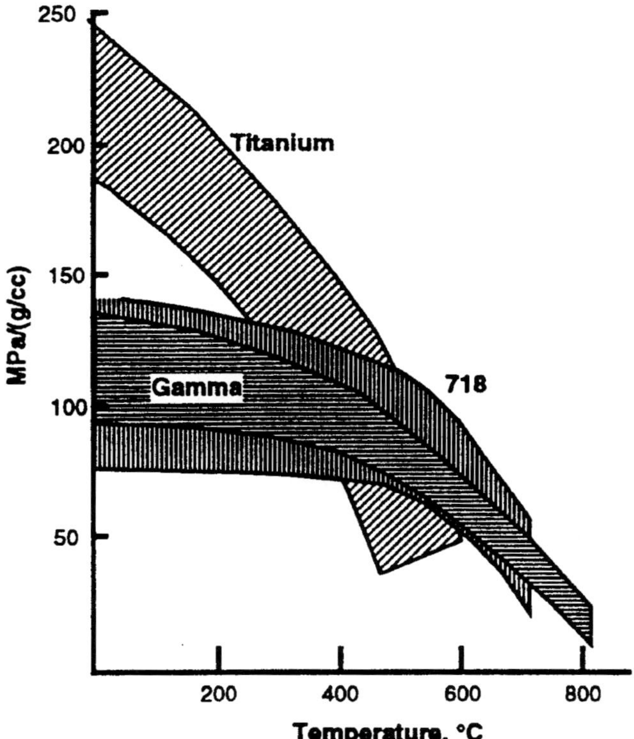

FIGURE 7. The temperature dependences of specific tensile strengths of gamma alloys, conventional titanium alloys, and the nickel base supperalloy IN 718. After ref. (23).

evaluated extensively: wrought processing and casting. Wrought processing relies on use of hot working to break up the ingot structure and, in some cases to produce a near-net shape. A basic problem with TiAl alloys is their tendency to exhibit severe solute segregation when cooled slowly after casting, such as in a large ingot. Homogenization of the ingot by heat treatment in the alpha-phase field results in a coarse-grained, fully lamellar microstructure which exhibit low ductilities even at hot working temperatures. If homogenization does not completely remove the segregated cast structure, microstructural banding in which bands contain lamellar grains, can persist even after hot working. Research and development efforts have been successful in coping with these difficulties, but considerable care (and expense) must be exercised in achieving a uniform microstructure with optimum properties.

An alternate, more direct and cost-effective method of achieving near-net shape in parts with complex shapes is casting. Most current efforts to employ gamma alloys in engineering applications rely on cast gamma, usually produced by investment casting. Major issues involve control of alloy composition, solute segregation and control of microstructure, notably microstructural refinement. The composition issue relates to the strong sensitivity of gamma alloy properties to composition, especially Al content. Further casting difficulties arise from Nb which enhances dendritic segregation; this effect can be removed by heat treatment. The successful applications of these procedures permit the retention of a fully lamellar, relatively coarse grain structure after casting and to utilize its good toughness and creep strength while accepting its poor tensile ductility (~1%). An alternate approach to microstructure control is to utilize "XD Gamma" alloys containing ~0.8 vol pct TiB_2 particles, which refine the grain size independent of section size. Cast XD gamma can exhibit room temperature tensile ductility in the range of 1.5% accompanied by a yield strength higher than "conventional" gamma; however, creep strength suffers as does fracture toughness.

In view of their promise, cast gamma alloys are being considered for both automotive engine and gas turbine applications. Automotive applications center around intake and exhaust valves in high rpm engines; cost remains a major factor here. Aerospace and power generation turbine applications are primarily non-rotating parts initially although applications to rotating parts are already being considered. Some examples include transition duct supports, seal housings and supports as well as low pressure turbine air foils and compressor blades.

REFERENCES

1. Sims, C.T. and W. Hagel (eds.). 1972. The Superalloys, John Wiley, New York.
2. Sims, C.T. 1984. Superalloys 1984, TMS, Warrendale, PA.
3. Sims, C.T., Stoloff, N.S. and W.C. Hagel (eds.). 1987. Superalloys II, John Wiley, New York.

4. Bradley, E.F. 1988. Superalloys, A Technical Guide, ASM International Materials Park, OH.

5. Martens, A. 1890. M.H.K. Tech. Versuchs-Anst. 8:216 - 225.

6. Beardmore, P., Davies, R.G. and T.L. Johnston 1969. On the Temperature Dependence of the Flow Stress of Nickel-Base Alloys. Trans. TMS AIME 245:1537-45.

7. Duhl, D.N. 1987. Directionally Solidified Superalloys. Ref 3: 189-214.

8. Koch, C.C., Liu, C.T. and N.S. Stoloff, (eds.). 1985. High Temperature Ordered Intermetallic Alloys, MRS, Pittsburgh, 39.

9. Stoloff, N.S., Koch, C.C., Liu, C.T. and O. Izumi (eds.). 1987. High Temperature Ordered Intermetallic Alloys II, MRS, Pittsburgh, 8.

10. Liu, C.T., Taub, A.I., Stoloff, N.S. and C.C. Koch (eds.). 1989. High Temperature Ordered Intermetallic Alloys III, MRS, Pittsburgh, 133.

11. Johnson, L.A., Pope, D.P. and I.O. Stiegler, (eds.). 1991. High Temperature Ordered Intermetallic Alloys IV, MRS, Pittsburgh, 213.

12. Baker, I., Darolia, R., Whittenberger, J.D. and M.H. Yoo (eds.). 1993. High Temperature Ordered Intermetallic V, MRS, Pittsburgh, 288.

13. Horton, J.A., Baker, I., Hanada, S., Noebe, R.D. and D.S. Schwartz (eds.). 1995. High Temperature Ordered Intermetallic VI, MRS, Pittsburgh, 364.

14. Pope, D.P., Liu, C.T. and S.H. Whang, (eds.). 1995. High Temperature Intermetallic, Mat'l Sci and Eng, A192/193.

15. Darolia, R., Lewandowski, J.J., Liu, C.T., Martin, P.L., Miracle, D.B and M.V. Nathal (eds.). 1994. Structural Intermetallics, TMS, Warrendale, PA.

16. Westbrook, J.H. and R.L. Fleischer. 1995. Intermetallic Compounds: Principles and Practice. John Wiley, New York.

17. Mishima, Y., Ochiai, S., Yadagawa, Y.M. 1986. Mechanical Properties of Ni3Al with Ternary Additions of Transition Metal Elements. Trans. JIM 27: 41-50.

18. Miracle, D.B. 1993. The Physical and Mechanical Properties of NiAl. Acta metall. mater. 41: 649-684.

19. Noebe, R.D., Bowman, R.R. and M.V. Nathal. 1993. Physical and Mechanical Properties of the B2 Compound NiAl. Int. Mat. Rev. 38: 192-232.

20. Darolia, R. 1994. NiAl for Turbine Airfoil Applications in Ref. 15: 495-504.

21. Y-W Kim. 1994. Ordered Intermetallic Alloys, Part III: Gamma Titanium Aluminides, Jour of Metals 46(7): 30 - 40.

22. Huang, S.C. Alloy ing Considerations in Gamma-Based Alloy s in Ref. 15: 299-308.

23. Austin, C.M. and T.J. Kelly. 1994. Development and Implementation Status of Cast Gamma Titanium Aluminide in Ref. 15: 143-150.

The Era of Materials. Edited by S.K. Majumdar, R.E. Tressler, and E.W. Miller © 1998, The Pennsylvania Academy of Science.

Chapter Twenty-Two

AQUEOUS CORROSION

HOWARD W. PICKERING

Department of Materials Science and Engineering
Steidle Building
Pennsylvania State University
University Park, PA 16802

INTRODUCTION

Great effort is expended every day in the USA and elsewhere in the developed world to control and reduce the corrosion of metals, alloys and other structural materials. These include the issues of cost, safety, environmental compatibility, conservation and the dynamics associated with technology's need for engineering materials that can withstand the ever increasing, more demanding applications within both the traditional and high tech industries. Thus, the chemical processing industries, automobile and other transportation industries, communication, energy conversion, electronics, chemical and metals industries, are all participants in a concerted effort to control corrosion. It is not totally surprising then that in the USA the annual cost of corrosion is enormous, approximately four percent of our gross national product or an annual expenditure in 1994 of 300 billion dollars (1)!

WHAT IS CORROSION?

Corrosion is the deterioration of a material that occurs when the material reacts with its environment (corrosive). When we refer to degradation of a bridge deck by corrosion of the steel reinforcing bar (rebar) in the concrete, the corrosive is an alkaline (pH 13) mixture of H_2O and other species (e.g., Cl^- from deicing salts or marine atmospheres) that can fill the pores of the concrete and came in direct contact with the steel rebar.

Corrosion can be classified as "wet" or "dry". It is the former when the environment is an aqueous liquid or moisture, and the latter when it is a dry gas typically at high temperatures. This chapter will focus on the fundamentals of "wet" corrosion and limit the discussion to metals and alloys. Aqueous liquids are stable between less than 0 and greater than 100°C at atmospheric pressure and to higher temperatures at higher pressures. An example of the latter is the pressurized water reactor with an operating temperature of approximately 289°C.

Corrosion is classified according to its appearance. Most forms of corrosion produce highly characteristic features that can be seen with the unaided eye or at a low magnification in a microscope. These forms of corrosion and their characteristic features are shown in Figure 1. Corrosion ranges from nearly uniform (Fig. 1a and 1b) to various degrees of nonuniform or localized (Figs. 1c to 1k).

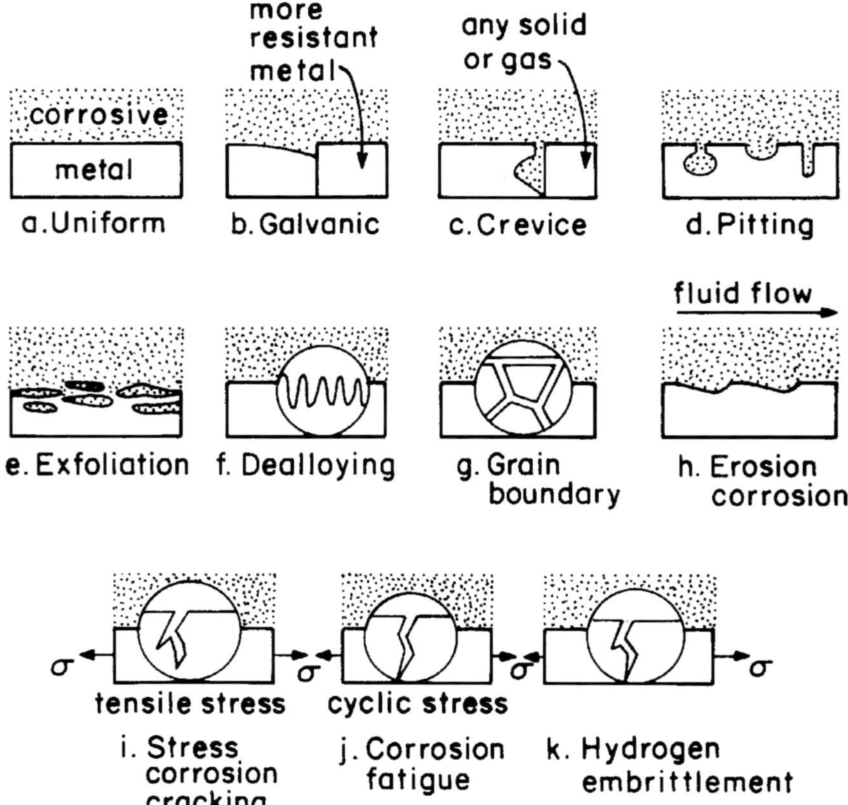

FIGURE 1. The appearance of corrosion determines its many forms.

In some cases defects in the alloy, such as nonmetallic inclusions, accelerate or cause the nonuniform corrosion. An example is MnS inclusions which can initiate pitting corrosion (Figure 1d). Another defect existing in metals and alloys is the grain boundary, which in certain metallurgical conditions (sensitized or segregated), can cause grain boundary corrosion (Figure 1g) or intergranular stress corrosion cracking (Figure 1i). These defective conditions of the alloy can be seen under a microscope at different levels of magnification. Non metallic inclusions and grain boundary precipitates that form during thermal-mechanical processing of the alloy can be seen at 10^2 magnification using an optical microscope. Figure 2 illustrates the sensitized condition at a grain boundary in stainless steel that forms during its thermal processing or subsequently during welding operations. On the other hand, segregation, involving minute amounts of a foreign element at the grain boundary, can only be "seen" with high resolution techniques that have atomic or nanometer resolution such as an atom probe field ion microscope (AP-FIM) or the transmission electron microscope (TEM). Techniques with low laterial but high depth resolution such as Auger electron spectroscopy can also be helpful in identifying segregation when the sample can be fractured intergranularly to allow for direct examination of the fracture surface.

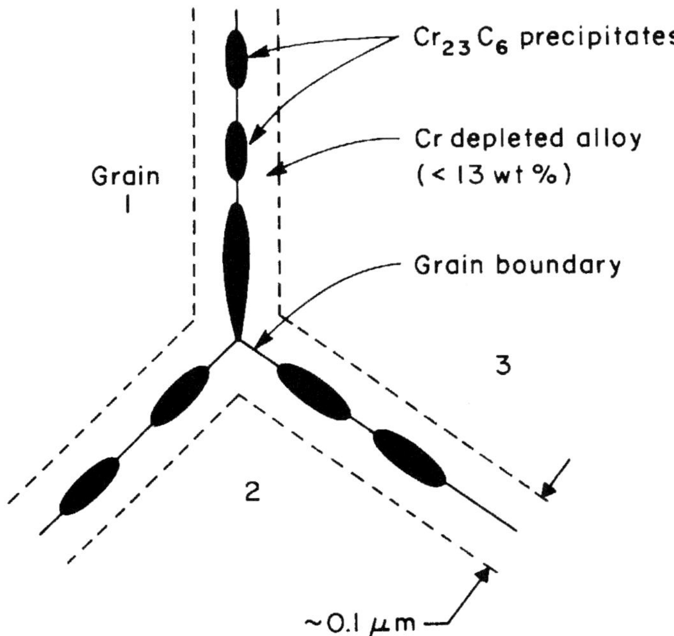

FIGURE 2. Schematic of the sensitized grain boundary structure of a stainless steel consisting of Fe, Cr (20 wt%) and C (0.05 wt%). The 0.1 um wide band of Cr depleted alloy along the grain boundary corrodes more rapidly than the bulk stainless steel or than the chromium carbide precipitate.

Because the corrosion occurs largely underneath the surface during the more severe forms of nonuniform corrosion, detection can be easily missed until it is too late and a penetration of the vessel wall or collapse of a structure occurs. This is a particular concern in stress corrosion cracking, corrosion fatigue, pitting corrosion and crevice corrosion where only one or a few corrosion penetrations are occurring over the entire surface. The lack of virtually any uniform corrosion on the outer surface can be very misleading when it comes to evaluation of the ultimate integrity of the system, e.g., in the case of the highly corrosion resistant stainless steels and nickel based alloys (Inconels) whose outer surfaces typically remain in the original brightly polished, lustrous metallic condition.

WHY DOES CORROSION OCCUR?

The great majority of the metallic elements are thermodynamically stable in the oxidized state as oxides, sulfides and other compounds rather than in their zero valent metallic state. Thus, the metal producing industries have for centuries been founded on the need to extract metals from their ores and then to process them into useful forms. These metals, either in the pure state or combined into alloys, will tend to revert back to their stable compound state when exposed to the atmosphere or other reactive environments. Only a few (noble) metals exist in nature in the metallic state, e.g., gold. The noble metals have intrinsic corrosion resistance because of their inertness in water, oxygen and other environments.

A primary task of the corrosion engineer is to slow down the corrosion rate of metals and alloys. There are a number of ways to do this, including the covering of its surface with a protective layer (e.g., paint or other more corrosion resistant metal). Other methods of protection include electrically biasing the metal's surface into a non-reactive condition (known as cathodic and anodic protection), and otherwise reducing a surface's reactivity by incorporating certain corrosion reducing species (inhibitors) at the metal/corrosive interface. In cathodic protection the metal is electrically biased to a potential that is thermodynamically unfavorable for the metal oxidation reaction,

$$M \rightarrow M^{n+} + ne\text{-} \tag{1}$$

In anodic protection the metal is biased in the more oxidizing direction into a potential regime where the driving force for metal oxidation is so high that instead of forming soluble metal ions, a dense solid metal oxide or hydroxide forms on the surface and protects the metal, causing the rate of metal oxidation to decrease by orders of magnitude (to the μA rather than $>mA\ cm^{-2}$ range). The latter is known as the passive state.

In the above thermodynamic discussion, corrosion occurs because the products of the corrosion reaction are more stable than the reactants, the metal being one of the reactants.

However, as with all chemical reactions, for a reaction to proceed at a finite rate, the conservation laws, including the conservation of charge, must be satisfied. Charge can only be conserved during an aqueous corrosion reaction when another reaction simultaneously consumes the electrons produced by the metal dissolution reaction typified by Reaction (1).

There are many such reduction or cathodic reactions that will spontaneously occur with Reaction (1), with the number of such reactions increasing as the reactivity of the metal increases. The standard EMF listing of reactions readily illustrates this concept. The reactivity of metals increases towards the bottom of the EMF series (larger negative equilibrium potentials). Thus, reactions higher in the series proceed spontaneously in the reduction direction when combined with a metal reaction that is lower in the series. Since Reaction 1 and those listed in the EMF series cannot proceed in either direction at a net rate by themselves, they are referred to as partial or half cell reactions. A small listing of half cell reactions in the EMF series is given in Table 1. Thus, the iron half cell reaction in Table 1 will proceed from right to left as an oxidation reaction, and the hydrogen reaction will proceed from left to right to form hydrogen gas as the only reduction reaction when iron is immersed in aqueous, (deoxygenated) hydrochloric acid. The sum of these two reactions, being mindful to conserve mass and charge, is the overall cell or corrosion reaction. The Cl^- ion is thermodynamically stable under these conditions and, therefore, does not undergo reaction. When any two (half cell) reactions are placed in contact with each other, one of the two directions will always be spontaneous. Hence, one has for the iron and hydrogen reactions from Table 1:

TABLE 1.
Standard Electrode Potentials

Electrode Reaction	25°C, NHE (v)
$Au^{3+} - 3e^- = Au$	1.68
$Cl_2 + 2e^- = 2Cl^-$	1.358
$O_2 + 4H^+ + 4e^- = 2H_2O$	1.229
$Ag^+ + 2e^- = Ag$	0.7996
$Hg_2^{2+} + e^- = 2Hg$	0.796
$O_2 + 2H_2O + 4e^- = 4OH^-$	0.401
$Cu^{2+} + 2e^- = Cu$	0.340
$2H^+ + 2e^- = H_2$	0.000
$Sn^{2+} + 2e^- = Sn$	-0.136
$Ni^{2+} + 2e^- = Ni$	-0.23
$In^{3+} + 3e^- = In$	-0.338
$Fe^{2+} + 2e^- = Fe$	-0.409
$Cr^{3+} + 3e^- = Cr$	-0.74
$Zn^{2+} + 2e^- = Zn$	-0.763
$2H_2O + 2e^- = H_2 + 2OH^-$	-0.828
$Ti^{2+} + 2e^- = Ti$	-1.63
$Al^{3+} + 3e^- = Al$	-1.71

Source: Handbook of Chemistry and Physics, 60th ed., CRC Press, Inc., Boca Raton, FL, 1979-1980.

reduction: $2H^+ + 2e^- = H_2$ 0.00 volts   (2)

oxidation: $Fe^{2+} + 2e^- = Fe$ -0.41 volts   (3)

so that the overall cell reaction and standard cell potential, E^0_{cell}, at open circuit or infinite internal resistance (zero net current flow) for the spontaneous direction are

$$Fe + 2H^+ = Fe^{2+} + H_2 \qquad 0.41 \text{ volts} \qquad (4)$$

Although the chloride ions are non reactive, they are necessary for charge neutrality of the aqueous solution so they are sometimes included in the formulation in which case Reaction (4) is written as

$$Fe + 2HCl \rightarrow FeCl_2 + H_2 \qquad 0.41 \text{ volts} \qquad (4a)$$

Identifying an overall reaction (like 4a) as having the special qualities of conduction through the metal and electrolyte phases and separate anodic and cathodic half cell reactions was a major breakthrough and is the foundation of modern corrosion science and engineering. The experiments by Wagner and Traud convincingly demonstrated it for the first time in 1938 (2).

The voltages in Reactions 2 to 4 are for the limiting case that Reaction (4) occurs at a zero (or infinitely slow) rate. Once Fe is in direct contact with the H^+ ions, the positive voltage indicates that Reaction (4) will spontaneously occur to the right. The magnitude of the voltage in Reaction (4) is a measure of the thermodynamic tendency for the cell reaction to go in the direction written being related to the standard free energy change of the reaction (ΔG^0) by

$$\Delta G^0 = -n \, F \, E^o_{cell} \qquad (5)$$

where n is the number of electrons in the reaction (per atom of iron dissolved or molecule of hydrogen gas formed), F is the charge on a mole of electrons, approximately 96,485 C mol^{-1}, and the negative sign reflects the convention that E^o_{cell} is positive for a cell reaction that is spontaneous in the direction written (left to right). When the substances in the above reactions are not at unit activity, Eq. 5 is written without the superscripts,

$$\Delta G = -n \, F \, E_{cell} \qquad (5a)$$

By adopting this convention, if one can measure the cell potential at open circuit (the equilibrium state of the two half cell reactions), this potential is equivalent to the Gibbs free energy.

Reactions (2) and (3) both occur simultaneously on the metal's surface but not typically at the same locations. When the anodic and cathodic reactions occur randomly over the surface with increasing time, the metal dissolves uniformly over the surface (uniform corrosion, Figure 1a). Separation of Reactions (2) and (3) on the surface means that charge must pass between these sites in both the aqueous and metal phases thereby creating an electrical circuit as shown in Figure 3. Recognition of this conduction process which was already experimentally demonstrated before the mid twentieth century (2) identifying corrosion as an electrochemical reaction, is essential for the application of corrosion control techniques by the corrosion engineer. Electrons flow in the metal from the anodic site where electrons are produced by Reaction (3) to the cathodic site where electrons are consumed by Reaction (2), while anions and cations flow (in the directions shown) to complete the electrical circuit.

These two flows of current through the metal and electrolyte conductive phases, respectively, require potential drops in both phases in accordance with Ohm's law, i.e., the higher the resistivity and the greater the length of the conductive paths, the larger will be the required voltage (IR) drops to maintain a given current flow in the circuit. To the extent an IR voltage exists in the circuit, since it is part of the cell voltage, less of the cell voltage is available for biasing (polarizing) Reactions (2) and (3) in their designated directions. Thus, the shorter the cell circuit and the lower the resistivity of the electrolyte phase (ignoring the metal's resistivity which is negligible in comparison to that of the electrolyte), the larger will be the cell voltage available to drive Reactions (2) and (3) and the higher will be the cell current. Hence, most of the metal dissolution process will

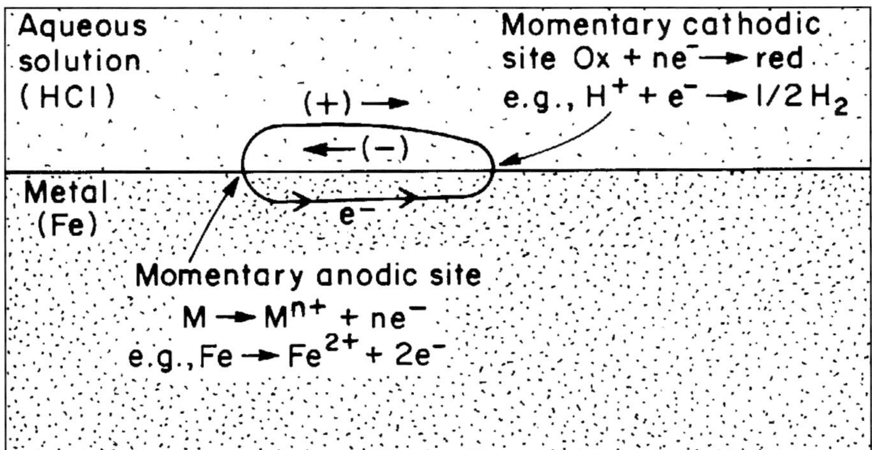

FIGURE 3. Schematic illustrating the electrical circuit existing in the metal and aqueous phases and, hence, the electrochemical nature of aqueous corrosion processes. At the anodic sites atoms of metal dissolve; at the cathodic sites solution species are reduced. In uniform corrosion the anodic and cathodic sites occur randomly over the surface resulting in a uniform rate of metal oxidation.

be the result of Reactions (2) and (3) operating close to each. Below, in conjunction with localized corrosion, we will see that if one or more of the required conditions for uniform corrosion is missing, the corrosion circuit becomes very large, e.g., mm rather than nm in length. Nevertheless, the driving forces are still high enough to cause high rates of localized corrosion, especially for the metals in the lower half of the EMF series. The most obvious of the necessary conditions for uniform corrosion and close proximity of Reactions (2) and (3) are uniform access of the reduction reaction to the metal surface and homogeneity (especially compositional) of the metal surface.

Since, Reactions (2) and (3) need to occur at the same rate to conserve charge in the corrosion circuit, if either reaction is slowed down so must the other slow down. This charge constrain is helpful in designing corrosion protection systems, e.g., by adding a substance to the corrosive which adsorbs on the metal surface and slows the rate of Reaction (2) (referred to as a cathodic inhibitor), one also slows Reaction (3) by the same amount. Or, if the metal is in a neutral aqueous solution where the rate of Reaction (2) may be very low but the corrosive contains some dissolved oxygen, corrosion of the metal occurs with the main cathodic reaction being

$$O_2 + 2H_2O + 4e^- \rightarrow 4OH^- \qquad 0.401 \text{ volts} \qquad (6)$$

Now, if a species (inhibitor) is added to the corrosive that scavenges the oxygen, the corrosion rate decreases by the amount that Reaction (6) decreases.

In the above, discussion of some of the half cell reactions in Table 1, the special case of standard conditions was assumed. In the general case the activities of the species are not unity. Therefore, the equilibrium potentials of the half cell reactions differ from those listed in Table 1. Their new equilibrium potentials can be obtained using the Nernst equation

$$E = E^o - \frac{RT}{nF} \ln \frac{a_p}{a_r} \qquad (7)$$

where R is the ideal gas constant (1.986 calories mole^{-1} K^{-1}), T is temperature in degrees Kelvin (K) and a is the thermodynamic activity of the products, p, and reactants, r. The E value obtained from Equation (7), like also the E^o values listed for the half cell reactions in Table 1, is the electrode potential of the half cell reaction measured against the normal hydrogen electrode ($E^o = 0$, see Table 1). Activity can be expressed as concentration times the activity coefficient, γ, making it obvious that the equilibrium potential of a half cell reaction is dependent on the concentrations of the oxidized and reduced species in the system. For example, Equation (7) for Reaction (3) becomes.

$$E = E^o + \frac{RT}{nF} \ln \gamma_{Fe^{2+}} [Fe^{2+}] \qquad (7a)$$

where the concentration of Fe^{2+} in mole liter^{-1} is indicated by the brackets and the metallic form of iron is pure so $a_p = 1$.

Reaction (3) shows that iron is oxidized to the divalent soluble ion in acid solution but other products are thermodynamically stable at higher pH and E. From the available thermodynamic data the stable species for many metals in water have been collected and presented in the form of E-pH (Pourbaix) diagrams (3, 4). The Pourbaix diagram for iron is shown in Figure 4. It shows which forms of iron are thermodynamically stable in the various E-pH regions.

A definition of the driving force, η, is the magnitude of the departure from the equilibrium potential of the oxidation or reduction reaction in question,

$$\eta = E_s - E \tag{8}$$

where E_s is the (open circuit) corrosion potential, E_{COR}, or an applied potential, E_{app}, both being measurable quantities, E is the equilibrium potential obtained from Eq. 7 and η is called the overvoltage. For the above example of iron immersed in HCl (which contains a finite concentration of Fe^{2+} ions), $\eta = E_{COR} - E$ where E is obtained from Eq. 7a.

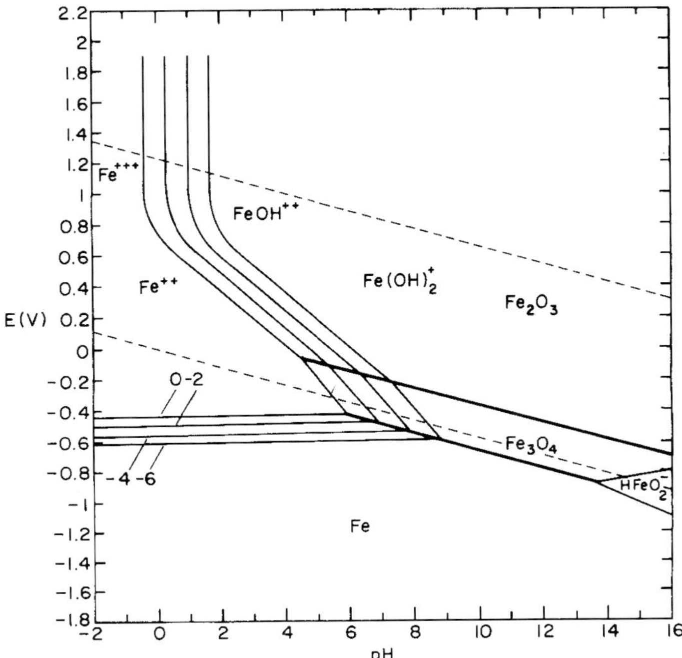

FIGURE 4. The iron-water E-pH diagram at 25°C. E is on the SHE scale (3).

THE CORROSION RATE

Under spontaneous corrosion conditions, Reaction (1) proceeds to the right causing a loss of mass and a reduction in thickness of the metal. However, in many corrosive environments, in particular neutral solutions, the metal corrodes by combining with water to form solid hydroxides and oxide films on its surface in addition to soluable metal ions. In this case the changes in mass and thickness are not simply related to the corrosion rate. In the former case for which Reaction 1 applies, as typically is the case in acids, one has

$$\delta = \Delta W / \rho \tag{9}$$

$$\delta = \frac{it}{nF} V_m \tag{10}$$

and

$$\Delta W = \frac{it}{nF} W_m \tag{11}$$

where δ is the reduction in thickness (cm), ΔW is the mass loss per unit area of the corroding metal surface (g·cm^{-2}), p as its density, V_m is its molar volume, W_m is its atomic weight, i (A·cm^{-2}) is the current density flowing for the time, t, that Reaction 1 occurs, n is the number of electrons in Reaction 1 and F is the Faraday constant. Expressing Eqs. 10 and 11 as rates one has

$$\dot{\delta} \ (\text{cm s}^{-1}) = \frac{i}{nF} V_m \tag{10a}$$

and

$$\Delta \dot{W} \ (\text{g cm}^{-2} \text{ s}^{-1}) = \frac{i}{nF} W_M \tag{11a}$$

Historically, other units than those given in Eqs. 10a and 11a have been used to express corrosion rates such as mils per year (mpy) and milligrams per square decimeter per day (mdd), respectively.

Constrains on the usefulness of Eqs 9 to 11 are that Reaction 1 occurs homogeneously over the surface, i.e., uniform rather than localized corrosion, and the difficulty in measuring the current in Reaction 9 to 11a, in particular the corrosion current, i_{cor}. For the directly measured current to be that of Reaction 1, Reaction 1 must be the only half cell reaction occurring on the surface of the metal. This requirement precludes direct measurement of i_{cor} since a cathodic reaction necessarily occurs simultaneously on the surface of the metal during a spontaneous corrosion reaction. Only when the corroding metal is biased to electrode potentials that are more positive or noble than the equilibrium potential of the oxidant (cathodic) reaction, does the measured (ammeter) current correspond to the rate of Reaction 1. This condition is illustrated in Figure 5. These bias-produced metal dissolution rates are typically much larger than the spontaneous corrosion rate. Nevertheless, they are meaningful since they can be used to estimate i_{cor} since, as mentioned above, i_{cor} is not directly measurable. The corrosion potential, E_{cor}, on the other hand is easily measured or can be determined from the directly measured polarization curves, being the potential in the polarization plot at which the measured current approaches zero from both directions (see Figure 5), i.e., at zero measured current all of the electrons produced by Reaction 1 are consumed by the cathodic reaction occurring on the same surface so that no electrons are available to flow through the ammeter of the external circuit. Actually, E_{cor} is more simply directly measured under open circuit (non biased) conditions using a reference electrode and high impedance voltmeter. For more details one can consult any of a number of books on electrochemistry and corrosion (5-7).

THE PASSIVE CONDITION

All corrosion resistant metals and alloys rely on what is known as the passive condition for their remarkable resistance to corrosion reactions. Most of these metals are highly reactive in a thermodynamic sense, being near the bottom of the EMF series. Yet, in spite of this very strong tendency for reaction, the reaction

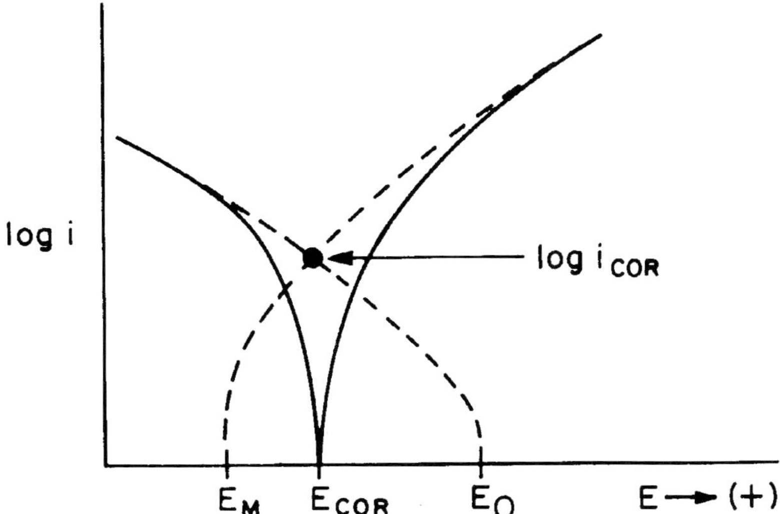

FIGURE 5. Schematic polarization curves (dashed lines) of the anodic dissolution of metal, M, and cathodic reduction of an oxidant, O, on the metal surface. Solid lines are the measured (ammeter) curves.

rate itself is often so low as to be inconsequential, i.e., < 10 μA cm^{-2} which from Eq. 10a is < 10 pm s^{-1} or < 10 mpy.

As described above the passive condition is established if the metal oxidation process leads to the formation of a very adherent, protective, solid, reaction product layer on its surface. A passive region in the anodic polarization curve is shown in Figure 6. At sufficiently positive electrode potential (E_{pass} in Figure 6), the rate of metal oxidation stops increasing with increasing potential and, instead, decreases precipitously to the very low passive current, i_{pass}. The relative sizes of the active and passive regions is a measure of the corrosion resistance of the metal or alloy. Hence, for carbon and low alloy steels, which are not known for their corrosion resistance, this ratio is large relative to that for a stainless steel, i.e., the addition of Cr to a carbon steel greatly suppresses the active peak (E_{pass} moves to the left in Figure 6) and also decreases the passive current. The corrosive, in particular its pH and the concentration of aggressive ions such as chloride ion which can cause local breakdown of the passive layer, also strongly influences the relative sizes of the active and passive regions. This is discussed more below in connection with localized corrosion.

For metal or alloys with small active regions the cathodic polarization curve typically crosses the metal dissolution curve in its passive region (Figure 6). The stronger the oxidant the more positive is the corrosion potential, E_{cor}. For this situation (E_{cor} is in the passive region) the protective corrosion product layer forms immediately. With weaker oxidants, i.e., those with not so noble an equilibrium potential, E_O, small exchange currents, i_o, or low concentrations (e.g., oxygen

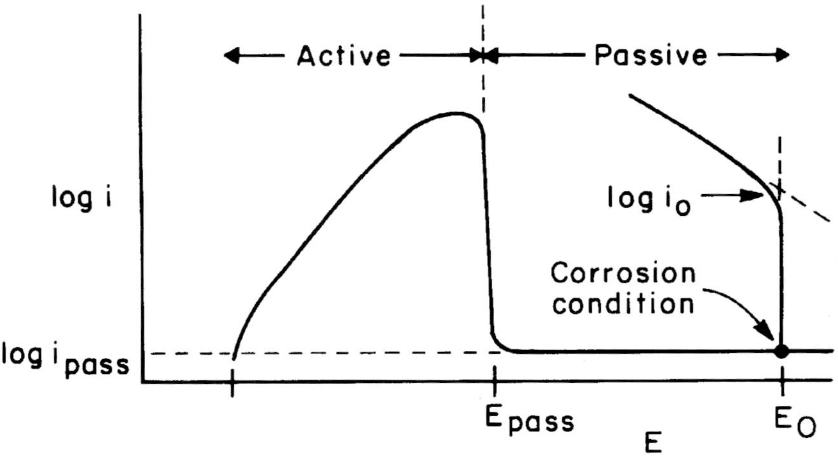

FIGURE 6. Schematic polarization curves of a metal exhibiting an active/passive transition and of the oxidant that has the equilibrium potential EO, and exchange current, io. The corrosion potential, E_{cor}, and corrosion current, i_{pass}, are established by the intersection of the two curves.

dissolved in water), the oxidant curve may also intersect the metal dissolution curve in its active region and cause high corrosion rates similar to the case above (Figure 5) with E_{cor} in the active region.

With E_{COR} in the passive region the metal dissolution rate is at the negligibly low i_{pass} value. Hence, the metal or alloy is functioning as a corrosion resistant material. However, in this corrosion resistant state the protective layer can under certain conditions, break down locally and produce one or more forms of localized corrosion (Figure 1c to 1j). An example is the pitting of passive stainless steels (E_{COR} is in the passive region) when the corrosive contains chloride ions.

THE IR VOLTAGE IN LOCALIZED CORROSION.

In uniform corrosion the anodic and cathodic reactions can and typically do occur very close to each other (nm scale). As a result the corrosion circuits described above are very small so that the IR drop in the circuit is negligibly small. Hence, the IR voltage has been ignored in the discussion to this point. In localized corrosion, however, the anodic reaction occurs inside the cavity at a large distance from the cathodic reaction. The latter typically occurs on the outer surface of the metal where there is a plentiful supply of the oxidant. The resulting corrosion circuits are very long (mm scale) and the IR voltages very large, approaching the total difference between the equilibrium potentials of the metal and oxidant. In this case the measured potential, again referred to as E_{COR}, is actually the potential of the cathodic reaction since the reference electrode used in the measurement is physically located at the outer surface where the oxidant is reduced. From potential theory the electrode potential of the anodic (metal dissolution) reaction which is occurring inside the cavity E_A, can be obtained from the measured E_{cor} value as

$$E_A = E_{COR} - IR \tag{12}$$

where I is the current flowing in the corrosion circuit and R is the total ohmic resistance of the circuit. As an approximation R can be taken as the ohmic resistance of the aqueous path since its resistance is orders of magnitude larger than that of the metal path even in the case of highly conducting electrolytes. As in the case of electronic conductors the value of R is determined by the length and cross sectional area of the electrolyte path within the cavity. For cylindrical and rectangular cavities the cross sectional area is constant along the length (depth) of the cavity. Since the IR voltage is dependent on distance into the cavity, one sees from Eq. 12 that the electrode potential along the wall of the cavity varies in the less noble direction with increasing distance into the cavity. Hence, on a polarization plot the abscissa is both the potential and distance into the cavity as shown in Figure 7. Consequently, for a cavity with a large enough aspect ratio (depth to opening dimension) the electrode potential on the cavity wall beyond the x_{pass} distance is less noble than the potential of the active-to-passive transition,

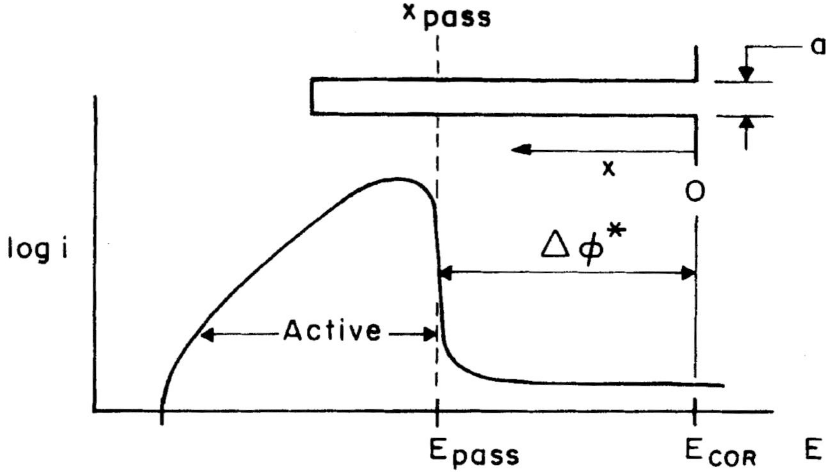

FIGURE 7. Schematic drawing showing that the local electrode potential, E (x) is less noble at increasing distance x into the cavity. Hence, at x < xpass, E (x) is in the active region of the polarization curve where order-of-magnitude higher metal dissolution rates occur.

E_{pass}, i.e., $E (x > x_{pass}) < E_{pass}$ is in the active region of the polarization curve where order-of-magnitude higher metal dissolution rates occur.

With $\Delta\phi *$ defined in Figure 7 as the voltage difference between the corrosion and passivation potentials, $\Delta\phi* = E_{cor} - E_{pass}$, the minimum requirement for localized corrosion is that the aspect ratio is large enough so that

$$IR > \Delta\phi* \tag{13}$$

at the deepest portion of the cavity. Since I and R are both functions of distance x, Eq. (13) is more rigorously expressed as

$$\int_{o}^{x} I(x)R(x)dx > \Delta\phi* \tag{14}$$

Some recent literature gives more details and describes the application of the IR mechanism of localized corrosion to specific metal/electrolyte systems (8-10).

CONCLUDING REMARKS

1. Examination of the definition of the $\Delta\phi*$ term in Figure 7 shows that an active

peak is required in the polarization curve for $\Delta\phi^*$ to have a finite value. It follows that by careful choice of corrosion resistant alloys with a view to minimizing the size of the active peak in the particular service environment, one may decrease or completely eliminate localized corrosion. This design approach, however, will not always be successful because of the change in solution composition that occurs with time in the stagnant corrosive that fills the corrosion cavity. The problem occurs when the change in solution composition produces or enlarges the active peak. An example is a neutral solution for which the solution becomes acidic through the occurrence of a hydrolysis reaction. Hence, as time increases an active peak forms in the polarization curve and increases in size so that the alloy becomes more susceptible to localized corrosion. The opposite is true for strongly acidic solutions in which the size of the active peak decreases with time leading to destabilization of the localized corrosion event as a result of deacidification and increase of the metal ion concentration within the cavity electrolyte (10).

2. From the above considerations it is clear that in all cavities which support anodic dissolution of the metal or alloy (i.e., pits, crevices, grain boundary grooves and corrosion fatigue or stress corrosion cracks), the local electrode potential changes in the more negative direction with increasing distance into the cavity, and the solution composition changes and its concentration increases with increasing time to new stable values. Although it is clear that consideration of these parameters is essential in determining a system's susceptibility to localized corrosion, more understanding is needed as the complexity of the situation increases. This is particularly true for the crack propagation forms of corrosion.

REFERENCES

1. Report, Battelle (Columbus, OH), April 1995.
2. C. Wagner and W. Traud, Z. Elektrochemie und Angewandte Physikalische Chemie, *44*, 52 (1938).
3. M. Pourbaix, Atlas d'Equilibres Electrochemiques at 25°C, Gauthier-Villars, Paris, p. 312 (1963).
4. M. Pourbaix, The Thermodynamics of Aqueous Solutions, trans. J.N. Agar, Arnold, London (1949).
5. A.J. Bard and L.R. Faulkner, Electrochemical Methods, John Wiley & Sons, New York, 1980, Chapter 2.
6. H. Kaesche, Metallic Corrosion, National Association of Corrosion Engineers, Houston, TX, 1985.
7. P.H. Rieger, Electrochemistry, Prentice-Hall, Inc., Englewood Cliffs, N.J., 1987.
8. B. DeForce and H.W. Pickering, JOM (J.Minerals, Metals & Materials Soc.), 22 (Sept., 1995).
9. H.W. Pickering, Materials Sci. & Eng., *A198*, 213 (1995).

Chapter Twenty-Three

HIGH TEMPERATURE CORROSION

GEORGE SIMKOVICH

Department of Materials Science and Engineering
Steidle Building
Pennsylvania State University
University Park, PA 16802

INTRODUCTION

There are many applications for metals at high temperatures. Examples are gas turbine engines in airplanes, turbines in power production, heating elements in many chemical process systems, metal housing for many high temperature processes and even the mundane muffler in your automobile.

The applications of metals at high temperatures has resulted from both empirical and theoretical studies. Such studies have been conducted primarily by the high temperature metallurgist. As may be surmised, the field of high temperature metallurgy is a subdivision of the broad field of materials science. Each sub-division of a profession has a particular general problem that confronts them. For example, the ceramist concerned with the mechanical properties of ceramics has, in general, the problem of a material that is brittle. This brittleness is of foremost concern in many applications but the resistance to high temperature corrosion is outstanding. Conversely, the metallurgist has a material, a metal, which displays outstanding strength and ductility but which suffers from the fact that it has the desire to become a ceramic when utilized at high temperatures in an air atmosphere.

More generally, one may say that all metals, Me, with the exception of gold, react with oxygen (and/or with many other oxidants) to form an oxide (and/or other compounds) in accord with the general reaction

$$Me + \frac{a}{2b} O_2 = MeO_{a/b} \tag{1}$$

proceeding spontaneously and irreversibly to the right. Thermodynamically (1) one may say that reaction (1) proceeds to the right at constant temperature and pressure because the Gibbs free energy change, ΔG, for the reaction is always negative. That is

$$\Delta G < 0 \tag{2}$$

holds for the normally encountered conditions. This reaction proceeds at low as well as at high temperatures; however, since the kinetics of the reaction tend to increase rapidly with temperature, frequently in a logarithmic fashion, the use of a metal at high temperatures relies upon decreasing the kinetics of reaction (1) in some manner.

Thus, in this chapter we shall present a bit of history relating to the published works which led present-day metallurgists to approaches that reduce the kinetics of reaction (1) and hence permit the use of metals in a wide variety of aggressive, corrosive atmospheres at high temperatures. Following this we shall discuss one aspect of high temperature superalloy behavior that has been and continues to be rather intriguing in a view of the fact that the alloying results, although most desirable, are not completely understood.

SOME HISTORY AND SOME THEORY

In 1920 Tamman (2) noted that the growth of a scale on a metal occurred in a parabolic manner as a function of time. That is, the thickness change (Δx) of the growing scale [or the gain in weight per unit area ($\Delta M/A$) of the oxidizing sample or the loss in metal thickness, $|\Delta y|$] decreases with time in the manner depicted in Figure 1A which is a schematic depiction of oxidation of a metal, say Ni, to NiO as a function of time.

A plot of the square of the ordinate of ($\Delta M/A$) against time (or the ordinate versus $t^{1/2}$) permits calculation of the parabolic rate constant, k, for the oxidation process. These constants may be converted to one another and also to Wagner's rational rate constant, k_r, which will be discussed, when the growing scale is a single compound.

Because diffusion processes also normally exhibit parabolic behavior it was hypothesized that solid state diffusion determined the kinetics of these corrosion processes.

For a pure metal forming only a single oxide which follows a parabolic relationship the diffusion process must occur in the growing oxide. The question then arises as to "what" is diffusing and more importantly, "how" is the diffusion process occurring. Fortunately during the 1920's and 1930's a number of investigators in the world's scientific community were interested in motion in ionic type compounds. This interest resulted in a number of nominal papers (3-6) which provided answers to these and many other questions related to solid state chemistry.

Frenkel's (3) paper in 1926 was the "ice breaker" for it postulated that a well-defined crystalline lattice need not be perfect and that the lattice could display point defects - vacancies and interstitials - which permitted motion through the lattice that did not require exorbitant energy expenditures. The papers by Wagner and Schottky (4) and Jost (5) basically described theoretically the possible combinations of point defects which might exist in solid ionic crystals. Thus the family of point defects in an electrically neutral crystal - vacancies, interstitials, electrons, electron holes and associates between these point defects - were "born" and have been proven to exist. The book by Kroger (6) describes these defects much more thoroughly.

With this information on point defects Wagner (7) in 1933 derived the elegant relationship

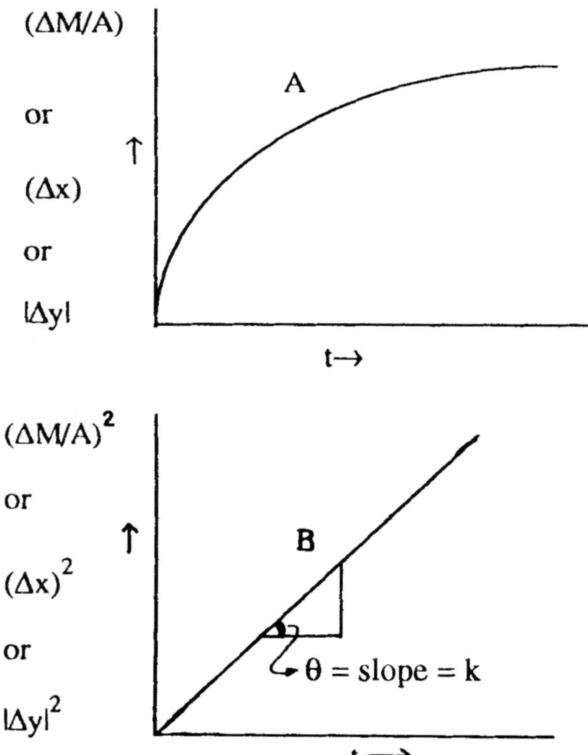

FIGURE 1 - A - Schematic behavior of a metal oxidizing parabolically at high temperature. - B - depiction of obtaining the rate constant of Figure 1-A by determining the slope from a plot of the square of the ordinate versus time, t. See the text for symbol description.

$$k_r = \frac{300}{96,500N_{Av}} \int_{\mu'}^{\mu''} \frac{1}{|z_2|} (t_1 +)t_2)t_3\sigma d\mu_x \qquad (3)$$

where k_r is the rational rate constant, N_{Av} is Avogadros number, μ_x is the chemical potential of element X, t's are the transport numbers, z's are the absolute electrical charges σ is the total electrical conductivity of the growing compound, superscript double dash ($''$) represents the value at the compound-gas interface, superscript single dash ($'$) represents the value at the metal-compound interface and the subscript numbers 1, 2, and 3 represent the cation, anion and electrons (or electron holes) in the compound. Thus, Wagner's concept may be depicted as in Figure 2.

For the case of a binary compound

$$t_1 + t_2 + t_3 = 1 \qquad (4)$$

When the growing binary compound is primarily an electronic conductor then t_3 is essentially one and making use of the Nernst-Einstein equation relating diffusion and electrical conductivities one may put equation (3) into the form

$$k_r = c_{eq} \int_{a_x'}^{a_x''} \left(\frac{z_1}{|z_2|} D_1^* + D_2^*\right) d \ln a_x \,|_{t_3=1} \qquad (5)$$

where $c_{eq} = z_1 c_1 = |z_2|c_2$ is the concentration of metal or non-metal ions in equivalents per cubic centimeter, D^*'s are self-diffusion coefficients and a_x's are the thermodynamic activities of x.

Equation (3) and its special case equation (5) permits one to control oxidation reactions in terms of the mechanisms of diffusion. Let us take several simple examples. Thus, we assume for case I that the growing compound is an ionic conductor, that is that $(t_1 + t_2)$ is almost equal to one, then from equation (3) and (4) we find that t_3 is rate controlling. Hence for an ionic conductor growing on a metal the motion of electrons or electron holes through the scale is rate determining. For case II we take that the growing compound is a semiconductor, $t_3 \cong 1$. Thus, equation (5) applies and we see that either cation diffusion and/or anion diffusion controls the rate of growth of the corrosion product.

There are a number of conditions that must be met in order that equation (2) and (4) hold. Among these are: (a) equilibrium is established at each interface; (b) the scale formed is adherent and continuous over the metal surface; (c) diffusion in the

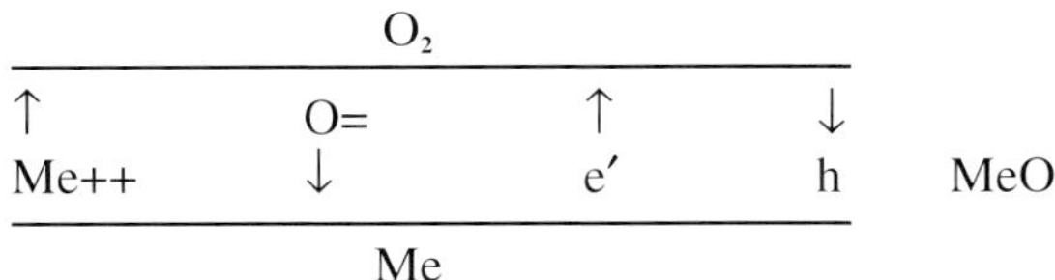

FIGURE 2 - A schematic sketch of possible diffusion processes in a growing MeO scale.

growing scale is bulk diffusion, i.e., no short circuit paths are followed and (d) no cracking or spalling of the growing scale occurs.

These relations led to an understanding of the growth processes that occur on pure metals - see Hauffe (8) for the description of many such pure metal cases - and they also served as guides to the development of many binary, ternary and multicomponent alloys.

Because the point defects in the growing scale were directly related to the kinetics the alloy development investigator searched to obtain scales that displayed low concentrations of the point defects primarily responsible for the growth process. The high temperature compounds that were developed either empirically and/or theoretically to serve as slow growing and adherent scales under many corrosive conditions were found to be Cr_2O_3, Al_2O_3 and SiO_2.

As an empirical example we take Ni-Cr alloys, e.g. Nichrome. If one measures a parabolic rate constant as a function of chromium at a given temperature and oxygen pressure one finds that this rate constant varies in the fashion depicted in Figure 3. From this depiction we see that, at low chromium concentrations, the rate constant first increases as chromium is added to the nickel and then maximizes followed by a decrease in the constant. This behavior may be explained via the relationship given in equation (5).

The analysis is as follows: NiO is a semiconductor, i.e. $t_3 \cong 1$ and $D^*_O << D^*_{Ni}$; Hence the diffusion of nickel via vacancies in the growing oxide scale is rate determining when only NiO is present as a scale. This is the case for low chromium concentrations since the Cr_2O_3 formed dissolves into the NiO in accord with the reaction

$$Cr_2O_3 = 2Cr^{\bullet}_{Ni} + V''_{Ni} + O^x_o \tag{6}$$

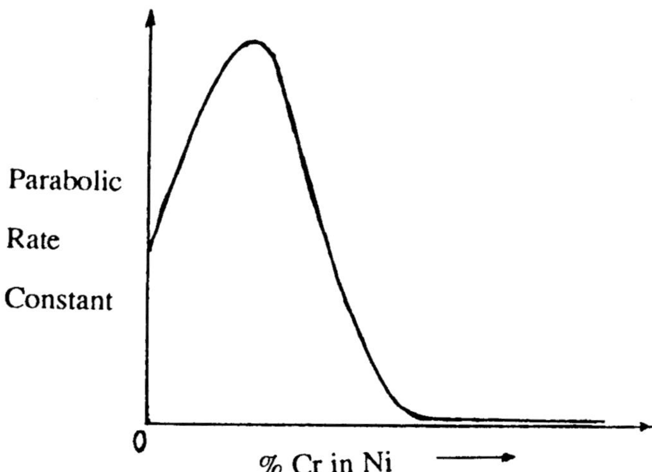

FIGURE 3 - Parabolic Rate Constant as a function of chromium content in nickel equation (5).

where, using the Kroger-Vink notation (6), Cr_{Ni} indicates a chromium ion substituting for a nickel ion at a nickel ion site, V_{Ni} indicates a vacancy at a nickel ion site O_O indicates an oxygen ion sitting at an oxygen ion site and the superscripts filled dot, dashed line and x indicate an excess positive charge, an excess negative charge and no excess charge respectively.

Thus the dissolution of Cr_2O_3 into NiO creates cation vacancies which increases the rate of oxidation since diffusion of nickel ions via these vacancies is rate determining. Such occurs in the scale until near the maximum where other oxides begin to appear - a spinel, $NiO \cdot Cr_2O_3$, and Cr_2O_3. Finally after additional chromium additions only Cr_2O_3 appears at the alloy-oxide interface and this slow growing compound then controls the kinetics of oxidation.

Nichrome, of this example, is an industrial alloy. Most industrially utilized metals are not pure metals - the major exception would be the noble metals. Thus the theory advanced for pure metals does not suffice for alloys employed in industrial applications although the concept of motion in the growing scale does generally hold.

BINARY ALLOYS

To aid in applications, a number of binary alloy theories have been advanced. These, as with the pure metal theory, were the product of Wagner (9,10). In these alloys one must also take into account diffusion in the alloy metal phase as well as in the growing scale.

Consider the case of the oxidation of a binary alloy, one component of which is noble. In this case we have the oxidation of the non-noble alloying element occurring while the noble metal remains in the alloy. As the alloy is oxidized the noble element concentration in the alloy at the alloy-oxide scale interface increases. Thus, the noble element tends to diffuse toward the bulk of the alloy while the active component diffuses toward the alloy-scale interface and then through the growing scale to form new scale. The rate of the growth of the scale, say k, and the rate of diffusion in the alloy, say D, then determine the actual rate of oxidation since, if diffusion is rapid in the alloy, one always has a supply of active element to be oxidized at the alloy-scale interface. Because the alloy cannot oxidize faster than the pure active element, say k°, one may surmise that the actual oxidization rate is strongly dependent upon the supply of active element at the scale-alloy interface, that is, upon the diffusion occurring in the alloy phase. Figure 4 depicts schematically the extreme behaviors feasible in such an alloy system.

When both components of an alloy oxidize one may obtain a single oxide scale of one or the other component or a mixture of the alloy oxides. This depends on a number of factors such as alloy diffusivity, parabolic rate constants of each of the oxides, oxygen potentials in the corrosive gas phase and temperature to list the major parameters considered in the advanced theory 10. It is possible by controlling alloy composition, temperature and oxygen pressure, to obtain single phase or multi phase scales. This is discussed rather thoroughly by Wagner (10).

It is possible to evaluate approximately the concentration needed to obtain a single oxide scale of one of the components in a binary alloy (10). Thus the concentration of B in a binary alloy A-B to obtain BO as a single scale oxide is given as:

$$N_B = \frac{V}{Z_B M_O}\left(\frac{\Pi k p}{D}\right)^{1/2}$$

(7)

where N_B is the mole fraction of B in the alloy A-B, V is the molar volume of the alloy Z_B is the valence of B, M_O is the atomic weight of oxygen, D is the diffusion coefficient of B in the alloy and kp is the parabolic rate constant of the growth of BO exclusively, This condition is necessary but may not be sufficient if factors not considered in the approximations leading to equation (7) become important in the growth of BO.

INTERNAL OXIDATION

In addition to oxidizing externally alloys may dissolve oxygen and the oxygen may then form an oxide in the bulk alloy. This is normally designated "internal oxidation." Theories covering this situation have been numerous and have been advanced by a number of authors (11-17). Let us assume that we have an alloy

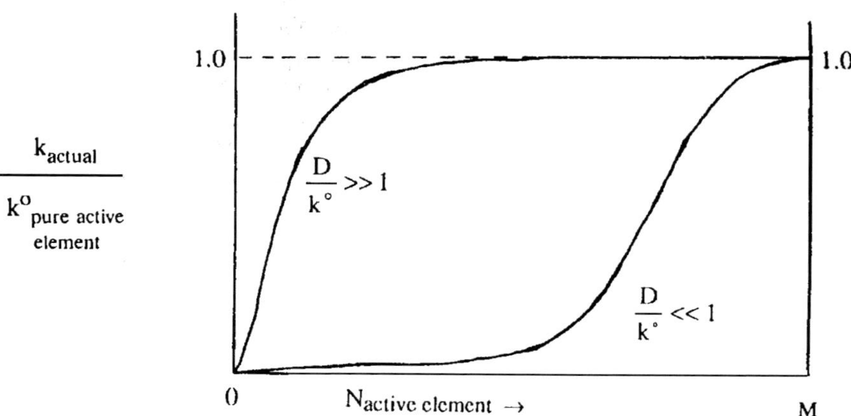

FIGURE 4 - Schematic plot of the ratio of parabolic rate constants, k actual of alloy, k° pure active element, versus the concentration of the active element in the alloy at various diffusion rates, D, in the alloy.

A-B and that B is an active element whose affinity for oxygen is much greater than is A's for oxygen, that is

$$\Delta G_{AO} >> \Delta G_{BO}$$

We also assume that the alloy dissolves oxygen in reasonable concentrations and that the concentration of B is less than that given by equation (7), that is, essentially no external scale of BO is formed.

Under these conditions the oxygen will dissolve in the alloy. Eventually the B and oxygen concentrations will reach their solubility limit, precipitating BO inside the alloy structure. Figure 5 depicts the structure obtained when the alloy is partially, internally oxidized and Figure 6 depicts the concentration profiles from the alloy interface into the alloy.

The internally oxidized alloy may be desired or conversely may degrade some desired alloy properties. A desirable property usually is the increase in strength conveyed to the alloy via the fine oxide particles. Additionally, one may wish to retain a specific property of the metal A, say the electrical conductivity, so that the internal oxidation process which tends to give a "pure" metal A in which particles are dispersed does drive the alloy towards the electrical properties inherent with the pure metal A.

Consider now the converse case, i.e. an alloy system where you wish to have BO as a protective oxide on an A-B-X alloy and hence where you do not wish to oxidize B internally. We consider several techniques to avail ourselves of an alloy where BO is formed as an external scale. The following are at least three methods of obtaining such an outer BO scale:

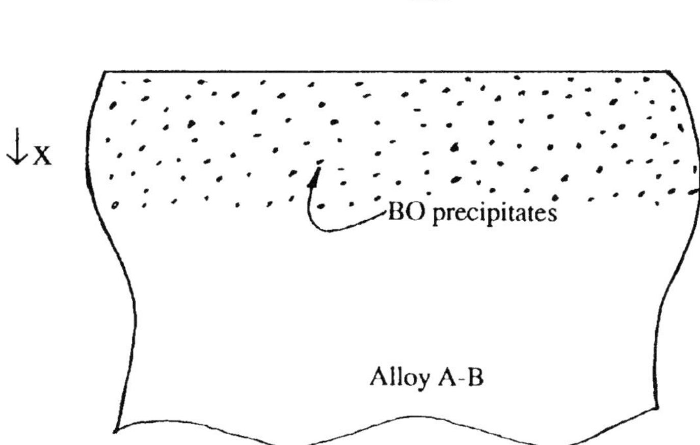

FIGURE 5 - Partial internal oxidation of an alloy in which BO is much more stable than AO.

(A)   One may increase the concentration of B to such an extent that one forces sole formation of a BO external scale - see equation (7) and Figure 6;

(B)   One may decrease the oxygen potential to such an extent that only B oxidizes in the alloy A-B. This may be deduced by viewing Figure 6 - lowering the oxygen potential reduces the concentration of dissolved oxygen in accord with Sievert's Law (1) and thus decreases the thickness of the internally oxidized zone until, at a sufficiently low oxygen pressure, no internal oxidation occurs; and,

(C)   One may introduce a third element, say x, into the binary alloy A-B whose affinity for oxygen is generally intermediate between that of A and B. Hence this element serves to lower the oxygen potential "felt" by the B component to an extent similar to that of lowering the potential in the gas phase as described in case (B) above.

This is called the "getter" effect and is displayed in Figure 7 for an alloy in the Fe-Al system to which is added Cr to give an Fe-Cr-Al alloy18. As may be seen, the effect of the Cr in the Fe-Al is most desirable.

The addition of alloying elements to a base metal usually is employed to not only obtain oxidation resistance but also to control other alloy properties, e.g., creep resistance and strength. Thus, few alloys are simple binary or ternary alloys. However, one may approach industrial high temperature alloy design utilizing the concepts presented.

Reactive element effects (REE)

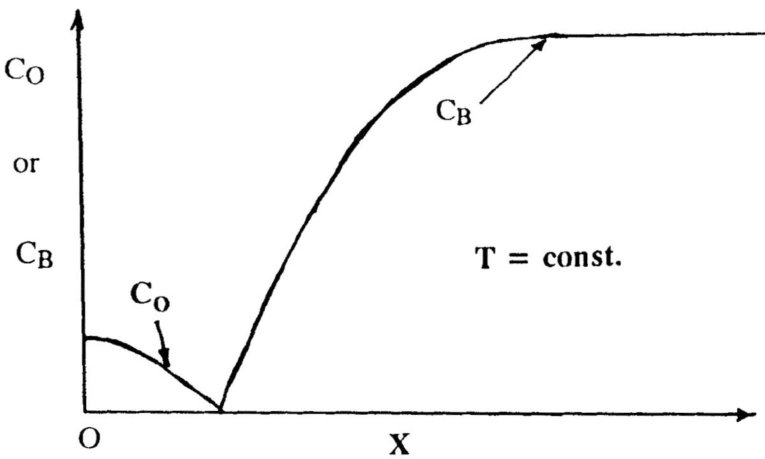

FIGURE 6 - The concentration of dissolved, uncombined oxygen and B (in alloy A-B) as a function of distance into the alloy.

We consider now the intriguing topic of the effect of the addition of trace amounts of reactive elements, or small quantities of reactive element oxides as a dispersed phase, to alloys which form either chromia, Cr_2O_3, or alumina, Al_2O_3, as their protective scale.

As was noted previously there are many requirements that a growing, high temperature scale must meet. Foremost among these requirements is that the scale is slow growing in the corrosive environment and that the scale remains adherent during any cyclic operation whether temperature or oxygen potential or other parameters.

In 1937 Pfeil (19) obtained a patent which was concerned with the addition of trace additions of oxygen active elements such as Y and Ce to alloys that form chromium or aluminum oxides scales at moderate and high temperatures. These additions conferred many desirable properties to such alloys in respect to their oxidation behavior. It was noted by Pfeil that the mechanism (s) responsible for the oxidation benefits derived by these trace additions was (were) not understood. This patent has led to a wealth of research studies and some clarification of the mechanisms involved has evolved from the studies undertaken.

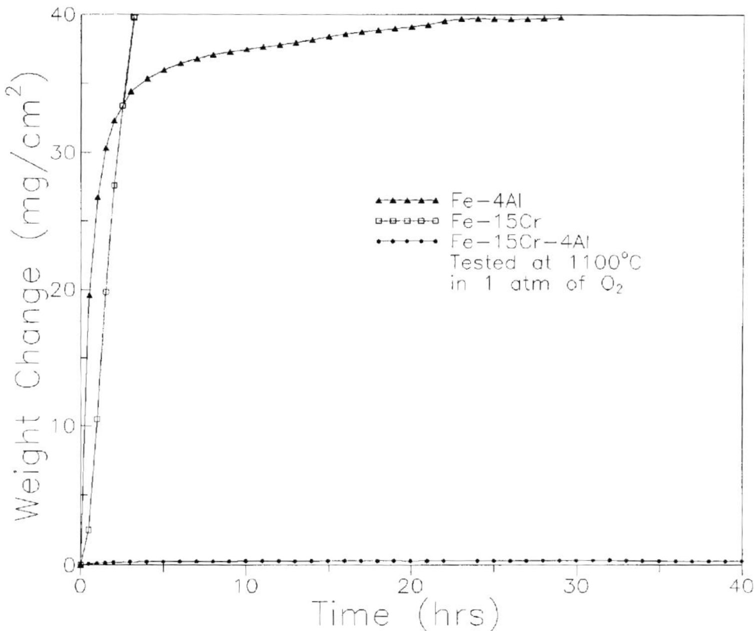

FIGURE 7 - $\left(\dfrac{\Delta M}{A}\right)$ vs time illustrating the effect of a chromium addition to an iron-aluminum alloy. Chromium displays the "getter" effect.

In the following, a summary of the benefits of the "reactive" element effect is given in respect to both Cr_2O_3 and Al_2O_3 formers.

Thus, the effects noted in oxidation behavior when traces of reactive metals (or dispersed particles of the reactive metal's oxide) are added to alloys are summarized as follows:

1. The adherence of the scale is immensely improved.
2. The rate of oxidation is decreased substantially.
3. The mechanism for scale growth, especially for chromia formers, is altered from predominantly cation motion to anion motion.
4. The concentration of chromium, and perhaps of aluminum, in the alloy needed to form the protective chromia (alumina?) scale is distinctly reduced; and
5. The grain size of the growing scale is markedly reduced.

A general review of these effects is given by Whittle and Stringer (20) and for alumina formers by Jedlinski (21). A 50 year anniversary publication recognizing Pfeil's patent was edited by King (22) and presents a number of recent papers concerned with REE.

A number of mechanisms have been proposed to explain the effects conveyed by the reactive elements or their oxides. However, no general mechanism appears to answer all of the observations, effects, noted. Hence, we present some selected mechanisms that have been advanced throughout the years.

The most important effect conferred by the reactive elements to alloys is the increased adherence of the scale during temperature cycling operations because temperature cycling is frequently encountered in practice. Initially it was hypothesized that the reactive elements oxidized, or were already present as oxides, and that these oxide particles served as "pegs" between the growing oxide scale and the alloy (23). Future studies did not confirm the presence at all times of these pegs and, therefore, serious considerations of the "pegging" effect are generally excluded even though some pegging may occur.

Another proposed mechanism for the increased adherence effect was that the reactive metal oxide with the growing oxide formed a graded seal between the alloy and the "pure" growing oxide (24). This concept has not been supported by further studies and findings.

The formation of voids at the growing oxide-alloy interface is virtually eliminated when reactive elements/oxides are added to alloys (21). This aids considerably in the bonding between the growing oxide and the alloy. This void elimination does relate to the increased adherence of the scale attained via the addition of the reactive metal or its oxide. Figures. 8A and 8B display quite clearly the effect on void formation that occurs on an Fe-15Cr-4 Al alloy, oxidized at 1100°C, when 1 Wt% Y_2O_3 is added to the alloy (18). Basically, the voids are eliminated in the presence of the Y_2O_3.

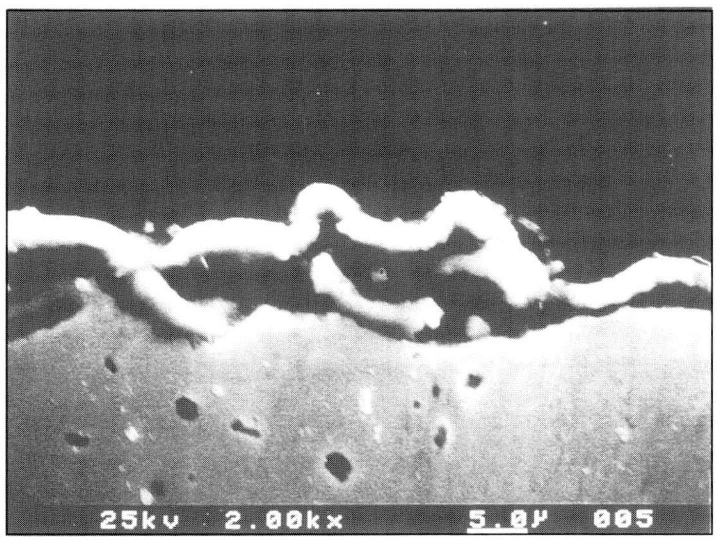

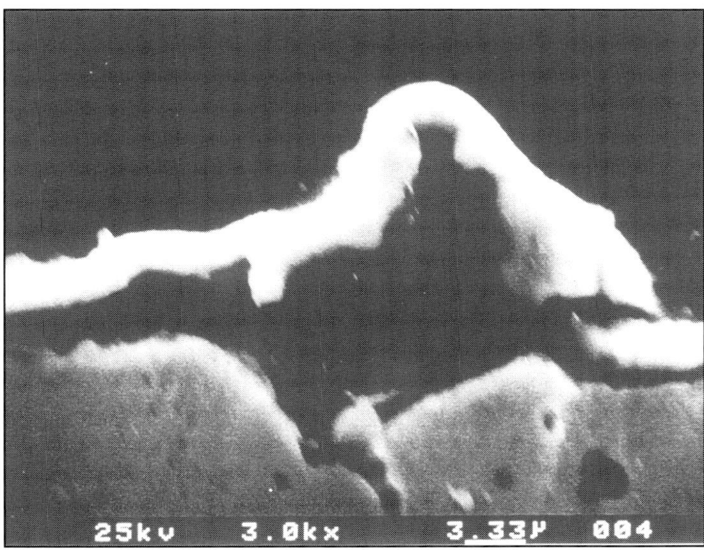

FIGURE 8 - Cross-section of an FeCr-4Al alloy without (A) and with (B) the addition of 1 wt % Y_2O_3. T = 1100°C; PO_3 = 1 atm - Reference (18).

Probably the recently proposed mechanism (25) that the reactive metals or their oxides act as "getters" for sulfur and other indigenous elements in the alloy serves best to explain the increased adherence conferred by the reactive elements/oxides. These indigenous elements adsorb at the scale-alloy interface and tend to weaken the bond between the scale and the alloy. Hence, reaction of these bothersome elements with the reactive metal/oxide removes this weakening effect and improves the bond strength between the scale and the alloy.

Figure 9 shows the effect of the addition of Y_2O_3 on the kinetics of oxide formation (18). It is obvious that the initial rate of oxide formation is decreased considerably by the Y_2O_3. This results primarily from the increased rate of formation of the protective oxides, Cr_2O_3 and Al_2O_3, in the Fe-Cr-4Al alloy when the Y_2O_3 is present.

Since the presence of the reactive metal or their oxides induces rapid formation of the protective oxides (18, 20, 21) the oxygen potential at the scale-alloy interface is reduced to such a low value that the base metal alloying component cannot form an oxide. Hence, the kinetics of oxidation are decreased while the amount of the alloying component, say chromium, and/or aluminum in the alloy, necessary to form the protective oxides is reduced. Such is desirable for a number of reasons but primarily because this lowering of the concentration of the active component of the

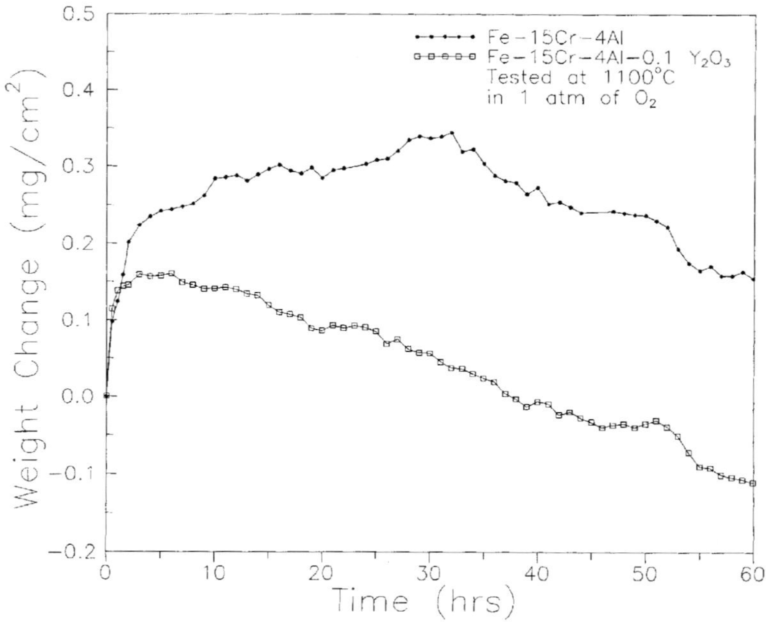

FIGURE 9 - Change in $\left(\frac{\Delta M}{A}\right)$ versus time obtained for an alloy exposed to 1 atm of oxygen at 1100°C without and with the presence of a dispersed Y_2O_3 phase.

alloy gives a broader range of useful concentrations to the alloy designer. Thus, the dependence of other alloy properties, e.g. ductility, etc., may receive more attention and may not be compromised in attaining oxidation protection.

Reduction in grain size of the growing scale may simply be attributed to the greater number of nucleation sites present on the fine reactive metal oxides that are present initially as a dispersed oxide phase or appear as initial oxidation products of the alloy which contains the reactive alloying element. Frequently, when the reactive metal oxide is added as a dispersed phase to the alloy the grain size of the alloy phase is also reduced in size. The fact that these phases, the alloy and the scale, retain fine sizes for long times at high temperatures indicates that the particles of the reactive metal oxides serve to prevent grain growth by pinning motion at the grain boundaries.

In respect to cation motion versus anion motion in the growing scale one notes that the chromia scale behavior is better documented than is the alumina scale (20). Thus it is known that chromium ions move from the alloy-scale interface to the scale-gas interface in the absence of reactive metal oxide particles (22) and that the kinetics are parabolic in nature. However, a number of other findings in relation to the influence of the reactive metal oxide particles are important. These are that motion in the Cr_2O_3 scale is actually through grain boundary regions primarily (26) and further that reactive metal oxide particles when present in the scale are found at the grain boundaries and are also adsorbed along the grain boundaries.

If one now considers the major ionic motion in the commonly employed or resulting reactive metal oxides one finds that oxygen ion motion is predominant (27). Thus, it was concluded (28) that the change in ionic motion from cations to anions is completely in accord with the motion in the grain boundaries, i.e., in "pure" Cr_2O_3 chromium ions are found to move outward but when reactive metal oxides are adsorbed at the oxide grain boundaries motion is in accord with that predominant in the reactive metal oxide-anonic motion.

SUMMARY

High temperature corrosion behaviors of metals and their alloys are presented in the article. The conduct of pure metals that corrode in a parabolic manner is rather well understood. Binary alloy actions become much more complicated than the pure metals and these actions vary tremendously with alloy composition.

A quantitative account of ternary and higher order alloys is not available at the present time. However, the behavior of such alloys may be guided by some of the quantitative principles advanced for pure metals and binary alloys. In addition a number of qualitative guidelines are available for these ternary and/or higher order alloys.

The rather unusual effects of minor additions of reactive elements (or their oxides) are discussed and some mechanisms advanced by various authors are presented. It does not appear that a single mechanism will answer the many observed effects, most of the effects are beneficial, promoted by such additions.

Finally, it is well to indicate that all materials tend to deteriorate at high temperatures. Thus, the ceramics formed from metals will also be degraded in corrosive atmospheres at high temperatures.

Much effort is still needed to overcome the many high temperature problems that confront science and industry.

REFERENCES

1. Darken, L.S. and R.W. Gurry. 1953. *Physical Chemistry of Metals,* McGraw Hill Co., New York, NY.
2. Tamman, G. 1920. *Z. anorg. u. allgem. Chemie,* 111:78.
3. Frenkel, J. 1926. *Zeit f. Physik,* 35:652.
4. Wagner, C. and W. Schlottky. 1930, *Zeit f. phys. Chemie,* (B), 11:163.
5. Jost, W. 1933. *J. Chem. Phys.,* 1:466.
6. Kroger, F.A. 1964. *The Chemistry of Imperfect Crystals,* John Wiley & Sons, Inc., N.Y., N.Y.
7. Wagner, C. 1933. *Zeit. f. phys. Chemie,* ((B), 21:25.
8. Hauffe, K. 1965. *Oxidation of Metals,* Plenum Press, New York, NY.
9. Wagner, C. 1952. *J. Electrochem. Soc.,* 99:369.
10. Wagner, C. 1952. *J. Electrochem. Soc.,* 103:627.
11. Rhines, F.N. 1940. *Trans. Metal. Soc.,* AIME, 137:246.
12. Rhines, F.N. 1942. W.A. Johnson and W.A. Anderson, *Trans. Metal. Soc.,* AIME, 147:205.
13. Darken, L.S. 1942. *Trans. Metal. Soc.,* AIME, 150:157.
14. Meijering, J.L. and M.J. Druyvesteyn, 1947. *Philips Res. Rep.,* 2: 81 and 260.
15. Wagner, C. 1959. *Elektrochem.,* 63:772.
16. Bohm G. and M. Kahlweit. 1964. *Acta Metall.,* 12:505.
17. Rapp, R.A. 1964. *Corrosion,* 21:382.
18. Yan, R-F. 1995. M.S. thesis, The Pennsylvania State Univ., University Park PA.
19. Pfeil, L.B. 1937. "Improvements in Heat Resistant Alloys," British Patent Number 459,848.
20. Whittle, D.P. and J. Stringer. 1980. *Phil Trans. R. Soc. London, A295.*
21. Jedlinski, J. 1992. *Solid State Phenomena,* 21 & 22:335.
22. "The Reactive Element Effect on High Temperature Oxidation," ed. by W.E. King. 1989. Trans-Tech Publications Ltd., Brookfield, VT, 05036, U.S.A.
23. Antill, J.E. and K.A. Pekall. 1967. *J. Iron St. Inst.,* 205:1136; and Allam, I.M., D.P. Whittle and J. Stringer. 1978. *Oxid Met.,* 12:35.
24. Pfeiffer, H. 1957. Werkstoffe and Korosion, 8:574.
25. Funkenbusch, A.W., Smeggil, J.G. and H.S. Bornstein. 1985. *Metal Trans. A.* 16A:1164.
26. Atkinson, A. and R.I. Taylor. 1985. In: G. Simkovich and V.S. Stubican (Eds.) *Transport in Non-stoichiometric Compounds* Plenum Press, N.Y., N.Y. 285.
27. Ando, K. and Y. Oishi in ref. 26, 203. 1985.
28. Simkovich, G. 1995. *Oxid of Metals,* 44:501.

Chapter Twenty-Four

TRANSPORT PROCESSES IN FUSION WELDING

TARASANKAR DEBROY

Department of Materials Science and Engineering
Steidle Building
Pennsylvania State University
University Park, PA 16802

INTRODUCTION

Welding is the most efficient and cost-effective joining technology available today. In a broad sense, it involves the formation of metallurgical bonds in joints but excludes joints made by mechanical means and adhesives. In the United States, welding is utilized in fifty percent of the industrial, commercial, and consumer products that make up the gross national product (1). The major economic impact of welding can be seen from its widespread use in the construction of buildings and bridges, and in the shipbuilding, aerospace, automotive, chemical, petrochemical, electronic, and power generation industries.

During the Second World War, there was a crucial national need to fabricate some three thousand cargo ships in a short time. Welding was extensively used for the rapid construction of the Liberty and the Victory ships. Welding also played a major role in the construction of the 800 mile long and 48 inch diameter Alaskan pipeline. The construction, maintenance, and continuing rebuilding of America's transportation links also require extensive use of welded components. Welding has also played an important role in the exploration of space. The fabrication of many critical components, ranging from the lunar module for moon landing in 1969 to the construction of all the space shuttles since then, has made extensive use of various welding processes. Today, designers of the next generation of fighter aircraft, the F-22, for the US Air Force must depend on advanced electron beam welding to build the fuselage.

In fusion welding, parts are joined by melting and subsequent solidification of adjacent areas of two parts. Welding may be performed with or without the addition of a filler metal. Depending on the thickness of the part, the weld may consist of one or more individual beads formed by each pass of the heat source along the joint. Fusion welding processes are classified according to the nature of the heat source. An electric arc is used in gas-metal arc welding (GMAW), gas-tungsten arc welding (GTAW), and submerged arc welding (SAW). In laser and electron beam welding, very high energy density beams are used as an energy source. Resistive heating is utilized in the electroslag welding process.

Figure 1 is a schematic diagram of the fusion welding process (2). Three distinct regions in the weldment are clearly observed: the fusion zone which undergoes melting and solidification, the heat affected zone which experiences significant thermal exposure and may undergo solid state transformation, but no melting, and the base metal which is unaffected by the welding process.

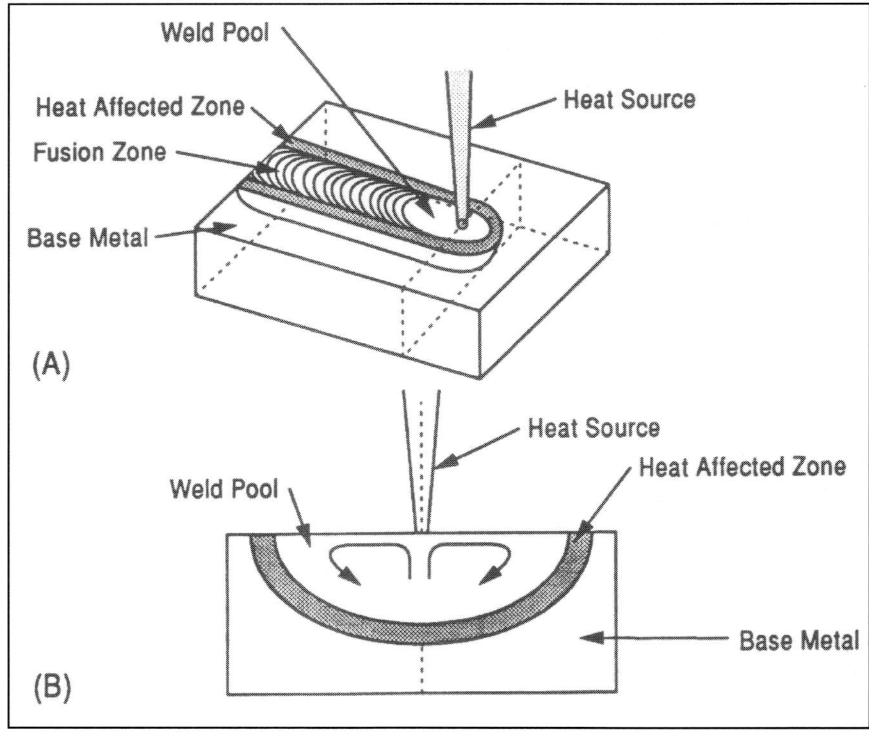

FIGURE 1. Schematic diagram (17) of the fusion welding process showing (a) the interaction between the heat source and the base metal, and (b) transverse section.

The interaction of the material and the heat source leads to rapid heating and melting of the base material. During welding, several types of forces act on the liquid metal. For example, at the weld pool surface, the spatial gradient of temperature and the resulting gradient of surface tension leads to a stress, known as the Marangoni stress. Furthermore, the spatial variation of density leads to buoyancy forces on the fluid. Similarly, when electric current is used, electromagnetic forces may become important depending on the strengths of the magnetic field and the current density distribution. As a result, molten metal circulates rapidly in the weld pool. The circulation helps in transporting heat from away from the region where the heat source interacts with the molten pool. The heat transfer and fluid flow affect the size and shape of the weld pool, the cooling rate, and the kinetics and extent of various solid state transformation reactions in the fusion zone and heat-affected zone. The resulting weld pool geometry influences dendrite and grain growth selection processes. Both the partitioning of nitrogen, oxygen, and hydrogen between the weld pool and its surroundings and the vaporization of alloying elements from the weld pool surface greatly influence the composition and the resulting microstructure and properties of the weld metal.

The composition, structure, and properties of the weld metal are affected by widely diverse physical processes. Therefore, improvements in the weld metal performance from a scientific viewpoint require an interdisciplinary approach. Major advances have taken place in welding science and technology over the last few decades. Because of this newly acquired knowledge, welding has evolved from almost an empirical art to a major interdisciplinary technical activity requiring synthesis of knowledge from various disciplines. This chapter reviews important advances in fusion welding through understanding of heat transfer, fluid flow, and mass transfer.

HEATING OF THE WORKPIECE

The outcome of the welding process depends on the amount of energy absorbed by the workpiece from the heat source. The consequences of the absorbed energy include formation of the liquid pool, the temperature-time cycle in the entire weldment, and the structure and properties of the weldment. Therefore, it is important to understand the physical processes involved in the absorption of energy during the welding processes. The physical phenomena that influence the energy absorption in the workpiece are unique to each welding process. For a given power source, the extent of the energy absorbed by the workpiece depends on the nature of the material, the type of the heat source, and the parameters of the welding process.

Let us consider a relatively simple process where a clean metal surface is heated by a laser beam. The absorption of infrared energy by metals depends largely on conductive absorption by free electrons. Therefore, for flat, clean metal surfaces, the absorptivity can be calculated from the knowledge of the electrical resistivity of the substrate. Arata and Miyamoto (3) demonstrated that the absorptivity of such surfaces is a linear function of the square root of the electrical resistivity of the respective metals based on experimental data. Bramson (4) related the absorptivity to the substrate resistivity and the wavelength of the laser radiation:

$$\eta(T) = 0.365[\frac{r}{\lambda}]^{\frac{1}{2}} - 0.0667[\frac{r}{\lambda}] + 0.006[\frac{r}{\lambda}]^{\frac{3}{2}} \tag{1}$$

where $\eta(T)$ is the absorptivity at a temperature T, r is the resistivity in ohm-cm, and λ is the wavelength in cm. Such calculations are accurate for clean and flat metal surfaces and when a plasma plume does not affect the absorption of the laser beam. All metal surfaces are highly reflective to the infrared radiation of a CO_2 laser at room temperature, and thus, the transfer of energy from the laser beam to the workpiece can be inefficient and, in some cases, lie in the 10 to 20% range.

Of course the welding process is much more complex than the heating of a clean, flat metal surface by a laser beam. During laser welding, the absorption of the laser beam by the workpiece is affected by several factors such as the wavelength of laser, the nature of the surface, joint geometry, the deformation of the liquid pool, and other factors. Since the absorption of the laser beam by metal surfaces is poor, how does one actually weld metals with a laser beam? When the power density is high, metals can vaporize rapidly from the molten pool, and a cavity in the shape of a keyhole can form because of the recoil force of the vapor on the liquid metal. When the keyhole forms, the energy absorption efficiency can improve dramatically, to a value much higher than that predicted by equation (1), owing to multiple reflections of the beam in the cavity. In many cases, during laser welding, a luminous gas plasma forms near the weld pool. A part of the laser beam energy can be absorbed by the plasma before the beam reaches the material. Thus, it is important to know how much of the laser energy is actually absorbed by the plasma. Theoretical treatments are available for simple systems (5).

A free electron traveling through the electric field in a plasma can absorb a photon and acquire additional kinetic energy. The free-free transition involving photon absorption is called inverse bremsstrahlung. The extent of absorption owing to this effect can be determined (6-9) if the electron density and electron energy in the plasma are measured or estimated. Rockstroh and Mazumder (9) conducted aluminum welding experiments using argon shielding gas. They found that the extent of inverse bremsstrahlung absorption was about 20% and 31% for welding with 5kW and 7kW laser powers, respectively. On the other hand, for welding of steels with low power (sub-kilowatt) CO_2 lasers, overall bremsstrahlung absorptions of less than 10% have been reported in the literature (6, 7). Thus, the attenuation of the laser beam energy by the plasma plume becomes important at high laser powers. Unless the keyhole forms, the absorption of a carbon dioxide laser by clean metal surfaces is poor. The absorption improves somewhat when shorter wavelength lasers, such as the Nd-YAG laser, are used.

For arc welding, the fraction of the arc energy transferred to the workpiece, commonly known as the arc efficiency, η, is given by (10):

$$\eta = \frac{q}{VI} = 1 - \frac{q_e + (1 - n)q_p + mq_w}{VI} \tag{2}$$

where q is the rate of heat absorption by the workpiece, I and V are the welding current and voltage respectively, q_e is the rate of heat transfer to the electrode from the heat source, q_p is the energy radiated and convected to the arc column per unit time, of which a proportion n is transferred to the workpiece, and q_w is the rate of heat absorbed by the workpiece, of which a portion, m, is radiated away. For a consumable electrode, the amount of energy transferred to the electrode is eventually absorbed by the workpiece. Thus, in the case of a consumable electrode, equation (2) can be simplified to:

$$\eta = 1 - \frac{(1 - n)q_p + mq_w}{VI} \tag{3}$$

Equations (2) and (3) are useful in understanding how the various types of heat loss affect arc efficiency. However, it is difficult to estimate the heat loss terms q_e, q_p and q_w accurately from theoretical considerations. Because of the importance of quantitative assessment of heat absorption, a large volume of literature exists on the arc efficiency and laser beam absorption coefficient. Correlations are available for the estimation of these parameters from welding variables. However, the available correlations may not always take into account the special circumstances of welding such as the nature of the workpiece surface, geometric orientation of the workpiece, the nature of the plasma plume, and other factors. Therefore, the experimental values of the arc efficiency and absorption efficiency should be used where such data are available for the conditions of welding. Only when the data are not available, theoretically calculated values should be used.

ORIGIN OF FLUID FLOW

Fluid flow and convective heat transfer are important in determining the size and shape of the weld pool, the weld macro- and microstructures, and the weldment properties. The flow is driven by surface tension, buoyancy, and, when electric current is used, electromagnetic forces. In some instances, aerodynamic drag forces of the plasma jet may also contribute to the convection in the weld pool. Buoyancy effects originate from the spacial variation of the liquid metal density, mainly because of temperature variations, and, to a lesser extent, from local composition variations. Electromagnetic effects are a consequence of the interaction between the divergent current path in the weld pool and the magnetic field that it generates. This effect is important in arc and electron beam welding, especially when a large electric current passes through the weld pool. In arc welding, a high velocity plasma stream impinges on the weld pool. The friction of the impinging jet on the weld pool surface can cause significant fluid motion.

The spatial gradient of surface tension is a stress known as the Marangoni stress. The variation of the surface tension at the weld pool surface may arise owing to variations of both temperature and composition. In most cases, the temperature variation is the main contributing factor. For such a situation, the weld metal can circulate at high velocities. DebRoy and David (2) have shown that the order of magnitude of the maximum velocity can be estimated from the following force balance:

$$u_m^{3/2} \approx \frac{d\gamma}{dT}\frac{dT}{dy}\frac{W^{1/2}}{0.664\rho^{1/2}\mu^{1/2}} \tag{4}$$

where u_m is the maximum velocity, γ is the surface tension, T is the temperature, y is the distance from the weld pool surface vertically downwards, W is the width of the weld pool, and ρ and μ are the density and viscosity of the metal, respectively. Let us consider an example where we calculate the order of magnitude of the maximum velocity, u_m. For a typical pool width of 0.5 cm, metal density of 7.2 gm/cm³, viscosity of 0.06 poise, temperature coefficient of surface tension, $(\frac{d\gamma}{dT})$, of 0.5 dynes/(cm °C), and spatial gradient of temperature of 600°C/cm, the maximum velocity is approximately 62 cm/sec according to equation (4). Considering that his weld pool is only 0.5 cm wide, this velocity is rather large. Computed values of the order of 100 cm/sec have been reported in systems dominated by Marangoni convection (11, 12). However, when the surface tension gradient is not the main driving force, the maximum velocities can be much smaller. For example, when the flow occurs owing to natural convection, the order of magnitude of the maximum velocity, u_m, can be approximated by the following relation (11):

$$u_m \approx \sqrt{g\beta\Delta Td} \tag{5}$$

where g is the acceleration due to gravity, β is the coefficient of volume expansion, ΔT is the temperature difference, and d is the depth. For the values of $\Delta T = 600°C$, $g = 981$ cm/sec², $\beta = 3.5 \times 10^{-5}/°C$, and $d = 0.5$ cm, the value of u_m is 3.2 cm/sec. In the case of electromagnetically driven flow (13) in the weld pool, the velocity values reported in the literature are typically (14) in the range of 2 to 20 cm/sec. In comparison, the magnitude of the velocities for both buoyancy and electromagnetically driven flows in the weld pool are commonly much smaller than those obtained for surface tension driven flows.

The velocities calculated from equations (4) and (5) can provide a rough idea of the flow of liquid metal in the weld pool. Solutions of the equations of conservation of mass, momentum, and heat are necessary for the calculation of temperature and velocity fields in the weld pool. Calculations of fluid flow and heat transfer in the weld pool are now routinely performed through numerical solution of the equations of conservation of mass, heat and momentum. In view of the complexities of the welding processes, attempts to understand welding processes through simulation must involve concomitant well-designed experimental work to validate the models. Trends in the computed velocities as a function of a welding variable such as the heat source intensity are more dependable than the exact computed velocity field values under a given set of welding conditions. Experimental determination of the velocity fields in the weld pool remains a major challenge in the field (15). In the absence of an adequate experimental technique, contemporary literature relies heavily on the available recourse of numerical calculations to obtain insight about weld pool heat transfer and fluid flow.

VARIABLE PENETRATION

Numerical calculations of heat transfer and fluid flow have provided detailed insight about welding processes that could not have been achieved otherwise. One example is the variable depth of penetration of welds (16-20) during welding of different joints made of the same grade of steels under identical welding conditions. Qualitative effects of surface active elements such as sulfur and oxygen in determining the weld pool geometry have been known (20) for well over a decade. In recent years it has been shown that unexpected, and even apparently anamalous results of weld penetration can be understood from calculations of heat transfer and fluid flow.

Cross sections of the welds formed in steel samples containing 20 and 150 ppm of sulfur and welded under different powers are shown in Figure 2. A comparison of these micrographs shows that, at a laser power of 1900 W, the pool geometries were similar in the two steels containing 20 and 150 ppm sulfur. However, when the samples were welded using laser powers of 3850 W and 5200 W, the weld geometries showed a pronounced dependence on the concentration of sulfur. The welds containing 150 ppm sulfur have a much greater depth of penetration than those containing 20 ppm sulfur at high laser powers. Thus, the concentration of sulfur may or may not have a significant effect on the weld geometry depending on the laser power and other welding variables when a laser is used as a heat source. Are these results anomalous? In general, for counter-intuitive results, possible lack of reproducibility of the data cannot be ruled out. However, in this instance, over eight carefully conducted experiments were analyzed and the results were reproducible. These results lead to an important question: how can we control weld penetration? A meaningful answer to this question requires discussion of numerically computed results. Therefore, the answer to the question will be deferred until the end of this section.

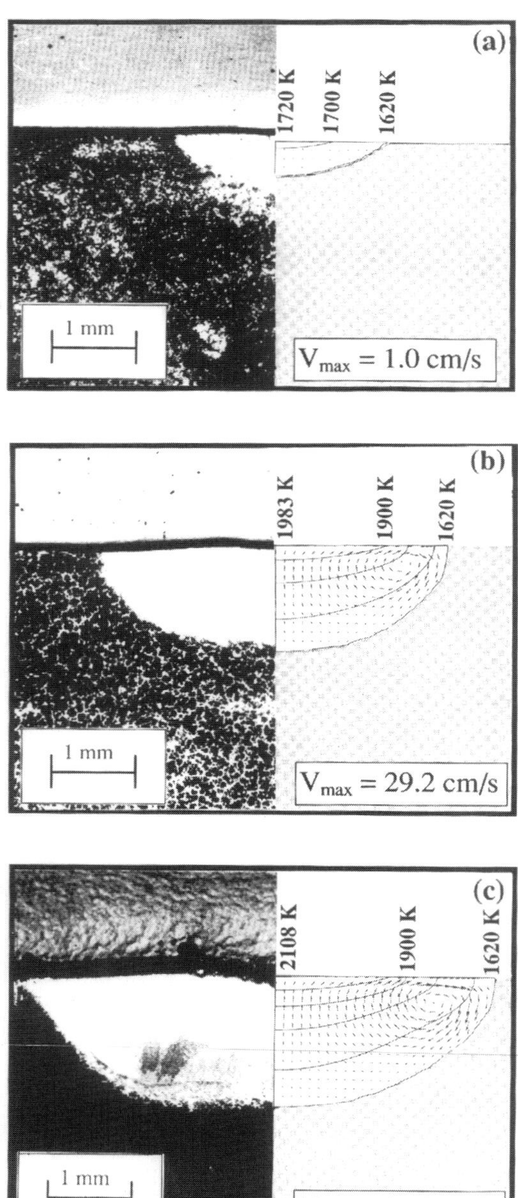

FIGURE 2. Comparison (18) of the predicted weld pool geometries with the experimental observations. for the steel containing 20 ppm sulfur for laser powers of (a) 1990 W, (b) 3850 W and (c) 5200W.

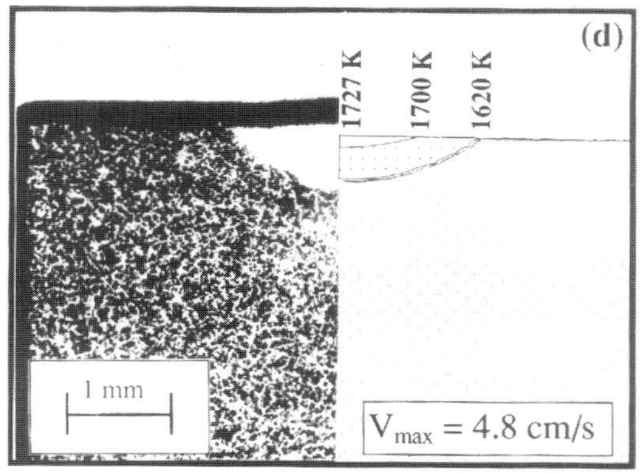

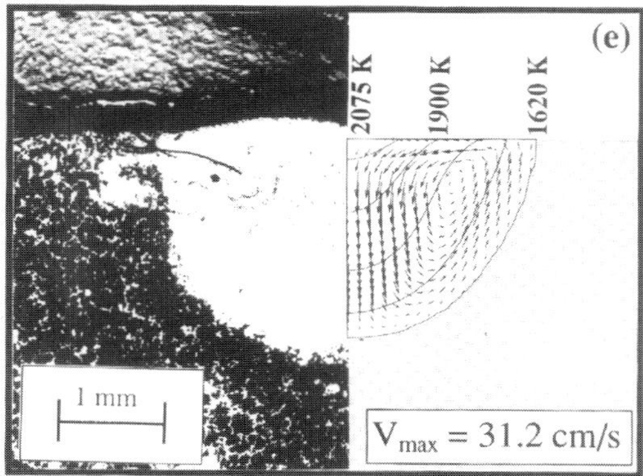

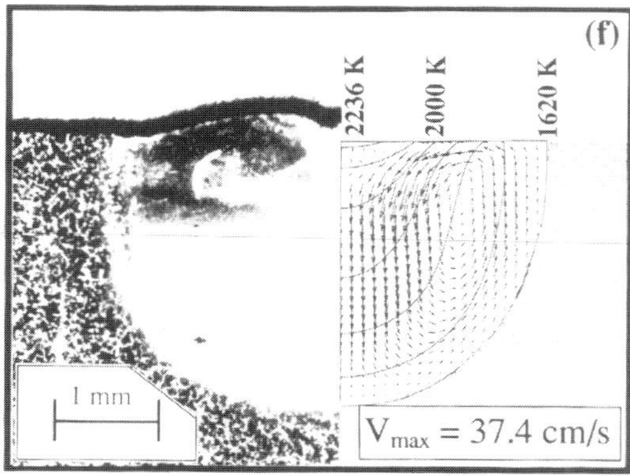

FIGURE 2 *(continued).* For the steel containing 150 ppm sulfur for laser powers of (d) 1900 W, (e) 3850 W and (f) 5200 W. Welding duration: 5s.

We start by examining the results (18) presented in Figure 2. The predicted weld geometries for the three powers for a welding time of 5 s are compared with the corresponding experimental data in Figure 2 for samples containing 20 and 150 ppm sulfur. It is observed that for a laser power of 1900 W, there is no significant difference between the computed weld geometries containing different sulfur concentrations. This insensitivity of weld geometry on sulfur concentration is consistent with the experimental observations. However, for laser powers of 3850 and 5200 W, the predicted geometries show that the weld penetration is deeper in the steel with 150 ppm sulfur than in the steel containing 20 ppm sulfur. Furthermore, it is observed that the predicted weld pool geometries are in good agreement with the corresponding experimentally observed values.

The similarity in the weld pool geometry observed for samples with 20 and 150 ppm sulfur, welded at a laser power of 1900 W, can be understood from the computed velocity and temperature fields presented in Figure 2. The results show that the peak temperature reached on the weld pool surface for both the cases is about 1720 K. The relatively low temperature gradients on the weld pool surface lead to low surface velocities and an insignificant effect of convective heat transfer on the weld pool geometry. The effect of convection on pool geometry can be examined from the dimensionless Peclet number for heat transfer, Pe. The Peclet number is a measure of the relative magnitudes of convective and conductive heat transfer and is given by:

$$Pe = V_{max} L/\kappa \tag{6}$$

where V_{max} is maximum velocity, κ is the thermal diffusivity of the liquid metal given by $k/\rho C_p$ and L is the characteristic length that can be taken as the depth of the weld pool. Using the experimental weld geometries and the maximum computed surface velocities, the Peclet number for the cases presented in Figures 2(a) and 2(d) are 0.18 and 0.91, respectively. These low values of Pe (<1) indicate that conductive heat transfer is more important than convective heat transfer in the development of the weld pool geometries in these two cases. As a result, fluid flow does not play an important role in determining the weld pool geometry in this case, and there is no significant difference between the weld pool geometries for steels containing 20 and 150 ppm sulfur for a laser power of 1900 W.

At high laser powers (3850 and 5200 W), the computed Peclet numbers are large (>200), and the convective heat transport has a pronounced effect on the weld pool geometry. For a sulfur content of 20 ppm, the temperature coefficient of surface tension (21), $d\gamma/dT$, is negative above 1700 K, as can be observed from Figure 3. Therefore, negative values of $d\gamma/dT$ prevail over much of the weld pool resulting in radially outward flow over all of the weld pool surface and shallow weld pools.

When the sulfur content is 150 ppm, the convective heat transport in the downward direction results in deep pools for both 3850 and 5200 W, as can be observed from Figures 2(e) and 2(f). Since $d\gamma/dT$ is negative at temperatures higher than 1980 K for the steel containing 150 ppm sulfur, it generates a radially outward secondary flow in a small region near the middle of the pool that opposes the radially inward flow at the pool periphery. However, the driving force for this secondary flow is mild and it is overpowered by the dominant radially inward flow which leads to a deep weld pool.

Careful assessment of the concentrations of surface active elements, such as oxygen and sulfur, and the use of a well-tested mathematical model would be useful for the solution of the variable penetration problem.

SIMPLE FEATURES OF SOLIDIFICATION STRUCTURE

The cooling rate at a given solidification front location is the product of the temperature gradient and the solidification growth rate. For several alloys, the secondary dendrite arm spacings in the newly formed solid have been experimentally correlated with the cooling rates. Therefore, using numerically computed cooling rates and the available experimental correlation,

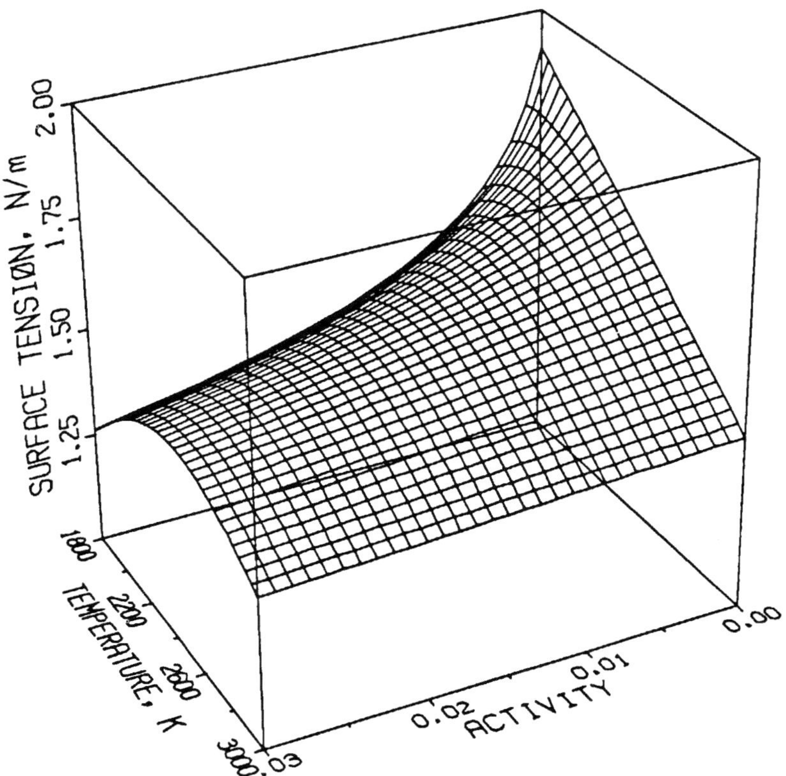

FIGURE 3. Variation of surface tension of Fe-S as a function of temperature and sulfur activity (21).

the secondary dendrite arm spacing can be estimated. The experimantal correlations between secondary dendrite arm spacing and the cooling rate obtained by Abdulgadar (22) and Bower et al. (23) are shown in Figure 4. Paul and DebRoy (12) calculated cooling rates at the edge of the weld pool from the numerical calculations of convective heat transfer. These cooling rates were used to determine the secondary dendrite arm spacings. For example, secondary dendrite arm spacings of 0.9 microns and 0.4 microns were obtained for computed cooling rates of 5 mm/sec and 31 mm/sec welding speeds, respectively. These values were in good agreement with the experimental results (12). The agreement between the predicted secondary dendrite arm spacing and the values obtained from independent data demonstrates that the calculated values of cooling rates are fairly accurate. In another investigation (24) of the effect of pulsed laser welding on the thermal response of 310 and 316 austenitic stainless steels, the cooling rates were theoretically calculated from fundamental principles of transport phenomena. The secondary dendrite arm spacings were determined from the microstructures. The results indicated excellent agreement between the measured and the expected secondary dendrite arm spacings based on the cooling rates. The investigations with both continuous (12) and pulsed (24) heat sources indicate that simple features of the solidification structure can be determined from the available numerical models of weld pool transport phenomena.

Prediction of microstructure and properties of the weldment from fundamentals is considerably more difficult than predicting simple features of the solidification

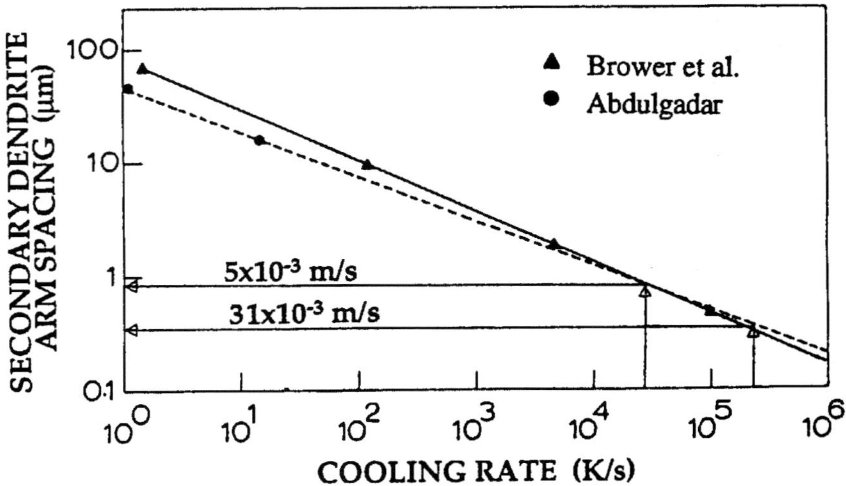

FIGURE 4. Plot of secondary dendrite arm spacing as a function of cooling rate for conduction mode laser welding of AISI 201 stainless steel (2).

structure. However, in recent years, significant progress has been made to understand the weld metal microstructure evolution from fundamentals of transport phenomena in the welding of low-alloy steels (25). The cooling rate calculated from the fluid flow and heat transfer model was coupled with an existing phase transformation model to predict the volume percentages of acicular, Widmanstatten, and allotriomorphic ferrites in the top bead of the multi-pass welds containing different concentrations of vanadium and manganese contents in Fe-C-Si low-alloy steel welds. Numerical calculations with ten weldment compositions predicted volume percentages of different ferrite morphologies that were in fair agreement with independent experimental data. The agreement between experimental and predicted measurements indicate a significant promise for predicting weld metal microstructure evolution in fusion welds, for any given low-alloy steel composition and welding process, from the fundamentals of transport phenomena.

Process models are an important component in achieving effective automation in welding. The comprehensive phenomenological models, embodying detailed description of the important physical processes in welding, cannot be used for real time welding applications, since they require extensive computer time. However, large phenomenological models can be used to calibrate and verify relatively simple process models, such as neural networks (26), which do not consider any of the physical processes in welding, and as a consequence, are computationally simple and can be used in real time.

VAPORIZATION FROM WELD POOL SURFACE

The weld pool surface temperatures are normally much higher than the melting points of the weld metals. Consequently, pronounced vaporization (2, 27-31) of alloying elements such as manganese in steels, and zinc in aluminum alloys takes place, especially when high energy density heat sources are used. Such losses often result in a change in the composition of the weld metal, affect weld properties and are a serious problem in the welding of many important engineering alloys.

Weld metal composition changes (28) due to laser welding of thin samples of various grades of high manganese stainless steels are indicated in Figure 5. The severe depletion of manganese in the weld zone in each case is clearly evident. From a mass balance, the decrease in the concentration of manganese, $\Delta\%\mathrm{Mn}$, can be expressed as follows:

$$\Delta\%\mathrm{MN} = \frac{100 r_{\mathrm{Mn}}}{\rho}\left\{\frac{A}{v}\right\} \tag{7}$$

where r_{Mn} is the vaporization rate of manganese per unit surface area, A is the weld pool surface area, ρ is the density of the weld metal, and v is the volume of the weld metal melted per unit time. The weld metal composition change depends on the vaporization flux and the surface to volume ratio of the weld pool, with the latter often being the dominant factor. For the conditions of the experiments presented in Figure 5, a small weld pool having a width of less than a millimeter was formed, and a significant composition change was observed. An increase in the power of the heat source does not result in a more pronounced composition change. This fact may appear counterintuitive at first. The composition change is most pronounced at low powers (29) because of the small size and, consequently, high surface to volume ratio of the weld pool, as can be observed from Figure 6.

A simple model to calculate the evaporation rate is given by the Langmuir equation (33):

$$J = \frac{p^o}{\sqrt{2\pi MRT}} \tag{8}$$

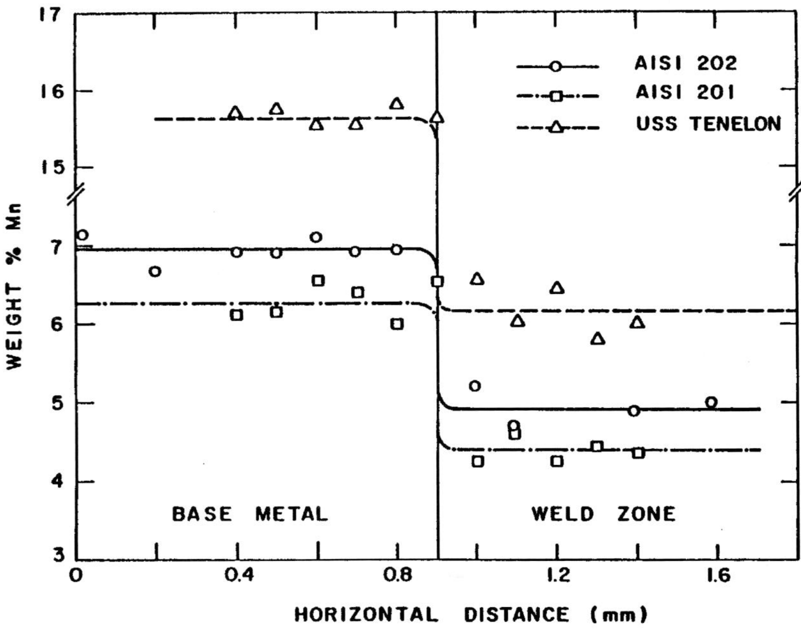

FIGURE 5. Concentration of manganese versus distance in the base metal and in the weld zone for continuous wave carbon dioxide laser welding (28). Laser power: 560 watts, welding speed: 3.5×10^{-3} m/s, shielding gas flow rate: 10^{-4} m³/s, and sample thickness: 7×10^{-4} m.

where J is the vaporization flux, p° is the vapor pressure, M is the molecular weight, R is the gas constant, and T is the temperature. The Langmuir equation can be used to calculate vaporization rates at very low pressure where significant condensation of the vapor does not take place. Experimental data indicate that the vaporization rate under most welding conditions is five to ten times lower than the rate predicted by the Langmuir equation. However, the equation has been used to determine relative rates of vaporization of various alloying elements (27, 29).

Quantitative understanding of the loss of alloying elements from weld pools involves numerical solution (30, 31) of the equations of conservation of mass, momentum, and translational kinetic energy of the vapor molecules near the weld pool surface. Furthermore, the weld pool surface temperature distribution is also required for the vaporization rate calculations. A key feature of the calculations is the consideration of the pressure gradient-driven mass transfer. At temperatures higher than the boiling point, the pressures in the vicinity of the pool are greater than the ambient pressure. This excess pressure provides a driving force for the vapor to move away from the surface. To include this effect, the velocity distribution functions of the vapor molecules escaping from the weld pool surface at various locations are used (31).

In Figure 7, the experimentally determined vaporization rates are compared (30) with the rates computed from the model and the values calculated from the

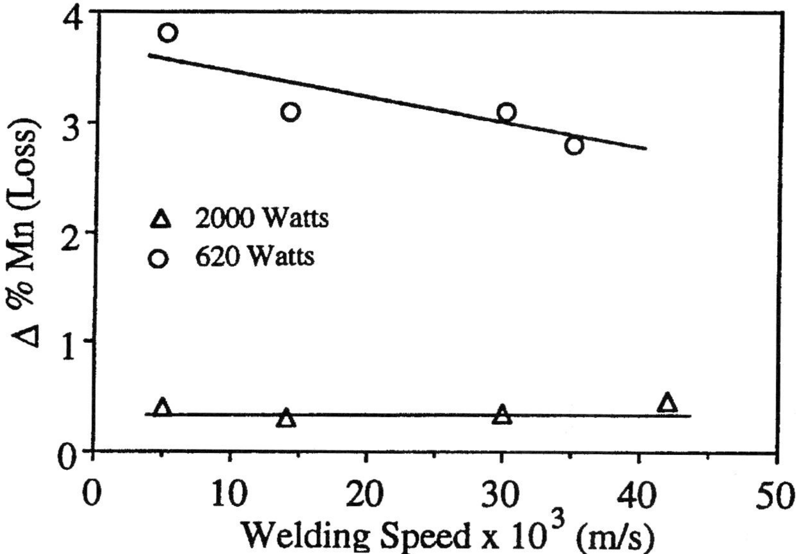

FIGURE 6. Effect of welding speed and power on the decrease in the manganese concentration of stainless steels resulting from laser welding (29). The symbol $\Delta\%Mn$ represents the difference between the manganese concentration in the base metal and the weld metal. Samples of AISI 202 stainless steels were welded with laser power of 620 W and AISI 201 samples were welded with laser power of 2000 W.

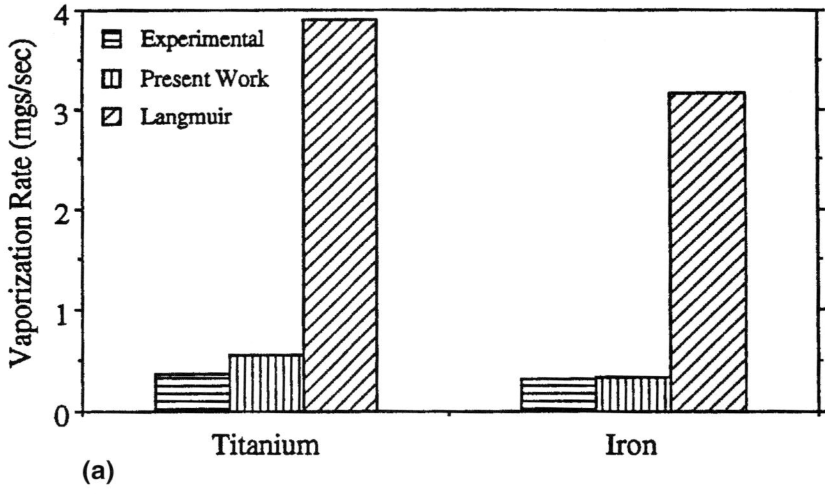

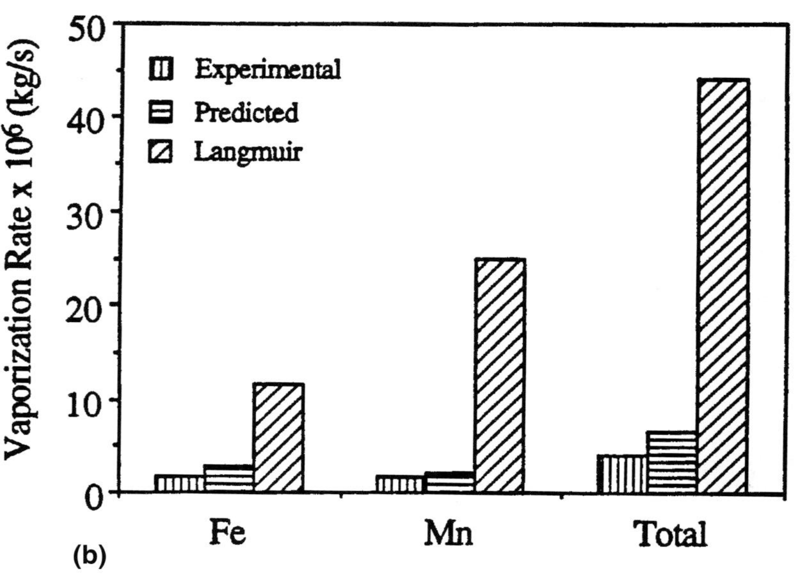

FIGURE 7(a): Comparison of the experimental vaporization rates of pure iron and titanium with the corresponding rates calculated from the Langmuir equation and the model (30). Laser power: 500 W, argon flow rate: 1 1/min. FIGURE 7 (b) Comparison of the vaporization rates calculated from the Langmuir equation and from the model with experimentally determined values for AISI 201 stainless steel (31). Laser power: 3000 W, welding speed: 15.24 x 10^{-3} m/s, argon flow rate: 5.5 x 10^{-4} m³/s.

Langmuir equation. It is observed that the experimentally determined vaporization rates are closer to the values predicted by the model than the rates calculated from the Langmuir equation.

The results show that higher accuracy is achieved by including more realistic and detailed descriptions of the physical processes, and consequently, by performing more complex calculations. Furthermore, the counterintuitive fact that higher power does not result in more pronounced composition change may be explained from the calculations of vaporization rates and the pool geometry.

GAS-WELD METAL REACTIONS

During welding, hydrogen, nitrogen and oxygen may dissolve in the weld metal, whereupon they subsequently form pinholes or porosity, or combine with elements in the alloy to form inclusions. In steels, hydrogen induces cracking, nitrogen increases the yield strength and the tensile strength but reduces the ductility, and oxygen promotes inclusion formation. Because of the important role of these species in determining the properties of weld metal, understanding their origin and science-based control remains an important goal in welding research.

In arc welding, the consumables can contribute to the weld metal oxygen and hydrogen concentrations, and slag-metal reactions have a strong influence in determining the concentration of oxygen. In laser welding, the concentrations of hydrogen, nitrogen and oxygen in the weld metal are affected by the partitioning of these species between the weld pool and the surrounding gas environment.

Shielding from these gases may be achieved by means of the addition of a flux, an inert gas, or a combination of the two, or from evacuation of the atmosphere, as in electron beam welding. In electroslag welding, the flux provides shielding of the metal from the atmosphere. In gas-metal arc, gas-tungsten arc and plasma welding, an external inert gas supply is used. In the welding of reactive metals, special shielding measures such as welding inside a gas-filled box are often used to protect the weld metal. In submerged arc welding, depending on the composition of the electrode coating, the evolution of CO_2 or other gases, such as H_2, can result. The flux helps in protecting the metal. Oxygen and nitrogen contents as high as 0.7 and 0.2 wt%, respectively, have been obtained in the weld metal during arc welding (33). These concentration levels were far greater than those in the base and filler metals and indicate the importance of dissolution of these species from the gas phase.

In the absence of a slag layer, the dissolution of a species from a diatomic gas environment can be represented as:

$$\frac{1}{2}G_2 = \underline{G} \tag{9}$$

The equilibrium concentration of a species such as hydrogen in a metal is given by Sieverts law which states that the concentration is proportional to the square root of the partial pressure at any given temperature.

$$C_{\underline{G}}^{d} = K_{eq}^{d} \sqrt{p_{G_2}} \tag{10}$$

where $C_{\underline{G}}^{d}$ is the species concentration in solution at equilibrium with the diatomic gas G_2, K_{eq}^{d} is the equilibrium constant for reaction (9), and p_{G_2} is the partial pressure of G_2. However, in most welding processes, the weld metal is exposed to a plasma environment. When a gas transforms to a plasma phase, its constituents may dissociate, ionize, or become electrically or vibrationally excited. All of these species have different equilibria with the weld metal. If a plasma contains a reactive monatomic gas, the equilibrium between the atomic gas and the weld metal must be considered. The equilibrium concentration of a species in a metal in a monatomic gas environment is given by:

$$C_{\underline{G}}^{m} = K_{eq}^{m} \, p_{G} \tag{11}$$

where $C_{\underline{G}}^{m}$ is the species concentration in solution at equilibrium with the monatomic gas G, K_{eq}^{m} is the equilibrium constant for dissolution from a monatomic gas, and p_{G} is the partial pressure of G. It should be noted that under the conditions of fusion welding, the monatomic gas is not in equilibrium with the diatomic gas in the plasma. Calculation of the equilibrium solubilities from equations (10) and (11) for different gases reveals that, in most cases, the equilibrium concentration of a species in solution is considerably higher in the monatomic gas environment than in the corresponding diatomic gas environment. To illustrate the difference, the computed equilibrium concentrations of nitrogen in iron at equilibrium with both diatomic and monatomic nitrogen are shown in Figure 8. It is observed that at 1850 K and 1.0 atm of diatomic nitrogen partial pressure the equilibrium solubility is about 0.045 wt% $\underline{N}$. However, at the same temperature, in monatomic nitrogen environment, the same equilibrium concentration can be achieved at a partial pressure of only about 1.0×10^{-7} atm. Several investigators (34-39) have indicated that the species concentration in the weld metal can be significantly higher than those calculated from Sieverts Law. The transformation of ordinary molecular species to excited neutral atoms and ions in the gas phase leads to enhanced solution of species in the metal. The presence of the partially dissociated diatomic gases and electrons in the gas phase introduces several special features of the system. Of these, three issues are of special interest in welding. What factors govern the extent of dissociation of a diatomic gas in the welding environment? How does the temperature affect the species concentration in the weld metal for different gases? How much of these dissolved species in the weld pool is still retained by the weld metal after cooling?

The answers to these questions are still evolving. Ouden and Griebling (38) proposed that under the arc, the weld pool interacts with atomic nitrogen and

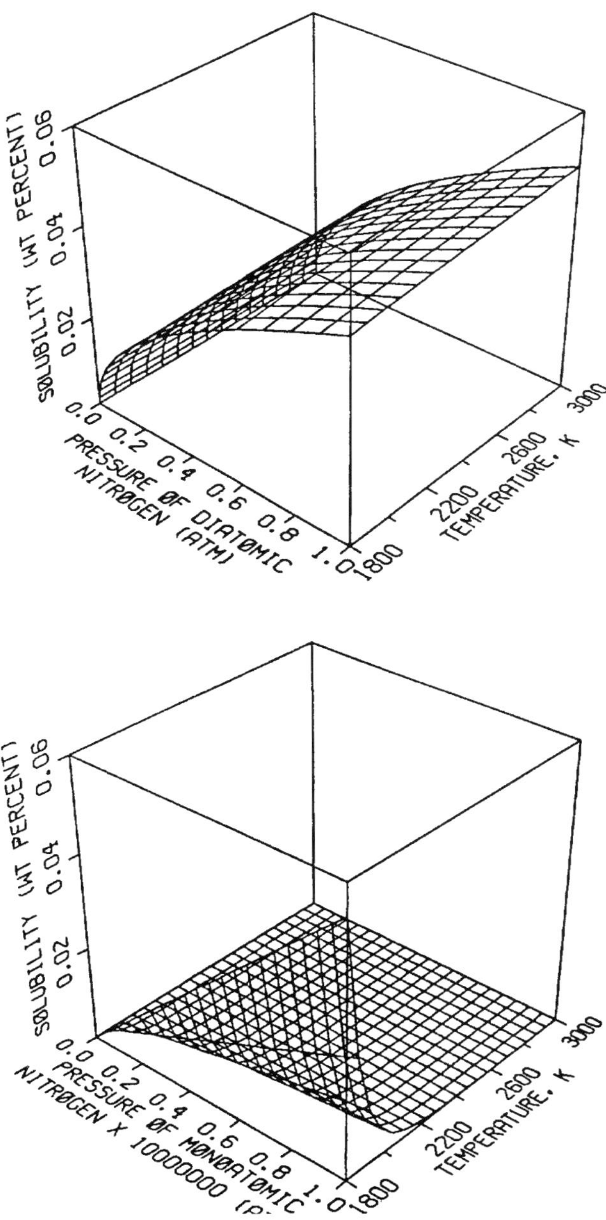

FIGURE 8. Computed equilibrium nitrogen concentrations in iron in diatomic and monatomic nitrogen gas environments (2).

receives an infusion of nitrogen. At the periphery of the weld pool where the pool is exposed to diatomic nitrogen, nitrogen is lost from the weld pool. The final weld pool nitrogen concentration is a balance between the absorption and the desorption processes. Mundra and DebRoy (34) proposed that the concentration of atomic nitrogen in the plasma is higher than what would be expected in equilibrium with diatomic nitrogen at the temperatures of the weld pool surface. They modeled the higher-than-equilibrium (superequilibrium) partial pressure of atomic nitrogen in the plasma by an equivalent thermal dissociation of diatomic nitrogen at a dissociation temperature which is about 200 K higher than the temperature at the weld pool surface. Their model was applied by Palmer and DebRoy (36) and by Cross, Hoffmeister, and Huismann (40) to examine nitrogen dissolution. While efforts have been made to examine, quantitatively, the dissolution of gases under the conditions of fusion welding, it is fair to say that quantitative prediction of partitioning of these elements between the weld metal and its surroundings still remains a major problem.

After the elements are dissolved in the weld pool, their concentrations change with time during solidification and cooling. Gedeon and Eagar (37) in their study of hydrogen dissolution, observed that a model for hydrogen dissolution in the weld metal must take into account the mechanism of hydrogen absorption in the weld pool, hydrogen rejection from the solidifying front, and the hydrogen diffusion away from the weld pool. Since the diffusivity of hydrogen is much higher than that of nitrogen or oxygen, the rate of diffusion of nitrogen or oxygen away from the weld pool is likely to be lower than that of hydrogen. A fundamental understanding of the rates of dissolution of hydrogen, oxygen and nitrogen in the weld pool, and their rejection from and transport in the liquid and the solid phases during solidification and cooling is just beginning. Because of the important role of these species in determining the properties of weld metal, understanding their origin and science-based control remains an important goal in welding research.

CONCLUDING REMARKS

In the last few decades, considerable progress has been made in understanding heat transfer, fluid flow, and mass transfer processes. Physical and mathematical modeling, which are commonly used tools in engineering science, have been used to probe contemporary issues in fusion welding. As a result of carefully planned experiments and concomitant modeling work, advances have been made in understanding the weld pool geometry, weld metal composition, and simple features of microstructure. However, the meaningful phenomenological modeling of a complex process, such as fusion welding, is computationally intensive, requires highly trained personnel, and is expensive. For these reasons, the modeling of fusion welding processes, while providing important insight, has been of limited practical use to practicing engineers. Neural network-based real-time models, calibrated by experimental results and these large phenomenological models, would be important in the pursuit of intelligent process control to produce defect free, structurally sound, and reliable welds.

ACKNOWLEDGMENT

The research was sponsored by the U.S. Department of Energy, Office of Basic Energy Sciences, Division of Materials Sciences, under grant number DE-FG02-84ER45158. Helpful comments by Mr. T.A. Palmer in the preparation of this manuscript are appreciated.

REFERENCES

1. American Welding Society Statement. 22 March 1992. *USA Today.*
2. DebRoy, T. and S.A. David. 1995. Physical Processes in Fusion Welding. *Reviews of Modern Physics.* 67(1):85-112.
3. Arata, Y. and I. Miyamoto. 1978. Technocrat. 11(5):33.
4. Bramson, M.A. 1968. *Infrared Radiation: A Handbook for Applications.* Plenum Press, New York.
5. ZelDovich, Y.B. and Y.P. Raizer. 1966. *Physics of Shock Waves and High Temperature Hydrodynamic Phenomena.* Academic Press, New York.
6. Collur, M.M. and T. DebRoy. 1987. Emission Spectroscopy of Plasma during Laser Welding of Stainless Steels. *Metall. Trans. B.* 20B:277-286.
7. Miller, R. and T. DebRoy. 1990. Energy Absorption by Metal Vapor Dominated Plasma During Carbon Dioxide Laser Welding of Steels. *J. Appl. Phy.* 68(5):2045-2050.
8. Rockstroh, T.J. 1987. *The Role of the Plasma during Laser-Gas and Laser-Metal Interactions.* Ph.D. Thesis, University of Illinois at Urbana-Champaign.
9. Rockstroh, T.J. and J. Mazumder. 1987. Spectroscopic Studies of Plasma during CW Laser Materials Interaction. *J. Appl. Phys.* 61(3):917-923.
10. Lancaster, J.F. 1980. *Metallurgy of Welding.* George Allen and Unwin, London, pp. 33-35.
11. Szekely, J. 1986. The Mathematical Modeling of Arc Welding Operations. In: S.A. David (Ed.) *Advances in Welding Science and Technology.* ASM International, Materials Park, OH, pp. 3-14.
12. Paul, A.J. and T. DebRoy. 1988. Free Surface Flow and Heat Transfer in Conduction Mode Laser Welding. Metall. Trans. 19B:851-858.
13. Woods, R.A. and Milner, D.R. 1971. Motion in Weld Pool in Arc Welding. *Welding Journal Research Supplement.* 50, 163s-173s.
14. Wang, Y.H. and S. Kou. 1986. Driving Forces for Convection in Weld Pools. In: S.A. David (Ed.) *Advances in Welding* Science and Technology. ASM International, Materials Park, OH, pp. 65-69.

15. Mazumder, J. 1993. Validation Strategies for Heat-Affected Zone and Fluid-Flow Calculations. *ASM Handbook.* ASM International, 6:1146-1152.

16. DebRoy, T. 1995. Interfacial Phenomena in Numerical Analysis of Weldability, In: H. Cerjak and H.K.D.H. Bhadeshia (Ed.) *Mathematical Modeling of Weld Phenomena II.* Institute of Materials, London, pp. 24-38.

17. David, S.A. and T. DebRoy. 1992. Current Issues and Problems in Welding Science, *Science.* (257):497-502.

18. Pitscheneder, W., DebRoy, T., Mundra, K. and R. Ebner 1996. Role of Sulfur and Wedling Variables on the Temporal Evolution of Weld Pool Geometry during Multi-kilowatt Laser Welding of Steels, *Welding Journal Research Supplement.* 75(3): 71-80.

19. Mills, K.C. and B.J. Keene. 1990. *International Materials Reviews.* 35(4): 185-216.

20. Heiple, C.R. and J.R. Roper. 1982. Mechanism for Minor Element Effect on GTA Fusion Zone Geometry. *Welding Journal Research* Supplement. 61:97s-101s.

21. Sahoo, P., DebRoy, T. and M.J. McNallan. 1988. Surface Tension of Binary Metal-Surface Active Solute Systems under Conditions Relevant to Welding Metallurgy. *Metallurgical Transactions,* B. 19B:483-491.

22. Abdulgadar, S.A. 1988. *Laser Welding of 200-Series Stainless Steels:* Solidification Behavior and Microstructural Characteristics. Ph. D. Thesis, Pennsylvania State University, p. 64.

23. Bower, W.E., Strachan, R. and M.C. FLemings. 1970. *AFS Cast Metals Research Journal.* December, 176.

24. Zacharia, T., David, S.A., Vitek, J.M. and T. DebRoy. 1989. Heat Transfer During Nd-YAG Pulsed Laser Welding and Its Effect on Solidification Structure of Austenitic Stainless Steels, *Metall. Trans.* 20A:957-967.

25. Mundra, K., DebRoy, T., Babu, S.S. and S.A. David 1997. Weld Metal Microstructure Calculations from Fundamentals of Transport Phenomena in the Arc Welding of Low Alloy Steels, *Welding Journal Research Supplement.* 76(4): 163-171

26. White, D.R., Carmein, J.A., Jones, J.E. and K. Liu. 1993. Integration of Process and Control Models for Intelligent Control of Welding. In: S.A. David and J. Vitek (Ed.) *International Trends in Welding Science and Technology.* ASM International, Materials Park, OH, pp. 883-887.

27. Khan, A. and DebRoy, T. 1987. Alloying Element Vaporization and Weld Pool Temperature During Laser Welding of AISI 202 Stainless Steel. *Metallurgical Transactions,* B. 15:641-644.

28. Collur, M.M., Paul, A. and T. DebRoy. 1987. Mechanism of Alloying Element Vaporization During Laser Welding. *Metallurgical Transactions,* B. 18B:733-740.

29. Khan, P.A.A., DebRoy, T. and S.A. David. 1988. Laser Welding of High Manganese Stainless Steels. *Welding Journal Research Supplement.* 67(1):1s-7s.

30. DebRoy, T., Basu, S. and K. Mundra. 1991. Probing Laser Induced Metal Vaporization by Gas Dynamics and Liquid Pool Transport Phenomena. *Journal of Applied Physics.* 70(3):1311-1319.

32. Dushman, S. 1962. *Scientific Foundations of Vacuum Technology.* Second Ed., John Wiley and Sons, New York.

33. Kou, S. 1987. *Welding Metallurgy,* John Wiley and Sons, New York, pp. 61-63.

34. Mundra, K. and DebRoy, T. 1995. Calculation of Weld Metal Composition Change in High Power Conduction Mode Carbon Dioxide Laser Welded Stainless Steels. *Metallurgical and Materials Transactions.* 26B:149-157.

35. Bandopadhyay, A., Banerjee, A. and T. DebRoy. 1992. Nitrogen Activity in Low Pressure Nitrogen Plasma. *Metallurgical Transactions, B.* 23B:207-214.

36. Palmer, T. and T. DebRoy. Physical Modeling of Nitrogen Partition between the Weld Metal and its Plasma Environment, submitted for publication in *Welding Journal Research Supplement.*

37. Gedeon, S.A. and T.W. Eager. 1990. Thermochemical Analysis of Hydrogen Absorption in Welding. *Welding Journal Research Supplement.* 69:264s-271s.

38. den Ouden, G. and O. Griebling. 1990. Nitrogen Absorption during Arc Welding. In: S.A. David and J.M. Vitek (Ed.) *Recent Trends in Welding Science and Technology.* ASM International, OH, pp. 431-435.

39. Uda, M. and S. Ohno. 1973. Effect of Surface Active Elements on Nitrogen Content of Iron Under Arc Melting. *Trans. Nat. Res. Inst. Metals.* 15(1):20-28.

40. Cross, C.E., Hoffmeister, H. and G. Huismann. June 1995. "Nitrogen Control in Hyperbaric Welding of Duplex Stainless Steel", Presented at the International Institute of Welding Conference in Sweden.

Chapter Twenty-Five

CARBON AND GRAPHITE MATERIALS

PETER THROWER

Department of Materials Science and Engineering
Steidle Building
Pennsylvania State University
University Park, PA 16802

INTRODUCTION

It is estimated that a little under 0.2% of the earth's crust is carbon, most of it existing as coal. In addition there are a variety of organic materials, naturally occurring graphite deposits, and metallic carbonates. The major sources of raw materials for the preparation of commercial carbons and graphites fall in this latter group, coal in itself being of minor importance. In a sense, all carbonaceous materials are prepared by a process of pyrolysis, or thermal degradation. The heating of the raw material tends to decompose it, releasing the other major elements, H, O, S, N, etc., in various gaseous forms, thus leaving a residue which is mainly carbon. Other impurities, mostly metals, may be boiled off at even higher temperatures. Because of the inherent simplicity of the carbonization process and the multitude of carbonaceous materials available, both naturally occurring and synthetically produced, there is an endless variety of forms of carbon and graphite. Indeed, there are more forms of carbon than of any other element. In each of the following sections concerned with different materials, reference will be made to many different precursors, or starting materials. The final product is not only a function of starting material(s), but also of the heat treatment temperature, heating rate to that temperature, heating time, and surrounding atmosphere.

In discussing carbons and graphites it would be helpful to have a clear distinction between the two terms. Unfortunately none exists. "Carbon" is the terminology used for the chemical element and also for the more disordered forms of the material. These are not necessarily totally amorphous forms, and in some respects may be quite crystalline.

The term "graphite" refers to the true, well-ordered single crystal with the structure shown in Figure 1, but is also used for polycrystalline materials and of some materials with quite a high proportion of very disordered regions. The use of the term "amorphous graphite" for some natural graphite deposits in Mexico, Korea, and other parts of the world may seem scientifically absurd, yet is quite common. The process of preparing a carbon or graphite involves heating a precursor to a temperature high enough to drive off the volatiles and leave a mainly carbonaceous framework, sometimes called a char. In this char the carbon atoms are mainly retained in their original arrangement, thus retaining some of the past history of the material. For example charcoal retains the pore structure of the original wood. Further heating of the char may, or may not result in the formation of a graphite. Carbons which graphitize are called "soft" carbons, while those which resist graphitization, even when heated to temperatures around 3000°C, are called "hard" carbons. The reason why chars retain the structure of the precursor is because of the low mobility of carbon atoms at all temperatures below 1000°C. In fact mobility which results in the systematic arrangement of atoms into crystals usually is not effective below a temperature of 2200°C.

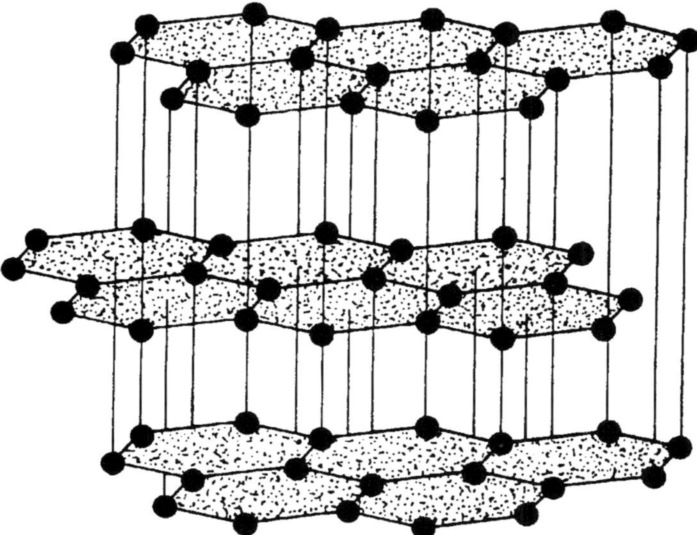

FIGURE 1. The crystal structure of graphite.

In order to destroy this randomized, or amorphous structure, there must be free mobility of the carbon atoms so that they can be rearranged into the graphite structure of Figure 1. However, there are limitations to the mobility of the atoms which are best understood in terms of the graphite lattice. In this structure the carbon atoms are bonded to three other coplanar carbon atoms, 1.42 A apart, the bonds oriented at 120° to each other. The carbon "layer plane", known as a graphene layer, thus consists of an arrangement of close-packed open hexagons, often referred to as a "chicken-wire" arrangement. In the normal hexagonal graphite structure, layer planes are arranged 3.35Å apart in an –ABABA– stacking sequence, with one half the atoms in one plane lying directly over atoms of the adjacent plane, while the other half lie over the centers of the hexagons. A graphite single crystal consists of over a million layer planes, up to several mm across, arranged in this ABABA stacking sequence. A char, on the other hand, may contain only very small volumes with such order, perhaps connected by very disordered regions. Within such a structure, carbon atoms may become mobile at high enough temperatures, but motion is normally confined to directions parallel to the layer planes and is not easy, because of the high binding energy (10.5eV) of the atoms within the layer planes. The forces which hold adjacent layer planes together are of the van der Waals type and there is therefore no great driving force for these layers to orient themselves perfectly. Another limitation in graphite manufacture is the fact that carbon cannot be melted, cast, and recrystallized as can metals. All transformations must therefore occur in the solid state.

Graphitization consists of the motion of carbon atoms so as to produce layer plane growth, the rotation of aligned adjacent graphene layers so that they can join to form one larger layer plane, and the reorientation of adjacent layer planes to form a crystal. Several studies of the graphitization process have been made and definitive reviews (1, 2) give different interpretations of the data. In any event it is well established that it is a kinetic process governed by an activation energy of 10^6 J/mole. As such, graphitization requires temperatures usually in excess of 2500°C. The following sections can only give a very brief summary of some of the many methods employed in making carbon and graphite materials. The wide variety of materials produced by the industry and the way processing variables may be adjusted to produce special properties have resulted in the nickname "black art" being applied to graphite manufacture. May of the specifics of the processes are proprietary and are probably based on empirical knowledge rather than a scientific understanding.

GRAPHITE

While graphite of a highly crystalline character is mined in various parts of the world, it is the production of bulk polycrystalline graphite which is the major industry. Blocks of material up to 1m in diameter and a few meters long have been produced, but most artifacts are much smaller. The vast majority of bulk graphites are made from at least two starting materials, a finely ground filler such as a coke, and a binder to hold the filler particles together. In some specialized applications, natural graphite or carbon black may be added to produce lubricative or abrasive qualities, and other adjuvants may sometimes be needed, such as MoS_2 to enhance lubrication in inert atmospheres or vacuum. The normal scheme of manufacture is shown in Figure 2. The various stages are as follows:

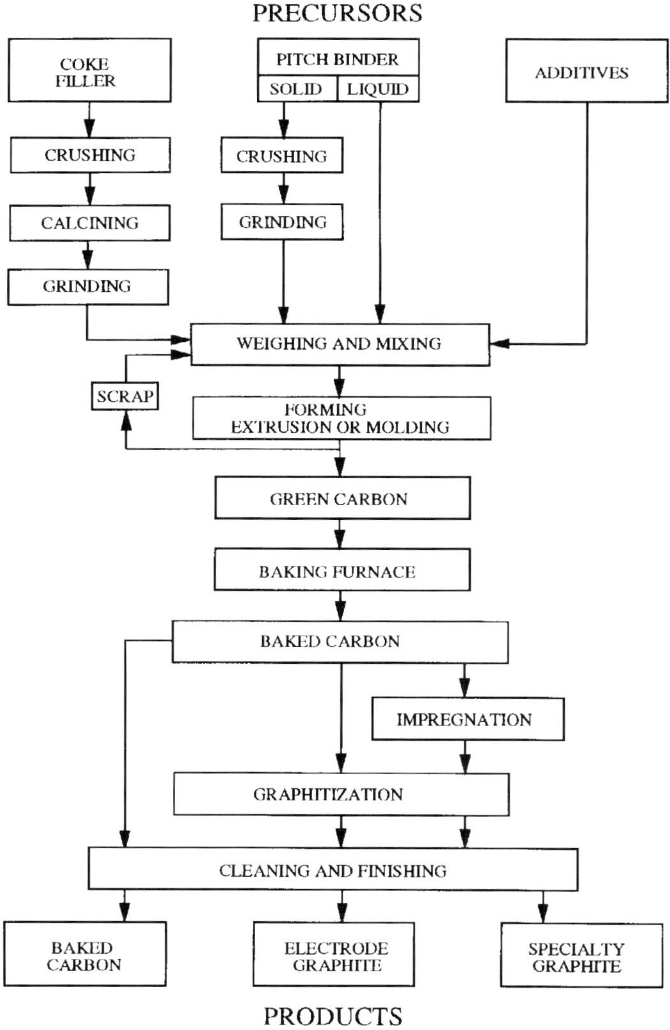

FIGURE 2. Simplified flow diagram for the production of some carbon and graphite materials.

Raw Materials and their Preparation

The starting materials are mainly derived from either petroleum or coal, the more usual being coal tar or petroleum pitch for the binder and petroleum coke as the filler. The binder materials are chosen because of their high carbon yield, usually in the range 40-60 wt %, their fluidity at mixing and forming temperatures, normally around 1000°C or slightly higher, and their ability to wet the filler particles so as to produce a carbon bond between them during the later baking operation. Its high carbon, low volatile, yield is essential because the binder is the one component which has not been carbonized prior to manufacture. Coal tar pitch is a by-product of the production of metallurgical coke and is produced by the distillation of tar from bituminous coal. Petroleum pitch is produced by the cracking of petroleum, and is used in relatively small quantities compared to coal-tar pitch. Although having a higher softening point (120°C vs 110°C) it has a lower viscosity up to 200°C and, most important, it contains no quinoline insolubles which normally account for 12-15 wt % of coal tar pitches. This latter property makes it an important impregnant for increasing the density of the baked carbon before graphitization. The brittle nature of solid coal tar pitch at room temperature allows it to be crushed or ground before mixing with the filler.

By far the major filler used in graphites is petroleum coke, produced as a by-product of the cracking of petroleum to make gasoline. The heavy residues produced by the cracking process are rich in polyaromatic compounds and may be coked in several ways, the most common of which is the delayed coking process. The production of coke from the pitch proceeds through a liquid crystal state known as the mesophase, which has been the subject of extensive recent research and the understanding of this process has thus been greatly improved (4, 5). The structure of the coke depends to a large extent on the precursor or feedstock. Of special interest are the needle cokes which are produced from aromatic feedstocks. The greater the aromaticity the more ordered the molecules in the mesophase, thus producing an ordered primordial layer structure in the coke, which is most susceptible to subsequent graphitization. The small ordered layer plane domains in needle coke are easily visible in an optical microscope.

Coke produced in a fluid bed by spraying residuum onto a hot coke particle is known as fluid coke. Because the coking process is almost immediate there is very little order to the crystallites formed and the final product has an onion-skin structure. Of course this limits layer plane growth during graphitization and imposes constraints on the coke particles (diameter ~ 500μm) which, in turn, restricts densification during later heat treatment. A property of fluid coke which can be taken advantage of in designing the final graphite is its inherent isotropy, due to its onionskin layer plane arrangement. A similar microstructure exists in gilsocoke®, a material produced from a naturally occurring deposit called gilsonite®, where the coke particles show multicoalescens of spherical particles (diam. -500 pm). The onionskin structure of this coke has higher layer plane order than that of fluid coke and the material has been used successfully in producing high quality isotropic graphites for use in the nuclear industry.

The raw or green coke usually has to be calcined before being used in graphite production. This consists of heating it to 1200-1400°C to remove volatiles, mostly methane and hydrogen, and to produce densification. The densities of delayed cokes can be as high as 2.1 g/cm³ after calcination, whereas fluid cokes have densities in the range 1.9-2.0 g/cm³. The calcined coke is then crushed, screened and milled to prepare it for mixing with the pitch binder. The size required depends on the final product desired. A high-strength graphite, or one on which a fine surface finish is required, must be fine grained. Such materials may be used in the nuclear and aerospace industries and for electrical discharge machining applications. On the other hand electrodes, such as used in steel electric arc furnaces, are usually much coarser, with coke particles up to and over 1 cm in diameter. In most processes the filler used is a mixture of size fractions chosen to maximize the packing efficiency. The art of mixing various size fractions is usually empirical and is aimed at minimizing the amount of binder required to produce an adequate high density. As a rule, binder requirements increase with decreasing size of filler particle.

Another factor to be borne in mind is the shape of the particles. Spherical particles such as gilsocoke will pack quite differently from the acicular needle coke. It is here that past experience is paramount. The blending of the coke fractions and their mixing with the binder is a crucial step in the manufacture of any graphite if the final product is to meet specifications.

Mixing and Forming

The blend of filler particles and the melted or pulverized binder (typically 25 wt %) is placed in a mixer to further blend the filler particles and to uniformly coat them with the binder. The mixer must be maintained at a temperature where the binder is relatively free flowing, and to accomplish this, steam jackets heated to 140-170°C are often used. It is essential that the binder is distributed evenly throughout the mix, otherwise the structural integrity of the final product will be unsatisfactory. After the mix has cooled to 100-110°C, a temperature slightly above the softening point of the coal tar pitch, it is ready to be formed into the desired final shape. Not only does this process shape the material but it also forces the filler particles into closer packing, thus increasing the bulk density.

There are two important forming methods, viz. extrusion and molding. The extrusion technique is used to form the majority of carbon and graphite products, which are in the form of cylindrical or rectangular logs. The extrusion press consists of a heated chamber, known as a mud chamber, which is filled with the binder/filler mix. On one side of the chamber is a hydraulic ram which forces the mix through a die of desired shape on the other side. It is during this processing step that the anisotrophy usually present in graphite materials is introduced. The elongated, acicular particles pass through the die with their long axes parallel to the extrusion direction. This produces a grain in the structure with the primordial layer planes parallel to the extrusion direction. In the final graphitized product this causes in-plane properties such as high thermal and electrical conductivity to predominate, while the cross-plane properties such as low thermal and electrical conductivity have a greater influence on properties perpendicular to the extrusion direction.

Molding is used to form nearly all small carbon and graphite products such as seals and brushes, but is also used to form large crucibles, etc. The hot binder/filler mix is simply compressed into a mold of the required shape using a plunger or ram. Pressures used vary widely depending on the mix, application etc. but are usually in the range of 15 - 210 MPa. During this process the coke particles are preferentially aligned with their primordial layer planes perpendicular to the pressing direction. High electrical and thermal conductivity in the final product is thus obtained perpendicular to the pressing or molding direction. This gives a graphite crucible the advantage of attaining uniform temperature relatively rapidly.

In some specialized applications where an isotropic product is required, an isostatic, or hydrostatic, molding technique may be used. The application of pressure in all directions tends to retain the filler particles in a random distribution of orientations, however the technique is usually only practical for smaller blocks and not for finished shape products.

After the extruded or molded part is cooled, it becomes hard and rigid as the temperature falls below the softening point of the binder. It may then be moved to the baking furnace.

Baking

The purpose of the baking cycle is to convert the binder to solid carbonaceous material which forms bonds between the filler particles and thus produces a strong product. During this process the binder passes through its thermoplastic state and eventually decomposes, causing considerable gas evolution. These processes impose two constraints on the baking operation. First, the green carbon must be supported in order to prevent slumping as the binder passes through its softening point. Second, heating rates must be slow enough to allow the evolved gases to escape without blowing up the structure. It is estimated that over 15 cm³ of gaseous products are released for every cubic meter of green carbon.

The green carbon is loaded into a baking furnace and packed around with ground coke and/or sand for support. The packing material also separates the pieces and stops them sticking together when they become plastic, and is porous enough to allow the evolved gases to escape. The temperature is then raised to 800-1000°C very slowly at first, the whole heating period usually being around 10 days. An approximately equal time is required for cooling to a temperature (~400°C) at which the material will not burn or be thermally shocked to fracture on exposure to air, so that a baking cycle takes 3-4 weeks. After unpacking, the baked carbon is scraped to remove packing material, and inspected for flaws before further processing. Alternatively it may be sold as a baked carbon.

Impregnation

This step in the processing may be omitted, with significant cost advantages. However for many high performance applications the final density of the finished graphitized product would be too low, and this is corrected by impregnation between the baking and graphitization stages. The surface of the baked carbon is first carefully cleaned to allow the impregnant access to the interior. It is then placed in an autoclave where it is heated to around 250°C, dried, and evacuated. Heated pitch is then introduced, and pressure is applied to hasten penetration of the binder to the inner pores of the structure. After removal and cleaning, the artifact will have an increased weight of around 15%. At this stage a second baking cycle is necessary before the material is either impregnated for a second time or taken to the graphitizing furnace. For nuclear and aerospace applications, the graphite may be impregnated several times. However, each impregnation shows diminishing returns in the improvement of properties and adds greatly to the cost. Remember that each impregnation-baking cycle takes several days.

Impregnation may use coal tar pitch, petroleum pitch, furfuryl alcohol, or some other organics; linseed oil has even been used. The advantage of petroleum pitch is that it contains no quinoline insolubles which reduce the penetration of the impregnant by blocking the pores near the outer surface of the artifact. A density increase of around 5% (1.6 to 1.7 g/cm³) in the final graphite is typical for a single impregnation, but the increases in strength and modulus are much larger, often in the 60-80% range, while electrical and thermal conductivities show slight improvements.

Graphitization

The final major step in the manufacture of graphite is heat treatment at around 3000°C, which causes carbon atom migration thus allowing an ordering, or crystallization, to occur. This migration begins to be significant at around 2200°C, and the term graphite is often loosely applied to materials heated above this temperature, even though little real graphite crystal development has occurred.

As mentioned earlier, there are some carbons which do not graphitize, but in those that do, crystal development follows the line of the primordial layer planes in the coke. The size and orientation of the crystallites can be varied by changing the starting materials, the forming process, and the time and temperature of graphitization. Most large graphite blocks are graphitized using the Acheson furnace which was introduced in 1895. This furnace can be enormous (15m long x 3m x 3m). The baked carbon is stacked inside the furnace between electrodes at the ends, and is packed and covered with coarsely ground metallurgical coke, with a final layer (~50 cm) of a blend of coke, silicon carbide, and sand. This layer acts as a thermal and electrical insulator and also serves to protect the product from oxidation. Heating occurs by passing large currents (>50,000 Amp) through the furnace resistive load. The thermal capacity of such a large structure obviously causes long heating and cooling times, respectively around 5 and 10 days.

After cooling, the graphitized blocks are removed from the furnace, cleaned, and inspected to determine whether property specifications have been met. After any required machining and finishing they are ready for delivery to the customer. For some materials the total manufacturing time maybe as much as five months. Other techniques involving resistive tube furnaces, induction furnaces, and even direct resistive heating of the components themselves are available for graphitizing smaller pieces of material. These batch processes require much shorter graphitization cycle times.

Other Techniques

Among the hundreds of grades of graphite offered for sale by the various manufacturers there are a number of specialty materials which have undoubtedly undergone some special proprietary processing. Bindless graphites (6) have been produced which have up to 95% theoretical density. These were produced by using finely ground, partially calcined coke which was heated in an autoclave under an isostatic pressure. The methane and other organics released from the coke during heating are pyrolyzed and probably form a thin pyrocarbon bond between the filler particles. The final product can be both highly isotropic and very strong. It is suspected that variations of this technique are used in the production of some of the fine grained, high strength, isotropic graphites commercially available today.

PYROLYTIC CARBONS

The term pyrolytic carbon can properly be applied to carbon filaments, carbon blacks and carbon films, as well as to the more massive deposits which are the subject of this section. Pyrocarbon materials are made by chemical vapor deposition (CVD) and may vary in density, properties, and structure as much as the bulk materials discussed in the previous section. A heated hydrocarbon gas decomposes into an entire series of molecular species with a wide spectrum of carbon contents and molecular weights (7). Within this pyrolyzing atmosphere "droplets" may be formed which pyrolyse and condense on a nearby surface. On the other hand large carbonaceous complexes may condense directly on the surface of the chamber. The former condition produces a fluffy, sooty, soft carbon, while the latter produces a hard solid carbon. It is the second of these materials which is of primary interest here. The structure of the carbon produced by the CVD process has been shown (8) to depend on: the type of hydrocarbon and its concentration, the pyrolysis temperature, the contact time, and the geometry of the pyrolyzing chamber. Of these, the pyrolysis temperature is perhaps the most important, but it is the nature of the chamber which conveniently divides the carbons produced into two distinct types.

Stationary Mandrels

This first experimental configuration is used to produce massive pieces of pyrocarbon such as for rocket nozzles and nose cones. It consists of a furnace containing a graphite mandrel on which deposition is to occur. Hydrocarbon gas such as methane, natural gas or propylene is diluted with an inert gas, introduced into the furnace containing the heated mandrel, and pyrocarbon is deposited on its surface. In this case difficulties arise because there is no mixing of the gas and there are large gradients in the temperature and composition of the gas. For example, as the gas progresses down the tube the concentration of hydrocarbon is depleted because of the carbon deposition, therefore the amount and structure of the deposit varies with position. The density of the pyrocarbon depends on the deposition temperature, showing a minimum at around 1700°C for both methane and propane precursors. This minimum may be eliminated by reducing the partial pressure of the hydrocarbon, thus lowering saturation levels in the pyrolyzing gas and preventing the formation of soot. It is the incorporation of this undesirable form of carbon into the deposit which causes the decrease in density.

Pyrolytic carbons deposited on stationary formers normally show well-defined growth structures which are nucleated by surface imperfections or by droplets nucleated on the surface, (Figure 3). These growth cones contain imperfect graphite basal planes lying preferentially parallel to the substrate. If the surface is very smooth, very few growth features are produced and the layer planes form a laminar structure. Materials such as these are known as substrate nucleated pyrocarbons. If hydrocarbon levels are increased to cause nuclei to form in the gas phase, these will be incorporated into the deposit, each nucleus acting as the origin of another growth cone. When growth cones are developed throughout the structure the material is referred to as continuously, or regeneratively nucleated. The preferred orientation of the defective small layer plane units parallel to the substrate gives the pyrocarbon very anisotropic properties even though the

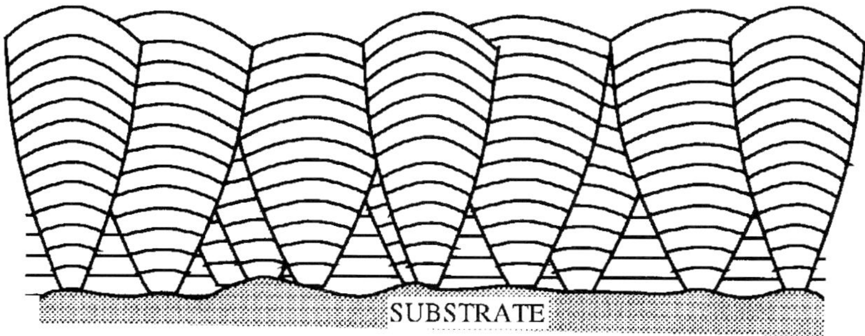

FIGURE 3. The microstructure of substrate nucleated pyrolytic carbon.

structure is nowhere near that of a perfect graphite crystal. Thermal conductivity perpendicular to the deposit is therefore very poor, which usually sets up a high temperature gradient across the deposit. This limits the thickness of the deposit which can be obtained, either because of the surface exposed to the gas becoming too cool for deposition to occur, or because of the differential strains in the material producing lamellar cracking on cooling.

Not only are the layer planes small and defective, they are also stacked with twists and tilts between them. This twisting of the layers with respect to each other gives rise to the term "turbostratic" carbon for these materials. The turbostratic, well oriented, substrate nucleated pyrocarbons are ideal candidates for graphitization and it is from these precursors that the highly oriented pyrolytic graphite (HOPG) used for x-ray and neutron monochromators is produced.

While simple thermal annealing at temperatures around 3000°C produces graphitization of these materials, and annealing at 3600°C gives materials whose basal plane properties approach those of the single crystal, there always remains an angular spread in the c-axes of the constituent crystallites of 2-3°. This mosaic spread is now reduced to as low as 0.2° by a process involving annealing to 3200-3600°C while applying a compressive stress perpendicular to the deposition plane. Pressures of up to 105 Pa have been reported, but actual pressures at temperature are probably unknown because of the plastic nature of the material. Not only can flat plates be produced, but also singly and doubly curved plates by using suitably shaped compression rams. Such materials are of extreme importance for producing focused, monchromated x-ray beams, and on a unit weight basis are probably the most expensive forms of graphite commercially available.

Fluidized Beds

In this configuration a bed of particles to be coated is held in a furnace while a mixture of hydrocarbon and inert gases is passed through it, thus causing fluidization of the bed. The hydrocarbon gas pyrolyzes in the bed, and carbon is deposited on the particles. The advantage of this method is the continuous mixing which occurs, producing a fairly uniform temperature and gas concentration throughout the bed. This results in a more controlled, uniform deposit. The incentive for the development of this technique was to produce coatings on nuclear fuel particles, which could retain the products of the fission reaction in a nuclear reactor. Depending on the reaction conditions, many types of structure may be formed within the deposit. These are classified as laminar, isotropic, and granular. The first two are respectively similar to the substrate and continuously nucleated materials already discussed. The granular carbons are produced at high bed temperatures (1700-2000°C) and low hydrocarbon gas concentrations (-5%), where the values in parentheses are for methane. One advantage of the fluidized bed technique is the ability to change the temperature and the type and composition of the gas quite easily. In this way layers of different structures may be laid down on top of each other. A porous, low density deposit followed by an impervious, strong, high density deposit is of great utility in retaining the fission gases in nuclear fuel particles. Another advantage is the ability to introduce other elements into the fluidizing gas for codeposition with the carbon, an important example being the codeposition of silicon to produce a fine dispersion of silicon carbide in the carbon. This material has been used in bioengineering for the coating of heart valves, etc.

GLASSY CARBON

When certain polymers are thermally degraded, a material resembling a black glass is produced. Its shiny surface and its brittle character have earned it the name vitreous or glassy carbon. Commercial production is very limited, and few of the uses proposed when it was first introduced in 1967 have ever come to fruition (11). Many details of manufacture are proprietary but the general principles seem well defined. Precursors for glassy carbon are polymers with a degree of crosslinking. These cross linkages are believed to remain in the carbonized material, allowing only limited layer plane growth and orientation. Indeed after heating to 2700°C the crystallite size in these materials is only around 5 nm parallel to the layer planes and 3 nm perpendicular. The interlayer spacing is then 0.35 nm, much greater than the 0.335 nm for a perfect graphite crystal. The difficulty in graphitizing the material is well demonstrated by transmission electron microscope studies which reveal the structure to be somewhat akin to stacks of ribbons of approximately ten graphite layer planes which are tangled and knotted together. There is therefore little opportunity for layer plane development to occur without a complete breakdown of the structure. Since liquefaction does not occur for carbon at normal pressures, the disorder is locked in, and the material is an excellent example of a hard carbon. Some of the precursors which have been used are cellulose, phenolic resins, and polyfurfuryl alcohol.

Cellulose carbon is made by dispersing cellulose powder in water and then centrifuging to obtain the shape of the part to be manufactured. When dried, this body can be machined so that it will produce the surface finish and thickness required. It is then heated to cause degradation, however because of the large volumes of gaseous products evolved during the early stages of decomposition, the heating rate must be extremely slow. Heating rates of 1 to 5°C per hour are often mentioned. The build up of gas pressure in the decomposing cellulose is balanced by an external gas pressure (14 MPa) which can be reduced as porosity is developed in the structure. When the temperature has reached around 500°C there is enough porosity for the material to be further heat treated in a conventional furnace under ambient pressure. The final heat treatment temperature may be anywhere in the range 1000 to 3000°C.

Furfuryl alcohol, a thermosetting resin, may be used to produce unusual shapes of glassy carbon artifacts by repeatedly painting the resin on a die of the required shape and size. After an adequate thickness has been obtained, the whole article may be slowly heated to produce carbonization. A manufacturing technique using 12:1 phenol hexamine has been developed (12) in which resin preheated to 300-330°C is crushed and compacted into discs under a pressure of 350 MPa. Sintering under a small load (105 Pa) produced a disc which maintained its shape during carbonization. Similar heating rates to those already mentioned were used during the carbonization stage.

Although the density of glassy carbon is only around 1.5 g/cm³, its permeability is ten to eleven orders of magnitude less than that of a bulk binder/filler graphite. The porosity which develops during the initial heating stages (to ~800°C) heals itself as the temperature is further increased. The mechanism for this is not understood. The shrinkage of the carbon from the precursor dimensions is greater than 25%, but experience allows manufacturers to machine the precursor to dimensions such that the finished product will be of the desired shape and size. This is particularly important because the resulting glassy carbon is very hard and almost impossible to machine. The large amounts of gas evolved during the processing limit final thicknesses to around 5 mm, however a cellular type of glassy carbon has been developed which is available up to 10 mm thick. The cell size can be varied to produce materials ranging from a fine foam to a very open three dimensional net. These materials are known as reticulated vitreous carbons.

CARBON FIBERS

Although carbon fibers can be made from glassy carbon, the fibers which are of most interest are those which have a more graphitic structure. This is because these fibers are designed to take advantage of the extreme strength and stiffness of the graphite layer planes. A graphite crystal has a very high Young's modulus (900 GPa) parallel to the layer planes. However a force applied a few degrees away from this direction causes shear of the layers over each other with a drastic lowering of the modulus (120 GPa at 10°). In order to obtain a high modulus fiber it is therefore necessary to have primary graphite layer planes aligned very close to the fiber axis. This orientation is introduced using three different methods, depending on the precursor (13, 14).

For cellulose, or rayon-based fibers (15) the production involves three distinct steps: heat treatment to ~350°C to form a thermally stable char, carbonization at 1000-2000°C, and graphitization at temperatures up to 3000°C. The heat treatment is usually performed in reactive atmospheres (O_2 Cl_2 or HCl) to prevent evolved tars from redepositing on the fiber, and also to improve process rates and weight yields. Carbonization to temperatures above 1000°C is very rapid (<1 min.) and is usually done in an inert atmosphere. At this stage there is no real order to the material, and heating to graphitization temperatures does little to improve matters. High strength, high modulus fibers may be produced by straining carbonized yarn at temperatures in the range 2800-3000°C. It is here that the layer plane alignment is produced. Increased stretching is accompanied by an increase in Young's modulus, with almost perfect alignment of the layer planes parallel to the fiber axis achieved with a strain of 180%. Strength also increases with increased strain and values of 2.8 GPa tensile strength and 800 GPa Young's modulus have been reported. The need to introduce the alignment by stretching during the graphitization step makes high modulus rayon based fibers very expensive.

Because of the high cost of rayon-based fibers and some inferior properties when they are incorporated into composites, most commercial manufacturers use polyacrylonitrile (PAN) precursors. There are again three distinct manufacturing steps: stabilization, carbonization, and graphitization. A continuous "tow" of between 3,000 and 10,000 fibers moves slowly through these three stages without interruption. During the stabilizing heat treatment at around 220°C for several hours in air, the thermoplastic PAN is converted into a non-plastic cyclic compound which does not melt at the higher carbonization temperature. By applying enough tension to prevent the 25% length shrinkage which would otherwise occur, the polymer chains are straightened and a high degree of alignment is introduced. The oxidizing atmosphere cyclizes the polymer chains to form a ladder polymer. At this stage the required orientation for a high modulus fiber is already present, and tension is not required during the later processing stages. The stabilized fibers are now heated in an inert atmosphere between 1000°C and 1500°C. Any remaining nitrogen, hydrogen and oxygen are driven off, and an essentially carbonaceous fiber is produced. During this stage, the cyclic molecular chains link up to form structures resembling the layer planes of the graphite structure. A final, brief, heat treatment at temperatures up to 3000°C increases crystallite size and improves orientation, thus increasing Young's modulus.

The properties of PAN fibers (~10 μm diameter) are very dependent on the final heat treatment temperature (HTT). Tensile strength reaches a maximum (3.5 GPa) at a HTT of around 1500°C, and then decreases, while Young's modulus continues to increase with increasing HTT with values over 480 GPa having been achieved. Because of this, three types of PAN fibers are recognized. Type I fibers have been subjected to an HTT of at least 2800°C, and have a high modulus. Type II fibers are high strength fibers with a final HTT of around 1500°C. Type III fibers are of inferior quality with neither the high modulus of Type I nor the higher tensile strength of Type II. Type III fibers are cheaper than either Type I and Type II PAN fibers or rayon-based fibers and are used extensively in sporting goods.

The third type of carbon fiber is the mesophase pitch-based fiber. The precursor is a coal tar or petroleum pitch which has been heated to convert it into the highly anisotropic liquid crystal or mesophase state. This mesophase is thermodynamically stable and does not convert to an isotropic liquid until heated above the mesophase-liquid transition temperature. However, before this happens carbonization occurs. The anisotropy of the mesophase is therefore retained in the carbonized product where it manifests itself as a preferred orientation of the primordial graphite layer planes. This is the same process that occurs on coking. In spinning a fiber from the mesophase, the forces involved result in a highly oriented structure, with the mesophase spheres usually elongated into cylinders parallel to the fiber axis. The spun fiber is then stabilized in an oxidizing atmosphere, carbonized, and graphitized as for the PAN fibers. A 345 GPa modulus, 2.0 GPa tensile strength fiber has been produced and is the cheapest, relatively high modulus carbon fiber available.

Depending on the geometry of the spinneret through with the mesophase pitch is squeezed, fibers of many different microstructures can be produced. In some the graphite layers are arranged as concentric cylinders with a structure similar to that of a leek or scallion. In others the layers radiate from the center like spokes on a wheel. Naturally, these configurations give quite different surface properties.

In recent years there has been considerable interest in fibers of cross-sectional geometries other than circular. By adjusting the shape of the spinneret, c-shaped, trilobal, and ribbon fibers can be produced (17). While the former two seem to offer the promise of some enhanced mechanical properties when used in composites, the ribbon fibers have been developed with a view to providing excellent thermal conduction (18).

Another interesting development in carbon fiber processing has been the development of special synthetic pitches for fiber spinning. Foremost among these is the naphthalene pitch developed by Mochida et al. in Japan (19, 20), using a HF/BF_3 catalyst for the pitch production. Fibers produced from this precursor exhibit easier processing characteristics and promise properties at least equal to those produced from other sources. Tensile strengths and moduli have been achieved with values as high as 4.0 GPa and 880 GPa respectively. The latter is over 80% of the value for a graphene layer, indicating a very high degree of layer orientation in the fiber. The pitch is now being made commercially available by Mitsubishi Gas Chemical.

CARBON/CARBON COMPOSITES

Carbon fibers are used almost exclusively in composite materials. Carbon fiber reinforced plastics (CFRs) are nowadays used extensively in many sporting goods, and carbon fiber reinforced cement has been taken advantage of in some modern architecture (21). The most complex and expensive are the carbon/carbon composites. Because the fibers only have strength along their length, composites are made using 3D-woven fibers (Figure 4) or laminates of 2-D woven fibers which are either impregnated and processed in the same manner as bulk graphites, or are infiltrated with pyrocarbon. In the latter case the pyrocarbon is deposited directly in the pores of the structure, thus producing a rigid 3D body. Difficulties arise because the pyrocarbon deposits preferentially at the entrances to the pores through which the pyrolyzing gases enter the structure. This is often solved (unsatisfactorily) by periodically machining away the outer surfaces.

The more usual manufacturing technique is to infiltrate the fiber mat with pitch, which then has to be pyrolyzed to produce carbon. Unfortunately the low carbon yields from the pitch, together with the effects of gas evolution during pyrolysis, mean that the process has to be repeated several times to produce a satisfactory material for some aerospace applications. With each impregnation the difficulty of

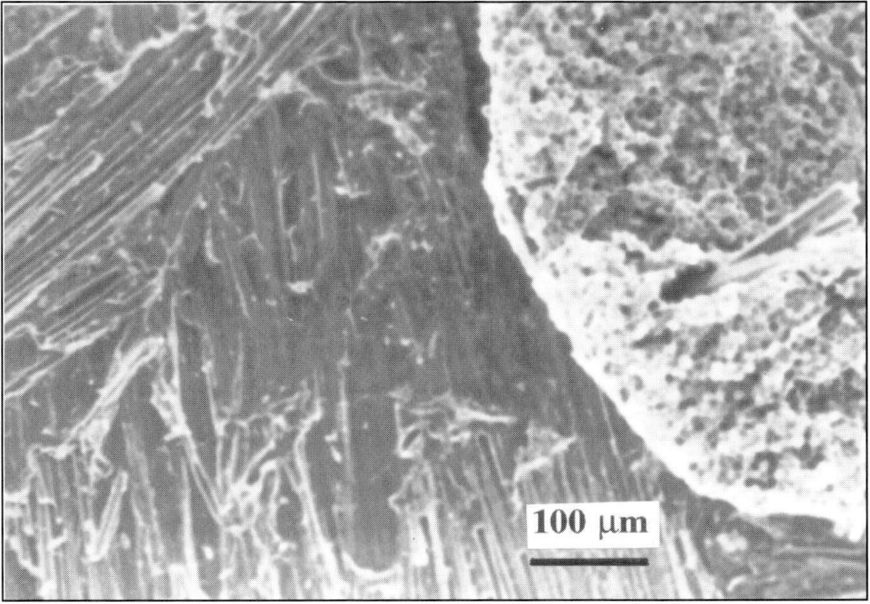

FIGURE 4. 3D Carbon/carbon composite. A fiber bundle perpendicular to the photograph (top right) is penetrating a woven fiber mat in the plane of the photograph.

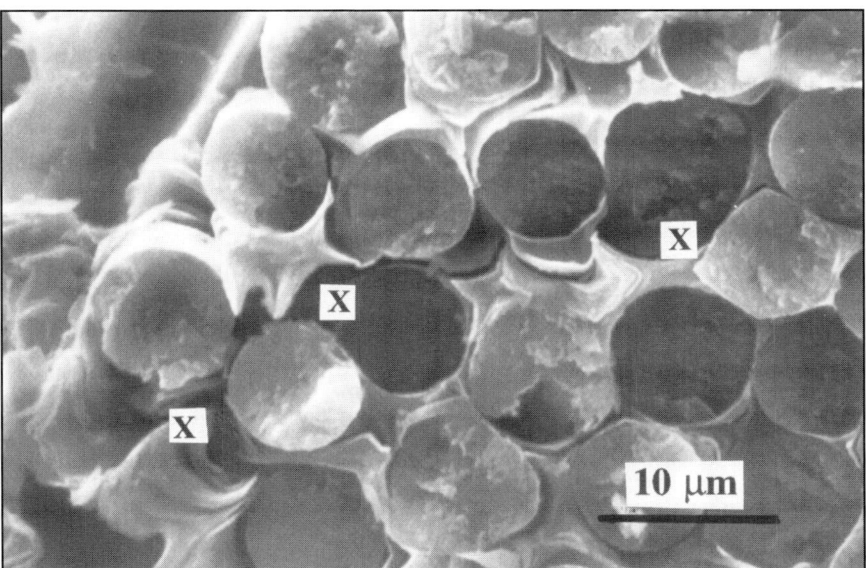

FIGURE 5. Detail of carbon/carbon composite showing separations (X) between fiber and matrix.

forcing the impregnating pitch into the pore structure becomes more difficult, and the amount of pore filling decreases. These problems can be somewhat solved by using less viscous impregnants in the later impregnations. In shrinkage of the pitch during pyrolysis also produces some loss of adhesion between fiber and matrix which sometimes results in a pore which is impossible to fill despite many repeated infiltrations. Such separations are evident in Figure 5.

As many as seven to ten successive impregnations may be necessary, each followed by a slow pyrolysis stage and a high temperature graphitization. Pyrolysis cannot be allowed to proceed at a rapid rate because otherwise the evolving gases would blow the piece apart. As a result materials can take several months to produce and, because of the large times and amounts of energy used in their production, they are some of the most expensive materials produced today.

Primary applications of carbon/carbon composite materials are in the aerospace industry, where in spite of their high cost, they can still be very cost effective. Using carbon/carbon composites for the brakes of the Concorde supersonic aircraft is reported to produce a weight saving of over 1000 lb, which translates into significant economies in operation.

ACTIVATED CARBONS

One of the more remarkable properties of graphite is the ability of the graphene layer to adsorb various molecular species on its edges. The total area of such edges in a carbonaceous material is known as the active surface area, and carbons with such areas of several hundreds of square meters per gram of carbon are readily available. These materials merit a separate chapter, such is their diversity. Almost any organic material can be used as the precursor for activated carbons, and in recent years there have been reports of the use of various woods, nuts and kernels (e.g. olive stones) being used for this purpose (22). Even spent coffee grounds have been investigated.

Each material has a unique pore structure and spectrum of pore sizes and, as such, serves a different purpose. A molecule is only adsorbed if it can penetrate the pores, with the result that different materials adsorb different molecules. It is this feature which even allows these materials to be used as molecular sieves for the separation of various molecular species, even isomers.

This adsorption property of activated carbons is used extensively in many large-scale industrial techniques nowadays to clean up various pollutants. The material has also been used in cigarette filters and in faucet-mounted household water purification devices. There is also much current interest in the use of these materials for the storage and transport of natural gas.

REFERENCES

1. Fischbach, D.B. 1971. In *Chemistry & Physics of Carbon,* Vol. 7 (P.L.

Walker, Jr., ed.) pp. 1-105. Dekker, New York.

2. Pacault, A. Ibid., pp. 107-154.
3. Rose, K.E. Nov. 7, 1971. *Hydrocarbon Process, 50*, 85.
4. Marsh, H. and C. Cornford. 1976. In *Petroleum Derived Carbons* (M.L. Devincy and T.M. O'Grady, eds.) pp. 266-281. American Chemical Society, Washington, D.C.
5. White, J.L. Ibid., pp. 282-314.
6. Chard, W., Conway, M. and D. Niesz. Ibid., pp. 155-171.
7. Palmer, H.B. and C.F. Cullis. 1966. In *Chemistry and Physics of Carbon*, Vol. 2 (P.L. Walker, Jr., ed.), pp. 265-325. Dekker, New York.
8. Bokros, J.C. 1969. In *Chemistry and Physics of Carbon*, Vol. 5 (P.L. Walker, Jr., ed.), pp. 1-118. Dekker, New York.
9. Diefendorf, R.J. 1960. *J. Chim. Phys.* (Paris), 7, 85.
10. Moore, A.W. 1973. In *Chemistry and Physics of Carbon*, Vol. 22 (P.L. Walker, Jr and P.A. Thrower, eds.), pp. 69-187. Dekker, New York.
11. Cowlard, F.C. and J.C. Lewis. 1967. *J. Mater. Science 2*, 507.
12. Jenkins, G.M. and K. Kawamura. 1976. *Polymeric Carbons.* Cambridge Univ. Press.
13. Goodhew, P.J., Clarke, A.J. and J.E. Bailey. 1975. *Mater. Sci. and Eng.*, 17, 3.
14. Diefendorf, R.J. 1976. In *Petroleum Derived Carbons* (M.L. Deviney and T.M. O'Grady, eds.), pp. 315-323. American Chemical Society, Washington, D.C.
15. Bacon, R. 1973. In *Chemistry and Physics of Carbon*, Vol. 9 (P.L. Walker, Jr. and P.A. Thrower, eds.), pp. 1-102. Dekker, New York.
16. Kotlensky, W.V. Ibid., pp. 173-262.
17. Edie, D.D., Fox, N.R., Barnett, B.C. and C.C. Fain. 1986. *Carbon 24*, 447.
18. Edie, D.D., Fain, C.C., Robinson, K.E., harper, A.M. and D.K. Rogers. 1993. *Carbon 31*, 941.
19. Yoon, S.-H., Korai, Y., Mochida, I. and I. Kato. 1994. *Carbon 32*, 273.
20. Fortin, F., Yoon, S.-H., Korai, Y. and I. Mochida. 1991. *Carbon 32*, 1119.
21. Inagaki, M. 1991. Carbon 29, 287.
22. Rivera-Utrilla, J., Ferro-Garcia, M.A., Mata-Arjona, A. and C. Gonzalez-Gomez. 1984. *J. Chem. Tech. Biotechnol.* 34A, 343.

Subject Index

V

W

Z